Economic Principles Applied to Space Industry Decisions

Economic Principles Applied to Space Industry Decisions

Joel S. Greenberg
Princeton Synergetics, Inc.
Princeton, New Jersey

Volume 201
PROGRESS IN
ASTRONAUTICS AND AERONAUTICS

Paul Zarchan, Editor-in-Chief
MIT Lincoln Laboratory
Lexington, Massachusetts

Published by the
American Institute of Aeronautics and Astronautics, Inc.
1801 Alexander Bell Drive, Reston, Virginia 20191-4344

American Institute of Aeronautics and Astronautics, Inc, Reston, Virginia

1 2 3 4 5

ISBN 1-56347-607-X

Table of Contents

Chapter 1

Introduction

THIS is not an economics book. It is a book about the application of economic principles and concepts in decision making related to space activities. The book is primarily a tutorial and elaborates upon concepts and methodology and their applications. Emphasis is placed upon applications with typical results of performed analyses presented to demonstrate concepts and methods.

The use of mathematical and simulation models serves as the underpinning for much of the presented material. A number of models are described with numerous applications presented. The specific models considered have been selected to demonstrate the role that a structured thought process can play in the decision process. Because most decisions relating to technology development, product design, capital expenditures, and, in general, investments involve uncertainty and risk, a number of the selected models, developed methodologies, and presented examples explicitly and quantitatively consider uncertainty and risk.

Unfortunately investment-type decisions have historically, with but few exceptions, been based upon analyses that consider single point estimates for a range of performance, cost, and other variables that are normally not known with certainty. In other words, even though there might be considerable uncertainty associated with technology development, product design, and related investment decisions, uncertainty and its consequence, risk, are but rarely quantified and taken into account in the decision process.[1–3] The lack of quantified assessments of uncertainty results in an incomplete characterization of technology development, product design, and investment alternatives. That is, an important product or investment attribute is left undefined and therefore not available as a dimension when comparing alternative courses of action. Sensitivity analyses are often used in lieu of the explicit consideration of uncertainty but are a poor and unreliable substitute. In addition, the lack of quantification of uncertainty often leads to the use of most likely values, which, even though mathematically incorrect, commonly are added and subtracted in arriving at various product or investment financial performance measures.

To compound the design and decision problem, the choice of design or "optimization" of performance is frequently accomplished at too low a level. For example, a product might be designed with the objective of minimizing mass or maximizing reliability, whereas a significantly different design or product choice might result from the maximization of the financial performance of the related business venture. For example, the Federal Aviation Administration (FAA) continues to contemplate the imposition of user fees for services provided to those applying for launch-related licenses, and the Department of Defense (DoD) is considering increasing fees for launch services provided to commercial users of federal ranges. Each is aimed at increasing the agency's budget (assuming the U.S. Congress allows the fees to be maintained by the agency and not returned to the U.S. Treasury). But increasing the cost of U.S. commercial space transportation services is likely to impact international competitiveness adversely by reducing U.S. market share for launch services and possibly associated satellites. A reduction in market share will effect exports, job creation, and Treasury revenue. Will these "costs" exceed the benefits (i.e., the increased user fees)? A method will be described that provides a means for assessing the likely impacts of service attribute (e.g., pricing) changes on market share and explicitly considers uncertainty and risk and end user decision processes in the estimation of changes in market share and related impacts on job creation and Treasury revenue.

This book addresses these and a number of other issues with emphasis placed upon applications rather than basic theory. A number of applications that explicitly use formal uncertainty characterizations and risk analysis are described. These have been selected both to demonstrate the breadth and scope of applications that have considered the quantification of uncertainty and the use of resulting risk measures in analyses and decision processes and to illustrate a number of data collection and analysis techniques that have been used with reasonable success.

Because government programs and policies can have significant impacts on private sector investment decisions, and many government programs and policies are undertaken with this as a specific objective, a methodology is described that allows these impacts to be evaluated in terms of changes in private sector investment decisions. These changes are developed in quantitative terms. Several examples are presented.

The objectives of this book are to put economic analysis into perspective with respect to real-world decision making in the space industry and to expand the perspective of the reader with respect to the type of tools and analyses that might be brought to bear on complex business and government problems.

I. Space Industry[4]

The commercial space industry is comprised of four primary segments: the satellite manufacturing sector that designs, manufactures, assembles, tests and sells satellites to both commercial and government customers; the launch sector that manufactures, assembles, and tests launch vehicles and provides launch services; the satellite services sector that leases transponders and provides subscription/retail services [such as direct to home television, satellite mobile telephone, satellite mobile data communications, very small aperture terminal (VSAT) services, and the sale of remote sensing imagery and value-added services (which includes the data processing sector that creates information products from space collected data)]; and the ground equipment sector that manufactures and sells major satellite-related hardware [such as gateways and satellite control stations, mobile uplink equipment, VSATs and ultra small aperture terminal (USATs), and consumer electronics, such as Direct Broadcast Service (DBS) dishes, village phone booths, handheld and notebook size phones, and global positioning systems (GPS) receivers].

In 1999, the commercial space industry generated direct revenues of approximately $68 billion. This is expected to increase to $199 billion by 2010, more than doubling over the next 10 years—a compound annual growth rate in excess of 10 percent per year. Although the revenues in all sectors are expected to increase, as illustrated in Fig. 1.1, revenues in the satellite services (subscription and retail) sector are expected to achieve a more dominant share of the total. This sector provides important technology and capability for the rapidly growing information technology sector, which, in turn, provides an important market for satellite services. Although launch service and satellite manufacturing revenues are expected to grow, these sectors' percentage of the total revenue will decrease.

U.S. federal space expenditures are another source of space industry revenues. Although the figures just cited include government procurement of commercial space industry services, significant federal space spending not related to the commercial space industry occurs. ["Revenue" includes revenues from government customers in those instances where an item or service was commercially procured and was relevant to the space industry; revenues from the creation, launch, and operation of nonsatellite spacecraft (such as science probes) are not reflected, nor are revenues derived from or related to any human space flight program.] In 1999, the DoD spent more than $6.8 billion on space-related procurement; space-related facilities and payroll constituted almost another billion in expenditures. NASA outlays exceeded $12.5 billion (procurement) in 1999, more than 92% of its budget authority. After considering that the DoD numbers do not include classified procurement programs and are therefore understated, the

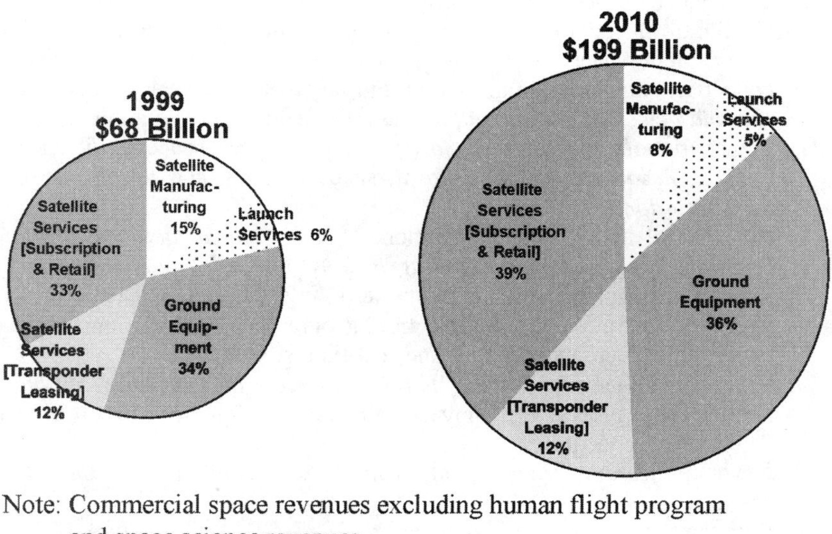

Note: Commercial space revenues excluding human flight program and space science revenues.

Fig. 1.1 Global commercial space industry revenues (Source: Ref. 4).

U.S. government alone spent more than $22 billion on space-related activities in 1999. Other world governments (excluding China and Russia) reported just over $7 billion in space spending in 1999. Combined, the global commercial space industry, the U.S. government, and foreign (excluding China and Russia) governments accounted for close to $100 billion in space spending and revenues in 1999.

The global space industry experienced significant growth during the 1990s as new commercial markets emerged to drive revenues upward. At the same time both civil and military government space spending first stagnated and then declined. As the industry adapted to its new environment, traditional space firms sought mergers to consolidate operations and shore up markets. In a further bid to adjust, these same firms both competed with and acquired new ventures in the emerging services markets. This context for the space industry, where commercial services markets drive the most significant growth, will shape the next decade of space business as it has shaped the last. New markets will continue to surface, presenting economic opportunity to those that step forward.

Today, space is used for a variety of missions, operations, and services. However, the potential for expansion is very high. Listed next are many current and potential uses of the space infrastructure. Most are not yet available, but these and many more missions and applications will become real in the coming decades, thanks to new technology. The following is a summary of current and potential space-based applications and missions with the latter indicated in italics:

1) Communications: fixed satellite services, direct broadcast services, mobile satellite services, and data collection and forwarding (By far, this is the dominant commercial space business sector.)
2) Government missions and operations: navigation (GPS), communications, intelligence, space science research, life sciences experiments, treaty verification, technology development and demonstrations, law enforcement, space debris management, search and rescue, weather data collection and forecasting, *human* and automated planetary exploration, *asteroid detection and mitigation*, and international cooperation
3) Transportation: space rescue, *fast package delivery*, space servicing and transfer, *hazardous waste disposal*, space tourism, and *ultrahypersonic transportation*
4) New orbital missions and operations: *space medical facilities, orbiting hospitals, space settlements and agriculture, and orbiting business parks*
5) Remote sensing: land-use surveying, crop infestation and disease, private weather observation and modification, geographical information services, upper atmosphere exploration, and real-time surveillance
6) Space manufacturing: *space mass production of consumer products*, microgravity processing, university research, biological products and laboratories, and space robotics
7) Novel applications: space burial, *theme parks, athletic events, and movie studios*
8) Extraterrestrial resources: *lunar-based* He_3 *and asteroid mining*
9) Advertising: *Orbiting billboards and space product demonstrations*
10) Space utilities: *Solar power satellites and artificial moons*

II. Breadth of Economic Concerns

There are many aspects to economics, all of which are aimed at providing information for clarifying and improving decisions concerning the commitment of resources and establishing government policies. Operationally, economic analysis includes financial and investment analysis, cost analysis, benefit/cost and cost effectiveness analyses, as well as traditional micro- and macroeconomic analyses. These methods, in turn, utilize many disciplines including mathematical, simulation, and econometric modeling, probability and statistics, and decision analysis. These and other aspects of space economic analysis and associated disciplines are discussed in the following pages with emphasis placed upon applications. Sufficient theory is presented as the applications are developed so that the methods and techniques described can be applied to the economic aspects of space applications beyond those chosen as examples in the pages of this book.

To appreciate the scope of economic analysis and its applications, a number of rather diverse, yet typical problems are indicated in the following sections, where economic and related analyses are likely to make important contributions in the decision-making process. A number of these are addressed in detail in the following pages. Both government and industry concerns are indicated.

A. Government Concerns

In general, government concerns relate to the efficient utilization of resources in order to achieve desired objectives and the implications of government policies and programs on private sector investment decisions and resulting economic costs and benefits. These concerns include

1) Development of private sector options for the financing of space business ventures (for example, business ventures based upon reusable launch vehicle technology) and establish the need for and appropriate role of the government in helping to achieve the necessary financing.
2) Establishment of the economic implications of loan guarantees, particularly when providing these guarantees for business ventures that are judged to have relatively high technology risk, thereby placing the government in a perhaps unintended risk-sharing environment.
3) Establishment of the need for, and alternatives to, government provision of third-party liability indemnification for space-launch-related activities.
4) Establishment of the economic impacts of regulatory actions (for example, licensing of space launch activities, establishing financial responsibility requirements for launch activities, establishing the need for and value and cost effectiveness of orbital debris remediation requirements) and identification and evaluation of alternative actions.
5) Development and justification of technology transfer activities and establishment of the value of technology transfer alternatives.
6) Establishment of the appropriateness and likelihood of success of privatization or commercialization of government assets or provision of services previously provided by the government.

7) From the government's perspective, establishment of the likely financing requirements for commercial space ventures and the consequent need for government assistance and the form of assistance.
8) Estimation of the cost for development and production of products including new satellites and launch vehicles.
9) Establishment of more realistic cost estimates and program plans for long-term multiphase technology programs (in support of initiatives such as solar space power, space exploration, space nuclear propulsion).
10) Development of the impact of government programs and policies on pricing policies for products and/or services in order to identify those that will likely maintain or improve competitive positions and achieve desired financial performance objectives (such as achieving a specified return on investment).
11) Formulation of efficient technology programs comprising multiple research and development (R&D) projects, each having multiple benefit attributes and resource requirements, the sum of which exceeds budgetary constraints.
12) Establishment of the likely impacts of the use of nonmarket economy goods and services on the U.S. economy and on specific organizations.
13) Establishment of the likely impacts on U.S. industry of making excess government assets available for general use.
14) Establishment of the appropriate tradeoffs between reliability, performance, and maintenance and sparing strategies aimed at minimizing life-cycle cost of space missions while achieving system performance objectives.

In the pages that follow, a number of the preceding areas are addressed directly, some indirectly, and others only by inference.

B. Industry Concerns

In general, business (industry) concerns relate to profitability and liquidity (cash flow and availability of required resources) and the efficient utilization of resources in order to achieve desired short- and long-term objectives. These concerns include the implications of government policies and programs on profitability and liquidity both in the short and long term. Many space businesses require large amounts of up-front capital funding, have long periods before payback, involve risks greater than those of other businesses (technological, market, and government intervention), and are subject to many external influences. Because of these characteristics, the space industry has historically relied upon a broad range of government supporting activities that encompass the government as a customer, developer of technology, subsidizer so as to maintain and/or improve international competitiveness, sharer of risk, source of capital, regulator, etc. Thus, the space industry's concerns are closely related to those of the government and include the following:

1) Development of private sector options for the financing of space business ventures (for example, business ventures based upon reusable launch vehicle technology) and establish the need for and appropriate role of the government in helping to achieve the necessary financing.

2) Establishment of the need for and the financial implications of loan guarantees and other incentives as a requirement for capital formation for new business ventures that are judged to have relatively high technology risk.
3) Establishment of the need for, and possible alternatives to, government provision of third-party liability indemnification for space-launch-related activities.
4) Establishment of the cost and other impacts of regulatory actions (for example, licensing of space launch activities, establishing financial responsibility requirements for launch activities, establishing the need for and value and cost effectiveness of orbital debris remediation requirements).
5) Assessment of the feasibility of privatization or commercialization of government assets or provision of services previously provided by the government.
6) Establishment of the likely financing requirements for commercial space ventures and the need for government assistance and the form of assistance.
7) Estimation of the cost for development and production of products including new satellites and launch vehicles.
8) Development of the impact of government programs and policies on pricing policies for products and/or services in order to identify those that will likely maintain or improve competitive positions and achieve desired financial performance objectives (such as achieving a specified return on investment).
9) Formulation of efficient technology programs comprising multiple R&D projects to achieve new and/or improved business goals.
10) Establishment of the likely impacts of the use of nonmarket economy goods and services on specific organizations.
11) Establishment of the likely impacts on U.S. industry of making excess government assets available for general use.
12) Establishment of the appropriate tradeoffs between reliability, performance, and maintenance and sparing strategies aimed at minimizing life-cycle cost of space missions while achieving system performance requirements.
13) Establishment of pricing policies that will achieve financial and other business objectives.

In the pages that follow, a number of the preceding areas are addressed directly, some indirectly, and others only by inference.

III. Organization of the Book

This is not an economics book. It is a book about the application of economic principles and concepts in decision making related to space activities. The book is primarily a tutorial and elaborates upon concepts and methodology and their applications. Typical results are presented, based upon performed analyses, to demonstrate the presented concepts and methods. An important aspect of the book is the use of multiattribute decision variables, and related databases, to characterize private sector investment decisions. This allows the impacts of government programs and policies to be developed quantitatively in terms of the change in the likelihood that the private sector will make investments, or take actions, that the government is trying to encourage. The methodology also allows the effects of various government and industry actions (i.e., pricing, performing research and technology development, etc.) to be established in terms of impacts on international competitiveness (i.e., changes in market shares).

The terms *uncertainty* and *risk* are used throughout this book. The author, based upon more than 35 years of experience with performing financial and risk analyses, has found the following definitions to be valuable. Risk is what the decision maker faces. Uncertainty is what the planner or analyst faces. Risk relates to the variability of decision metrics that are the consequences of technology, market, cost and other uncertainties, probabilistic actions and/or events that are, to varying degrees, beyond the control of the planner or analyst, and less than perfect reliability. Risk can be characterized as the probability that a decision metric will exceed various values (for example, there is a 20% chance that the return on investment will exceed the firm's cost of capital, or conversely, there is an 80% chance of not achieving or exceeding the cost of capital) or simply as the standard deviation of the probability distribution of the metric. This is particularly useful (as will become apparent in the following pages) when the resulting probability distribution is near normal.

Uncertainty is associated with the perceived range of variability of quantities or variables used in the determination of the values of the decision metrics. Uncertainty can be characterized in terms of minimum, most likely, and maximum values (i.e., the triangular probability distribution associated with the value of a variable such as unit sales, market share, etc.), expected values and associated standard deviations, and specific probability distributions. A particularly useful method for characterizing uncertainty perceptions is described in Sec. I of Chapter 4 and is the primary method considered in this book.

A. Introduction (Chapter 1)

The scope and magnitude of the space industry was discussed in Section I, and the breadth of economic concerns was described in Section II. The various sectors comprising the space industry (i.e., space transportation, communication, remote sensing, and future possibilities such as space power, tourism, etc.) were described (both qualitatively and quantitatively) and historic and projected levels of activity presented. Both U.S. and worldwide levels of activity and forecasts were presented so that following discussions can be viewed with appropriate perspectives.

The book considers economic principles and concepts relating to both public and private sector investment decisions and government policies and programs that might influence private sector investment decisions. The role of economics in decision making is described through the presentation of a number of examples. Examples include the following: financial analysis of reusable launch vehicle business ventures to identify the need for and desired form of government supporting initiatives (such as loan guarantees, anchor tenancy, R&D investments, tax policy); planning and evaluation of an efficient R&D program; determining cost-effective satellite sparing and maintenance strategies; establishing an appropriate orbital debris regulatory regime; analyzing the economic impacts of government policies and programs on international competitiveness in terms of job creation and changes in Treasury revenue (i.e., taxes and unemployment transfer payments).

A number of typical investment decisions are described together with useful metrics. (The importance of appropriate metrics is emphasized.) This is followed by an introduction to business planning, financial analysis, capital budgeting and assessing the impacts of government programs and policies on international competitiveness and measured in terms of changes in job creation and Treasury revenue. This introductory section is followed by detailed discussions of economic issues and decisions relating to reusable launch vehicles, space operations (including launch-on-demand, on-orbit service/repair, and launch campaigns), licensing and regulatory issues with emphasis on financial responsibility (i.e., establishing insurance requirements) and orbital debris. To illustrate the scope of economic concerns, the final section discusses the application of economic and decision analysis in the areas of energy conservation and casino gaming and demonstrates the similarity (with space-related decisions) of useful analytic techniques.

B. Investment Decisions (Chapter 2)

Investment decisions are normally concerned with committing current resources in order to achieve future rewards. Several different types of decisions are considered, namely those relating to unconstrained choices (i.e., evaluating and selecting investments based upon the individual investment's value), constrained choices (i.e., selecting a project set within resource constraints), and forced choices (i.e., selecting the best alternative to implement a prior decision). The reusable launch vehicle (RLV) investment decision is illustrative of the first, planning an R&D program is typical of the second, and selecting a launch vehicle (given the decision to launch a communication satellite) is typical of the latter. The characteristics of these types of decisions are described together with typical metrics and related decision predictive databases. The role of government actions in influencing private sector investment decisions is also described, as are the resulting economic impacts. A number of detailed examples are presented.

Analysis, assessment, and evaluation of alternatives imply the use of metrics for comparison purposes. Both private and public sector metrics are discussed. The impact of different public sector actions/initiatives on private sector metrics is also discussed. The development and use of appropriate metrics are discussed and illustrated. Examples used to demonstrate the importance of appropriate metrics include space transportation operability/resilience and economic impacts of orbital debris remediation policies.

Because many metrics are the result of financial analysis and because most private sector investment decisions are based upon the results of financial analyses, basic concepts underlying financial analysis are discussed. Variables such as profit, cash flow, payback period, return on investment, risk, and net present value are defined, and their importance/significance in decision making is discussed. A geosynchronous communication satellite business financial planning model is described and used to illustrate the financial analysis and evaluation of a typical communication satellite business venture. In addition, the model is used to demonstrate the sensitivity of financial performance to schedule delays, launch vehicle reliability, and satellite design life.

To establish financial impacts, it is customary to develop a business plan. The business plan serves as the basis for communication between those seeking the commitment of resources and those having the ability and authority to commit resources. The business plan provides a description of the proposed projects as well as a justification of the requested resources. It also provides an indication of the value to be received from the commitment of resources. The financial analysis, in the form of a financial plan, is a component of the business plan. Other components of the business plan include a description of the market, competition, sales forecast, and staffing plan. These and other components of the business plan are described.

Capital budgeting decisions relate to making resource investments, in the near term, which will produce benefits in the long term. Evaluations of these investment opportunities involve the quantification of multiple performance measures that then serve as the basis of decisions to commit resources. Typical performance measures are described together with a multiattribute decision predictive database. Predicting the influence of government programs and policies on private-sector investment decisions is described through the use of this database. A number of different government initiatives such as anchor tenancy, tax policies, guaranteed loans, research, development and demonstration programs are discussed together with their likely influence on decision variables such as return on investment, risk, payback period, and magnitude of investment at risk.

The use of the decision database is described in terms of the comparison of alternatives and the selection of a launch vehicle for a geosynchronous communication satellite mission. First, the comparison and selection is considered from the point of view of the customer. Next, the comparison is considered from the point of view of a provider of space transportation services. The objective is to forecast market share based upon likely decisions that will be made by the user of the transportation services. The predictive decision database is a key element in this analysis. The database describes likely choices in terms of relative expected value and relative perceived risk. The unique method for collecting the data contained in the database (i.e., the tradeoff between expected return on investment and perceived risk) is described. Finally, decisions are considered from the point of view of the government. These decisions relate to the evaluation of fair pricing of foreign competitive space transportation services and the planning of an R&D program that can affect the attributes of U.S. space transportation services and explicitly and quantitatively considers the impact on international competitiveness.

A number of typical R&D program formulation methods are discussed. These include methods applicable to the planning of a focused (or correlated) technology program, methods applicable to the planning of a technology program considering many project undertakings but in a resource constrained environment, and finally the ranking and selection of a set of technology projects from amongst a larger set utilizing a benefit/cost ratio rank ordering approach The differences between these approaches and their appropriateness is discussed. Finally, a method is described for the public sector planning and evaluation of a technology program (using space transportation as an example) that considers the implications of the program on international competitiveness, through changes in market share, and consequent economic impacts (for example, job creation and Treasury revenue).

The impact of government programs and policies on private sector investment decisions is discussed. A number of government programs and policies are described, and their impact on private sector financial performance measures (i.e., expected return on investment, risk, payback period, and magnitude of investment at risk) is described. This is followed by a discussion of the effect of altering financial performance measures on likely investment decisions. To demonstrate the developed concepts, a change in pricing of government launch services provided to commercial users of federal launch ranges is considered and the likely consequences of government actions developed first in terms of changes in transportation pricing, then on international competitiveness (i.e., market share), and finally in terms of changes in exports/imports, job creation and Treasury revenue.

The implications of changes in international competitiveness are developed through the consideration of space transportation. When foreign suppliers capture launch business, they gain many advantages. First of all, just as loss of business harms U.S. manufacturers, gains in foreign business enable these suppliers to gain market share and hence become more competitive by expanding the base upon which they write off fixed costs. For the U.S. there is a net loss of jobs with its effects in the economy. There is a loss in revenue base to the launch industry that could, in part, have formed the basis for R&D that would improve the technology and make it more competitive in the future. And, more important, there is a loss of domestic access to space. The economic implications of foreign competition to the launch and satellite communication industries are developed through the use of a mathematical model that considers both direct and indirect effects for the loss (or gain) of a launch and its associated payload to foreign competition. The model develops both direct and indirect impacts on job creation, exports/imports, and Treasury revenue. This model is further discussed (in a following section) in terms of the economic impacts that can result from the successful development of a reusable launch vehicle that will reduce transportation prices and thereby affect demand. The economic impacts that can result from changes in launch fee pricing (for commercial users of government launch facilities) are developed.

Government frequently undertakes activities that are aimed at transferring the results of its research and development activities to the private sector. The objective of these activities is to influence investment decisions so that the R&D results are utilized beneficially. If investment decisions are not changed, there is no value obtained from the investment in technology transfer programs. A number of government programs have been specifically aimed at influencing investment decisions by performing experiments that would demonstrate technology and the ability to provide new and/or improved services. Two examples are the NASA flight experiments program and the Advanced Communication Technology Satellite (ACTS) program.

These programs are used to illustrate assessments and evaluations of technology transfer activities. The former is considered from the point of view of determining a priori which flight projects should be selected for funding when limited resources are available, specifically taking into account the potential impacts of the experiment in terms of the decision making processes of the target audience. The latter is considered as an example of a lost opportunity to measure the value of a demonstration program in terms of a priori perceptions and measured experiment results and likely impacts on investment decisions.

C. RLV Economics (Chapter 3)

The development of new space transportation systems is characterized by relatively large capital requirements, perceived high risk, and relatively long payback for the initial investment. Risks relate to achievable technical performance, markets (i.e., the demand for transportation services as a function of price, time, and competition) and government short- and long-term policies. Because commercial development of new transportation services is the goal, the role and perceptions of those responsible for financing the commercial space transportation ventures (i.e., the capital markets) are of concern as are the possibilities of altering these perceptions through the judicious application of government programs and related policies. From the government's point of view, the benefits of these programs and policies must outweigh their costs. Thus private and public sector decisions are entwined and must be considered simultaneously. The discussion of RLV economics considers these underlying issues relating to public policy involving space station, space shuttle and expendable and reusable space transportation services and public sector initiatives and associated legislative requirements. In addition, methods for analyzing RLV financing requirements and potential implications of government actions are described together with typical analysis results obtained to date. Pricing policies are considered as well as the likely consequences of alternative government policies and programs on the U.S. economy (measured in terms of job creation and Treasury revenue). Finally, RLV insurance-related issues are discussed, and a method for ensuring the long-term availability of insurance at affordable prices is described.

A broad range of policy issues relating to the government support of a reusable launch vehicle program is discussed. In particular, the rational for government investment in the development of RLV technology is discussed as is the relationship of the space station and the RLV and the relationship between the space shuttle and its improvement to the RLV. The issues relating to United Space Alliance (USA) commercialization/privatization of the space shuttle and the role of the space shuttle in launching commercial payloads are also discussed. Finally, the potential role for Congressional action (tax policy, technology investment, participation in Research, Development, Test, and Evaluation (RDT&E) funding, loan guarantees, etc.) is discussed, as are concerns with impacts on a competitive environment. Setting the stage by reviewing the Space Act and its relationship to space transportation and the RLV program precedes this discussion.

Prior to embarking on a discussion of the RLV, lessons that might be learned from the space shuttle economic analysis are reviewed. The last major space transportation program undertaken in the U.S. was the space shuttle transportation system (STS). This section is concerned with looking back at the environment within which the STS program and policy decisions were made and the economic and related analyses performed for and the resulting program and policy decisions made for the STS. As a result of reviewing the STS program, a number of "lessons learned" are identified in the hope that pitfalls can be avoided for the development of new space transportation systems based upon reusable launch vehicles. The intent is to identify shortcomings of the STS analyses and resulting program and policy decisions in order to improve the program and policy decision process relating to the current RLV-related activities. The issues and associated lessons to be learned range from underlying Executive Branch and Office of Management and Budget (OMB) policy decisions within which the STS program was forced to function to limitations of analyses performed by NASA and its contractors.

The general situation of financing space-related business ventures is reviewed with specific consideration given to large capital-intensive risky programs having reasonably long payback periods. The reusable launch vehicle development and the creation of a business venture employing a fleet of reusable launch vehicles is used as an example and used to set the stage for describing the various aspects of the capital markets and the magnitude of resources available from its various segments together with capital market requirements (in the form of anticipated financial performance, control concerns, and characterization of the various forms of risk). Various important financial performance measures are described as are their importance in the evaluation and decision process. This is followed by a description of how the financial community evaluates large capital-intensive, risky and long payback business ventures, methods that might be appropriate for reducing risk perceptions (and/or improving other financial performance measures) and the appropriateness of different funding sources as the business venture proceeds through time. Differences between the financing of different types of business ventures [for example, RLVs vs. geosynchronous orbit (GEO) and low Earth orbit (LEO) communication satellites] are also discussed.

Pricing of new launch services in a competitive environment is of obvious concern. Thus the general theory behind product/service pricing is briefly discussed. The general theory is then used to develop a number of pricing policies that might be of concern in the planning and evaluation of RLV business ventures. A number of pricing policies are developed including pricing to maximize profit but with prices constrained by competitive service pricing, pricing such that the RLV price is a specified fraction of the anticipated competitive price, pricing in a competitive market so as to achieve a specified market share, pricing at a specified fraction of anticipated competitive pricing then changing pricing to maximize profit, setting price as a constant times operating costs, and two-tier pricing (a government price for a specified duration with commercial pricing to maximize profits). These pricing schemes are considered in terms of price elasticity and the development of an RLV financial planning model.

Methods and analyses are also described for analyzing the implications of various government actions and policies on the financial performance of typical reusable launch vehicle business ventures and on the U.S. economy. A financial planning model that allows the impacts of various government actions (such as various tax policies, anchor tenancy, participation in RDT&E, loan guarantees) to be established in terms of impacts on the financial performance of typical RLV business ventures and on government costs and benefits is described. The financial planning model takes into account many factors including pricing policies, market segmentation, and price elasticity as a function of market segment, uncertainty and risk, etc. and develops typical business proforma documentation as well as government costs and benefits and economic impacts. Typical results of analyses performed to date are presented including the impact of various government initiatives (such as various tax policies, anchor tenancy, participation in RDT&E, loan guarantees) on financial performance, government costs, and benefits, and job creation and Treasury revenue. Implications of market size and market share are discussed, as is the effect of space shuttle costs and costs that can be forgone, on government costs and benefits. The implications of significant price changes are discussed in terms of price elasticity and impacts on financial performance and costs and benefits.

Future deliveries of payloads to orbit are likely to utilize space transportation services that employ a fleet of reusable launch vehicles. To achieve this capability will require the investment of billions of dollars. Such sizable investments will require the formulation of detailed business plans that demonstrate the financial viability of the proposed business and that the business can function within an acceptable level of risk. Besides coping with the normal business risks associated with achievable level of technical performance, demand for launch services, competition, etc., the business must also cope with the risks associated with transportation system failures and delays. This requires the development of a viable risk management plan. A risk-management scenario that can demonstrate and provide an assurance is described, at the time that investment decisions must be made, that the necessary insurance coverage at affordable prices will be available over the long term. Without a long-term risk management plan including insurance, financing from capital markets is unlikely.

D. Space Operations (Chapter 4)

Space operations involve the placement of satellites into orbit and the maintenance of services that they provide over a period of time. Topics to be discussed include a risk-based approach for comparing advanced space transportation architectures, the selection of a launch vehicle for GEO communication satellite missions, the selection of launch vehicles for LEO missions (both initial placement and replacement), determination of launch and satellite sparing strategies that involves the consideration of cost risk, life-cycle cost and availability analyses, implications of achieving a launch-on-demand capability on satellite sparing strategies and space transportation supply/demand (resiliency/operability) considerations, on-orbit service and repair, and the implications of various government actions on international competitiveness of U.S. space transportation services. Underlying all space activities is the consideration of reliability and the taking (or not taking) of insurance as a means of managing risk.

Advanced space transportation architectures are currently being researched, resulting in a large number of alternatives that bear consideration for undertaking enabling research and development. Because not all research activities can be funded, a high-level risk-based methodology has been developed that can be used to identify those architectures which should be investigated in further detail, thereby setting aside those that are less attractive. This methodology is described and considers performance and cost uncertainties associated with the development and operations of multistage vehicles and their payload interactions.

The decision process concerned with the selection of a launch vehicle, or more appropriately, transportation service, for a GEO communication satellite mission is complex because of the many attributes of launch vehicles that affect financial decision variables. Launch vehicle attributes include price, payment schedule, reliability, lift capability, final orbit placement accuracy, insurance rate, scheduling (including possible delays), and other factors. These influence decision variables such as return on investment and risk. A financial planning model is described that transforms the transportation system attributes into decision variables. A typical analysis is described with results (including sensitivity analyses) presented. Using a decision database (described in a preceding section) together with the described financial analyses, a method is developed for forecasting launch vehicle market share in a multilaunch vehicle competitive environment.

Launch campaign modeling is concerned with the determination of the mix of launch vehicles and their scheduling for LEO satellite placement and replacement missions. Methods and approaches for analyzing launch campaigns are discussed with emphasis placed upon the consideration of reliability, failures, and resulting time delays and their effect on overall mission availability and cost. This discussion is then followed by a discussion of methods for taking into account cost uncertainty in the determination of costs via a detailed work breakdown structure. Both uncorrelated and correlated cost items are considered. The result is the determination of the probability distribution of both nonrecurring and unit recurring costs from which different levels of cost risk can be established. The developed cost risk data are then considered in the context of development of a satellite placement and sparing strategy aimed at minimizing the present value of mission life-cycle costs while meeting both short- and long-term availability requirements. A number of strategies are considered including on-orbit and ground spares, launch on failure, launch in anticipation of wear out failure, launch according to a schedule, and mixed strategies. Explicit consideration is given to launch vehicle reliability, scheduling uncertainty and time delays resulting from failures, and satellite random and wear out failure characteristics.

Recently, the possibility of achieving a launch-on-demand capability has received attention. Launch on demand refers to the ability to place satellites into orbit as and when required. Being able to launch on demand has significant ramifications for satellite on-orbit sparing, satellite design and storage, launch vehicle scheduling, and the robustness of the space transportation architecture. The implications of launch on demand are discussed with particular emphasis placed upon the determination of the launch-on-demand capability that must be achieved in order to eliminate the need for on-orbit operating spares, the methodology that is required in order to establish the robustness of a transportation architecture which will have the ability to recover from adversity (i.e., a launch failure) and to keep up with a steady-state demand for placing satellites into orbit, the ability to respond to the requirement to launch a satellite when the need arises unexpectedly either because of an unexpected (random) on-orbit failure that requires replacement or because of the sudden recognition of an unanticipated requirement, and the ability to recover from adversity (i.e., a launch failure) during the placement of a constellation into orbit. Appropriate metrics are developed for these situations.

Methods and procedures for analyzing the economics of on-orbit service and repair missions are described including the results of resent studies. Service at a satellite location, repair at a transportation node (in orbit or on the Earth), and simple placement with replacement are discussed. Mission scenarios are described that then serve as the basis of the economic analyses.

Orbital debris will increase the costs of space missions through costs incurred to conform to debris mitigation policies and/or losses that result from debris impacts. It is likely that orbital debris mitigation policies will, in the long term, reduce mission costs (below those that would result from pursuing a policy of no orbital debris mitigation) but are likely to increase costs in the short term. This is the typical investment dilemma: Is it desirable to make an investment in the near term in order to achieve long-term returns? An orbital debris economic analysis methodology is described together with the results of economic analyses of the potential impacts of orbital debris on the life-cycle costs of performing typical low-Earth-orbit missions. The regulatory implications and possible regulatory regimes are discussed in a following section.

E. Licensing and Regulatory Issues (Chapter 5)

Licensing and regulatory issues and processes relating to commercial space launches and commercially operated launch sites are discussed. This is accomplished by first describing the regulatory agencies, their roles and authorities, followed by a description of formal analyses required in support of a new regulatory action. The requirements for and the scope of regulatory economic impact analyses are then described. This is followed with a description of the economic impact analysis performed in support of U.S. Department of Transportation (DoT) financial responsibility regulation that establishes insurance requirements for commercial launches and describes the so-called government indemnification of licensed launches. Finally, orbital debris is revisited, and the regulatory implications and associated economic impact analysis are discussed. To conclude this section, various aspects of space-related insurance are described with particular emphasis placed upon third-party liability insurance and the so-called "indemnification."

The general regulatory regime associated with space launches is discussed. The agencies involved, together with their roles and their statutory authority, are described. The formal analyses required in support of a new regulatory action are described with major emphasis placed upon the need for and form and scope of supporting economic impact analyses. The basic concepts of establishing "current practice" from which the costs and benefits associated with alternative actions are measured, and the need for and considerations of alternatives are described.

To demonstrate regulatory economic impact analysis, DoT's rule making relating to financial responsibility is described. This rule making is in response to a statutory requirement for establishing insurance requirements for commercial space launches. The insurance requirements cover both insurance to cover potential damage to government facilities and potential third-party losses. The statutory indemnification requirements and interparty waiver of claims are discussed as are the establishment of current practice, the use of maximum probable loss in the setting of insurance requirements, the form of and role of risk assessment used to establish the probability of and associated levels of damage, the setting of threshold probabilities (i.e., the acceptable levels of risk for third-parties and for government facilities), and the identification of alternatives and the development of economic impacts (i.e., costs and benefits). Also discussed are industry identified issues associated with the rule making, in particular, the definition of third parties.

As the number of Earth-orbiting objects increases, so does the potential for generating orbital debris with the consequent increase in the likelihood of impacting and damaging operating satellites. Various debris remediation approaches are being considered, which encompass both in-orbit and return-to-Earth schema and have varying degrees of operations, cost, international competitiveness, and safety implications. Because of the diversity of issues, concerns, and long-term impacts, there is a clear need for the setting of government policies that will lead to an orderly abatement of the potential orbital debris hazards. These policies might require the establishment of a supportive regulatory regime. This section describes the various agencies that might be involved in orbital debris regulatory actions, the myriad of issues concerning the need for establishing government policies relating to orbital debris remediation, and proposed technical solutions and their economic and safety implications. Particular emphasis is placed upon addressing cost effectiveness and economic analysis methods and techniques as they relate to economic impact analysis in support of regulatory impact analysis.

The insurance industry has, for many years, provided an array of insurance products for the space industry. These products have included facility insurance, third-party liability insurance, launch and satellite insurance, on-orbit insurance, as well as other products. As the space industry evolves, new products and concerns emerge. The need to provide insurance cover for the anticipated large number of LEO communication satellites that will be launched on both existing and new launch vehicles with different numbers of satellites per vehicle coupled with closely spaced missions (in orbit) has caused new issues and concerns and insurance products to surface. Recent insurance industry history and product offerings are described. Insurance products and their availability and pricing are also discussed. Finally, the future of the space insurance industry is discussed in terms of likely changes in product offerings, particularly as they are likely to be affected by the needs of LEO communication satellites, the development and operational use of reusable launch vehicles, and the ever-menacing orbital debris environment as well as the possible effects of "space weather" (meteor showers, solar flares, etc.). Special emphasis is placed upon third-party liability indemnification issues and the need to develop a long-term course of government action within a competitive environment.

F. Beyond Space: Energy and Gaming (Chapter 6)

To illustrate the scope of economics and the similarity of decision making in a competitive environment and related analytic techniques, the application of economic and decision analyses and associated computer simulation techniques in the areas of energy conservation and casino gaming are considered. In the energy category, two specific applications are described. The first considers government decisions concerning investing in technology programs and the setting of policies that effect manufacturer and end-user decisions relating to building envelope and equipment with the objective of achieving energy conservation goals. The emphasis is on a stochastic market penetration model that relates government technology programs and policies to new product attributes that then affect end-user multiattribute and multiproduct replacement and purchase decisions. User decisions are considered as probabilistic. The second application considers manufacturer product design, and pricing and investment decisions in the area of manufactured housing (i.e., panel construction). Multiproduct decisions that are aimed at achieving investment objectives and are influenced by related government investments and policies whose goal is to enhance energy conservation are considered. Stochastic supply-side and demand-side models are utilized with probabilistic product choices considered in the demand-side model. These two models are then utilized to establish supply–demand equilibrium.

The discussion of gaming industry applications focuses upon the table games of Blackjack, Craps, and Baccarat with the games described and issues, resulting analyses, and decisions discussed briefly. This is then followed by a discussion of high-roller play and the use of computer simulation and decision analysis discussed in terms of awarding complementaries and negotiating with high rollers to obtain their play by offering appropriate incentive packages. The discussion of the table games first considers decisions that rely upon the long-term statistics of the games (for example, setting of game rules and procedures). This is then followed with a discussion of the short-term statistics of the games (for example, the long-term statistics of coin tossing indicates a probability of 0.5 of obtaining a head on the next coin toss; the short-term statistics are concerned with developing the probability distribution of the number of heads obtained as a function of the number of coin tosses where the number of tosses is relatively small). The importance of considering the short-term statistics in the development of money-management strategies is discussed in terms of the probability distribution of player bankroll (and the lack thereof, i.e., ruin) as a function of time. The game of Baccarat is used as an example to demonstrate, through the use of simulation, the effects of decision strategy, bet-to-bankroll ratio and money-management technique on the probability distribution of player bankroll as a function of time.

Finally, both are combined by considering the problem of real-time negotiations with high rollers in order to develop an incentive package that will result in the high-roller play but with the casino being able to achieve profitability objectives within specific risk constraints. To obtain high-roller play in a competitive environment, casinos negotiate incentive packages (complementaries, discount of losses, and bonuses) in return for player commitments (minimum and maximum bet size and duration of play). Tradeoffs are made between casino profit and risk objectives and incentive packages, taking into account player wager and decision strategies in determining necessary player commitments (i.e., minimum and maximum bet size and duration of play). The short-term simulation and mathematical modeling techniques to accomplish this and typical decision tradeoffs are described, and typical results are presented. Comparisons are made between the casinos offering incentives so as to alter player decisions and government offering of incentives to influence industry investment decisions. Both situations employ similar analytic techniques, and both situations involve actions taken, within acceptable performance regimes, in order to influence investment decisions.

References

[1]Schlaifer, R., *Analysis of Decisions Under Uncertainty*, Krieger, Malabar, FL, 1978.
[2]Edelman, F., and Greenberg, J. S., "Venture Analysis: The Assessment of Uncertainty and Risk," *Financial Executive*, Vol. 37, No. 8, 1969, pp. 56–62.
[3]Hertz, D.B., "Investment Policies that Pay Off," *Harvard Business Review*, Jan.–Feb. 1968.
[4]"Colorado's Strategic Plan for Space," Space Foundation, Colorado Springs, CO, 5 Sept. 2000.

Chapter 2

Investment Decisions

INVESTMENT decisions are normally concerned with committing current resources in order to achieve future rewards. Several different types of decisions are considered, namely those relating to unconstrained choices (i.e., evaluating and selecting investments based upon the individual investment's value), constrained choices (i.e., selecting a project set within resource constraints), and forced choices (i.e., selecting the best alternative to implement a prior decision). The reusable launch vehicle (RLV) investment decision is illustrative of the first, planning a research and development (R&D) program is typical of the second, and selecting a launch vehicle (given the decision to launch a communication satellite) is typical of the latter.

A decision implies that there are choices, and choices imply that there are alternatives that must be considered, analyzed, and evaluated in order to make the best possible decision. Comparison implies that there is at least one yardstick, or metric, that can be used as the measure for comparison and evaluation purposes. The following paragraphs describe various metrics that can be used in decision making processes; it is shown that usually multiple metrics are relied upon for taking a decision and that the appropriate metrics vary from decision to decision. Both private and public sector decisions and associated metrics are discussed. The impact of different public sector actions/initiatives on private sector metrics is also discussed. The development and use of appropriate metrics are discussed and illustrated. Examples used to demonstrate the importance of appropriate metrics include investment in space transportation, space transportation operability/resilience, and economic impacts of orbital debris remediation policies.

Because many metrics are the result of financial analysis and because most private sector investment decisions are based upon the results of financial analyses, basic concepts underlying financial analysis are discussed. Variables such as profit, cash flow, payback period, return on investment, risk, and net present value are defined, and their importance/significance in decision making is discussed. A geosynchronous communication satellite business financial-planning model is described and used to illustrate the financial analysis and evaluation of a typical communication satellite business venture. In addition, the model is used to demonstrate the sensitivity of financial performance to schedule delays, launch vehicle reliability, and satellite design life.

To establish financial impacts, it is customary to develop a business plan. The business plan serves as the basis for communication between those seeking the commitment of resources and those having the ability and authority to commit resources. The business plan provides a description of the proposed projects as well as a justification of the requested resources. It also provides an indication of the value to be received from the commitment of resources. The financial analysis, in the form of a financial plan, is a component of the business plan. Other components of the business plan include a description of the market, competition, sales forecast, and staffing plan. These and other components of the business plan are described.

Capital budgeting decisions relate to making resource investments, in the near term that will produce benefits in the long term. Evaluations of these investment opportunities involve the quantification of multiple performance measures that then serve as the basis of decisions to commit resources. Typical performance measures are described together with a multiattribute decision predictive database. Predicting the influence of government programs and policies on private sector investment decisions is described through the use of this database. A number of different government initiatives such as anchor tenancy, tax policies, guaranteed loans, and research, development, and demonstration programs are discussed together with their likely influence on decision variables such as return on investment, risk, payback period, and magnitude of investment at risk.

The use of the decision database is described in terms of the comparison of alternatives and the selection of a launch vehicle for a geosynchronous communication satellite mission. First, the comparison and selection is considered from the point of view of the customer. Next, the comparison is considered from the point of view of a provider of space transportation services. The objective is to forecast market share based upon likely decisions that will be made by the user of the transportation services. The predictive decision database is a key element in this analysis. The database describes likely choices in terms of relative expected value and relative perceived risk. The unique method for collecting the data contained in the database (i.e., the tradeoff between expected return on investment and perceived risk) is described. Finally, decisions are considered from the point of view of the government. These decisions relate to the evaluation of fair pricing of foreign competitive space transportation services and the planning of an R&D program that might affect the attributes of U.S. space transportation services and explicitly and quantitatively considers the impact on international competitiveness.

A number of typical R&D program formulation methods are discussed. These include methods applicable to the planning of a focused (or correlated) technology program, methods applicable to the planning of a technology program considering many project undertakings but in a resource constrained environment, and finally the ranking and selection of a set of technology projects from amongst a larger set utilizing a benefit/cost ratio rank ordering approach. The differences between these approaches and their appropriateness are discussed. Finally, a method is described for the public sector planning and evaluation of a technology program (using space transportation as an example) that considers the implications of the program on international competitiveness, through changes in market share, and consequent economic impacts (for example, job creation and U.S. Treasury revenue).

The impact of government programs and policies on private sector investment decisions is discussed. A number of government programs and policies are described, and their impact on private sector financial performance measures (i.e., expected return on investment, risk, payback period, and magnitude of investment at risk) is described. This is followed by a discussion of the effect of altering financial performance measures on likely investment decisions. To demonstrate the developed concepts, a change in pricing of government launch services provided to commercial users of federal launch ranges is considered and the likely consequences of government actions developed first in terms of changes in transportation pricing, then on international competitiveness (i.e., market share), and finally in terms of changes in exports/imports, job creation and U.S. Treasury revenue.

The implications of changes in international competitiveness are developed through the consideration of space transportation. When foreign suppliers capture launch business, they gain many advantages. First of all, just as loss of business harms U.S. manufacturers, gains in foreign business enable these suppliers to gain market share and hence become more competitive by expanding the base upon which they write off fixed costs. For the U.S. there is a net loss of jobs with its effects in the economy. There is a loss in revenue base to the launch industry that could, in part, have formed the basis for R&D that would improve the technology and make it more competitive in the future. And, more important, there is a loss of domestic access to space. The economic implications of foreign competition to the launch and satellite communication industries are developed through the use of a mathematical model that considers both direct and indirect effects for the loss (or gain) of a launch and its associated payload to foreign competition. The model develops both direct and indirect impacts on job creation, exports/imports, and Treasury revenue. This model is further discussed (in a following section) in terms of the economic impacts that can result from the successful development of a reusable launch vehicle that will reduce transportation prices and thereby effect demand. The economic impacts that can result from changes in launch fee pricing (for commercial users of government launch facilities) are developed.

Government frequently undertakes activities that are aimed at transferring the results of its research and development activities to the private sector. The objective of these activities is to influence investment decisions so that the R&D results are beneficially utilized. If investment decisions are not changed, there is no value obtained from the investment in technology transfer programs. A number of government programs have been specifically aimed at influencing investment decisions by performing experiments that would demonstrate technology and the ability to provide new and/or improved services. Two examples are the NASA flight experiments program and the Advanced Communication Technology Satellite (ACTS) program. These programs are used to illustrate assessments and evaluations of technology transfer activities. The former is considered from the point of view of determining a priori which flight projects should be selected for funding when limited resources are available specifically taking into account the potential impacts of the experiment in terms of the decision making processes of the target audience. The latter is considered as an example of a lost opportunity to measure the value of a demonstration program in terms of a priori perceptions and measured experiment results and likely impacts on investment decisions.

I. Metrics

When shopping in a supermarket for chicken soup, many brands are encountered. The items frequently have different actual and unit prices. The "actual price" is the price that must be paid for the selected item upon checkout. The "unit price" is the price per unit of measure (for example, dollars per unit volume or dollars per unit weight). When the different brands have different volumes or weights, both actual and unit prices are likely to be important. The former is important in an absolute sense (i.e., is the price reasonable and does it fit within the budget?), and the latter is a relative measure (i.e., getting the biggest bang for the buck). When all brands offer the same size can of soup, it is normal to consider purchasing the brand that offers the lowest unit price (given that the volume or weight of the product is sufficient and not excessive). However, if the shopper has experience with brand A and not with brand B, and the unit price of brand B is the lesser of the two, will brand B be purchased? The shopper, from prior experience, knows the taste of brand A but does not know if the taste of brand B will be acceptable. Thus, there is a measure of risk, and the shopper's decision as to whether to select A or B takes into account risk avoidance preferences. Thus even relatively simple decisions involve multiple attributes and the relative values placed upon these attributes by the shopper.

Nearly all decisions concerned with the commitment of resources (whether the decision is related to the purchase of the can of soup or the investment in a technology research and demonstration program aimed at developing a space transportation service that utilizes a fleet of reusable launch vehicles) consider, either explicitly or implicitly, multiple product or service attributes and their impacts on multiple decision measures or metrics. This is developed in the following pages.

A metric is a measure, or yardstick, against which alternatives are measured. The purpose of a metric is to establish a common basis from which the value of alternatives can be measured and compared. It should be immediately apparent that although it is possible to identify a very large number of possible metrics only those metrics that are of significance in the decision making process have value. It therefore follows that the first step in the decision process is to identify the decisions that are to be taken and then to identify those metrics that are required in order to make informed decisions.

Before becoming too immersed in the subject of metrics, several fundamental issues need to be addressed that have significant impact on the selection of appropriate metrics and resulting decisions. These concerns are addressed in the following sections.

A. Consideration of Uncertainty and Risk

Investing in the development of space transportation technology and related transportation services is considered as high risk. The risk, as measured by the perceived variability of the value of a metric or performance measure such as cost or return on investment or measured as the chance of being worse off as a result of a decision (a particularly useful concept when comparing alternatives), is the result of technology, cost, market, and other uncertainties. Not explicitly considering uncertainty and resulting risk implies that there is no difference in the level of risk associated with investment alternatives, and decisions should be made to select the alternative that appears, for example, to minimize cost (of achieving a specified set of objectives). This is illustrated in Fig. 2.1a. When uncertainty and risk are not considered, alternative A appears to be preferable to alternative B. When uncertainty and risk are explicitly and quantitatively considered, the situation shown in Fig. 2.1b can develop. In this case most decision makers would likely select alternative B because it is less risky (i.e., the costs are more predictable) than alternative A. The comparison of alternatives considering risk is illustrated in Fig. 2.1c, which indicates that preferences must ultimately be made between achieving lower expected costs but having higher risk.[1,2]

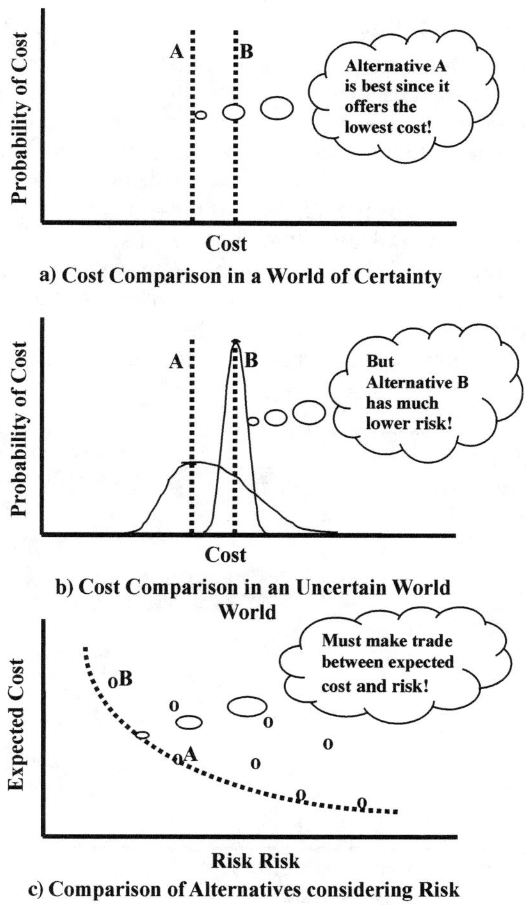

Fig. 2.1 Comparison of alternatives.

B. Consideration of Expected Values (Not Most Likely Values)

It is correct to add and subtract expected values. It is mathematically incorrect to add and subtract most likely values (except in the case where expected values and most likely values are the same). Using most likely values in analyses will lead to fallacious results. This is a fact that is all too frequently overlooked or ignored.[3] [The NASA analysis of Space Solar Power (SSP) was based upon the use of most likely values and not expected values. Thus, obtained results and related decisions are likely to be seriously flawed.] A simple method for ensuring that expected values are considered is illustrated in Fig. 2.2. Estimates are obtained that indicate the minimum, most likely and maximum values that are anticipated for each variable. The expected value (or mean) of the resulting triangular uncertainty profile (i.e., perceived uncertainty profile) is simply the sum of the three values divided by three. It is obvious, as illustrated in Fig. 2.2, that there can be significant differ-ences between the most likely values and expected values.

C. Consideration of Multiple Metrics

It is a known fact that investment decisions are normally taken after due consideration of multiple metrics. (This is discussed in considerable detail in Secs. IV and VI.) The consideration takes into account, either implied or explicitly stated, tradeoffs or preferences amongst the multiple decision metrics and their estimated values. In the private sector the frequently considered decision metrics include return on investment, payback period, risk, net present value, relative magnitude of investment, annual profit and cash flow, time to first profitable year, etc.[4] The specific mix considered and their relative importance depends upon the type of investment and decision makers (corporate decision makers view the world somewhat differently than venture capitalists). Figure 2.3 illustrates the relationship between expected return on investment, risk, payback period, and

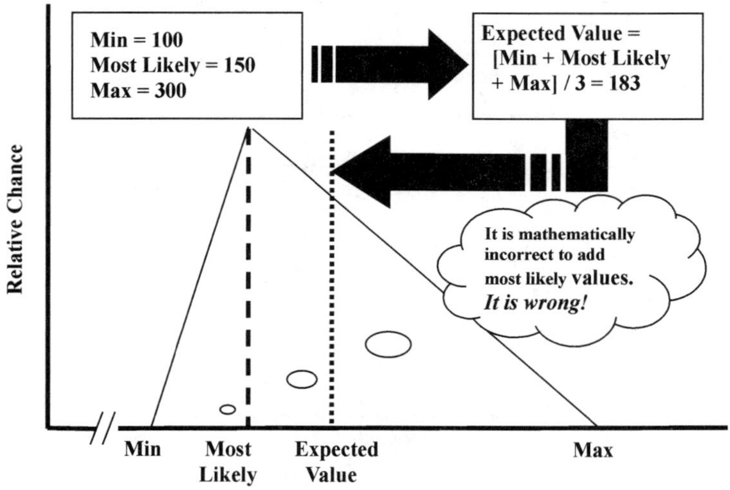

Fig. 2.2 Importance of utilizing expected values in computations.

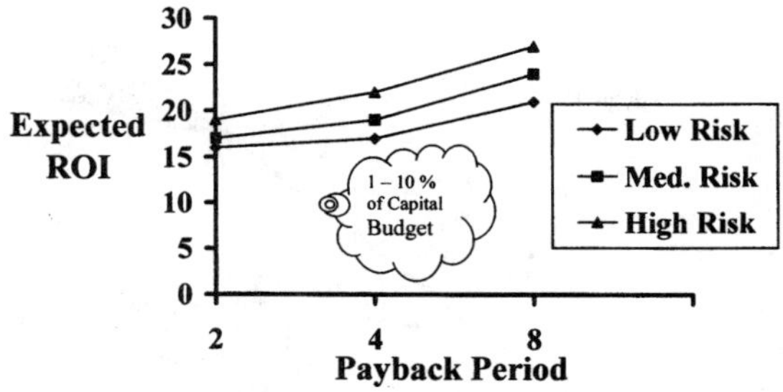

Fig. 2.3 Expected ROI required to achieve a 50% chance of obtaining funding in terms of normal business risk.

magnitude of investment on the likelihood of obtaining financing. Government investment decisions increasingly consider both the expected and risk dimensions of costs and benefits.[5] The space shuttle economic analysis included considerations of nonrecurring and unit recurring cost risks.[6] The point is that neither in government nor industry are decisions based solely on single attributes or metrics.

D. Need for Simplicity

There usually exist tradeoffs between simplicity (both in data collection and in analytic structure), reality, and the development of meaningful information. For example, the consideration of uncertainty and risk implies the need to increase the quantity of data utilized in analysis but tends to ensure that expected values rather than most likely values are considered and that quantitative measures of risk are developed. Analytic tools become somewhat more complex when the risk dimension is to be considered. But not adding these complexities is tantamount to saying that all alternatives [for example, single-stage-to-orbit (SSTO) and two-stage-to-orbit (TSTO) transportation architectures] have the same level of risk. Simplicity has its place but not at the sake of developing information that can lead to improved investment decisions, particularly relating to space transportation systems where investments can be in the order of billions of dollars. It is very disconcerting and somewhat alarming to be told by NASA engineers that the consideration of uncertainty and risk is too complicated and data intensive to be utilized on a regular basis.

E. Importance of Demand Forecast

Unless comparisons of alternatives are limited to the consideration of only nonrecurring cost, and possibly first unit cost, in one way or another it will be necessary to take into account forecasts of the demand for space transportation services. There is no getting away from the fact that this adds considerably to the uncertainty surrounding the development of annual recurring cost and all financial performance metrics, including price, return on investment (ROI), payback period, etc. The sensitivity of alternative transportation investment metrics to demand forecasts is an important consideration and can be a significant factor in the risk

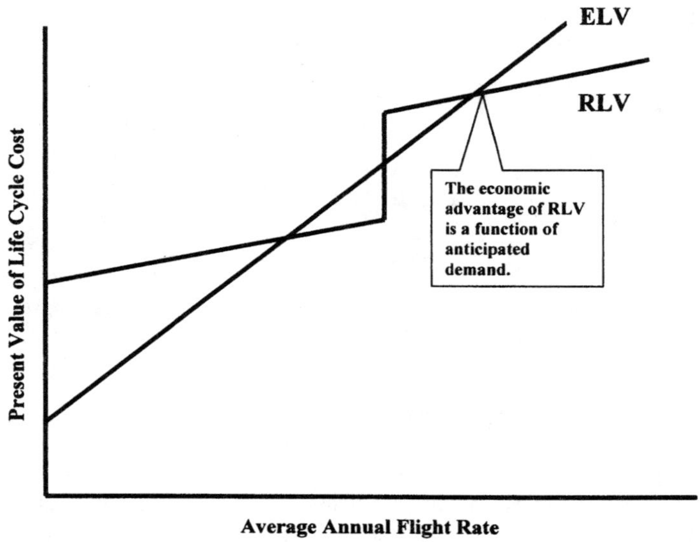

Fig. 2.4 Present value of life-cycle cost for expendable and reusable space transportation systems.

metric of alternatives under consideration. For example, Fig. 2.4 illustrates the general form of the present value of the life-cycle costs for expendable launch vehicle (ELV) and RLV based transportation architectures. ELV architectures can be tailored more easily to actual demand or short-term demand forecasts, whereas RLV architectures must consider long-term forecasts when establishing fleet requirements. This has an impact on risk.

In summary, metrics developed to aid in the decision making process should take into account the following facts of life:

1) It is mathematically incorrect to utilize most likely values; it is mandatory to develop and utilize expected values.
2) Risk is an important consideration when choosing between investment alternatives.
3) Decisions are normally based upon the consideration of multiple metrics.
4) Reality should be of more concern than simplicity both in data collection and in analysis. (Simplicity is good as long as it does not materially distort reality.)
5) Demand forecasts play an important role and cannot be neglected.

F. Example: Space Transportation Investment Decisions (Cost- vs Price-Based Metrics)

Historically, cost and cost per pound of delivered payload have been the metrics relied upon for a wide range of space transportation investment decisions. Recently, and as a result of the potential private-sector investment in the development of reusable launch vehicles, a much broader perspective has emerged that involves metrics such as return on investment, capital requirements, risk, and pricing. The difference between the two being that the former (i.e., cost-based metrics) seeks cost minimization, whereas the latter (i.e., price-based metrics) seeks profit maximization. Knowledge of cost is always important, but clearly there are situations where other metrics are of considerable import in the decision making process.

The importance of the choice of space transportation cost- and/or price-based metrics must be judged in terms of the decisions to be taken. Cost and/or price have significance only when decisions might be affected and the appropriateness of the use of cost or price or other metrics must be judged in terms of the correctness of resulting decisions.

Space transportation decision makers include the users of space transportation services, the providers of space transportation services, and government agencies that support technology development and formulate space transportation policy. Their decisions are concerned with a broad range of issues including user decisions relating to the selection of an appropriate launch vehicle, provider decisions relating to the improvement of existing launch services and the development of new launch services, and government research and development initiatives aimed at providing a technology base for future space transportation services.

The appropriateness of cost- and/or price-based metrics for space transportation related investment decisions is a function of the decisions that are to be taken. Because most decision processes consider multiple attributes (for example, magnitude of investment, risk, timing, etc.), it seems rather inappropriate to place emphasis on a single metric such as cost per pound or price to achieve a specified rate of return. Therefore, in the following paragraphs emphasis is placed upon the consideration of multiattribute decision processes that explicitly and quantitatively include risk as an important attribute. [Unfortunately, it is likely that risk considerations were not adequately taken into account in government decisions relating to investment in technology to support SSTO (i.e., the X33 program) or TSTO transportation architectures.] In addition, a metric relating to investment likelihood is considered for those decisions that involve second- and third-party decisions that are beyond the direct control of the first party (for example, establishing government programs and policies that are aimed at altering private sector investment decisions).

Unfortunately there appears to be no simple metric such as cost per pound or price per pound that has universal application. In fact, appropriate metrics are a function of the decision that is to be taken. This is the subject of the following paragraphs.

1. Cost Considerations

Cost estimating relationships (CER), are frequently used for developing cost estimates during the early planning and development phases and providing the cost information that is a necessary component for decision making. Cost estimates of nonrecurring and recurring costs are required for making informed decisions and must include estimates of all costs that will influence the decisions to be taken. These costs include those relating to technology development, inventory, fleet acquisition, ground operations, flight and mission operations, program administration and system management, insurance, etc. These costs and related expenses and data [such as depreciation lives, cost of capital, financing scheme, (a generalized financing scheme is discussed in Chapter 3, Sec. IV) etc.], together with revenue forecasts, should be of sufficient completeness to make possible the establishment of cash flow projections and the development of decision metrics.

In the early planning and evaluation stages of a program, heavy reliance is normally placed upon the use of cost estimating relationships. The CERs must be used carefully if uncertainty of input data is to be transformed into cost risk information. Unless all functional relationships between performance parameters, and between performance parameters and cost are linear, a simple graphical or analytical approach cannot be used to transform input variable (for example, mass, power, specific impulse, etc.) uncertainty into resulting cost risk. The required and correct transformation process can be accomplished by utilizing a mathematical model that employs Monte Carlo techniques that perform random sampling of input data, which is then utilized in integrated mathematical functional relationships (i.e., a cost model integrated with a performance model) to transform all sampled data into the resulting cost probability distributions. In other words, if uncertainty and risk are to be considered the CERs should be encompassed within the framework of a mathematical design model that configures the launch vehicle based upon the random sampling of the input data with the resulting design data then utilized to enter the CER. This random sampling of input data, when performed a large number of times, results in the probability distribution of the cost of the launch vehicle under consideration.

CERs provide a functional relationship between parameters such as mass and power and cost. The CERs are derived by fitting curves through a set of historic data points (the actual cost of previously developed products in terms of specific product attributes) utilizing least-mean-square-error techniques. As such, they provide estimates of expected cost when expected mass, power, or other attribute is specified. All too frequently the estimates of mass, power, etc. are most likely values. When this is done, the resulting cost estimates are not expected values, but most likely values and the costs of different subsystems are not additive.

2. *Pricing Considerations*

Pricing considerations play a major role in private sector investment decisions because it is the interrelationship between price, capital expenditures, expenses, and costs that lead to the determination of the set of financial performance metrics upon which decisions are taken. As a result, a number of pricing schemes warrant consideration during the planning and evaluation process leading to space transportation investment decisions. These include pricing to maximize profit but with prices constrained by competitive pricing; two-tier pricing (a government price for a specified duration with commercial pricing to maximize profits); pricing such that the transportation service price is a specified fraction of the anticipated competitive price; pricing in a competitive market so as to achieve a specified market share; setting price at a specified fraction of anticipated competitive price, then to maximize profit (analogous to product introduction pricing); and pricing as a constant times operating cost. The first two are of primary importance with the former of concern because it is similar to current practice (A mathematical model for this price determination is presented in Sec. IV of Chapter 3.) and the latter because it is a way of providing an assurance that the government can receive cost savings as a result of its investment in technology and the provision of incentives (i.e., justification for government investment).

Whatever the pricing scenario, the consideration of price together with costs, expenses, and other factors, results in the quantification of financial performance metrics such as annual profit, cash flow, indebtedness, payback period, and risk. These are the metrics that are important in the decision process with price, cost, and other factors acting as intermediaries.

3. *Decisions/Decisions/Decisions*

Keeping the preceding discussion of cost, price, uncertainty, and risk in mind, metrics are now discussed in terms of several investment situations.

Government investment in a focused transportation technology program. Government investment in the development of space transportation technology might be aimed at the development of a transportation system primarily for achieving government objectives (for example, providing space station transportation services or providing transportation for manned Mars exploration) or for assisting industry in the development of new and reduced price transportation services. In the former case government has the ability of specifying a set of requirements that must be satisfied by alternative technology programs and resulting transportation systems. The approach has been to select the alternative that minimizes the cost (hopefully the present value of cost and hopefully all costs are expected values) of achieving the stated objectives. This is not adequate when the alternatives being considered have differing levels of cost and performance uncertainties and therefore different levels of risk, for example, the probability of achieving the stated objectives. Usually, performance is assumed to be achievable, and the uncertainty is with respect to the cost of achieving the desired performance. Therefore, metrics should include both expected cost and the standard deviation of cost.

When the resulting level of performance is uncertain, the objective should include a statement of the desired probability of achieving performance objectives, and the cost should then be the cost (expected value and standard deviation) associated with meeting the objectives at the stated level of acceptable risk. (A methodology has been developed for planning and evaluating long-term multiphase technology programs that explicitly considers the uncertain nature of technology development, cost, and performance and the likelihood of achieving overall program objectives. This is described in Ref. 7. A methodology has also been developed for screening alternative transportation system investments and includes the consideration of both expected costs and the standard deviation of cost. This is described in Ref. 5.) It is inappropriate to compare alternatives based upon a single measure such as program cost or cost per pound of delivered payload. Because risk is of concern, explicit and quantitative risk information must also be provided for the decision process. In addition, it is inappropriate to compare the cost of alternative development programs unless they are all configured so as to achieve the same set of objectives. Technology programs configured to support SSTO and TSTO architectures cannot be compared on a cost basis unless an equal capability of resulting architectures is considered. The probability distribution of the cost to achieve the same level of performance will differ with the risk dimension being an important component of the comparison metrics.

In the latter case when government initiatives are undertaken specifically to effect private sector investment decisions, it is necessary to assess the impacts of these initiatives on private sector investment decisions, over which the government has little or no control. Therefore, from the government's point of view it is only possible to predict these decisions in terms of the probability, or likelihood, that specific decisions will be made given a set of financial performance measures. The results of government analyses of private sector business ventures, and the impacts of government programs and policies on these ventures, should develop the performance measures, or metrics, that will result in the estimation of the likelihood of private sector investment. Because surveys have been conducted[4] that have resulted in the development of investment likelihood functions in terms of expected return on investment, risk, expected payback period, and magnitude of investment, these parameters become important metrics that, taken together with the survey investment likelihood data, result in the estimation of investment likelihood as a function of government initiatives. The financial performance measures are intermediaries and are needed for the determination of the change in likelihood of investment brought about by each of the considered government initiatives. If a government program is undertaken to effect private sector investment decisions, benefits result only if decisions are altered; there are no benefits if decisions are not effected. Thus, a useful metric should relate to the change in the likelihood of private sector investment that results from the various considered government programs and policies.

Because different launch vehicles can be optimized for different market segments (placement of LEO payloads, placement of GEO payloads, space tourism, space-station support, space solar power system support, achieving a launch-on-demand capability, etc.) and can utilize different technologies, government decisions relating to technology development must go beyond the consideration of costs and must consider the benefits that might result from its actions.

User selection of a launch vehicle. The decision process concerned with the selection of a launch vehicle, or more appropriately, transportation service, for a particular mission is complex because of the many attributes of launch vehicles that affect financial decision variables. Launch vehicle attributes include price, payment schedule, reliability, lift capability, final orbit placement accuracy, insurance rate, scheduling (including possible delays), and other factors. These influence financial decision variables, such as return on investment and risk, which are based upon an analysis of the business or mission utilizing each of a number of alternative launch vehicles and estimates of the transportation service attributes.

Users of space transportation services are obviously concerned with the cost of the provided services. However their cost is the provider's price and not the provider's cost per pound.

Industry development of a new launch vehicle. Industry development of a business venture based upon the development and use of a new launch vehicle must surmount a number of financial hurdles, including the uncertainty of future demand, the high level of initial investments, potential competition from existing and future expendable and reusable launch vehicles, obtaining sufficient third-party liability and launch and payload insurance at reasonable rates, and the difficulties of adjusting current policy mechanisms to new government–private cooperative arrangements. Technology risk must all but be eliminated in order for the capital markets to provide funding at reasonable rates. This implies that the development cost will be a function of the technology risk associated with a transportation architecture. The cost of risk reduction therefore becomes a significant consideration, and all alternatives must be considered at the same risk level.

Because it is likely that industry will look to government for assistance, it is important to have an understanding of the implications of differing government policies on promoting venture financing. This understanding must be developed prior to industry investment decisions and requires the government to have insight into the likelihood of different policies affecting investment decisions and the reasonableness of alternative business scenarios. After all, the government will be playing the role of an investment banker (although without the degree of accountability) and, as such, should seek to maximize the return from its resources.

As a result of various surveys and analyses, it has become clear that investment decisions are not normally based upon the consideration of a single financial performance measure. Investment decisions, particularly those involving large resource commitments, consider a broad range of information. Of particular importance are such financial performance measures as ROI, magnitude of investment, payback period, and risk. Risk represents the variability of financial performance measures that results from uncertainty associated with inputs to the analysis. Risk is most frequently considered as the variability of ROI (and indicated as the standard deviation of ROI). In any event the development of the decision metrics requires the consideration of nonrecurring and recurring costs, demand forecasts, competitive pricing, and many other factors. Cost estimates are important but are not the metric used for decision-making—cost plays an important role in the development of the decision metrics.

Summary. Different decisions utilize different metrics. In all cases metrics should be based upon expected values and standard deviations. This is necessary because we live in an uncertain world with the result that all decisions must be taken with risk being a key consideration. Decisions to be taken and reasonable metrics are summarized in Table 2.1. Government decisions to invest in a focused technology program aimed at achieving government objectives (i.e., the development of a transportation system primarily for government use) should utilize expected cost (present value of nonrecurring and/or present value of life-cycle cost) and associated risk measures (in the form of the standard deviation of forecasted costs), and/or the expected value and standard deviation of the cost to achieve a specified likelihood of achieving stated objectives. Use can also be made of expected cost and risk for screening of alternatives at an early stage of planning.

Government actions aimed at influencing private sector investment decisions must consider the likely effect of these actions on private sector decisions and the

Table 2.1 Decisions and decision metrics

Decision to be taken	Decision metrics
Govt investment in a focused technology program aimed at:	
Achieving government objectives	Expected development cost Standard deviation of development cost Expected present value of life-cycle cost Standard deviation of present value of life-cycle cost Probability of achieving objectives
Effecting private-sector investment decisions	Expected ROI Standard deviation of ROI Magnitude of investment Payback period Likelihood of investment Expected government investment Standard deviation of government investment Expected government benefits
User selection of a launch vehicle	Expected ROI Standard deviation of ROI
Industry development of a new launch vehicle	Expected ROI Standard deviation of ROI Magnitude of investment Payback period

effect of these decisions on benefits (such as government cost savings, job creation, and Treasury revenue). (In addition, the cost of government action on the government needs to be considered, for example, government guaranteed loans have a cost to the underwriting agency based upon Office of Management and Budget (OMB) determination of risk of loss.) The metrics are therefore related to private sector investment decisions and include expected ROI, risk, magnitude of investment, and payback period. These, in turn, can be used to establish the likelihood of investment and expected government benefits.

User (for example, a provider of communication satellite services) decisions concerned with the selection of a launch vehicle normally place emphasis upon expected ROI, variability or standard deviation of ROI that results from the utilization of alternative launch vehicles in the business venture, and the schedule/availability of the launch vehicles.

Industry decisions relating to the development of a new launch vehicle normally consider metrics such as expected ROI, risk, magnitude of investment, and payback period. This is not to say that other metrics are not considered, but the aforementioned are extremely important. Nonrecurring cost, recurring cost, and other cost and expense elements are all encompassed in these metrics together with demand forecasts, competitive pricing, and other normal business factors. These are not decision metrics but are necessary for the determination of the decision metrics.

G. Example: Transportation Architecture Resiliency/Operability Decisions

Decisions concerning the development and use of space transportation systems have, in the past, placed emphasis primarily upon cost. The lack of other than cost considerations, as just discussed, in particular the lack of consideration of unreliability and its consequences, was a major factor in the inability to launch payloads for a significant period after the loss of Space Shuttle *Challenger*. This demise of the U.S. space transportation capability in the late 1980s strongly suggests that other than cost considerations are important. (Failures of U.S. expendable launch vehicles, in addition to the loss of *Challenger*, resulted in an extended period of time during which space transportation was unavailable, except for the use of foreign launch vehicles.) It has been suggested that "resiliency" might be an important consideration in defining robust future architectures.[8,9] Resiliency is defined as the ability to recover readily from misfortune and diversity—in other words, resiliency is concerned with the ability to recover from transportation system failures as manifested in the ability to deliver payloads to orbit. Resiliency can also be thought of in much broader terms that make it synonymous with "operability." It can be thought of as the overall ability to place payloads into orbit as and when required considering all of the realities including less than perfect reliability and consequent failures, standdown time resulting from failures, and constraints such as procurement cycles, number of vehicles maintained in inventory, and many other factors.

Methods for analyzing resiliency and operability are described in Sec. X of Chapter 4. Suffice it to say at this point that the primary measure of transportation system operability should be based upon the objective of the transportation system; the placement of payloads into orbit. Thus the primary measure of operability should be concerned with payload queue or backlog as a function of time. Long queues or queues that increase over time indicate that the transportation system is not responding to payload needs.

It is now understood that future transportation architectures should be planned and evaluated explicitly considering the ability to place payloads into orbit as and when required. This requires the consideration of and planning for the recovery from failures resulting from less than perfect reliability. It is shown in Sec. X of Chapter 4 that the consideration of less than perfect reliability and consequent failures, standdown time resulting from failures, and constraints such as procurement cycles, number of vehicles maintained in inventory, etc., results in the need to include metrics that relate to the probability distribution of payload inventory (i.e., queue length) as a function of time, the probability distribution of launch vehicle (by stage) inventory as a function of time, the probability distribution of the percentage of the time (annually and across the planning time horizon) that the transportation system is operating in a surge mode, and the expected value and standard deviation of the number of launch successes, launch failures, total launches, and desired number of launches (as determined by payload

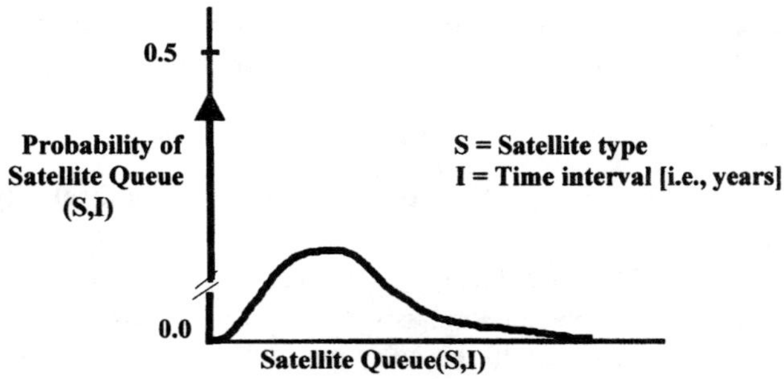

Fig. 2.5 Typical form of annual payload (satellite) queue statistics.

arrival rates and average number of payloads per flight) annually and the expected value across the planning time horizon. These descriptive statistics, though somewhat complicated, are necessary because simply using expected values is meaningless, as illustrated by the annual payload queue statistics summarized in Fig. 2.5.

Performance measures can be developed that describe the responsiveness of the transportation architecture to meeting satellite delivery (to orbit) needs. Because the degree of nonresponsiveness of an architecture is represented by the shape and magnitude of the probability distribution of satellite queue given that a queue exists, a reasonable performance measure can be the expected value of the satellite queue given that a queue exists (normalized and expressed as a percentage of total satellite demand). This effectiveness measure is of the form

$$\text{Effectiveness}(\%)$$
$$= 100 - \frac{100}{\sum_{i=1}^{T} \text{Desired Sats}(i)} * \sum_{i=1}^{T} \left\{ \frac{1}{p[Q(i,0)]} * \sum_{j=1}^{N} p[Q(i,j)] * Q(i,j) \right\}$$

where

$\text{Desired Sats}(i) =$ demand for satellites as a function of time

$p[Q(i,j)] =$ probability of satellite queue length $Q(i,j)$

$Q(i,j) =$ satellite queue length j at time i

$T =$ number of years in time horizon

This effectiveness measure represents the degree to which the expected value of satellite queues (given that the queue is greater than zero) exceeds zero, that is, the expected queue when the architecture cannot meet satellite delivery demands. Long and/or high probability satellite queues represent a nonresponsive architecture. A highly responsive architecture will have an effectiveness near 100% with less responsive architectures having lower values. The effectiveness measure partially takes into account the risk dimension by considering both the magnitude of the queue and the associated probability distribution. (It should be noted that additional emphasis can easily be placed upon the risk dimension by multiplying by the square of the satellite queue rather than by the satellite queue.)

H. Example: Capital Investment Decisions

An investment decision involves the commitment of resources with the hope of obtaining future rewards. Many different metrics have been developed and are in common use for establishing the value, both absolute and relative, of an investment decision. The specific metrics of importance depend upon the decision maker. As a result of various surveys and analyses,[4,10] it has become clear that investment decisions are not normally based upon the consideration of a single financial performance measure. Investment decisions, particularly those involving large resource commitments, consider a broad range of information. Of particular importance are such financial performance measures as ROI, magnitude of investment (the maximum value of indebtedness), payback period, and risk. Risk represents the variability of financial performance measures that results from uncertainty associated with inputs to the analysis. Risk is most frequently considered as the variability of ROI (and herein considered as the standard deviation of ROI). These and other metrics and their determination are discussed in detail in Sec. II.

Because the use of multiple metrics is the rule, it is obvious that tradeoffs are necessary between established values so that a final decision, that is, to commit or not, can be taken. Such tradeoffs between metrics can be, and have been, developed. A typical tradeoff between metrics is illustrated in Fig. 2.3, which indicates longer payback periods require greater ROIs in order to achieve the same chance of obtaining funding.

I. Example: R&D Program Investment Decisions

Government R&D programs (in the context of this discussion, include applied research, development, and demonstration programs) are undertaken to develop and demonstrate technology that will 1) reduce government costs and/or increase its capability (for example, NASA's program aimed at improving the capability of its space transportation system), 2) improving the international competitiveness of existing U.S. businesses, or 3) transferring the technology to industry resulting in cost reductions, increased capability of existing products/services, or the provision of new products/services. Similarly, private sector R&D programs are aimed at improving competitiveness and developing new products/services.

The purpose of metrics is twofold: first to determine if the development/demonstration of the technology has merit (i.e., do the benefits, given success, exceed the costs?); and, given that the benefits exceed costs, to identify the specifics of the R&D program alternative that maximizes net benefits (often within stipulated constraints such as maximum allowable annual budgets). Government metrics typically include cost savings, change in consumer and producer surpluses, change in gross national product (GNP), change in Treasury revenue, and job creation. Of course these are all the result of the government's investment in the technology program, and, of course, all are established as at the very least, expected values.

Other metrics need to be considered such as the likelihood that the R&D program will achieve its objectives and the consequences if the objectives are not achieved or are exceeded. At the very least it is necessary to multiply the benefits (as indicated in the preceding paragraph) by the likelihood of achieving objectives. When the R&D program is undertaken to influence private-sector investment decisions, it is further necessary to establish the likelihood that the private sector will make the desired decisions given different outcomes of the R&D program are obtained. The consideration of the probability of success and the probability of private-sector investment are discussed in more detail in following sections.

Also of concern is the formulation of long-term multiphase technology programs, such as might be undertaken in support of a Mars mission or the development of a commercial space solar power system. The metrics associated with such an activity and the necessary methodology are discussed in Sec. VII.

Private sector metrics are somewhat narrower in scope. Instead of being concerned with global measures such as consumer surplus and producers surplus, they are concerned with changes in profit (both near-term and long term), cost savings, cash flow, indebtedness, return on investment, present value of cash flow, risk, maintenance and/or increase in market share, etc. Again, all values should at least be expected values. As with government programs, the likelihood of achieving objects should be developed and its consequences considered.

J. Final Word About Metrics

At this point it should be obvious that the metrics of import depend upon the decision to be taken. Metrics have value only when they can influence a decision that is to be taken. Metrics should be developed to the extent that they are useful and important to the decision maker. For example, identifying the anticipated first year of profitability of an RLV business venture is likely to be important when funding from the capital markets is being sought, but is likely to be of no concern to a government agency planning a technology program aimed at developing the technology upon which a commercial RLV could be based.

Selection of appropriate metrics should concentrate on identifying those factors that can separate alternatives under consideration. When alternatives under consideration are perceived as being risky, then it should be immediately apparent that whatever metrics are developed, quantitative measures of risk should be included. When relatively large capital investments are being considered, it is not adequate to argue that the development of specific metrics, particularly risk-based metrics, is too difficult or that it is not possible to quantify uncertainty.

II. Financial Analysis

An investment decision involves the commitment of funds with the hope of future benefit. Many investment decisions are concerned with transforming an idea or emerging technology into a business venture. This transformation process requires the use of capital and labor resources to develop and provide products/services that satisfy market needs in such a manner that market-determined rates of return are provided on employed capital. To obtain necessary capital (from investment banks, venture capital firms, corporations, and other parties), it is necessary to prepare a business plan. A major objective of the business plan is to demonstrate to the investment community (including corporate management) the need for and the magnitude and timing of resources to be employed, products/services to be developed and provided, markets to be satisfied, and the financial performance that will result from the planned business venture. The business plan serves as the justification for the capital and other forms of investment. It also, upon commitment of resources, serves as both the initial plan for conducting business and the means for measuring performance. The business plan requires continuous updating as status and information available change.

The business plan normally consists of several supporting plans and analyzes the specifics and details of which are functions of the anticipated source of capital. These can include (as necessary to obtain financing) a market analysis and resulting sales forecast, a competitive analysis including rationale for achieving forecasted market shares, a detailed product/service cost analysis, a financial analysis that develops proforma income, cash flow and balance sheet statements for the considered planning horizon, a market/sales plan that identifies the distribution channels, sales force (and training), maintenance requirements, an advertising and promotion plan, a staffing plan for obtaining required personnel, and a management plan including identification of required and available skills (including identification of key individuals). These are all mutually supporting and feed the financial analysis.

The role of the financial analysis is to integrate all of the supporting analyses and plans into a framework that is reasonably standardized and readily understandable by potential investors. It provides the *raison d'etre* for the investment decision in the form of anticipated financial performance measures such as annual profit (loss), annual and peak cash requirements, payback period, net present value of cash flow, and various rates of return such as return on sales, return on assets and return on investment, and quantitative measures of risk (i.e., the possible variability associated with key financial performance measures).[4]

Financial analysis is concerned with integrating the results of other analyses, forecasts, and plans and establishing their financial consequences.[11] The financial consequences are usually expressed in terms of after-tax profit (revenue less cost and less taxes), cash flow (after-tax profit less capital expenditures, plus depreciation and less changes in other balance sheet items), indebtedness (negative of cumulative cash flow to any point in time with the maximum indebtedness being indicative of the maximum funding requirement), payback period (the time at which indebtedness becomes negative, that is, the time when net cash inflow is equal to net cash outflow), return on sales (ratio of after-tax profit to net revenue), return on assets (ratio of after-tax profit to value of assets less accumulated depreciation), net present value of cash flow [the discounted—at the firm's cost of capital—value of the cash flow stream suitably adjusted so as to eliminate financing costs (Not eliminating financing costs from the present value computation would amount to double counting.)], and discounted return on investment (the discount rate that results in a net present value of cash flow equal to zero). These parameters are described in more detail in the following paragraphs.

Before-tax profit is the difference between revenues and expenses. After-tax profit (ATP), adjusts this amount to account for federal and other taxes and takes into account carry-forward losses and available tax credits. [ATP is given in simplified form by

$$\mathrm{ATP}_t = \left[1 - \frac{\text{Tax Rate}}{100}\right] * \left[\sum \text{Revenues}_t - \sum \text{Expenses}_t - \text{Depreciation}_t\right]$$

where the subscript t indicates the time period (i.e., years).]

Capital expenditures are not explicitly included in the profit computation, but occur indirectly (and in any one year only partially) through the depreciation expense. Cash flow (CF), on the other hand, reflects the flow of funds through the business. CF is given by

$$\begin{aligned}\mathrm{CF}_t = {} & \mathrm{ATP}_t + \text{Depreciation}_t + \text{Change in Payables}_t - \text{Change in Inventory}_t \\ & - \text{Change in Receivables}_t - \text{Capital Expenditures}_t\end{aligned}$$

The cash flow computation includes the magnitude and timing of the inflow and outflow of funds. It includes such measures as after-tax profit, depreciation, increase in payables (i.e., due but not yet paid), decrease in inventories, and decrease in receivables (i.e., due but not yet received) as cash inflows (i.e., sources of funds), and such measures as losses (negative values of ATP), capital expenditures, decrease in payables, increase in inventories, and increase in receivables as cash outflows (i.e., uses of funds). Cash flow (which includes profit and loss as a component), not profit, is the most important determinant of the value of a venture. Profit is an accounting artifact; cash flow is a basic measure. A profitable business venture can fail because of cash flow problems. The significance of profit cannot be overlooked, however, because it is a key consideration when evaluating the availability of funds from the financial community. (Stock prices are usually measured in terms of price-earnings ratios.)

Figure 2.6 illustrates, in simplified form, the process for establishing annual profit and cash flow. Indebtedness is defined as the negative of the cumulative cash flow to any point in time and

$$\text{Indebtedness}_T = -\sum_{t=1}^{T} \text{Cash Flow}_t$$

where T is the point in time at which indebtedness is to be measured. Figure 2.7 illustrates typical profit, cash flow, and indebtedness patterns.

The cash flow and profit streams normally start off as net cash outflows and losses, respectively, as a result of R&D expenditures, engineering efforts, and initial operating or startup costs, which precede revenue from sales. Maximum annual net cash outflow decreases, eventually becoming a net cash inflow. The maximum funding requirement, that is, exposure, is indicated by the peak of the indebtedness curve. When the indebtedness is positive, the total investment has not been recovered, and the cumulative cash outflow exceeds the cumulative cash inflow. When the indebtedness is negative, the cumulative cash inflow exceeds the cumulative cash outflow. The indebtedness decreases to zero when sufficient cash

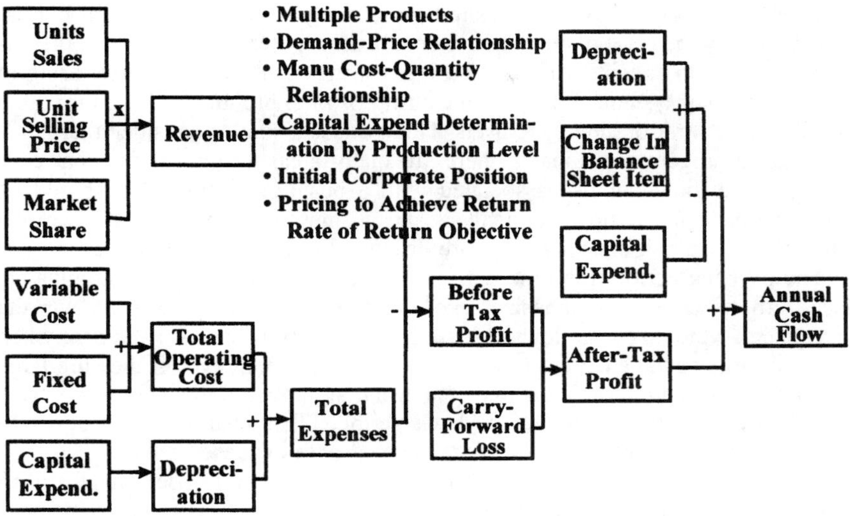

Fig. 2.6 Simplified profit and cash-flow structure.

has been generated to "pay off" the total investment. The time for this to occur is referred to as the "payback period."

As indicated, the financial analysis appears to be rather straightforward but in real life is complicated by many factors including the need to consider many time periods, multiple products that might be interrelated, price and manufacturing cost elasticities, alternative pricing policies ranging from specifying annual prices to

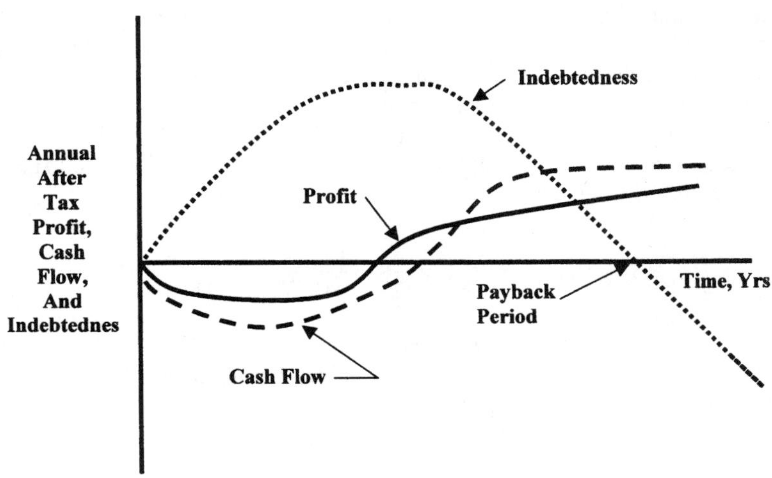

Fig. 2.7 Typical profit, cash flow, and indebtedness patterns.

establishing prices that provide desired gross margins or rates of return but are constrained by the price of competitive or substitute products. In addition, there is usually a need to consider different business structures such as manufacturing of components, subsystems, etc., manufacturing of final products selling through a distributor network or directly to users. There is also a need to consider alternative financing arrangements. Finally, there are many areas of uncertainty that contribute to the financial risk associated with a business venture that should be conveyed through the financial analysis and business plan to potential investors. Uncertainty and risk considerations are discussed in the following pages.

The payback period (the time until cumulative cash inflow equals cumulative cash outflow) criterion emphasizes exclusively the time required to recover the investment. Note that cash flows that are anticipated to occur after the time when net cash inflow equals net cash outflow are totally ignored even if they are large and persist for long periods of time. The usual argument is that shorter payback periods are preferable to longer payback periods. This results from a desire to avoid risk. Longer payback periods are equated with increased risk.

Another performance measure, somewhat less frequently used, is return on assets. It is the ratio of after-tax profit to asset value where asset value includes cash, receivables, inventory and book value (capital expenditures less accumulated depreciation) of capital items.

Present value or discounted cash-flow analysis seeks to adjust cash flows occurring in future time periods in a way that allows for their timing. The rationale behind the adjustment is that a dollar received in the future is worth less than a dollar received today, because the dollar in hand today could be put to work to create additional earnings. The adjustment process, known as discounting, establishes a present or "now" value of future cash flows. Shifting a proposed project's future cash flows to their now equivalent makes it possible to arrive at a single figure representing the value of the project. This in turn allows fair comparisons of alternative projects with different cash flow patterns over time and payback periods.

The computational approach is to reduce the cash flow occurring in a particular future period by a discount factor such that the discounted amount is the one that, if invested at the discount rate from the present to the corresponding future time, would be equal to the unadjusted value. In a sense this process is the complement of compounding interest on a savings account. Net present value (NPV) refers to the present value of a future cash flow stream. [NPV of a cash flow occurring in a given time period t is given by

$$\mathrm{NPV}_t = \frac{\mathrm{CF}_t}{(1 + r/100)^t}$$

where r is the discount rate (percent) and it is assumed that the cash flow occurs at the end of the time period. The net present value, that is, the algebraic sum of all present value contributions, is given by

$$\mathrm{NPV} = \sum_{t=0}^{n} \frac{\mathrm{CF}_t}{(1 + r/100)^t}$$

where n is the planning horizon and is finite. CF_0 implies the cash flow at the beginning of the first period.]

The NPV of a project depends on the magnitude of the cash flows, their timing, and the discount rate. (If cash flow is in current dollars, then the discount rate should include an inflation component. If the cash flow is in constant dollars, then the discount rate should not contain an inflation component and is the "real" discount rate.) This is illustrated in Fig. 2.8, where a typical cash flow pattern is indicated along with its NPV as a function of discount rate. If the discount rate is low, the NPV is high, because future cash flows are not heavily discounted. If the discount rate is large, the NPV normally becomes negative because of the heavy discounting of future cash flows (in fact, as the discount rate becomes very large, the net present value approaches asymptotically the value of the initial cash flow component).

Central to the use of the NPV criterion is the choice of the appropriate discount rate. Although economists have generally agreed that the firm's adjusted weighted average cost of capital (taking into account the specific debt/equity situation of the firm) is the appropriate rate to be used for the private sector, controversy still exists about the appropriate rate for use in government decision making. Some have maintained that the intermediate or long-term government bond rates are the most appropriate (depending upon the duration of the project under consideration). Others have maintained that the rate should be no lower than the typical rate of return achieved by investments in the private sector. The OMB set a rate of 10% for use in evaluating government projects on an equitable basis.[12] [This seems to be somewhat excessive when costs and benefits are expressed in terms of constant dollars. In this case a rate of perhaps 3 to 5% is more appropriate (i.e., the real cost of capital that does not include the inflation rate component).]

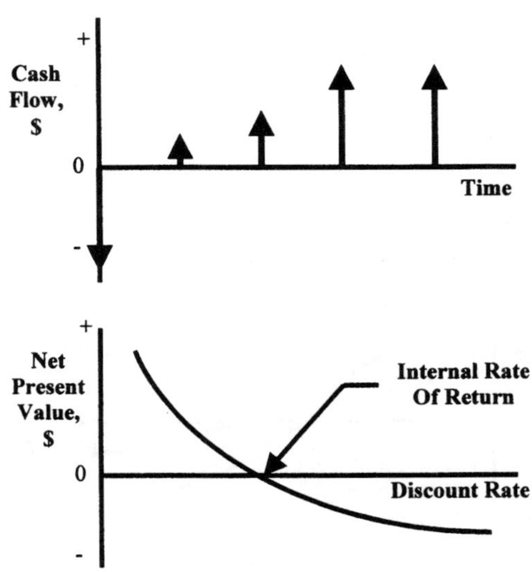

Fig. 2.8 Present value and internal rate of return.

NPV is a widely used criterion, though normally not the sole criterion, in the process of evaluating and selecting investments.[4] Theoretically, in a perfect world of certainty and unconstrained budgets all worthy projects with an NPV greater than zero should be undertaken, projects with an NPV less than zero should not be undertaken, and one can be indifferent about undertaking those projects with an NPV equal to zero. However, resources are seldom sufficient to undertake all projects passing this test, and resource rationing is said to exist. This lack of resources for all projects that an organization would like to undertake leads to the problem of choosing the set of projects or investments expected to maximize the net present value of the firm. This is commonly called the "portfolio selection problem." Optimum solutions can be developed utilizing linear or integer programming techniques and reasonably good or near optimum solutions can be obtained utilizing heuristic techniques.[13,14] This is discussed in Sec. VII.

The discount rate at which the NPV of a proposed project would be zero is known as the internal rate of return (IRR) of the investment. This is also referred to as the discounted return on investment or just return on investment. In Fig. 2.9 internal rates of return for hypothetical projects A and B are indicated. The IRR, in effect, represents the maximum rate of return that might be paid for funds borrowed to make the investment. The IRR computation, which can be performed by computing the net present value at several different discount rates and interpolating to establish that rate which yields a net present value of zero [or by solving the NPV equation for the real positive root(s) that yields NPV = 0], attempts to avoid the issue of fixing a single correct discount rate. Each project has its associated discount rate. When using IRR as an investment criterion, projects should be undertaken as long as the IRR exceeds the currently accepted threshold or "hurdle rate." The hurdle rate would be the same discount rate used in the net present value computations. Note that a determination of discount rate (the cutoff or hurdle rate) is still necessary. The hurdle rate is normally somewhat greater than the firm's cost of capital in order to compensate for risk.

An interesting situation is illustrated in Fig. 2.9. The net present value of two projects is shown as a function of discount rate. The cost of capital to the firm is r_0.

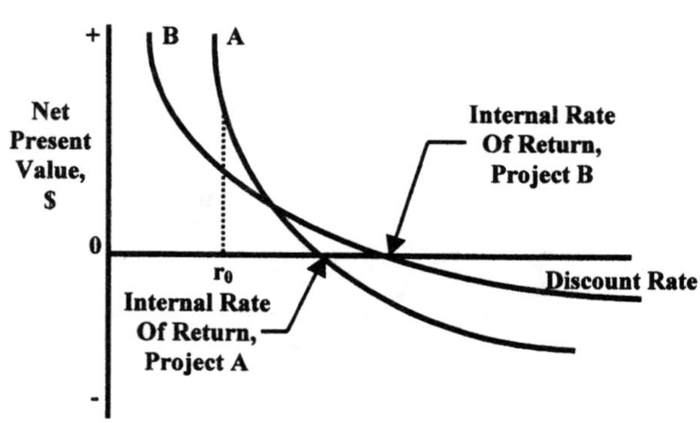

Fig. 2.9 Comparison of internal rate of return for hypothetical projects.

At this cost of capital, the net present value of project A exceeds that of project B, and, according to the NPV criteria, project A ranks ahead of project B. On the other hand, the internal rate of return of project B exceeds that of project A (both exceed r_0). Therefore, according to IRR criteria project B ranks ahead of project A. In general, the mix of projects selected will depend on the criteria used for ranking. This conflict can only be resolved by considering other projects and determining which portfolio of projects maximizes the net present value of the entire portfolio of projects within specified cost constraints.

Because investment decisions are concerned with the commitment of funds with the hope of future benefits and because there is uncertainty about the exact course of future events, the specific results of an investment decision should be described in terms of the probability of achieving different outcomes. In other words, uncertainty about the exact course of future events creates risk—fluctuations in the resulting costs, benefits, and cash flow patterns. Decision making should explicitly take into account these uncertainties. This requires the quantification of risk. The technique that makes provision for the explicit quantitative consideration of uncertainty and its effects, that is, the risk, in the analysis of a business situation is called risk analysis and is described in the following paragraphs. Chapter 4 describes several applications of risk analysis techniques and the associated computer simulation models.

Mathematical models used to simulate business situations and to develop risk measures frequently utilize Monte Carlo simulation techniques. Monte Carlo entails the simulation of a situation many times (perhaps 1000 or more), each time selecting the input data in a random fashion according to weighting provided by specified uncertainty profiles. The creation of uncertainty profiles that quantify risk perceptions is discussed in detail in Sec. I of Chapter 4. The result is the establishment of risk profiles of appropriate performance measures as well as the usual expected value financial performance measures. Figure 2.10 illustrates the concept of financial analysis with the explicit consideration of uncertainty and resulting risk.

Basic input data consist of deterministic and probabilistic data. Examples of deterministic data are the number of time periods to be considered, the discount rates, and tax rates. Probabilistic data consist of the probability density functions, here referred to as "uncertainty profiles," associated with the variables whose values cannot be predicted or known in advance with a reasonable degree of

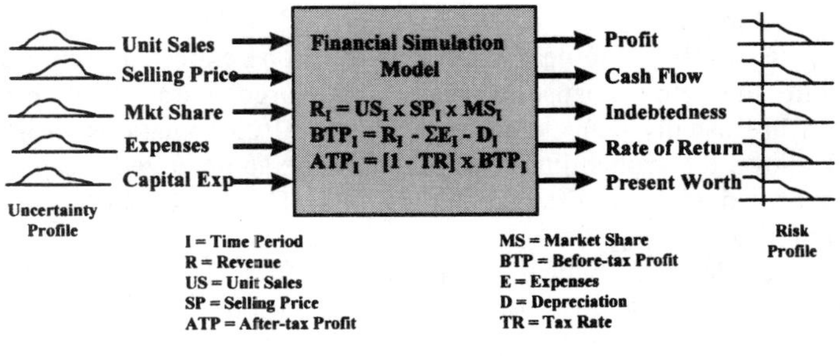

Fig. 2.10 Financial planning model: Monte Carlo concept.

certainty. Typical uncertainty variables include demand for products/services, market share, expense items, and capital expenditures. Uncertainty profiles are subjective estimates that describe the range and form (shape) of the uncertainty. For many ventures, particularly those relating to space, there is another type of variable that must be described in probabilistic terms and that can significantly affect risk. Variables relating to reliability are in this class. Reliability measures such as probability of success and mean-time-to-failure lead to variable or unpredictable timing of events (i.e., time of failure) that affects both the magnitude and timing of capital expenditures and related revenues and expenses.

These data are placed into a financial simulation model that represents the real-world situation being evaluated. The illustrated model (Fig. 2.10) states that revenue at a given time I is equal to the product of unit sales, selling price, and market share; before-tax profit is equal to revenue less the sum of all expense items less the depreciation expense; after-tax profit is the before-tax profit multiplied by one minus the tax rate.

Risk analysis is performed by random sampling of input data (according to weighting of the uncertainty profiles), performing computations contained within the simulation model, saving the results, then repeating the process. This process is repeated many times (i.e., Monte Carlo) until a reasonable set of histograms can be developed from the saved output. These histograms are worked into the desired form to indicate the variability of performance measures, such as profit, cash flow, indebtedness, rate of return, and net present value. A convenient form of displaying performance measures is a "risk profile," which indicates the chance of a performance measure exceeding specific levels (i.e., the complement of the cumulative probability distribution).

It is important to recognize that the same financial computations need to be performed with or without the formal consideration of uncertainty and risk. The primary difference is in the description of the input data and the form of the developed results.

Uncertainty makes annual cash flow probabilistic. Therefore, NPV must also be characterized by a probability distribution that represents the chance of achieving each of its possible levels. The probability of NPV, ROI, and other metrics exceeding specified levels can be obtained by measuring the area under the probability distribution curve for all of the values greater than the specified level—the risk profile of NPV, ROI, etc. Fortunately, and as expected by the central-limit theorem, these risk profiles tend to be normally distributed and can be summarized in terms of their expected values and standard deviations, thus making comparisons of alternatives relatively straight forward. [The NPV, ROI, and other probability distributions are near normal as would be expected according to the central-limit theorem because a reasonably large number of probability distribution are usually considered in the financial computations. Approaching normality allows the probability distributions to be summarized meaningfully in terms of expected values and standard deviations. (This has been done in the NPV results summarized in Fig. 2.11).]

Figure 2.12 illustrates typical risk profiles of NPV. The vertical scale represents the chance or probability p of exceeding the various levels of NPV, indicated by the horizontal scale. In general, the steeper the curve, the lower the risk (i.e., variability). When comparing alternatives, it is important to compare the

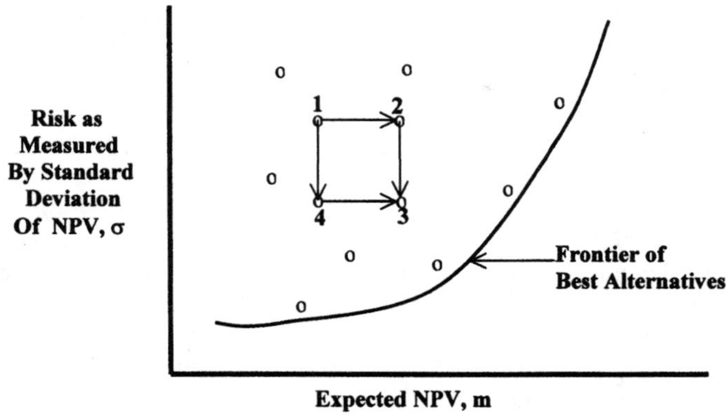

Fig. 2.11 Risk comparison of alternatives.

expected or "most likely" net present values. Comparing risk levels is equally important.

Figure 2.12 illustrates the NPV risk profiles for hypothetical alternatives A and B. A decision maker performing a conventional analysis without the explicit and quantitative consideration of uncertainty and resulting risk usually evaluates quantitatively only the most likely present value. To this uninformed decision maker, alternative A and B look alike because they show equal ($p = 0.5$) expected and most likely values.

In the absence of performing a risk analysis, the decision maker will try to pick the alternative yielding maximum NPV. In risk analysis the selection process has more information and in a sense is made more difficult. Tradeoffs must be made between alternatives possessing different expected present values and associated levels of risk. When the risk dimension is added, the decision maker finds alternatives A and B in Fig. 2.12, for instance, quite different. Alternative A assumes greater risk (variability) than alternative B. Thus, a conservative decision maker (averse to risk) would probably select B (if there were no other unquantified pressures to select alternative A).

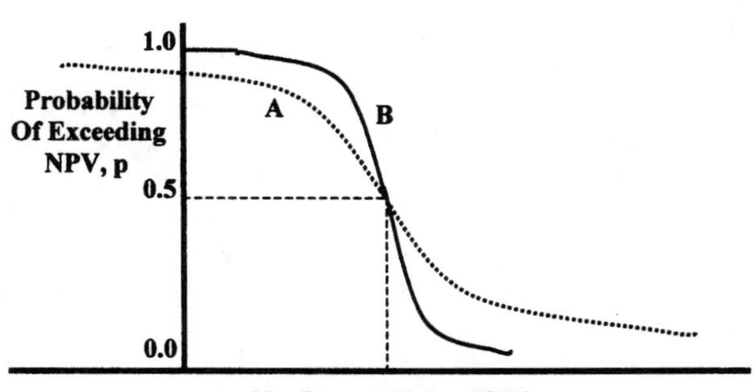

Fig. 2.12 Typical risk profiles of net present value.

The decision maker will usually identify many alternatives necessitating this kind of choice, but the task will be eased somewhat by the fact that the NPV probability distributions usually take the form of the familiar bell-shaped curve. That is, it takes the normal distribution or Gaussian form. In this case the distributions can be fully characterized by their expected values or means, m, and standard deviation σ, and a point on the m-σ plane can represent each alternative. For example, Fig. 2.11 illustrates two alternatives (1 and 2) having the same level of risk ($\sigma_1 = \sigma_2$), but with the expected NPV of alternative 2 greater than that of alternative 1. Therefore, alternative 2 is preferable to alternative 1. Similarly, it can be argued that alternative 3 is preferable to alternative 4; that alternative 3 is preferable to alternative 2, because both have the same expected NPV but alternative 2 is riskier; and so on through the whole list of alternatives.

If this process is continued, the decision maker can chart a frontier of best alternatives, each choice on the chart differs in both risk and expected NPV. This frontier might be called the "class of best alternatives." The best alternative can then be selected based on the decision maker's risk judgment. That is, the decision maker must make the tradeoff between an increase in expected NPV and an accompanying increase in risk. (This is discussed in more detail in Sec. IV.)

Risk analyses have been performed for many different situations, such as capital budgeting within large corporations, evaluation and planning of startup situations, evaluation of alternative technologies, and analyses of the potential impacts of government policies on private sector business ventures. Many different risk analysis models have been developed and are in current use—from those requiring the analyst to write short programs (as part of the input data stream) that tailor the model to the specific venture being analyzed, to generalized models that can simulate a wide variety of situations.[15] In Sec. I of Chapter 4, a general financial planning model (*DOMSAT*) for communication satellite business ventures is described along with a number of its applications. This model explicitly considers uncertainty and unreliability and the resulting risk consequences. Other models utilizing risk analysis techniques and the application of these techniques to a wide variety of decisions are also discussed in following sections.

The following paragraphs describe a financial planning model *PSI-RISK* that was developed as a general financial planning tool that specifically addresses uncertainty and risk in the business planning and evaluation process and has been used in a number of the following applications. The model can be used to address a wide variety of financial planning and evaluation situations ranging from the planning and evaluation of either a new or ongoing business venture to the evaluation of pricing policies that aim to achieve a desired level of financial performance.[16–20] (The original version of the model was developed more than 30 years ago to assist RCA corporate management in the capital budgeting process by explicitly considering uncertainty and risk. At about the same time, Dupont developed a similar capability. These models and capabilities have been utilized successfully over the years.) Through the provision of input data, the model can be used to consider an ongoing business and to establish product pricing in terms of rate of return objectives, production decisions, and competitive product pricing. The *PSI-RISK* model, which explicitly and quantitatively considers uncertainty, is a financial planning software tool that allows a business (such as a manufacturer of factory-produced panelized wood-frame construction products) to be modeled based on data provided to it and develops a proforma income statement, cash flow statement, and balance sheet. These are the basic financial documents that form the financial basis for decision making and are a major part of a business plan.

PSI-RISK is a general venture simulation model (operational in the DOS environment) that is capable of simulating the financial aspects of a wide range of multiproduct business ventures including both regulated and nonregulated. The basic cash flow structure of these computations is illustrated in Fig. 2.6. A number of the key structural aspects of *PSI-RISK* that enable it to simulate a very diverse range of business ventures are highlighted here: 1) multiple time periods considered, 2) trends considered for following time periods, 3) large number of stored uncertainty profiles, 4) multiple products, 5) product interrelationships, 6) demand–price relationships (by product), 7) manufacturing cost-quantity relationships (by product), 8) choice of depreciation method (by capital expenditure), 9) inter-period correlation, 10) capital expenditure determination by production requirements, 11) consideration of initial corporate position, 12) interest rate a function of debt/equity ratio, 13) pricing policy established to achieve desired rate of return (with market constraints), 14) flexible tax structure, 15) financial computation for different business situations, 16) part of existing corporation, and 17) separate corporation.

Data describing the projected market and competitive prices for as many as 20 products over a period of up to 10 years, manufacturing cost of sales and its elasticity, fixed and variable components of expenses, capital expenditures (both specifically identified and those determined to be required as a result of production demands) and depreciation methods, business structure and pricing policy, tax structures and balance sheet items are entered into a menu-driven spreadsheet data input system. Data need only be provided within the ranges set by variables such as the number of years to be analyzed, number of products to be considered, number of capital expenditures to be considered, etc. Thus the data requirements are a function of the complexity of the business being analyzed. A FORTRAN-compiled program that calculates and prints proforma income, cash flow and balance sheet statements for the overall business, and individual income summaries for each product then acquires the data set. These statements can be reviewed to assess the prospective financial viability of the business modeled, product pricing and achievability of desired financial performance goals, etc., and necessary modifications can then be made to the business plan.

Although it can be used as a simple deterministic financial model, the *PSI-RISK* Model is more complex and powerful than a simple financial planning model in several ways. This model is designed as a stochastic simulation model using Monte Carlo simulation techniques to convert a combination of deter-ministic and probabilistic input data describing a business venture into risk profiles indicating the probability that various financial performance measures such as profit, cash flow, and return on assets will exceed different levels. The same model utilizing the same set of equations can be used to also perform deterministic analyses. If a stochastic analysis is required, certain of the data can be entered in the form of a "range of uncertainty," whereby a minimum and maximum value are specified and an uncertainty profile describing the probability distribution of the data within the range of uncertainty are furnished. [To minimize data entry, the model contains a large number of stored uncertainty profiles (i.e., probability distributions) that can be referenced or others can be created as the need arises.]

Another useful feature of the *PSI-RISK* Model is that it can either utilize specified product prices or it can derive product prices that will result in specified rates of return (either return on sales, assets, or gross margin) within the constraints of competitive or substitute product prices. This feature plays a prominent role in the manufactured housing application. For each product an iteration and curve fitting process is utilized to derive that price which will allow the desired rate of return to be achieved, subject to the market price constraints. Because of this capability, the model can represent a price setting firm or a price taking firm. In the case of a price setting firm, product-selling price is computed to achieve a specified rate of return (either return on sales, assets, or gross margin). With a price taking firm, the firm must charge competitive prices (those that have been established by the price setting firm or those current in the market) and rates of return for the firm are computed based on these prices.

The *PSI-RISK* Model allows for the consideration of product interdependency relationships and relationships between unit sales and required production levels and associated capital expenditures. This allows the effects of variations in independent variables on dependent parameters to be automatically calculated by the model. Variations in prices for, or sales of a product, can affect sales of that or other products and therefore production levels. Changes in production level, in turn, can affect capital expenditures and manufacturing cost per unit.

These relationships are communicated to the *PSI-RISK* Model via the input data. Product interrelationships can be specified whereby the sales of one product may be indicated as a function of the sales of another product sold by the firm (as would be the case with complimentary goods, such as razors and razor blades). Demand-price (or price elasticity) relationships for each product and manufacturing cost-quantity relationships (sensitivity of unit manufacturing cost to changes in production level) can be specified to the model. Capital expenditures can be specified as production independent and production dependent, which allows the model to establish the magnitude and timing of expenditures depending on production level.

The *PSI-RISK* Model can represent various business situations, including a startup situation or a venture that is part of an existing corporation. In this latter case the venture can be considered either as a startup situation or as a continuation of current operations. The model has a flexible tax structure that calculates carry-forward losses and tax credits.

The following input data can be specified to the *PSI-RISK* Model via the set of spreadsheet data screens:

1) General data
    (a) Time horizon of business
    (b) Business type (separate entity or part of a larger organization)
    (c) Number of products
    (d) Discount rates
    (e) Timing of product introduction
    (f) Uncertainty profiles
    (g) Number of simulation runs

   (h) Tax rates
   (i) Investment tax credit rate
   (j) Type of debt/equity computation (i.e., maintain constant debt/equity ratio)
   (k) Initial debt/equity position
   (l) Interest rates (cost of capital at different debt/equity ratios)

2) Revenue related data
   (a) Unit sales
   (b) Market share
   (c) Distributor markup
   (d) Product interrelationships
   (e) Rate of return objectives and market constraints
   (f) Price of competitive products
   (g) Allowed or target rates of return
   (h) Price elasticity data
   (i) Uncertainty specifications (if desired) for unit sales, market share, unit price, or price of competitive products

3) Cost of sales and expense data
   (a) Cost of sales
   (b) Sensitivity of cost of sales to production level
   (c) Fixed and variable expense data items
   (d) Uncertainty specifications (if desired) for cost of sales and fixed and variable expense items

4) Capital expenditure data
   (a) Unit independent capital expenditures
   (b) Timing
   (c) Depreciation type and life
   (d) Applicability of investment tax credit
   (e) Uncertainty specifications (if desired) for magnitude of expenditures
   (f) Unit dependent capital expenditures
   (g) Production threshold level (production level at which a new production-level-related capital expenditure is required to be made)
   (h) Incremental production capability
   (i) Expenditure to achieve incremental capability
   (j) Depreciation type and life
   (k) Uncertainty specifications (if desired) for magnitude of expenditures

5) Balance-sheet-related data
   (a) Debt/equity ratio
   (b) Cash requirements
   (c) Level of receivables
   (d) Level of payables

(e) Level of inventory
(f) Other assets and liabilities
(g) Book value of existing (i.e., initial) capital assets
(h) Depreciation type and life for existing capital assets

After data are entered into the spreadsheet data program, a transfer file is created and sent to the FORTRAN-compiled program with a few simple keystrokes. Proforma income, cash flow and balance sheet statements for the business are generated, which display after-tax profit, return on sales, return on assets, total assets and liabilities, debt, equity, debt/equity ratio, net cash flow and indebtedness for each year of the analysis, as well as net present value (of annual cash flow) at several discount rates.

The consideration of production-dependent expenditures allows the *PSI-RISK* Model to establish the magnitude and timing of expenditures that can be driven by level of production. Another important feature is the ability to consider the initial position of the firm being modeled. *PSI-RISK* allows the initial corporate position to be specified in terms of current balance sheet items. This includes cash, receivables, inventory, book value of capital assets and remaining depreciation life and type of depreciation so that depreciation expenses can be computed, and debt/equity ratio. The model allows debt/equity ratio to be maintained as a constant over the time frame analyzed or allows equity to be maintained as a constant. Positive cash flows are used to reduce debt. When additional investment is required, it can be funded through additional debt (keeping equity constant) by increasing both debt and equity and maintaining a constant debt/equity ratio. When a constant debt/equity ratio is to be achieved, equity is not repurchased as positive cash flows occur, but distributions of earnings occur in a manner so as to maintain a near constant debt/equity ratio so that the cost of borrowing increases as the firm's financial position deteriorates.

III. Business Plan

A business plan is a formal document that describes a business and its operating environment, its short- and long-term objectives, and the intended means and methods for achieving those objectives. It is a management tool for the development, expansion, and operation of a business and is essential when it is necessary to use outside sources of capital. A properly prepared business plan serves both as the means of communicating with and convincing investors to make available the resources that are necessary to accomplish the short- and long-term objectives and as the initial plan for business operations. It also provides a basis for measuring performance.

Most business plans share many common features including a description of the business including its products and services (both existing and those proposed for development), an industry analysis, a market analysis, a description of the proposed marketing strategy, a description of the organization and its management, an implementation plan, a discussion of potential risks and pitfalls, a financial analysis including financial statements and projections, and sensitivity analyses. When resources are sought from the government, or specific government actions are required for the success of the business, it is also necessary to demonstrate the resultant economic impacts to the government usually in the form of cost savings, changes in imports/exports, job creation, and net treasury revenues.

When the objective of the business plan is to obtain financing from outside sources, it is necessary to provide sufficient detail and supporting analyses so as to convince the potential investor(s) that the rewards are of sufficient magnitude to justify the investment. Because in the space business, it is likely, though not always the case, that capital requirements will be large, risk will be perceived as being high, and payback period long, it is necessary to develop a business plan that directly addresses these issues. This is the purpose of the overall plan, particularly the financial plan. The financial plan must describe all of the assumptions that have been made, as they are likely to effect the overall financial performance. The financial plan includes financial documentation in the form of proforma income, cash flow, and balance-sheet statements. Because of the perception of the high risk associated with space business ventures, it is likely that prospective investors would require a formal risk analysis. The risk analysis should consider the uncertainties associated with each of the inputs to the financial plan (in particular, the uncertainties associated with technology, development and test, and manufacturing schedules, associated costs and achieved level of performance, demand for launch services and market share as a function of time, and reliability) and develop risk profiles of annual profit, cash flow and indebtedness, payback period and return on investment. Of particular importance is the chance that provided capital will be exceeded. (In other words, the chance that provided capital will be insufficient for the business to be successful.) The financial plan should specifically address the financing requirements and the use of provided financing. When resources are sought from the government, an interactive private-sector and public-sector analysis should be performed and government costs and benefits developed.

The following paragraphs describe the business plan from the perspective of a business venture attempting to develop a RLV. The RLV business plan should first describe the overall business that is being planned, including the development, demonstration, and the operation and maintenance of a fleet of RLVs. The specific services to be provided must be described together with the general aspects of the business that are immediately of concern to an investor. These aspects include cost and pricing, reliability, fleet size, availability, resiliency/operability, competition (during the time frame of the proposed business venture), short- and long-term availability of insurance, magnitude of markets for provided services, capital requirements as a function of time, etc.

A number of the specific plans that together make up the business plan are indicated in Fig. 2.13. Note that they all feed into the financial plan that is the unifying or cohesive force of the business plan. The financial plan utilizes quantitative data obtained from each of these plans to develop financial documentation, as summarized in Figs. 2.14–2.16, that serves to communicate to the investment community (and to the government) the resource needs of the business (as a function of time) and the rewards that can result to those who provide the needed resources. In addition, the financial plan can, and should, quantitatively indicate the level of risk associated with the provision of the resources. The risk can be communicated by indicating the chance that the

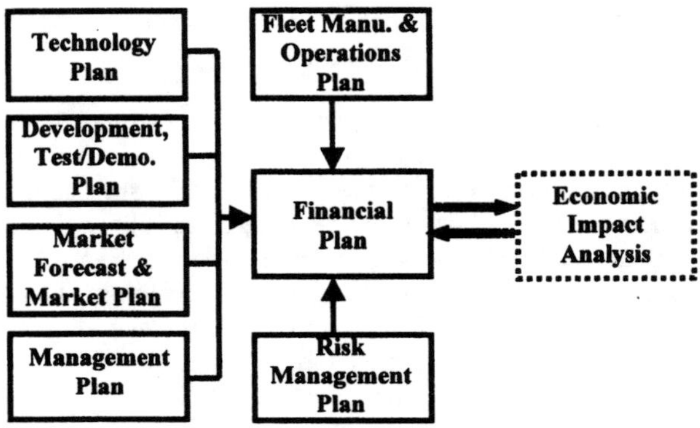

Fig. 2.13 Financial planning is at the heart of the business plan.

	Year					
	'02	'03	'04	'05	'06	'07 ....
No. of Flights						
Price						
Revenue [Government]						
No. of Flights						
Price						
Revenue [Commercial]						
Total Revenue						
P/L Integration						
Other Operations						
Launch Insurance						
Liability Insurance						
Launch Fees						
Maintenance/Repair						
Launch Operations						
DDT&E						
Total Operations Expenses						
General & Administrative Expense						
Facilities						
Fleet						
Depreciation Expense						
Debt Service [Interest]						
Allowance for Uninsured Loss						
Continuing Engineering/Development						
Before Tax Profit						
Carry-Forward Loss						
Gross Income Tax						
R&D Tax Credit						
Investment Tax Credit						
Tax-Holiday Credit						
After Tax Profit						
Return on Sales [%]						
Return on Assets [%]						
Debt Coverage						

Fig. 2.14 RLV transportation business proforma income statement.

	Year						
	'02	**'03**	**'04**	**'05**	**'06**	**'07**	**........**
After Tax Profit							
Decrease in Receivables							
Increase in Payables							
Decrease in Cash							
Depreciation							
Debt Financing							
Paid-In Capital [Equity]							
Total Cash Inflow							
Loss							
Increase in Receivables							
Decrease in Payables							
Increase in Cash							
Facilities							
Fleet [Original]							
Fleet [Replacement]							
Capital Expenditures							
Dividends							
Loan Repayment							
Stock Buy-Back							
Total Cash Outflow							
Annual Cash Flow							
Indebtedness							

	Discount Rate [%]										
	0	**5**	**10**	**15**	**20**	**25**	**30**	**35**	**40**	**45**	**≥50**
NPV Contribution [within planning horizon]											
NPV Contribution [infinite planning horizon]											
Total Net Present Value											

Fig. 2.15 RLV transportation business cash-flow projection.

business will be successful (for example, by indicating the chance that different levels of ROI will be exceeded). To accomplish this requires the use of a Monte Carlo financial simulation model that includes integrated models of the technology, manufacturing and operations, development test/demonstration, market forecast, and risk management models. Each of these is discussed briefly in the paragraphs that follow.

The technology plan is concerned with presenting a description of the technologies that are necessary to achieve the RLV performance upon which the overall business is based. The status of the various technologies must be communicated in terms of the cost and schedule to achieve required levels of performance for those technologies that are critical for achieving the overall RLV performance. Simulation modeling techniques such as the R&D Simulation Model (RADSIM)[21] should be employed for this purpose and result in probability distributions of cost and schedule that can then be utilized in an overall financial risk analysis performed as part of the financial plan. (It cannot be emphasized strongly enough that cost estimating relationships are not appropriate when there is

	Year					
	'02	'03	'04	'05	'06	'07 ....
Cash						
Accounts Receivable						
Prepaid Expenses						
Facilities Investment						
Less Accumulated Depreciation						
Book Value of Facilities						
Fleet Investment						
Less Accumulated Depreciation						
Book Value of Fleet						
Total Assets						
Accounts Payable						
Prepayments [Deferred Income]						
Long Term Debt						
Total Liabilities						
Corporate Funding [% Ownership]						
Equity Purchase [% Ownership]						
Paid in Capital						
Retained Earnings						
Total Stockholder Equity						

	Discount Rate [%]										
	0	5	10	15	20	25	30	35	40	45	≥50
ROI [Discounted Rate of Return on Corporate Investment]											
ROE [Discounted Rate of Return on Equity Investment]											

Fig. 2.16 RLV transportation business balance sheet projection.

uncertainty with respect to whether or not a project can be successfully completed. CERs are based upon the cost of successfully completed activities.) For those technologies that are in hand, it is necessary to indicate their history of utilization thereby convincing potential investors of level of maturity.

The development, test, and demonstration plan is concerned with describing how the various available and developed technologies will be utilized and integrated to create a flight demonstration model. The plan is also concerned with describing the test and demonstration program, indicating the various possible outcomes of the test and demonstration program, the likelihood of each outcome, and the consequences given each outcome. (This might actually be considered as part of the risk management plan.) The consequences might be procurement of additional demonstration models, additional testing, etc., all of which are likely to affect costs and schedules. Again, costs and schedules are uncertain and should be described by probability distributions. In all cases it is necessary to disaggregate costs into capital expenditures and expenses.

The fleet manufacturing and operations plan consists of a description of the launch vehicle manufacturing process that will lead to the procurement of the necessary fleet. It describes the supply chain as well as possible failures in the chain and methods for recovery. (This can also be considered as part of the risk management plan.) The plan also describes the method for, and resulting cost estimates for manufacturing the first and following units, schedules for manufacturing replacement units should a replacement be required, and the annual maintenance costs and their relationship to flight schedules. All cost and schedule estimates should be expected values, and associated uncertainties should be described in quantitative terms (i.e., probability distributions). These estimates will be used in the financial risk analysis.

The marketing plan consists of a description of the existing and future markets, including appropriate market segmentation, the competition for products/services to be provided, and a rationale for achieving the market share. Market share will depend upon the attributes of the transportation service that will result from the investment and the attributes of competitive space transportation services. These attributes include pricing, cost and availability of insurance, lift capability, transfer stage capability and availability (when direct payload placement is not possible), schedule/availability, and reliability. A method for estimating the effect of pricing and other attributes on market share for launching communication satellites is described in Sec. II of Chapter 4. The marketing plan should also describe likely competitive reactions (for example, if the RLV service reduces transportation price, can/will the competitive services also reduce price?) and what are likely RLV transportation service reactions to changes in competition. The marketing plan should also describe the price elasticity by market segment for provided transportation services (discussed in Sec. IV of Chapter 3). Existing and new markets should be discussed separately because there is likely to be a considerable difference in risk perceptions between the two. A description of the requirements for new markets to develop should be discussed as well as how the RLV will meet these requirements. The timing of market development and market penetration rates by market segment also needs to be described. Finally, the uncertainties associated with all factors that are likely to effect market development need to be described together with their implications on market share and total markets as a function of time.

The risk management plan is concerned with minimizing both the things that might go wrong and their effects, coping with the things that might go wrong, and recovering when things do indeed go wrong. Insurance plays a significant role in risk management (i.e., after minimizing the risk, transferring the remaining risk, for a price, to others who are in the business of sharing risks). The risk management plan should describe the manner in which insurance will be used to reduce financial risk. Of particular importance is the consideration of insurance short- and long-term cost and availability. The cost and availability of insurance for RLVs having replacement cost significantly greater than the cost per flight are of particular concern. For example, if cost per flight is to be less than 100 million when utilizing a vehicle that has a replacement cost of 1 billion, it is necessary to achieve insurance costs on the order of 1–2% of the replacement cost. (This is more than an order-of-magnitude reduction from current expendable launch vehicle insurance rates.) The risk management plan should describe how this will be achieved and in a manner that can convince an investor that this low cost insurance will likely be available over the long term. A method for accomplishing this reduction in insurance cost and ensuring long-term availability is discussed in detail in Sec. VI in Chapter 3. The risk management plan should also discuss the method used for the determination of fleet size and the possible use of other vehicles as backup. The resiliency/operability (see Sec. X of Chapter 4) of the transportation service should be described from the perspective of the RLV business (including consideration of the supplier chain for both the initial fleet and any replacements that may become necessary in the future) and from the perspective of customers for the provided transportation services.

The management plan is concerned with describing the management organization and structure of the business with the goal of convincing potential investors the management team is capable of organizing, controlling, and operating the business so as to achieve the financial results detailed in the financial plan. The management structure as a function of time serves as a major input in the determination of the general and administrative expense that is indicated in the financial plan.

Finally, if commitments are sought from the government such as funding, anchor tenancy, low-interest loans, tax relief, etc., then it is necessary to perform an analysis from the government's perspective. In other words, it is necessary to demonstrate that the benefits to the government outweigh the cost to the government. It is necessary to describe the details of what is required from the government in order to make the business a success and what the benefits are likely to be. The benefits should be demonstrated in terms of the present value of government cost savings, change in imports and exports, job creation, and net treasury revenues.

The business plan should be based upon the development and use of several interactive Monte Carlo simulation models as illustrated in Fig. 2.17. The *financial planning model* contains all of the necessary structure for performing the financial computations and developing the required set of financial reports. The model should accept inputs direct from the other models as well as direct-analyst-provided data (including all appropriate uncertainty estimates). The model should have sufficient structure so that various pricing policies can be considered such as establishing pricing so as to yield a desired rate of return in a competitive launch service environment. The developed financial reports should also contain quantitative risk information either in the form of risk profiles or summarized as expected values and associated standard deviations.

The determined pricing should feed the *Demand Model* that considers price elasticity of demand for different market segments. The output of the demand model and the determined pricing, together with RLV and competitive transportation service attributes, would be used in a *Market Share Model* that estimates the RLV transportation service market share as a function of time. The output of the market share model is the determination of the desired launch time for the next RLV flight. This would be used in the *Resiliency and Operability Model* that establishes, based upon prior launch and related failures, when the next launch can take place. The time of the next launch is then utilized in the *Launch and On-Orbit Operations Model* that simulates the launch and on-orbit activities taking into account reliability estimates. The model establishes the success, failure and recovery characteristics of the launch. The model establishes annual event statistics that include the number of launch successes and failures, refurbishments, fleet replacement requirements, etc. This information is then used in the financial planning model to establish all operations-related expenses and expenditures. The

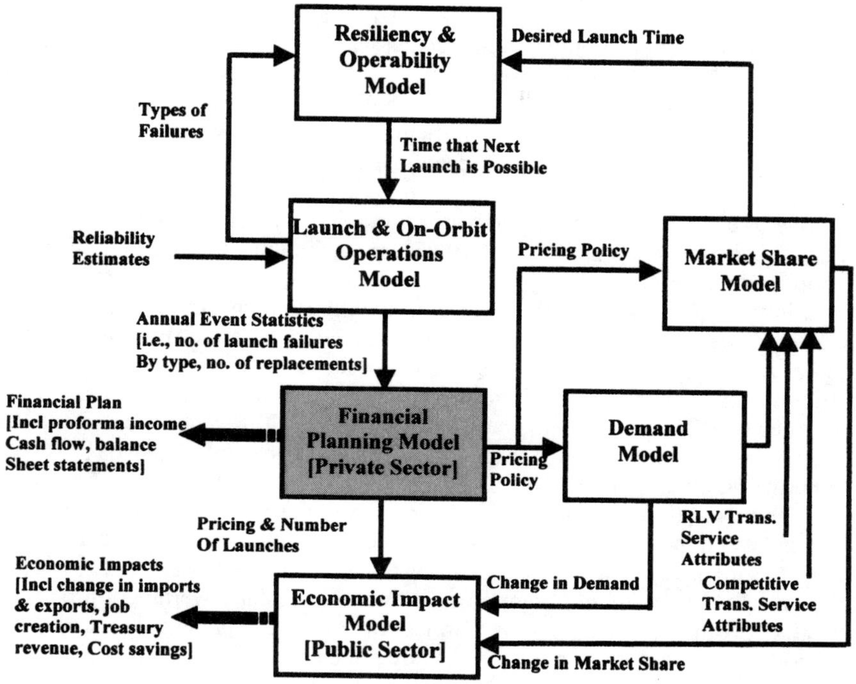

Fig. 2.17 Interactive planning models.

structure and interactions of these models ensures all computations are linked together and that changes in one or more estimates are appropriately considered throughout the system.

Finally, the outputs of the financial planning, demand, and market share models feed an interactive *Economic Impact Model* that establishes economic impacts to the U.S. economy in the form of the change in imports and exports, job creation, and Treasury revenue and the cost savings that would result to the government through the use of the RLV transportation services. Care must be taken to appropriately take into account reduction in market share for other U.S. provided space transportation services.

IV. Comparison of Alternatives (Decision Database)

When considering investment alternatives that involve relatively small amounts of investment capital, it is reasonable to make decisions based upon the value of a single metric such as present value of cash flow, payback period, return on investment, cost savings, etc. However, in most real-world situations, unfortunately, complications exist that force the consideration of multiple metrics or investment attributes. These complications include differences in the magnitude of the capital that will be placed at risk if an alternative is selected, differences in net present value that is likely to be achieved if an alternative is selected for investment, differences in the risk associated with each alternative, etc. It should immediately be apparent that the comparison of alternatives having different values associated with multiple attributes implies that it is necessary to consider the relative value of the different attributes to the decision maker and to convert the multiple attributes into one which can be used for ranking and selecting an efficient set of investments.

Private sector investment decisions are normally based upon the consideration of a number of financial performance measures including expected ROI, payback period, capital requirements, and the consideration of risk as indicated by the variability of ROI or variability of other measures. In general, private sector investment decisions are concerned with evaluating potential business opportunities and considering tradeoffs between such factors as expected ROI, risk, and magnitude of investment, which must fall within acceptable ranges, for example, the expected ROI must exceed a cutoff rate of return. However, this cutoff rate of return or hurdle that must be exceeded is related to the level of perceived risk with the hurdle increasing to offset increased risk. In addition, both the expected rate of return and risk hurdles are normally also related to the magnitude of the required investment and the payback period. Thus, government programs can affect or alter private sector investment decisions through their impact on perceived risk (i.e., through risk reduction) and required investment. Efficient government programs aimed at affecting private sector investment decisions are aimed at those specific aspects of an investment opportunity that are impeding the investment—programs to increase ROI if return on investment is perceived as too low, programs to reduce risk if risk is perceived as too high. This is discussed in detail in the following section.

To provide government planners with insight into the effect that their policies and programs might have on investment decisions, a relationship between the key financial performance measures of expected ROI, risk (as measured by the standard deviation of ROI; see Sec. II, expected payback period, and magnitude of investment at risk (relative to the budget under the decision-maker's control), and the likelihood of investment was developed. In essence, a multi-attribute utility function was developed relating the likelihood of investment to several of the more important financial performance measures. The selected financial performance measures are those that are readily available from a financial risk analysis.[15,22,23] Thus,

$$\text{Likelihood of Investment} = f\{m, \sigma, \text{PB}, \text{IND}\}$$

where m is the expected ROI, σ is the standard deviation or variability of ROI, *PB* is the expected payback period, and *IND* is the indebtedness but is not utilized directly; a proxy is used in its place. The proxy is the ratio of the magnitude of the investment to the budget under the control of the decision maker, expressed as a percentage.

To establish the desired functional relationship, the key to establishing the effect of risk and other financial performance measures on investment decisions, a survey was conducted under the auspices of the American Management Associations (AMA).[4] Persons queried generally had the titles of vice president

(Finance), controller, or director of corporate planning. The questionnaire at first glance appeared rather complex and imposing, yet considerably different than most questionnaires. The AMA mailed approximately 3000 questionnaires to their members selected at random but having the preceding titles. To these were added a number of specifically selected individuals in decision-making positions. From all of these, in excess of 300 useful responses were obtained. A very large fraction of those who responded identified themselves and indicated that they would be available for further discussions. A number of these were queried to clarify responses. Nearly all of the returned forms were filled out correctly and were internally consistent. Thus only a very small number had to be discarded.

The questionnaire consisted of two basic parts, with the former being of the standard type, collecting information on size of organization, type of business, investment decision criteria, and cost of capital. It is this part that provided specific information on the important decision criteria. The latter part of the questionnaire was concerned with the risk avoidance attitudes.

Figure 2.18 illustrates the instructions that were provided for completing the risk-related aspects of the survey. The basic approach for gathering the risk data was to present a series of pictures as in Fig. 2.19 (a page from the questionnaire). Each picture contains four risk profiles of ROI. Within each picture the four risk profiles all have the same expected ROI, but each has a different standard deviation of ROI. The pictures are arranged in columns with five pictures per column. Each picture or set of four risk profiles has a different expected ROI. These are shown relative to the firm's cost of capital R. Thus, expected ROIs of R, $1.1 \times R$, $1.2 \times R$, $1.4 \times R$, and $1.8 \times R$ are considered. Each column has associated with it a statement of payback period (2, 4, or 8 years) and a statement of the size of the investment (less than 1% of the capital budget, 1–10% of the capital budget, and greater than 10% of the capital budget). This approach standardizes the questionnaire to the per-ceptions of the respondent, using his or her own data as the general frame of reference. With each picture is a query pertaining to whether or not it is anticipated that the firm would make the necessary investment given the risk profile (A, B, C, and D), the payback period, and the size of the investment. It is these data that lead to the determination of the investment likelihood functions.

The collected risk-related data are summarized in Figs. 2.20–2.23 and indicate the effect of investment magnitude on investment likelihood in terms of expected ROI (relative to the cost of capital) and risk as measured by the standard deviation of ROI (relative to the cost of capital). Figures 2.20–2.22 illustrate the effect of payback period (2, 4, and 8 years, respectively) on the investment likelihood function. Figure 2.23 illustrates the effect of payback period on investment likelihood in terms of expected ROI (relative to cost of capital) and risk as measured by the standard deviation of ROI (relative to the cost of capital). Figure 2.23 indicates the manner in which investment attitudes change when the investment being considered is small compared to the capital budget and payback varies from 2 to 8 years. There is a rotation of the investment likelihood curves in a counterclockwise direction as payback period increases indicating the general preference for investments with short payback periods.

To review, the collected risk-related data (for the particular combination of 4-year payback period and an investment that is between 1 and 10% of the capital budget) are summarized in Fig. 2.24, which indicates investment likelihood (i.e.,

It is intended to establish the liklihood of private sector investment in terms of expected (average) ROI and risk, where risk is a measure of the variability or unpredictability of ROI. The variability of ROI is illustrated in the form of graphs which indicate the chance that ROI will exceed specific values for different business situations. These situations will be described in terms of expected payback period and the maximum expected investment relative to your capital budget. In answering the following questions, please assume that the business venture is consistent with the objectives and general goals of your firm and that the basic skills, facilities, etc., are all available and that all basic non-financial criteria are satisfied. It is the intent of this questionnaire to consider only the impact of financial criteria upon the investment decision assuming all other criteria are satisfied.

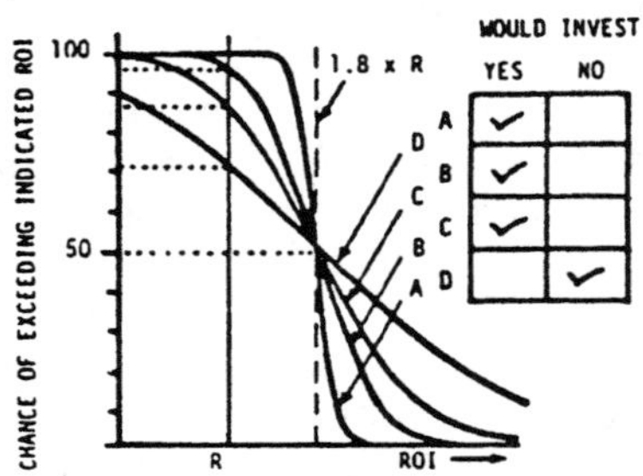

Investment decisions are to be stated in terms of return on investment (ROI) which is described in the following pages in the form of risk profiles. This is illustrated where the chance of ROI exceeding different values is indicated. Of specific importance is the vertical line indicating the cost of capital (R) of your firm. In each picture in the following pages, a family of risk profiles (A → D) is presented. Each member of the family has the same expected value, but each has a different variability. For the case illustrated, the expected ROI is 1.8 times your cost of capital (i.e., if your cost of capital is 10% then the expected value is 18%. Given this information and a statement of payback period (for example, 4 years) and a magnitude of investment (for example, less than 1% of your annual capital budget, do you think that your firm would invest given situation A?, situation B?, etc. There is a 100% chance that situation A will yield an ROI greater than your cost of capital (R) and a 0% chance that it will yield an ROI less than your cost of capital. There is a 73% chance that situation D will yield an ROI greater than R and a 27% chance that it will yield an ROI less than R. These percentages are obtained as the intersection of the risk profiles with the vertical line through R.

In the facing page an example is presented to further clarify the approach. In following pages, please check the appropriate "yes/no" box for the "would invest" decision for each of the risk profiles (A → D) and for each of the families. There are five (5) families for each of the three (3) "situations" on each page. Each page is for a different payback period; 2, 4 and 8 years. Each page therefore requires responses to 15 "would invest in A, B, C, D" questions. The "would invest" question refers to your estimate as to whether or not your firm would invest under the specified conditions.

Thank you.

Fig. 2.18 Instructions for responding to survey questions.[4]

the chance that an investment will be made given a specified set of conditions) in terms of the expected ROI and level of risk. The scales are normalized to the firm's cost of capital. Thus, the vertical scale represents the expected ROI as a fraction of the cost of capital (1.0 indicates that the ROI is equal to the cost of capital). The horizontal scale represents the risk or standard deviation of ROI as a fraction of the cost of capital. The curves represent contours of equal likelihood or probability of investment.

The line marked 0.5 indicates that those investments that are characterized by points (m, σ) that fall on this line have, a priori, a 50% chance of receiving funding. The zero likelihood line is based upon the assumption that, except in special

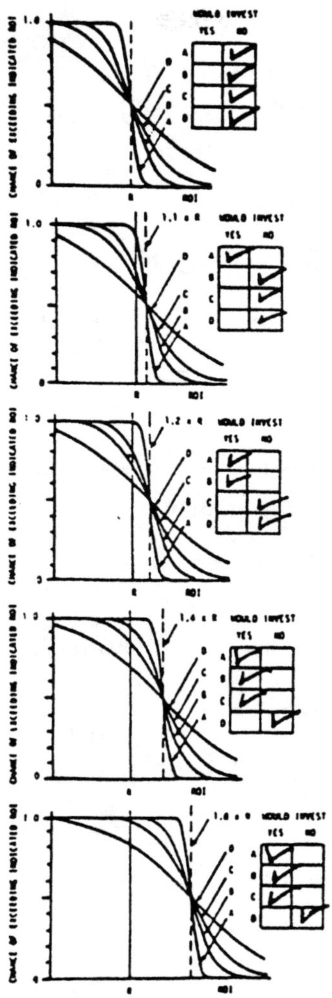

SCENARIO	CHANCE THAT ROI IS GREATER THAN COST OF CAPITAL (R)
A	50%
B	50%
C	50%
D	50%
A	90%
B	67%
C	60%
D	55%
A	100%
B	80%
C	70%
D	61%
A	100%
B	95%
C	83%
D	70%
A	100%
B	100%
C	97%
D	82%

Fig. 2.19 Example used to clarify survey questions.[4]

circumstances, it is unlikely that an investment would be made when the expected ROI was less than the firm's cost of capital. Because all points that fall on the same contour have an equal chance of being funded, the contours indicate the general risk avoidance preferences of an industry sector. For example, there is a 50% chance of an investment when the decision makers perceive an expected ROI that is 1.73 times the cost of capital and a level of risk (i.e., standard deviation of ROI) that is 0.5 times the cost of capital. This is indicated in Fig. 2.24. Also indicated in Fig. 2.24 are the points B, C, and D that represent three different alternatives that

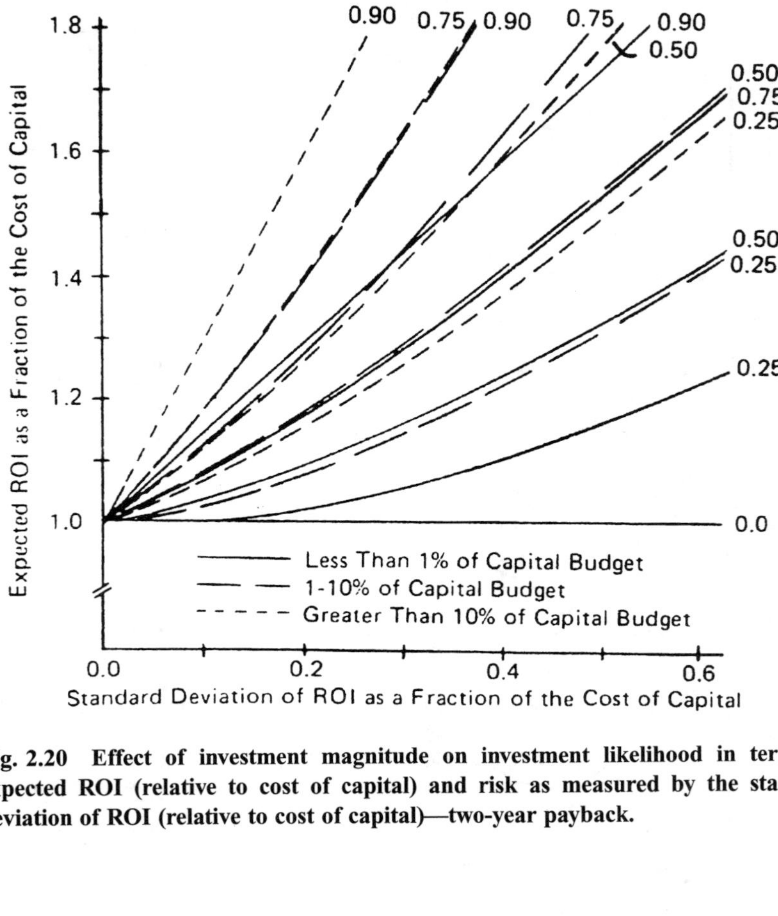

Fig. 2.20 Effect of investment magnitude on investment likelihood in terms of expected ROI (relative to cost of capital) and risk as measured by the standard deviation of ROI (relative to cost of capital)—two-year payback.

have the same likelihood of receiving financing. Figure 2.25 indicates the ROI risk profiles for these three investment opportunities.

The important point of the investment likelihood curves is that no matter what the set of developed financial performance measures (i.e., expected ROI, risk, payback period, and magnitude of investment), the resulting investment decision can only be predicted as a probability or likelihood of investment. Reducing risk perceptions will increase the likelihood of investment as will an increase in expected ROI, reduction of payback period, and reduction in the magnitude of investment. It is apparent that government policies and programs can alter these financial performance measures. The effects of government policies and programs on private-sector investment decisions can be measured in terms of changes in investment likelihood. For example, government tax policy will affect expected rates of return, whereas government R&D can alter risk perspectives, both of which can alter investment decisions. This is the subject of following sections.

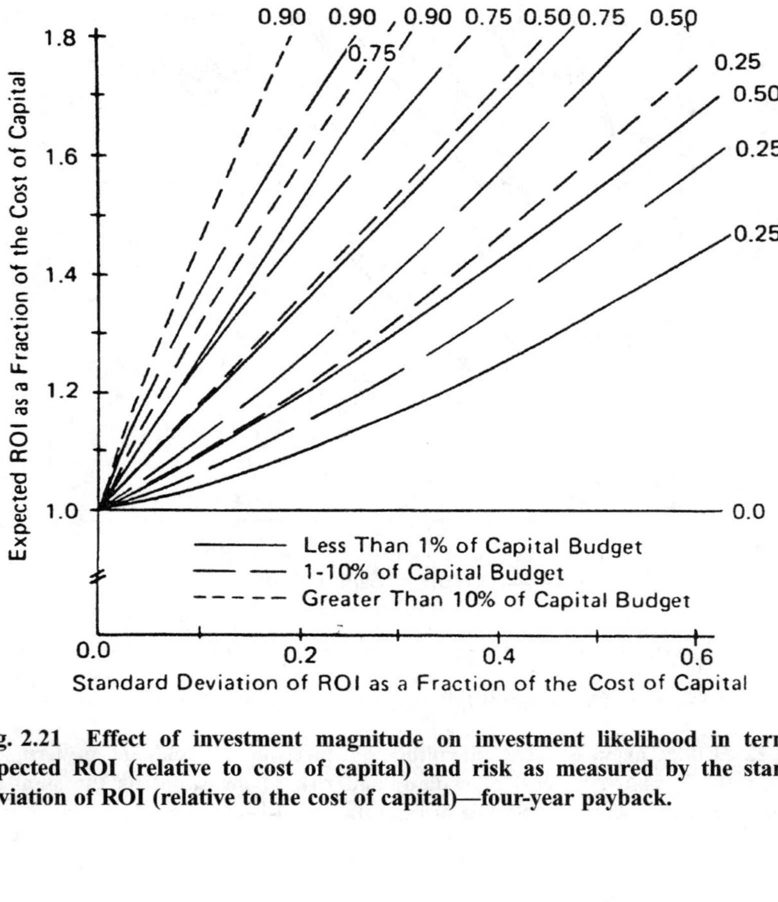

Fig. 2.21 Effect of investment magnitude on investment likelihood in terms of expected ROI (relative to cost of capital) and risk as measured by the standard deviation of ROI (relative to the cost of capital)—four-year payback.

V. Government Influence on Private-Sector Decisions

Both government, the public sector, industry, and the private sector, undertake programs that are aimed at influencing private sector investment decisions. These investment decisions encompass those of commercial organizations as well as those of individuals. For firms, these investment decisions range from corporate strategy decisions such as whether or not to make long-term technology investments (for example, develop on-orbit optical communications technology), enter a new business area (for example, become a provider of satellite communication services), to capital investment decisions (such as which satellite launch service provider to utilize). For individuals, these capital investment decisions range from equipment replacement (replete with considerations of tradeoffs between higher nonrecurring cost and lower recurring costs and fuel choice) to the acquisition of new products and/or services made possible by the introduction of new/advanced technology.

The private sector undertakes research, development, demonstration, and advertising/promotion projects that lead to increased producer and consumer

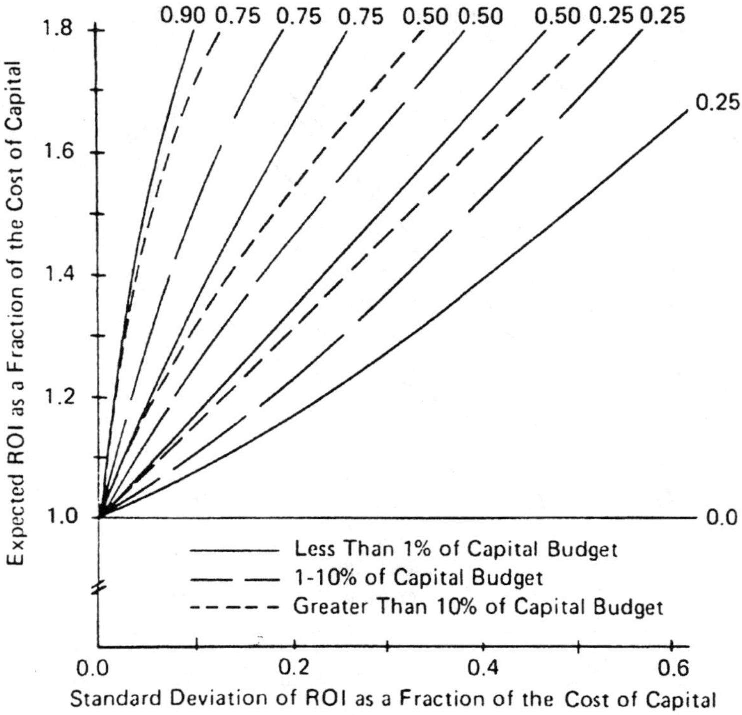

Fig. 2.22 Effect of investment magnitude of investment likelihood in terms of expected ROI (relative to cost of capital) and risk as measured by the standard deviation of ROI (relative to the cost of capital)—eight-year payback.

investments and to the formation of commercial ventures. To a large extent, these programs are undertaken with the expressed intent of maintaining or increasing the value of the firm. A vast array of government-sponsored programs are undertaken with the specific objective of developing technology and/or creating the environment, which will lead to increased private sector investment, and the formation of commercial ventures, which are in the public interest. These encompass research, development and demonstration programs as well as incentive, regulatory, standardization, and other programs.

The National Space Policy issued by the White House on 16 November 1989, and the policies of the previous administration, sought to encourage and foster private-sector investment in space as a major goal of the federal government. Several approaches for achieving this goal have been developed and are being applied. One approach is to encourage federal agencies to buy commercially available space products and services, another is to avoid conducting government activities that compete (or potentially compete) with private sector products and

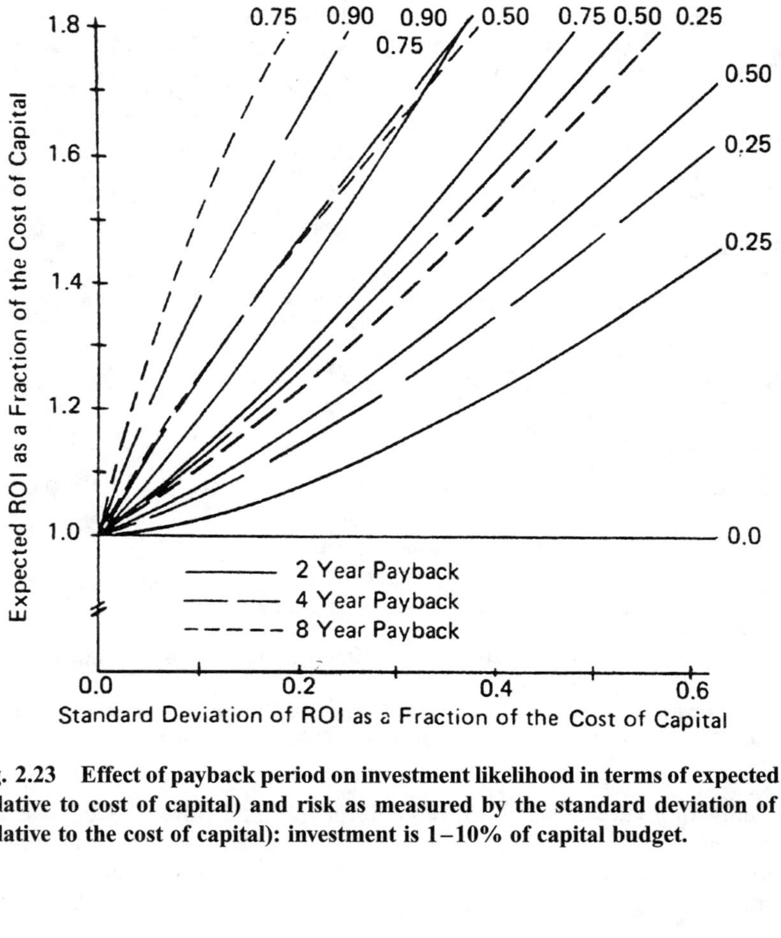

Fig. 2.23 Effect of payback period on investment likelihood in terms of expected ROI (relative to cost of capital) and risk as measured by the standard deviation of ROI (relative to the cost of capital): investment is 1–10% of capital budget.

services, and another is to develop technology and perform demonstrations that would not normally be undertaken by the private sector but can have long-term impacts on the U.S. economy.

Government assistance is often sought by the private sector when contemplating investments in space business ventures. How should the government evaluate the need for such assistance, and what forms of assistance should be considered? In particular, government assistance might be required in order to encourage and achieve private sector financing for the development of new private sector space transportation businesses, particularly those based upon a fleet of RLVs. Various forms of government "participation" or initiatives have been and are being considered and affect investment decisions in different ways; for example, some directly affect the magnitude of the private sector investment, some affect long-term rates of return, some affect risk perceptions, and some affect several of the private sector decision concerns. Because the creation of such a business venture must surmount a number of financial hurdles, including the

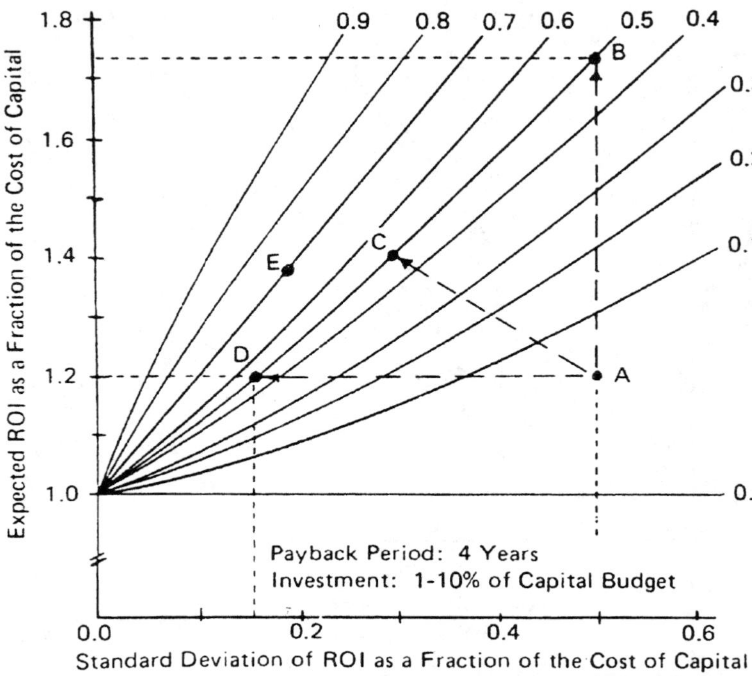

Fig. 2.24 Investment likelihood in terms of expected ROI (relative to cost of capital) and risk as measured by the standard deviation of ROI (relative to cost of capital).

uncertainty of future demand, the high level of initial investments, potential competition from existing and future expendable launch vehicles, and the difficulties of adjusting current policy mechanisms to new government–private cooperative arrangements, it is not clear whether or not such an endeavor is likely to be achieved without supportive government actions.

Government programs and policies attempt to alter private sector perceptions of risk and return that determine the extent of private sector innovation and investment relating to the commercial development of space. Space business ventures are normally characterized by large capital investment, long payback period, and high risk. The large capital investments will be made by the private sector only if the private sector perceives that adequate rates of return are likely to be achieved at acceptable levels of risk. Thus, a potential role of government relates to reducing these impediments to investment thereby increasing the likelihood of private sector investment with benefits to the U.S. economy.

In general, the government can encourage the continued commercial development of space by selectively affecting perceived levels of risk (business and technical) and by improving the return on investment potential of space-related ventures. The government can accomplish this by

1) creating new opportunities for investment (for example, through the development of new infrastructure),

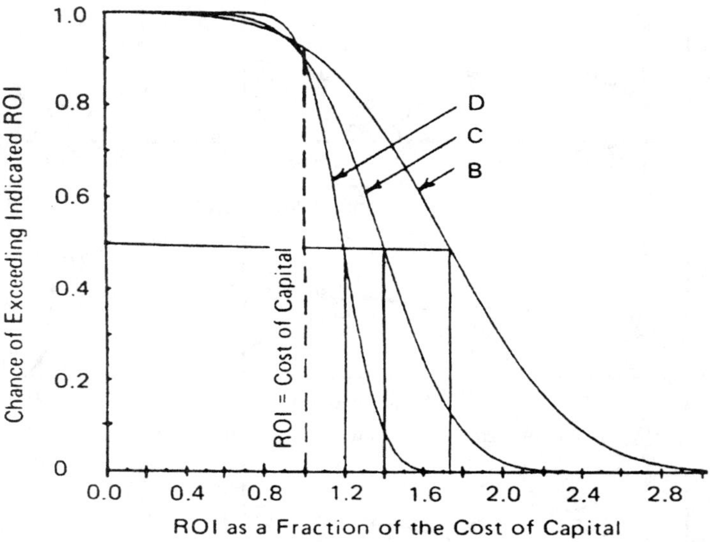

Fig. 2.25 Typical risk profiles of ROI that yield the same likelihood (50%) of investment. (Payback is four years, and investment is 1–10% of capital budget.)

2) reducing private sector perceived risk (for example, by undertaking technology development and demonstration programs),
3) increasing the potential for return on investment by reducing the initial capital burden associated with high technology ventures by the private sector and/or increasing product/service markets, and
4) creating an awareness of the opportunities, technology, and policies that are favorable to private sector investment.

To illustrate the effects that public sector programs can have on private sector investment decisions through the reduction of private sector perceived risk and by shifting the burden of funding from the private to the public sector, consider the example indicated in Fig. 2.26. This example illustrates the impact of a government demonstration program on the private sector. Figure 2.26 illustrates a common situation (for example, the ACTS) wherein the private sector is contemplating the investment in a business venture that requires a significant development effort culminating in a feasibility demonstration. Figure 2.26a illustrates the effect of full investment, including the demonstration program, by the private sector. The degree of variability of exposure (maximum of the indebtedness curve—negative of the cumulative cash flow), payback period, and ROI is indicated. A large part of the variability is as a result of the uncertainty of the cost and outcome of the demonstration program.

Figure 2.26b illustrates the impact on the private sector of NASA undertaking the development and demonstration program (for example, the development and demonstration of the Earth Observation Satellite, Landsat, and the ACTS) with no

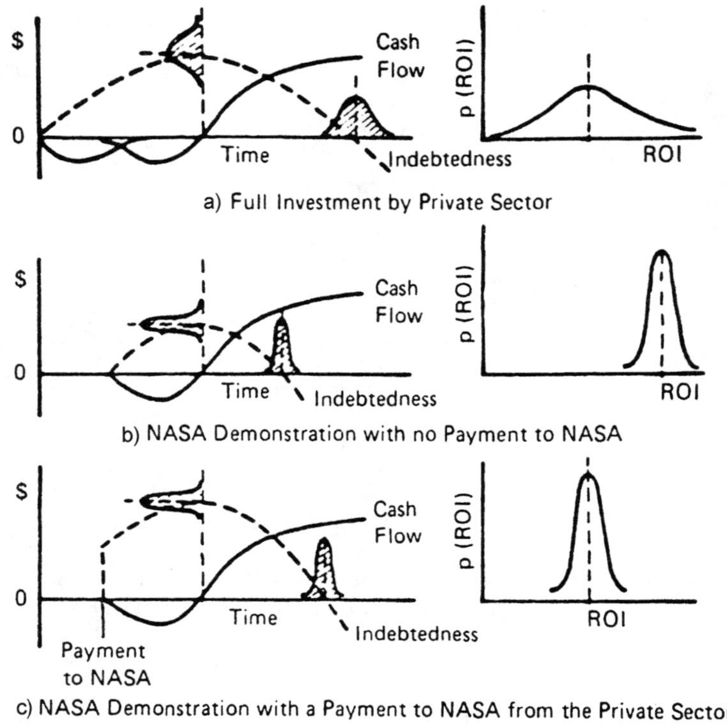

Fig. 2.26 Effect of a public-sector demonstration program on the private sector.

transfer payment from the private sector to NASA. Note that private sector exposure is reduced, as is the variability of exposure, payback period, and ROI. At the same time the expected ROI is increased. This is because of the uncertainty in the cost and outcome of the demonstration program being eliminated (private-sector decisions can await the outcome of the NASA program) as well as the private sector demonstration program funding requirement being eliminated.

Figure 2.26c indicates the impact of a NASA demonstration program when a payment is made to NASA from the private sector to cover the cost of the demonstration, for example, repayment of the cost in return for using a NASA-funded demonstration satellite (i.e., Landsat, ACTS) in an operational system. The impact of the payment is to increase expected exposure and payback period and to reduce expected ROI. Note that there is no change (relative to Fig. 2.26b) in the variability of the performance measures. It is obvious that the business venture under scenario B is more desirable than under scenario C, which is more desirable than under scenario A. It is also obvious that the course of government action can affect the likelihood of the private sector undertaking the hypothetical business venture. Because NASA or public sector benefits are a function of private sector investment decisions, the benefits will differ for each of the scenarios.

The three cases illustrated in Fig. 2.26 are superimposed, in Fig. 2.27, upon the investment likelihood curves developed in Sec. IV in order to illustrate the effects that public sector programs can have on private sector investment decisions. Point A indicates the private sector perceived consequences of pursuing a program without government support; point B indicates the same business situation but with the government funding the demonstration program; and point C indicates the same business situation but with the government funding the demonstration program but requiring recoupment of its investment from future proceeds of the business. These points demonstrate the role of the public sector in affecting private sector investment decisions through perceived risk reduction and shifting the burden of funding from the private to the public sector. The effect of a recoupment policy (i.e., payback to NASA for services rendered) is to drive point B toward point C and thus reduce the likelihood of private sector investment. It is, of course, desirable for a government agency to be paid for services rendered, but this must be tempered by its effect on total benefits—the true objective should be maximization of benefits.

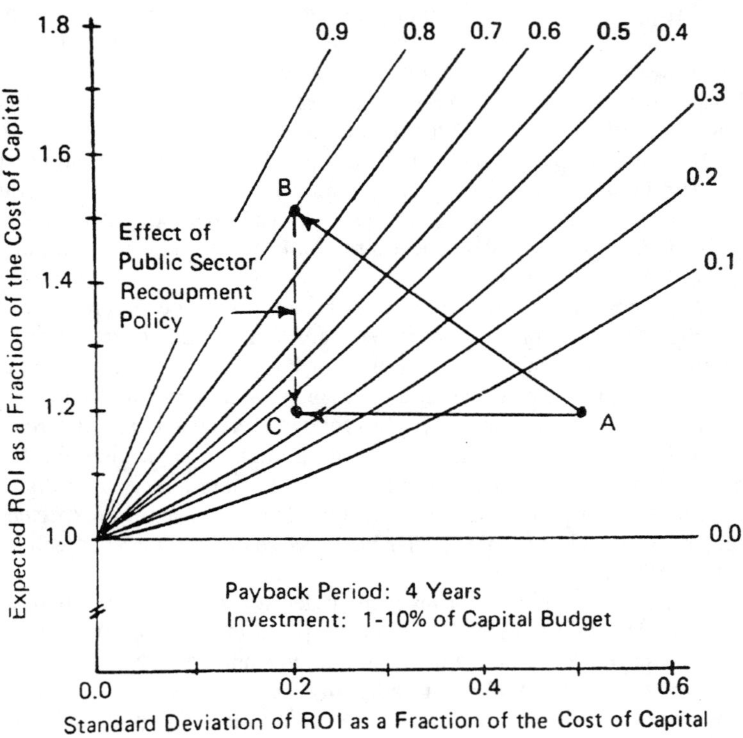

Fig. 2.27 Investment likelihood in terms of expected ROI and risk—effect of a public sector demonstration program.

The significance of the investment likelihood curves is that they provide the mechanism, the linkage, for indicating and measuring, in quantitative terms, the impact of government programs and policies on private sector investment decisions. The value of these programs and policies results from altering private sector decisions and stems from the benefits that would be foregone if the government program were not undertaken. The expected benefits from a public sector program undertaken to influence private sector investment decisions are given by

$$\text{Benefits} = \alpha_b \times \text{NPV}_b - \alpha_a \times \text{NPV}_a - \text{PVC}$$

where NPV_b and NPV_a are the expected public-sector benefits with and without the government program (i.e., development, demonstration, incentive, or other form of government program), respectively. The values of NPV_a and NPV_b can be determined using benefit-cost techniques to establish public sector benefits. [This is discussed in Secs. II and IV of Chapter 4 in terms of the effect of government programs on market share for U.S. provided launch services. The change in market share, combined with number of total launches and the economic impacts per launch lost (or gained), as described in Sec. IX, provide insight into the potential economic impacts of government policies and programs.] *PVC* is the expected present value of the cost of the public sector program. α_b and α_a are the probabilities of private sector investment with and without the public sector program, respectively, as obtained from the investment likelihood curves. It is immediately apparent that the difference between α_b and α_a has a major effect on public sector benefits. It is also evident that the value of α should not arbitrarily be taken as 1.0 with the public sector program and 0.0 without the public sector program. The appropriate value of α can only be established by government planning and evaluating private sector business ventures explicitly taking into account quantitative estimates of uncertainty and the effect of the public sector program on these uncertainties.

The RLV program is of particular interest because it is a current activity aimed at influencing private sector investment decisions and has the general attributes of large capital investment, high risk, and long payback period. Because the goal of the program is to have the private sector develop, build and operate a fleet of RLVs and to provide space transportation services for commercial and government users, private investment and financing are critical. A number of leading investment organizations have indicated that investment capital will not likely be forthcoming until the technology risk has been eliminated. (This has been the experience of those seeking to develop a commercial business based upon SSTO technology, such as VentureStar.)

As just indicated, private-sector investment decisions are normally based upon the consideration of a number of financial performance measures including expected ROI, payback period, capital requirements and considerations of risk as indicated by variability of ROI or variability of other measures. In general, private sector investment decisions are concerned with evaluating potential business opportunities and considering tradeoffs between such factors as expected ROI, risk and magnitude of investment that must fall within acceptable ranges, for example, the expected ROI must exceed a cutoff rate of return. However, this cutoff rate of return or hurdle that must be exceeded is related to the level of perceived risk with the hurdle increasing to offset increased risk. In addition, both the expected rate of return and risk hurdles are normally also related to the magnitude of the required investment. Thus, government programs can effect or alter private sector investment decisions through their impact on perceived risk (i.e., through risk reduction) and required investment. Efficient government programs aimed at affecting private sector investment decisions are aimed at those specific aspects of an investment opportunity that are impeding the investment—programs to increase ROI if return is perceived as too low, programs to reduce risk if risk is perceived as too high.

For example, if risk is perceived as too great then programs that are specifically aimed at risk reduction (which might be more specifically related to market uncertainty, technical performance uncertainty, schedule uncertainty, and so on) are normally required rather than those aimed at increasing expected rate of return sufficiently to offset the perceived risk. In such a case the impact on the private sector is through a reduction in private sector perceived risk and/or exposure with the increased likelihood of the private sector developing and marketing beneficial products and/or services.

Clearly, benefits can be achieved as the result of government actions altering private sector investment decisions. The benefits would result, in most cases, only if investment decisions are altered and the desired investments are made and ventures are ultimately funded and the ventures prove successful. Thus, government, in allotting its scarce resources, must consider the likelihood of altering business investment decisions and the viability of the resulting business.

The investment likelihood functions just described have been developed to provide government planners with insight into the effect that their policies and programs might have on investment decisions. The investment likelihood function is, in essence, a multiattribute utility function that relates the likelihood of investment to several of the more important financial performance measures. Because those concerned with establishing government policies and programs lack direct control of private sector investment decisions, it is important to have measures or metrics that can be used for anticipating private sector response to government policies and programs. The multiattribute survey results can be used for this purpose.

There are many government initiatives that individually and/or collectively can effect private sector investment decisions and can be used to achieve public sector goals/objectives and include the following.

Subsidized loans are loans provided at below market rates with the government paying the difference between the loan rate and the market rate. This has the effect of reducing expected annual expenses, increasing expected annual profit and cash flow, and reducing expected indebtedness. Depending upon the debt/equity structure of the business venture and the duration of loan subsidization, this can have a significant impact on the firm's projected expected discounted rate of return (IRR). However, loan subsidization will have little or no effect on perceived risk, defined in terms of the variability of financial performance measures. The cost to the government relates to the difference between the market and loan rates, the magnitude of the loan as a function of time, and the change in Treasury revenue resulting from a change in the likelihood of investment.

Loan guarantees[24] are loans provided by commercial institutions where the loan interest and repayment are guaranteed by the government. The government repayment guarantee is aimed at reducing the cost of the loan because of risk reduction (to the lending institution). The impacts on financial performance of loan guarantees are similar to, but of lesser magnitude than, those resulting from subsidized loans. Loan guarantees are aimed at reducing a firm's overall cost of capital with resulting impacts on expected financial performance measures. However, loan guarantees place the government at risk for the magnitude of the outstanding loan in the case of a default. [This has an effect on budget scoring with the likelihood of significant impacts on agency budgets at the time of the loan guarantee. (The Congressional Budget Office would normally assess the level of risk incurred by the government. The level of risk is multiplied by the amount at risk to establish a required budget offset.)] This latter attribute of loan guarantees is significant because, for all practical purposes, it eliminates technical risk as a concern to the investment community. If a firm does not achieve the appropriate level of performance, the firm has the option of defaulting with the government "left holding the bag" and the lending institution being unaffected by the lack of achievement of desired technical performance.

Tax policy (*Tax holiday*) implies foregoing the payment of taxes during a specified period of time. It is assumed that the tax holiday would be associated with income derived only from a space-related business venture (for example, an RLV business venture). Tax holidays have consequences when taxes would be due on profits. Typically, in a new business situation there is a period of technology and product development, followed by a startup period leading to ongoing business operations. During the former and initial stage of the latter periods, losses will be incurred, and no taxes will be due. These losses can normally be carried forward to offset profits from operations. It is thus likely, in the case of an RLV business, that tax holidays will play little or no role for an extended period of time. Tax holidays are likely to have no impact on the maximum magnitude of indebtedness (i.e., required level of investment), no effect on risk, and possibly some effect on payback period. Depending on the duration of the tax holiday, tax holidays can have a significant impact on internal rate of return. But this will be the consequence of events that will occur well into the future and are likely to have less significance in the decision making process than to impacts in the near term.[25]

Tax policy (*Tax credits*) are offsets against taxes because of investing in R&D, increased hiring, etc., and could possibly be passed through to offset non-RLV profits. This would eliminate some of the disadvantages of tax holidays because impacts would be more in the near term. Even though the magnitude of tax credits would likely be less than that resulting from tax holidays (when the venture is profitable), tax credits might be of more significance in the decision process because they can occur in the near term. Tax credits will have more impact on indebtedness and payback period than will tax holidays but lesser impact on internal rate of return and no impact on risk.

Anchor tenancy implies that the government enters into a long-term agreement for the utilization of products and/or services. This was utilized successfully for Spacehab and could have significant impacts on decisions relating to the investment in an RLV launch services business venture. This would, or could, have significant implications on market uncertainty perceptions but would require legislation to enable this to occur. Because market uncertainty is a major contributor to perceived risk, anchor tenancy could have a significant impact on risk reduction (and the form of available financing), but little impact on expected indebtedness, payback period, and return on investment.

R&D and RDT&E participation implies the government sharing in the costs associated with technology development, demonstration and testing and has the effect of the government sharing in the nonrecurring cost (front-end financing) of commercial ventures. This sharing has the effect of reducing industry's investment and consequently the expected maximum magnitude of the indebtedness. Because of the reduced indebtedness, expected payback period is also reduced, and expected internal rate of return is increased. Risk may be reduced through the cost sharing arrangement. If RDT&E participation is to be recovered through favorable pricing or other mechanism, then the impacts on payback period and internal rate of return can be significantly reduced.

Insurance guarantor is a role for the government. Because there is not likely to be a long-term commitment by the space insurance industry regarding the availability and pricing of third-party liability insurance or launch insurance, there will be uncertainty with respect to being able to recover from a catastrophic situation. The government, acting as an insurer of last resort or as a provider of short-term resources in support of an insurance provider, could have a significant effect on risk perceptions.

Debt/equity participation can include the government. The government can participate in the debt/equity structure of a firm (i.e., an investor) as it did in the initial creation and operation of COMSAT Corporation, which was the U.S. participant in the provision of international communication services. Direct investment participation will reduce funding requirements (from capital markets) and can effect risk perceptions via the government participation.

The preceding list include government programs/policies, as depicted in Fig. 2.28, that are aimed at altering private-sector investment decisions through risk reduction and capital formation. Table 2.2 summarizes the relationship between a number of the preceding government initiatives and decision variables with the number of Xs used to indicate their relative importance in achieving their objectives.

To evaluate government initiatives, it is necessary to assess their impacts on private sector investment decisions, over which the government has little or no control. Therefore, from the government's point of view it is only possible to predict these decision in terms of the probability, or likelihood, that specific decisions will be made given a set of financial performance measures. The results of government analyses of private sector business ventures, and the impacts of government programs and policies on these ventures, should develop the performance measures, or metrics, that will result in the estimation of the likelihood of private sector investment. Because surveys have been conducted that have resulted in the development of investment likelihood functions in terms of expected return on investment, risk, expected payback period, and magnitude of investment, these parameters become important metrics, which taken together with the survey investment likelihood data result in the determination of investment likelihood as a function of government initiatives. The financial performance measures are intermediate and needed for the determination of the change in likelihood of investment brought about by each of the considered government initiatives.

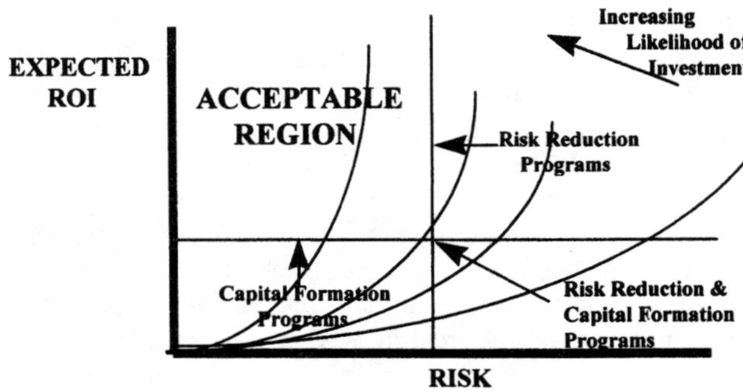

Fig. 2.28 Risk–return relationship and impact of government programs. (The shape of the curve is also a function of expected payback period and magnitude of the investment that is at risk.)

In summary, there are two types of metrics that need to be considered: those that are important to financial planners engaged in the planning of space-related business ventures and those that are important to government planners concerned with setting policies and allocating scarce resources in support of commercial space business ventures. In the former case, metrics include ROI, net present value, payback period, and quantitative measures of risk are important in establishing value and viability. However, because of the magnitude of many space-related investment undertakings there continues to be concern with the feasibility of financing space business ventures that are considered to be large-scale risky endeavors. The developed financial performance measures, particularly including quantitative measures of risk, can be used to assess the likelihood of obtaining financing and the form of capital structure that will increase the likelihood of financing.

Table 2.2 Summary of relationship between government initiatives and private-sector decision variables

	Decision variable		
Government initiative	Expected ROI	Risk	Required capital
Research and development	—	XXXXX	—
Tax policy	XXX	—	—
Cost sharing	XXXXX	—	XXXXX
Loan guarantees	X	XXX	X
Low-interest loans	XXX	—	XXX
Anchor tenancy	—	XXX	—
Anchor tenancy (with termination liability)	—	XXXXX	—

X···X indicates the relative magnitude of impacts on decision variables.

In the latter case, similar metrics (i.e., expected rate of return, standard deviation of rate of return, expected payback period, and expected magnitude of investment) can be utilized by the government to assess the impact of individual and combinations of selected government initiatives on effecting private sector space transportation investment decisions. However, the government must go beyond this and establish the benefits and costs that can result from the various considered policies and programs. This implies that metrics need to be developed which indicate the change in the likelihood of investment, government cost savings, and economic impacts in the form of job creation and Treasury revenue (resulting both from a change in market share and a change in demand for launch services).

Before leaving this subject of government impacts on private sector investment decisions, it is interesting to examine information developed by NASA and used in its analysis of the potential impacts of government programs and policies on RLV business performance. Figure 2.29 is a reproduction of a NASA-generated figure[26]

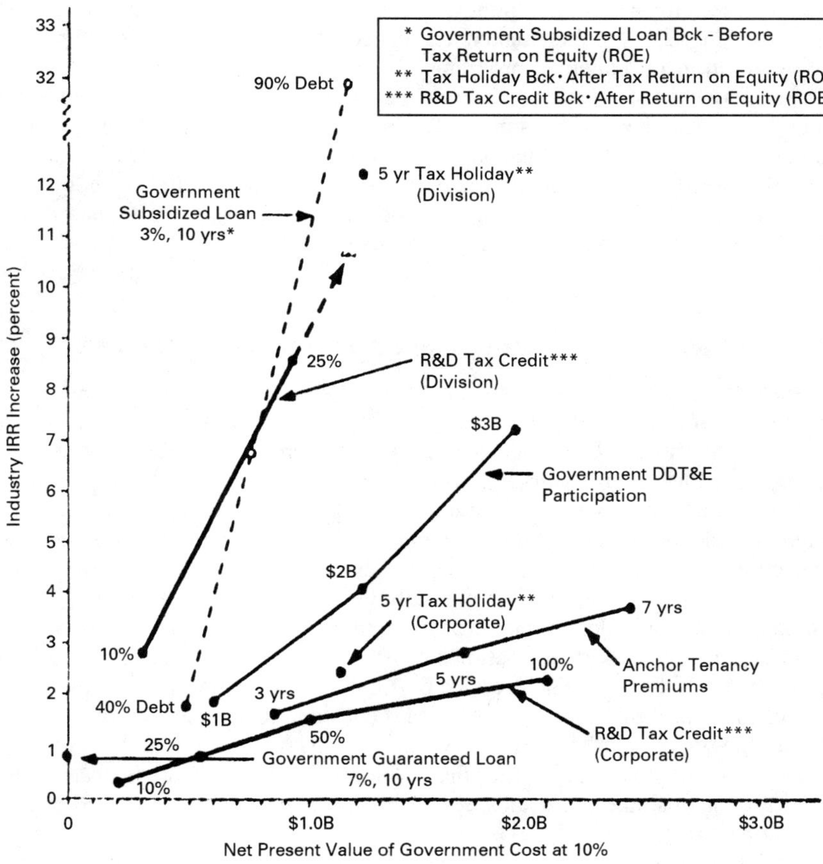

Fig. 2.29 Efficiency of RLV incentives (as shown in Ref. 26). Don't always believe what you see!

that purports to show the efficiency of RLV incentives. Figure 2.29 illustrates the impact of various government programs and policies on industry IRR in terms of the net present value of government cost of the program/policy. Illustrated in this manner are the impacts of programs dealing with government-subsidized loans, government RDT&E participation, tax holidays, anchor tenancy, R&D tax credits, and government-guaranteed loans. By placing the impacts of the various programs and policies on the same chart, the tacit implication is that these programs and policies can be compared in terms of impacts on IRR and government NPV. The implication is that the program/policy that has the maximum impact on IRR at the lowest cost is most preferable. This, unfortunately is not true because, as already discussed, programs/policies such as those relating to tax credits and loan guarantees affect expected values, whereas those relating to anchor tenancy affect risk perceptions and those relating to government RDT&E participation can affect both expected values and risk perceptions. Their impacts cannot, and should not, be depicted as in Fig. 2.29 because the risk dimension is not shown. Thus, tax-related incentives are indicated as being preferable to other policies such as anchor tenancy that affect primarily the risk dimension and are likely to have a greater impact on decisions than tax policy. As a result of this type of economic comparison (Fig. 2.29), policy makers can be led down the wrong path and public funds spent on inconsequential programs.

Comparisons must take into account the impacts of the various programs/policies on private sector investment decisions that consider multiple decision attributes: expected ROI, risk, payback period, and magnitude of investment. As already discussed, an important metric is the change in investment likelihood that is the result of the government program/policy. If investment decisions are not altered, then there is no value to the government program/policy.

VI. Comparison of Alternatives (Space Transportation Alternatives)

A decision database was described in Sec. IV that provides the means for assessing the impact of government programs and policies on private sector investment decisions. The collected data considered individual investment decisions and the general situation where the investments under consideration were independent and *not* mutually exclusive, that is, funds were considered to be available for the investments and the selection of one investment did not preclude the selection of another investment. There are, however, many situations that arise where investment opportunities are mutually exclusive. This situation arises, for example, when a provider of satellite communication services selects a launch service for launching a communication satellite. The selection of launch service A precludes the selection of launch service B. (Of course when multiple satellites are to be launched and a mix of launch services is considered, the selection of one mix precludes the selection of another mix.) These alternatives are said to be mutually exclusive, and their analysis requires the development of a decision database. The method for establishing this database, and obtained results, are described in the following paragraphs.

To understand better the characteristics of the demand for space transportation services for launching communication satellites, it is necessary to understand the

INSTRUCTIONS FOR RESPONDING TO SURVEY

Suppose that your firm has decided to launch one or more communications satellites. Taking this decision as a given, you must choose between two possible launch vehicle/service alternatives. Although there are many attributes which are associated with a given system, suppose again that these have been translated into financial performance measures, as represented by a histogram describing a probability distribution of return on investment. This value is expressed as your normalized return on investment; i.e., ROI divided by your cost of capital. This normalized value is refered to as R. Your responsibility, then, is to choose the better of two systems by evaluating their respective R curves.

Through the tip of each curve a dashed line has been drawn, indicating the expected, or most likely, normalized return on investment. The spread, or width, of the curve represents the associated risk. Thus if a curve is tall and narrow, it indicates a rather sure choice, with little chance of large deviations, for better or for worse, from the expected return. A shorter, wider curve, on the other hand, represents a riskier choice, with a greater chance that the realized R will be more or less than anticipated.

On the following pages are shown a series of pictures, each made up of histograms of return on investment, normalized to your cost of capital. For each column, there is a top histogram, labeled "A", which will always serve as one of your choices. Listed below are five other pictures, each labeled "B". Within each of the "B" pictures are four curves which you will individually compare to the "A" curve. You might notice that each "B" picture contains the same four histograms of R, except that they have successively higher expected values. The "A" curve is drawn in gray on each picture to simplify comparison.

The horizontal axis measures R. At the point labeled "1", your return on investment is equal to your cost of capital. At the point labeled "2", your return on investment is double your cost of capital, and so on. Each tic mark in between thus represents a change of 0.2 in R.

Each set of graphs also has an associated list of probabilities. These indicate the chance that a particular curve "B" will produce a higher return than curve "A". Note that this is only one factor in your decision, and not the only criteria. Some other factors you may wish to consider:

* What is the relative difference in expected value between the two curves?

* Does either curve extend beyond the "1" line? i.e. Could the firm lose money on either investment?

For each picture depicting the "B" alternatives, please indicate by placing an "X" in the appropriate boxes, whether you believe your firm would, based upon the R information shown, select alternative "A" or "B". This should be done for each of the four curves within each of the five "B" alternatives relative to the "A" curve at the top of the column, and for each of the columns on the following pages.

To the right of this text is an example to illustrate the procedure. Please make sure that only one "X" is placed in each of the four indicated rows (that is, the "A" and "B" alternatives are mutually exclusive).

Thank You.

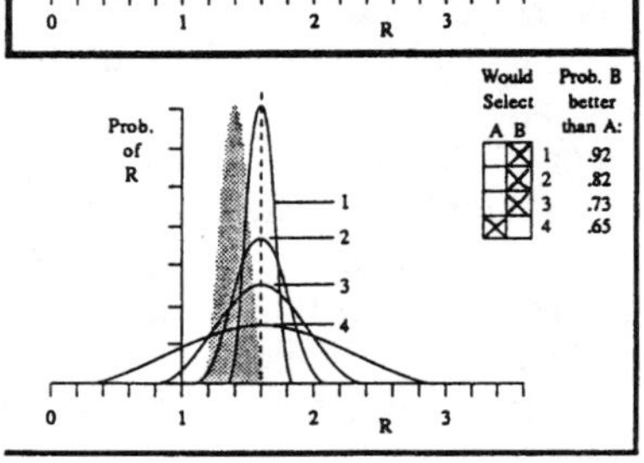

Fig. 2.30 Instructions for responding to survey.

implications of financial performance measures in selecting alternatives. The basic generic problem is as follows: Given that a decision has been made to commit resources to a communication satellite business venture and that there exist multiple implementation alternatives (for example, the selection of a specific space transportation service such as Ariane, Atlas, Long March, etc.), what is the chance that different alternatives will be selected in terms of perceived differences in resulting financial performance measures?

Clearly in a riskless environment and with all other factors being equal, the choice would be made to select that alternative which maximizes net present value of cash flow or return on investment. Enter risk and the problem becomes more complex because it is necessary to understand the tradeoffs between perceptions of risk and other financial performance measures. In other words it is desired to establish the likelihood of selecting alternative A rather than alternative B. This is equivalent to establishing expected market share for alternatives A and B. This is discussed in more detail in Sec. II of Chapter 4.

The method used to evaluate the likelihood of selecting one alternative vs another, that is, market share, separates perceptions of launch vehicle and launch service attributes from the decisions that are likely to be made based upon the consideration of financial performance measures. In other words, each launch vehicle and service has an associated set of attributes such as price, reliability, final payload placement accuracy, schedule, and likely delays. User perceptions of these attributes will affect the proforma financial performance projections of the resulting business venture with each launch service choice resulting in a different set of financial performance measures. (The *DOMSAT* financial planning model, described in Sec. I of Chapter 4, was specifically developed to transform launch vehicle and service attributes into business financial performance measures, which then establish the likelihood of selecting alternative A rather than alternative B.) The likely choice of a particular launch service is assumed to be a function of the resulting set of financial performance measures and its relationship to investment attitudes that relate to financial performance measures. It is with this latter subject (i.e., investment attributes relating to financial performance measures) that data collection is concerned.

The objective was to establish the likelihood of investing in one alternative relative to other alternatives given that a decision has been made to pursue a business venture (e.g., the decision has been made to launch one or more communication satellites and the problem is to select from amongst several alternatives for the launch vehicle/service that will be utilized). This was accomplished via a survey in which a number of histograms or probability distributions of normalized return on investment (i.e., ROI divided by the cost of capital) of a number of hypothetical alternatives were presented, compared, and decisions for investment noted. The consideration of the probability distributions allowed both the expected value and risk dimensions of the compared alternatives to be developed and tradeoffs made relative to risk avoidance preferences. The risk dimension is represented by the spread or standard deviation of the probability distribution of the normalized return on investment R.

The selection of the launch vehicle must consider the risk dimension because launch vehicle/service attributes such as reliability, placement accuracy, scheduling, and delays contribute significantly to differences in the overall business risk. Therefore, a number of business scenarios were described in the survey in terms of their R histograms, and choices were made by the respondents relating to these hypothetical scenarios. The developed decision database is based upon the return of approximately 15 useful survey questionnaires.

Figure 2.30 indicates the instructions that were provided to the survey participants. All participants were contacted via telephone prior to distribution of the survey and followup calls made after receipt, but prior to return of the survey. All survey questions related to the selection of one alternative (A) relative to other alternatives (B), of which there were four (B1–B4). Each column in the questionnaire allowed for the comparison of A and B, with each comparison vertically down the page having increasing expected values for the Bs. Each column on the page considered the same A alternative but with different expected values. The questionnaire contained four pages with each page having a different A alternative (different risk characterization). Thus, 60 different comparisons were provided in order to develop the likelihood function. Figure 2.31 illustrates one of the four questionnaire pages.

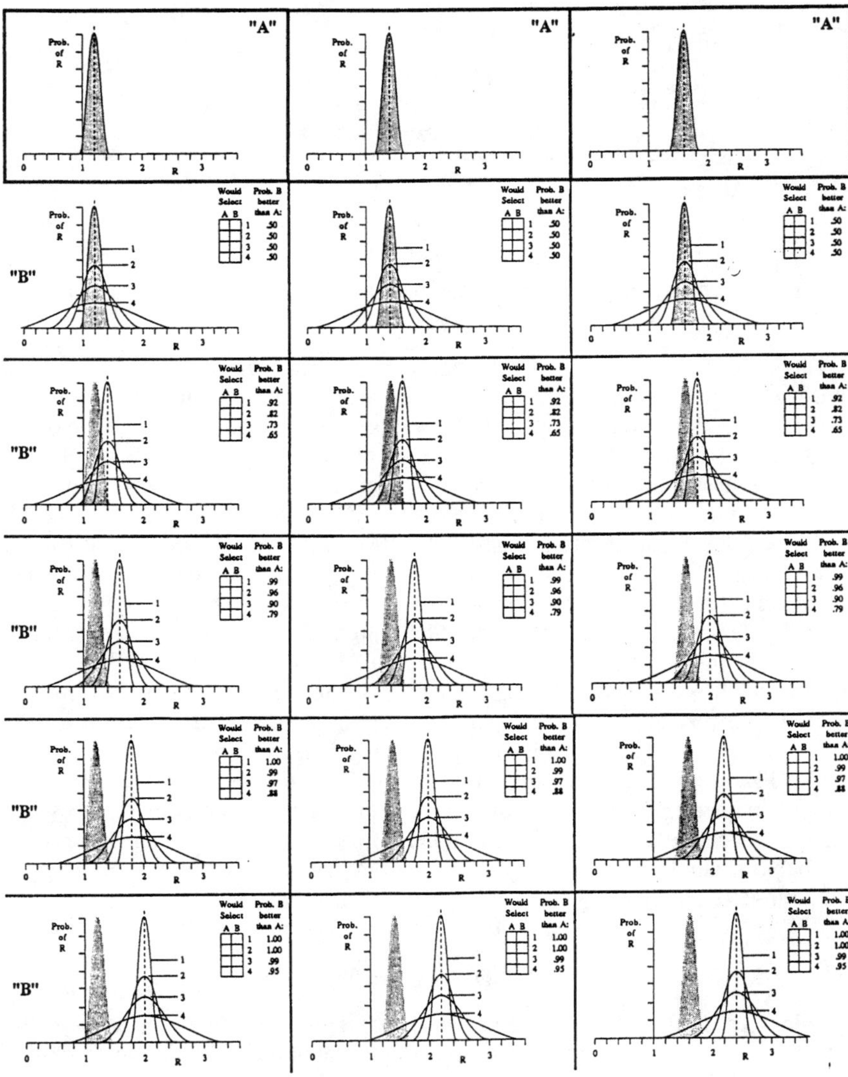

Fig. 2.31 One of the four questionnaire data collection pages.

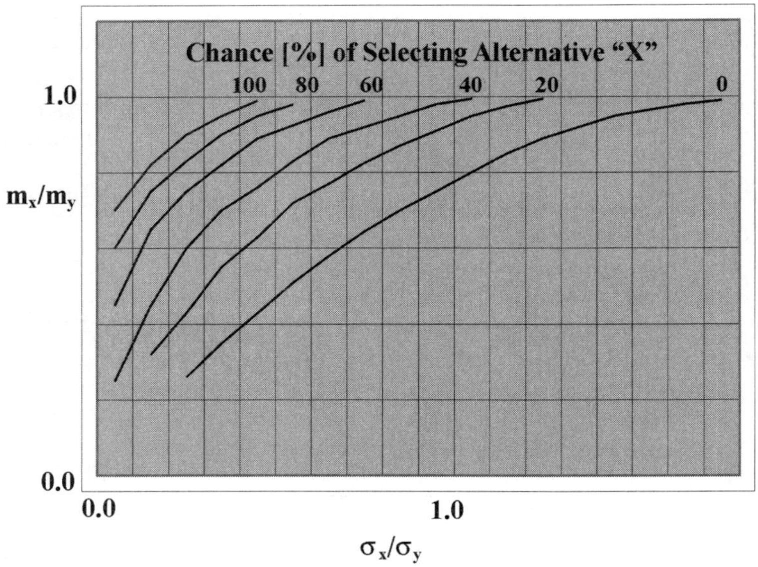

Fig. 2.32 Investment likelihood function for mutually exclusive alternatives.

The collected data resulted in the decision database summarized in Fig. 2.32. Figure 2.32 indicates the chance of selecting alternative X (relative to alternative Y) in terms of the ratio of expected ROI of alternatives X and Y and the ratio of the standard deviations of ROI of alternatives X and Y. For example, the likelihood function indicates that when the ratio of expected ROI of alternatives X and Y is 0.8 and the ratio of the standard deviations of ROI of alternatives X and Y is 0.5, there is a 40% chance that alternative X will be selected for investment (or the market share of launch service X is 40%). If a government program were undertaken that decreased the perceived risk associated with alternative A such that ratio of the standard deviations of ROI of alternatives X and Y were reduced to 0.4, then the market share of alternative X would be increased to 50%. The benefit that results from a government program that alters market share is clearly a function of the change in market share. If there is no predicted change in market share, then clearly there is no benefit from the government program or policy.

The likelihood function can also be utilized in the same manner by those engaged in the development and/or provision of space transportation services. For example, a change in a competitive bid price (for the provision of launch services) can be evaluated in terms of the change in the likelihood of winning the competitive procurement.

VII. R&D Program Planning

This section discusses but one of many approaches to the planning and evaluation of R&D programs. These approaches encompass simulation modeling, linear and integer program modeling, and the use of specific algorithms for ranking and project selection. Simulation modeling, such as RADSIM,[7,27] should be used for the planning of long-term, multi-phase R&D programs particularly where project success is not certain. It is notable that NASA chose not to do this for their current low-cost space transportation and space solar power initiatives, each requiring a long-term multiphase effort and each aimed at achieving significant technology advances where success of projects is less than certain. The linear and integer programming approach attempts to establish, through the use of mathematical algorithms, the mix of projects that maximizes program benefits given a set of cost and/or other constraints. Specific algorithms have been used such as establishing a benefit–cost ratio for each project of a possible portfolio of projects and selecting projects, from highest to lowest benefit–cost ratio until resource constraints are encountered. This latter approach is relatively simple and works reasonably well when projects are relatively short term and independent and project management has limited comprehension of more powerful approaches.

Technology programs are underway, and government policies are being evaluated and revised so as to enable the introduction of new lower-cost space transportation systems. Because of the high cost of, and long time required for, the development of new space transportation systems, their useful life time must be extended to the maximum extent practicable. This has been (and most likely will continue to be) accomplished by investing in research and technology programs that lead to improvements in launch vehicle attributes. These improvements have allowed the U.S. space transportation systems to remain reasonably competitive in international competitions and have led to cost savings for those missions utilizing the improved capability space transportation systems. Because resource requirements for possible research and technology projects normally exceed available resources, it is necessary to select, in a rational and defensible fashion, a set of projects from amongst the identified possible projects that can be undertaken with available resources.

Technology projects have the goal of altering launch vehicle attributes such as cost, reliability, lift capability, and scheduling (turnaround time). Changes in launch vehicle attributes directly affect the value of missions utilizing the improved launch vehicles (cost savings for civil and military missions, increased net present value of cash flow for commercial missions, and improved safety for all missions). In addition, changes in launch vehicle attributes will be reflected, by changing user decisions regarding the choice of a launch vehicle, in changes to market share. An increase in market share, resulting from altering choices because of improved user economics (resulting from improved transportation service attributes) will result in increased U.S. launch vehicle sales with benefits in the form of increased exports, increased employment, and increased Treasury revenue (the government's domain of concern when considering government investment in technology development and demonstration), and net present value of incremental (as a result of the improved capability launch vehicle relative to the otherwise competitive choice) cash flow of the U.S. commercial users of the improved capability launch vehicle and/or net present value of incremental cash flow of the provider of the launch services. (The split in benefits between the users and providers depends upon whether or not the launch provider alters pricing to offset the improvement in value derived from improvements in other launch vehicle attributes.)

Thus, the value of a technology program can be measured in terms of its combined impact on cost savings (net present value) for civil and military programs, increase in net present value of Treasury revenue resulting from improved cash flow for commercial missions, and the benefits from increased market share as measured in terms of net present value of increased Treasury revenue (or increased employment or increased exports). The former two areas can be thought of as equal capability (i.e., no change in demand for U.S. launch services) benefits and the latter as increased capability [i.e., resulting from a change in total demand (the result of price elasticity) and the demand for U.S. launch services (due to both the price elasticity and the change in market share)] benefits. Typical impacts resulting from improved competitiveness are illustrated in Table 2.3 in terms of an additional Atlas or Delta launch and the fourth-quarter 1994 impacts (i.e., if the U.S. captured the foreign launches). The development of these impacts is described in Sec. IX.

The following paragraphs briefly describe a resource allocation method appropriate for decision making relating to investments in R&D aimed at improving space launch vehicle attributes. Emphasis is placed upon relating research and technology projects to anticipated changes in launch service attributes, and then converting launch service attributes into equal capability and increased capability benefits with the increased capability benefits being related to user decisions (as effected by changes in perceived financial performance measures) and resulting changes in market share. (Changes in demand that result from price reductions have not been taken into account in the analysis but can be readily included.) The changes in market share are then related to changes in exports, employment, and Treasury revenue with both direct and indirect effects considered. An algorithm is then identified that can utilize obtained benefits and costs to establish an efficient technology investment portfolio. Although the emphasis of the following paragraphs is placed upon government investment decisions, the same process is applicable for the consideration of private-sector investments but with a somewhat different interpretation of benefits. This is discussed in following paragraphs.

Table 2.3 Summary of economic impacts of launch competition (relative to what would have been if U.S. firm won competition)

	Typical per launch impacts		
Impact Areas	Atlas	Delta	4th Quarter 1994 impacts
Reduction in exports, M$	100–165	45–94	445
Reduction in Employment			
Direct, person-yrs	2550–5120	1060–2930	13,362
Indirect, person-yrs	3580–7000	1510–4010	18,389
Total, person-yrs	6130–12,130	2580–6940	31,751
Reduction in federal revenue, M$	50–110	20–63	281

When resource requirements exceed available resources (as is frequently the case), a formal process is generally developed and followed in order to allocate resources in a rational and defensible manner with the goal of maximizing the value (benefit) that can be obtained from the allocation of the scarce resources (costs). Many different resource allocation schemes/algorithms have been developed including those based upon the use of formal R&D simulation models, mathematical models based upon linear and integer programming techniques, and heuristics.[28]

The R&D simulation approach attempts to capture the structure of, and uncertainty associated with, alternative R&D programs comprised of multiple projects undertaken both in parallel and in series—in other words, a multiphase program where decisions can be made at the completion of each phase and the projects, through obtained results and decisions, is highly correlated.[7] The objective is to identify the set and arrangement of multiple time-phased projects with decision rules concerning the starting and termination of projects aimed at maximizing the probability of overall program success while minimizing costs. This approach is most appropriate when dealing with long-term multiphase programs having considerable uncertainty and risk and where individual project results are highly interrelated when considering overall program results.

The mathematical modeling approach seeks to utilize formal linear and/or integer programming methods to maximize benefits given individual project benefits and costs and associated overall resource constraints. Multiple benefit, cost, and resource categories can be considered and can be functions of time. Finally, the heuristic approach attempts to develop a rational approach that is likely to come close to maximizing the benefits that can be achieved from a portfolio of R&D projects.

The heuristic approach is considered for the problem at hand. It is assumed that each technology project can be characterized in terms of costs and benefits; more specifically, the net present value of benefits (NPV) that are anticipated to result from the project and the present value of the cost of the project (PVC). (If the results of technology projects are highly correlated and/or the resulting benefits cannot be considered to be independent, a group of projects can be considered together as a single or focused project.) The heuristic approach consists of the following steps:

1) Calculate the present value (necessary when costs will be spread over multiple years) of the cost of each project [PVC(P)], where P is the project identifier.
2) Calculate the net present value (present value of project benefits less the present value of the costs that result from the implementation of the technology project results) that is anticipated to result from the implementation of the project.
3) Establish project benefit cost ratio [$B/C(P) = NPV(P)/PVC(P)$].
4) Identify project costs that will be consumed (in the planning year) from the available resources.
5) Rank order the projects by the calculated B/C ratio.
6) Select projects according to the rank ordered list until the available budget is consumed.

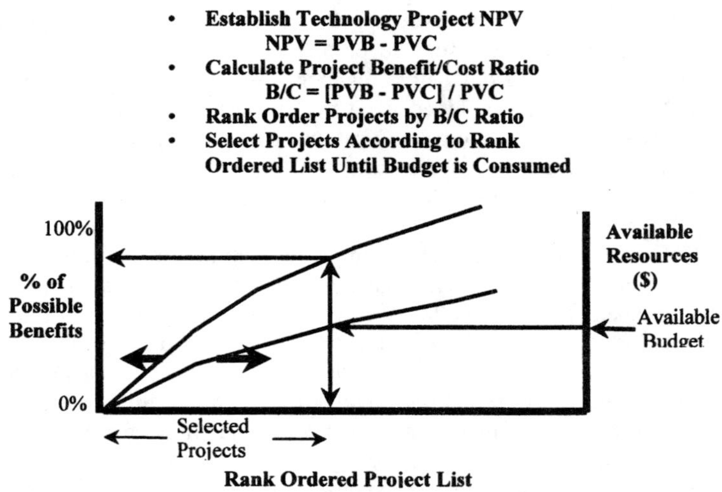

Fig. 2.33 Resource allocation algorithm.

This process is summarized in Fig. 2.33 where a rank ordered (by B/C ratio) list of projects is presented on the horizontal axis and the percent of possible benefits [the ratio of the benefits from the selected set of ranked projects relative to the benefits (NPV) that would result if resources existed to undertake all of the identified projects] is presented on the vertical scale. Also on the vertical scale is the level of available resources. If resources are allocated to projects, according to the ranked (by B/C ratio) project list, until available resources are consumed, the percentage of possible benefits can be obtained. The results of the heuristic approach can be modified to reflect specific needs not captured in the B/C approach, and changes in percent of possible benefits can be observed from the "forced" solutions.

The following paragraphs are concerned principally with the development of benefits and costs that can be used in such a heuristic approach and the inclusion of other qualitative benefits (for example, improvement of safety).

Figure 2.34 presents an overview of the resource allocation process associated with technology projects undertaken to alter launch service attributes (for example, reliability, cost, lift capability, etc.). The *technical-oriented analysis* results in the definition of the technical projects, their costs, and their anticipated impacts on launch vehicle service attributes. An *economic-oriented analysis* transforms the launch vehicle service attribute impacts into mission impacts and associated benefits. The benefits and costs are then used in the resource allocation algorithm, together with resource constraints, to prioritize and rank projects.

The interface between the characterization of launch vehicle technology projects and the economic (benefit/cost) analysis is illustrated in Fig. 2.35. This characterization consists of identifying the specific launch vehicle attributes that can be affected by the technology program, the nominal values associated with these attributes (i.e., the values of the attributes if the technology projects were not

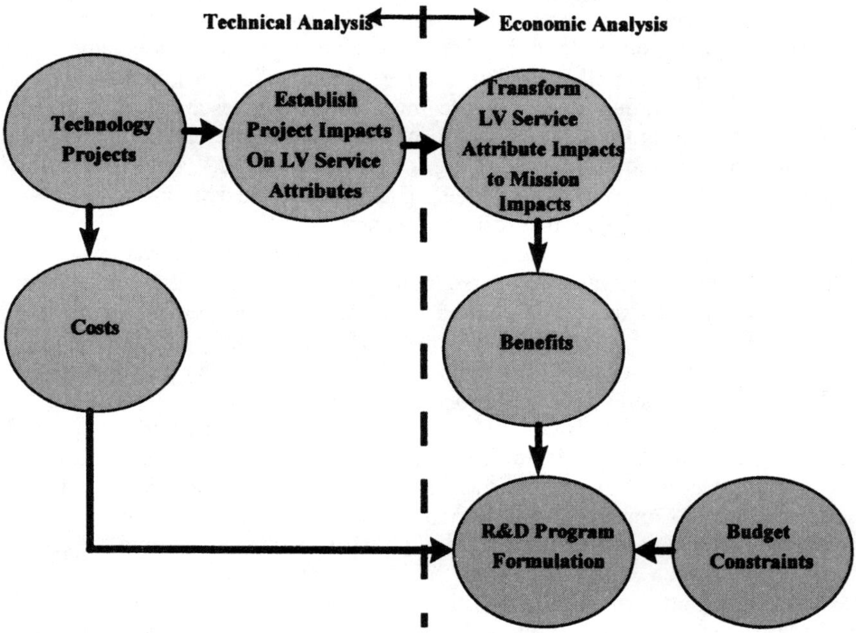

Fig. 2.34 Overview of resource allocation process.

undertaken), and the goals of the technology projects in terms of the percent change that they will cause on the nominal attribute values. Because projects might not be independent (i.e., impacts might not be linearly additive), combined impacts might also be considered.

The launch service attributes can be transformed into mission value through the use of the *DOMSAT* and *SATCAV*, or similar, simulation models. (These models are discussed in Sec. I of Chapter 4.) A number of typical missions (for example, fixed

Launch Vehicle Attribute	Nominal Value	Technol Proj Impact [% Change]				Combined Impact [%]
		Proj. A	Proj. B	.....	Proj. N	
Cost [M$] Reliability [Stage 1] Reliability [Stage 2] LEO Delivery Capability [kg] GEO Delivery Capability [kg] Placement Accuracy [m] Availability [Days] :						
Technology Project Cost [M$]	X					

Fig. 2.35 Characterizing technology project impacts on launch service attributes.

satellite communication service, direct broadcast communication service, GPS, etc.) need to be considered and the change in value (representing the value of the technology project) established relative to the mission value (Net present value of cash flow for communication satellite missions can be established using the *DOMSAT* model, and the present value of life-cycle costs for government and other missions can be established using the *SATCAV* model.) using the nominal values of the launch vehicle attributes. The form of the anticipated results is illustrated in Fig. 2.36 in the form of sensitivity coefficients that in certain instances might not be linear functions. These sensitivity coefficients are used to establish the economic value of the technology project for an individual mission. The obtained results must be modified to account for the number of missions (of each considered type) annually and over time properly adjusted through discounting. This is described through an example in following paragraphs.

The preceding paragraphs described the basic approach for estimating the equal-capability benefits or cost effectiveness. In addition, the added-capability benefits must also be considered. The added-capability benefits are the result of changes in international competitiveness that can result from the technology projects. That is, improving the launch service attributes will influence user-perceived financial performance that might alter decisions concerning the choice of a launch service, causing changes in market shares. Changing market share associated with international competitions will not only improve the profitability of the launch service provider but will impact exports/imports and resulting direct and indirect income, job creation, and Treasury revenue. (This is illustrated in Table 2.3).

The process of transforming launch service attributes into market share impacts is illustrated in Fig. 2.37. The *DOMSAT* model [as has been used for the Federal Aviation Administration (FAA)/Department of Transpotation (DoT) to provide insights into the implications of launch service attributes on communication

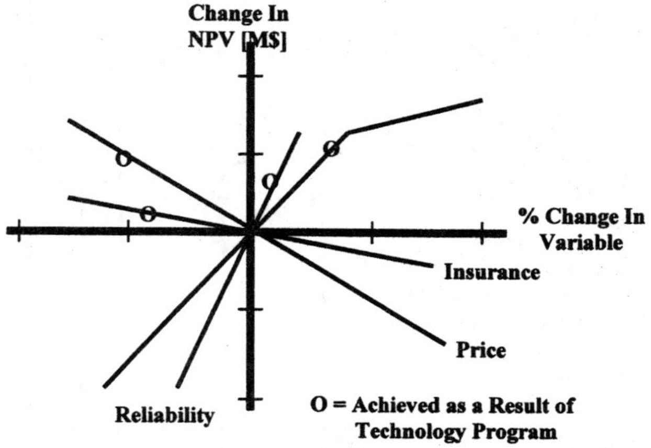

Benefits = NPV[Price] + NPV[Rel] + NPV[Life] + NPV[Insur]

Fig. 2.36 Transforming launch service attributes into mission value.

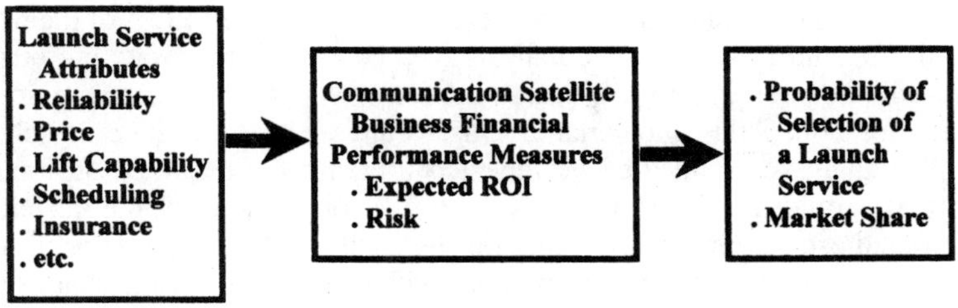

Fig. 2.37 Transforming launch service attributes into market share impacts.

satellite business financial performance and fair pricing policies] can be used to establish financial performance measures (expected return on investment, ROI, and risk as measured by the standard deviation of ROI) of representative communication satellite business ventures using the nominal values of launch service attributes of both U.S. and foreign launch services. The analyses must then be repeated utilizing the changed (because of the technology projects) launch service attributes and will result in changes in the perceived financial performance measures. These changes lead to changes in market shares.

The survey, and resulting database, described in Sec. VI, was aimed at developing an understanding of the implications of different financial performance measures on the choice of a particular launch service. It was found that relative (between competing launch services) expected return on investment and perceived risk were important indicators of the likelihood of selecting a launch provider. Figure 2.38 illustrates the type of results obtained from the survey, namely, the

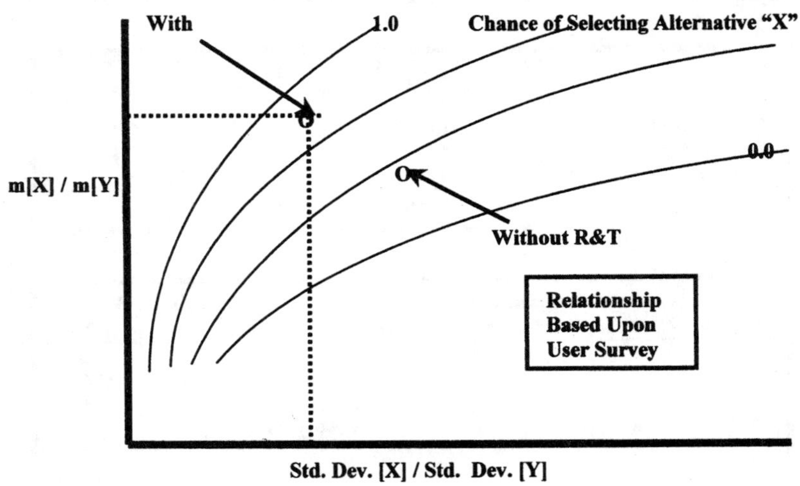

Fig. 2.38 Change in launch service market share (relationship of choice to financial performance). (Note that an R&T program aimed at alternative X can be met by a response within the scope of alternative Y. If this is the case, then the attributes associated with Y must be modified accordingly.)

likelihood of selecting alternative X relative to alternative Y in terms of the relative expected ROI and risk. Thus, an analysis of a communication satellite business venture utilizing, for example, Atlas and Ariane, would first utilize the nominal attributes for both Atlas and Ariane. This would yield the chance of selecting Atlas. The analysis would then be repeated utilizing the modified attributes of Atlas (that would result from the successful completion of the technology program) and establishing a new value for the chance of selecting Atlas. Because the chance of selecting Atlas is synonymous with market share,[29] the change in the likelihood of selecting Atlas is equivalent to the change in market share that is the result of the technology program. Thus, a mechanism is available for relating market share to launch service attributes and launch service attributes to technology programs with the result that the technology program can be related to changes in market share. By utilizing the data illustrated in Table 2.3, the number of launches per year, and the changes in market share that result from the technology program, it is possible to establish the value of the added-capability benefits.

Figures 2.39 and 2.40 present hypothetical examples that illustrate the method for establishing the benefit/cost ratio for technology projects with Fig. 2.39

Launch Vehicle Attribute	Nominal Value	Attribute Change [%] from Tech. Project			Total [%] Impact
		a	b	c	
* Cost [Price] [M$]	70	-5			-5
* Reliability	0.85	+2			+2
* Lift Capabil [Sat. Life]	12				
Technology Project Cost [M$]		8			8

	Sensitivity Coefficient		Change in NPV [M$]	
	Mission A	Mission B	Mission A	Mission B
* Cost [Price] [M$]	-1.0	-1.0	5.0	5.0
* Reliability	+1.5	+1.2	3.0	2.4
* Lift Capabil [Sat. Life]	+2.0	+1.5		
Total Change in NPV			8.0	7.4

Mission	No./Year	Change In Market Share [%]	Change In Treasury Rev. [M$]	Savings [M$]	Benefits [M$]
A	2			16.0	16.0
B	4			29.6	29.6
C	5	5	75*5*0.05		18.8
D	5	3	75*5*0.03		11.3
				Benefits [M$]	75.7
				B/C Ratio	9.5

Fig. 2.39 Hypothetical example of project evaluation.

Launch Vehicle Attribute	Nominal Value	Attribute Change [%] from Tech. Project			Total [%] Impact
		a	b	c	
* Cost [Price] [M$]	70	-5	-5		-7
* Reliability	0.85	+2		+4	+5
* Lift Capabil [Sat. Life]	12		+4		+4
Technology Project Cost [M$]		8	15	2	25

	Sensitivity Coefficient		Change in NPV [M$]	
	Mission A	Mission B	Mission A	Mission B
* Cost [Price] [M$]	-1.0	-1.0	7.0	7.0
* Reliability	+1.5	+1.2	7.5	6.0
* Lift Capabil [Sat. Life]	+2.0	+1.5	8.0	6.0
Total Change in NPV			22.5	19.0

Mission	No./Year	Change In Market Share [%]	Change In Treasury Rev. [M$]	Savings [M$]	Benefits [M$]
A	2			45.0	45.0
B	4			56.0	56.0
C	5	5	75*5*0.05		18.8
D	5	3	75*5*0.03		11.3
				Benefits [M$]	131.1
				B/C Ratio	5.2

Fig. 2.40 Hypothetical example of project evaluation (correlated projects).

considering a single project and Fig. 2.40 considering three correlated projects. The only difference is that in the former the impacts of the individual projects are linearly additive, whereas in the latter they are not [see the total (%) impacts]. Both examples consider two generic government missions (A and B) that generate cost savings and two generic missions (C and D) that involve international competition for the provision of launch services. Nominal values of attributes are shown together with the anticipated effects of the technology projects. Also indicated are hypothetical sensitivity coefficients for the net present value of mission life-cycle costs (developed utilizing the *SATCAV* Model) for missions A and B. The total attribute impacts multiplied by these sensitivity coefficients yield the change in mission net present value for each mission. Also indicated is the number of such missions per year. (The indicated examples consider only a single year. In practice, multiple years and associated discounting should be considered in the determination of benefits.)

Because C and D are assumed to be commercial communication satellite missions, the change in market share likely to result because of the technology projects is developed utilizing the results obtained from the *DOMSAT* Model and the launch service decision database. If three different launch services (a, b, c) are considered having expected values of ROI and risk as given by $\{m_a, \sigma_a\}$, $\{m_b, \sigma_b\}$, and $\{m_c, \sigma_c\}$ determined through the use of the *DOMSAT* Model, and resulting values of $p_{a/b}$, $p_{a/c}$, etc., as determined from Fig. 2.38, it is possible to establish market share for launch service a, S_a, as

$$S_a(\%) = \frac{100 * [p_{a/b} + p_{a/c}]}{[p_{a/b} + p_{a/c} + p_{b/a} + p_{b/c} + p_{c/a} + p_{c/b}]}$$

Multiplying the change in market share by the number of competitions and the likely change in Treasury revenue that result from a competitive "win" results in the determination of the added capability benefits. For the government missions the benefits are equal to the cost savings. Summing the benefits and dividing by the technology project cost yields the project benefit/cost ratio that is to be used in the prioritization algorithm.

From the launch provider's point of view, selecting a technology investment portfolio is similar but for the consideration of benefits. The launch provider is concerned with the firm's value and not the flow of Treasury revenue. Thus, benefits must be considered in terms of changes in the firm's net present value (of cash flow) and risk. This can be approximated as the change in net present value of government missions that are likely to utilize the launch vehicle with the presumption that part or all of the savings can be captured by the launch provider through price changes to offset the potential customer value obtained as a result of the technology investments. Commercial missions must also be considered. Here, changes in market share must be converted into changes in the launch provider's net present value (of cash flow).

In summary, a methodology was described in the preceding paragraphs for formulating efficient technology programs (comprising many technology projects) to improve the competitiveness of space transportation services. The objective is to improve financial or economic performance as a result of technology investments and government policies. Improving competitiveness implies increasing market share. Thus, the developed methodology seeks to relate change in market share to the specifics of the technology projects under consideration. Because market share is determined by user choices that depend upon perceptions of transportation service attributes and resulting user financial performance, the methodology takes into account the relationships between technology programs and launch service attributes and resulting user financial performance and resulting user choices. The preceding paragraphs describe a method for relating market share and user choices to perceptions of transportation service attributes, impacts on proforma financial decision variables, and financial preferences relating to the tradeoffs between the decision variables. Transportation services are described in terms of service attributes such as reliability, availability, price, etc., that are transformed into decision variables. Market share changes are related to benefits and costs that can then be used for the allocation of resources. The end result is the determination of technology project benefits and costs. The described method is aimed at the selection of technology projects for funding based upon benefit/cost ratios.

VIII. Assessment of Economic Impacts

According to the White House's National Space Transportation Policy (released 5 August 1994), "the U.S. Government is committed to encouraging a viable commercial U.S. space transportation industry." The policy also calls for the creation of a "plan that will focus on measures to foster an internationally competitive U.S. launch capability."

Competitiveness relates to pricing and market share. Industry viability relates to achieving an adequate return on investment. Government policies relating to the establishment of quantity restrictions and pricing policies of foreign launch services, pricing of U.S. Government-provided launch services, use of commercial spaceports, use of non-U.S. components in U.S. launch vehicles, etc., will have an effect on either or both pricing and market share, as well as industry viability. [It is possible that encouragement of commercial spaceports might lead to higher U.S. launch prices, thereby adversely impacting competitiveness and the U.S. economy: this would be directly counter to overall goals of fostering an internationally competitive U.S. launch capability. To improve competitiveness requires that the *value* of the U.S. launch service be increased through a combination of changes in pricing, scheduling, safety (and its impact on insurance), etc. Current U.S. Government policy relating to the provision of launch services is based upon the concept of incremental cost recovery for services provided to commercial users of U.S. ranges. Fixed costs are not recovered. (This may or may not be thought of as a government subsidy.)

Commercial spaceports cannot operate with an incremental cost-related pricing policy. Ultimately, all costs (fixed and variable) must be recovered plus a fee (leading to a reasonable return on investment). The implication is that prices will likely increase (because of the need to recover fixed costs) for the use of commercial spaceports unless offset by other actions such as increased operations efficiency, improved scheduling, etc. If the pricing of services provided by commercial spaceports exceeds that of comparable services provided by government ranges, it might be necessary to increase prices at government ranges because the U.S. Government is not supposed to compete with U.S. industry. This would then lead to increased costs for U.S. commercial launch services. An interesting potential dilemma! U.S. Government policies will affect launch competitions and the number of launches provided by U.S. providers of space transportation. When considering the setting of policies that will affect market shares, it is important to understand the economic implications of the considered policies and the importance of the results of launch competitions to the U.S. economy. The focus of the following paragraphs is thus on establishing the economic impacts that can result from the loss (or gain) of a launch to foreign competition. It is hoped that the developed metrics will be of value when setting policies in support of fostering an internationally competitive U.S. launch capability and in establishing a viable commercial U.S. space transportation industry.

The U.S. has led the major space activities of the western world and has been a principal provider of access to space. Indeed, the U.S. remains the only western-world provider of manned access to space via the space shuttle. But Europe, Japan, China, and Russia now also offer launch services to a market that is indeed international in scope. (India has recently demonstrated its capability to place satellites into geostationary orbits.) In addition, this competition is mounting at just the time that the U.S. DoD and other federal users are seeking ways to reduce their expenditures on space. Reductions could undermine the tenuous competitive position of U.S. commercial launch vehicle manufacturers and the providers of space transportation services. Although the U.S. manufacturers have received considerable subsidies in the development of their launch vehicles, it is important that production volume be maintained at a sufficient level to enable fixed costs to be amortized at a competitive rate and to provide a base for continuing product development. Reductions in federal markets could, therefore, seriously impact the competitive position of these manufacturers with respect to foreign competition.

When foreign suppliers capture launch vehicle business, they gain many advantages. To begin, just as loss of business harms the U.S. manufacturers, gains in foreign business enable these suppliers to become more competitive (i.e., gain market share) by expanding their base upon which to write off fixed costs. To the U.S., there is a net loss of employment (i.e., jobs, or more appropriately, person-years) with its ripple effects in the economy. There is a loss in revenue base that could, in part, form the basis for research and development that would improve the technology and make it more competitive in the future. And, more importantly, there is a loss of domestic access to space. In the near term use of foreign launch vehicles to place U.S. manufactured satellites into orbit can increase the competitiveness of U.S. manufactured satellites and reduce end-user fees. However, in the longer term loss of access to domestic launch vehicles by U.S. suppliers of satellite services could eventually result in losses to domestic satellite manufacturers. At the least one could expect that foreign providers of launch services would give priority to the launching of their domestically manufactured satellites. And in the extreme case, where foreign suppliers dominate the launch vehicle market, access to space could be denied to enterprises that do not grant the launching state access to the satellite technology or, worse, to the manufacture of at least a significant portion of the satellite.

It has long been clear that the ability of a nation to develop space-based business depends on that nation's continued and ensured access to space. It follows that the loss of ensured access to space threatens a nation's entire prospect for the development of space. Yet international agreements are being made (with respect to the establishment of "fair" pricing policies and quantity restrictions) that are likely to impact the viability of the U.S. space industry and affect the overall U.S. economy.[30,31] It is the objective of the following paragraphs to place a quantitative (where possible) value on the economic cost of foreign competition to the U.S. commercial space transportation industry by developing the economic cost of a launch lost to foreign competition.

In the following paragraphs the economic costs are developed for the loss of a U.S. launch of a foreign-manufactured satellite to foreign competition, the loss of a U.S. launch of a U.S.-manufactured satellite to foreign competition, the loss of a U.S. launch and its associated satellite to foreign competition, and the use of a foreign launch vehicle from U.S. facilities to launch U.S. and foreign satellites. First, a taxonomy of losses is presented. This is followed by a discussion of the methodology employed to estimate impacts. Finally, typical results are presented in terms of changes in balance of trade, income, and loss in jobs (i.e., expressed in terms of person-years) and tax revenues. Both direct and indirect effects are considered.

A. Taxonomy of Impacts

In general, losing a competitive launch procurement to foreign competition can have far-reaching effects. These effects can be observed as both costs and benefits, can linger over an extended period of time, and can differ over time. In addition, the magnitude of the effects will depend upon the specific U.S. launch vehicle involved in the competition and the implications for U.S. satellite procurements. In general, the costs resulting from a lost launch will outweigh the associated benefits particularly if the procurement of a U.S. satellite is also lost in the process.

The economic impacts of a lost launch fall into a number of categories, namely 1) income opportunity, direct and indirect; 2) employment, direct and indirect; 3) federal revenue impact; and 4) other. "Direct income" is income received (or, in the case under consideration, income lost as a result of competition) by an aerospace firm or organization from sales of services related to an export item. "Indirect income" is income generated (or lost) in the community (e.g., among restaurants, doctors, service stations, auto dealers, retailers, etc.) as a result of expenditures by employees of the firm that has sold the basic product or service (in the case under consideration, space transportation services). In other words, the loss of a launch to foreign competition will reduce direct income, which in turn will reduce purchases made in surrounding communities (reduction in indirect income).[32]

Direct and indirect employment are defined in the same manner, where "direct employment" is the consequence of procurement of products and/or services relating to the construction and provision of space transportation services and "indirect employment" is the consequence of money being spent in a community by the direct employees. The following paragraphs deal with the impacts of a launch lost to foreign competition. This implies that, unless specifically stated otherwise, direct and indirect income refers to a reduction in income (i.e., a loss) relative to having won the launch competition, and direct and indirect employment refers to employment (measured in person-years) associated with the launch, the satellite if applicable, and employment associated with the spending of income earned in the production of the launch (the indirect effects). These impacts are relative to that which would have been if the U.S. firm were the winner of the launch competition. (Consideration of winning a launch competition implies simply changing all of the signs.)

Federal revenues will be affected by taxes that are levied on direct and indirect income as well as by transfer payments (unemployment compensation that might have to be paid for a time to unemployed workers).

In addition to the preceding, there are other impacts of a launch lost to foreign competition. These impacts include increased costs as a result of reduced reliability caused by reduced flight experience and loss of competitiveness capability resulting from increased unit costs (i.e., the result of a smaller production base). In addition, there will be long-term impacts associated with specific uses of revenue. A firm utilizes the revenue from the sale of a product or service for a number of different purposes: labor, material, overhead, and profit (ultimately all of these go to labor). However, the specific uses differ and have different consequences on the economy. For example, a part of the firm's expenditures can be traced to R&D. The use of funds for R&D has different long-term consequences than the use of funds for manufacturing.[33–37] R&D expenditures tend to increase productivity in the long term (more than other types of expenditures) and therefore result in a long-term multiplier effect—in other words, dollars spent on R&D today will generate several times their amount in the long term. Thus R&D expenditures require separate consideration.

These short-, medium-, and long-term impacts of a competitively procured launch lost to foreign competition are summarized next. The short-term impacts are as follows: 1) direct (i.e., aerospace sector and its suppliers) income loss; 2) indirect (i.e., nonaerospace sectors) income loss; 3) reduced direct employment (i.e., aerospace sector and its suppliers); 4) reduced indirect employment (i.e., nonaerospace sector); 5) reduced exports; and 6) reduced tax revenues (from both direct and indirect income) and increased transfer payments. The medium-term impacts (in addition to those identified as short-term impacts) are as follows: 1) loss of competitive capability as a result of higher unit costs; 2) failure to provide a base for research and development to upgrade systems; 3) lower reliability resulting from less flight experience; 4) potential loss of satellite procurements and technology; 5) potential failure to maintain world-class facilities if demand is reduced; and 6) reduced opportunity to increase productivity (i.e., caused by reduced investment in R&D). The long-term impacts (in addition to those identified as both short- and medium-term impacts) are as follows: 1) failure to develop space activities to their fullest potential; 2) potential loss of satellite procurements, technology, and manufacturing base; and 3) failure of the U.S. to remain a leader in the development of space.

The type and magnitude of the impacts of the loss of a competitively procured launch to foreign competition are a function of the circumstances associated with the lost launch. For the sake of completeness, the following situations are considered: 1) loss of a U.S. launch of a foreign satellite, 2) loss of a U.S. launch of a U.S. manufactured communication satellite (for providing domestic service) to foreign competition, and 3) loss of a U.S. launch of a U.S. manufactured communication satellite (for non-U.S. service) to foreign competition.

These situations are considered for both Atlas and Delta launch vehicles. In addition, the use of foreign hardware as part (or all) of a launch vehicle to launch U.S. and foreign satellites from U.S. facilities is also considered. For these, the impacts of switching from the use of a U.S. launch vehicle to the foreign launch vehicle are considered:

1) reduction in direct income,
2) reduction in indirect income,
3) reduction in direct employment (person-years),
4) reduction in indirect employment (person-years),
5) reduction in tax revenue (direct and indirect) and increase in transfer payments,
6) reduction in R&D expenditures (with long-term impacts on productivity),
7) reduction in capital expenditures,
8) loss of competitive capability resulting from increased unit costs (i.e., the result of smaller production base),

9) lower reliability as a result of less flight experience, and
10) loss of a U.S. launch and U.S.-manufactured satellite to foreign competition.

The consequences of a loss in revenue that results from the loss of a launch competition, and possibly from the loss of an associated satellite, to foreign competition are summarized here.

These impacts are summarized under the headings of "Lost launch revenue" and "Lost satellite revenue" as entries in Table 2.4. Table 2.4 presents a number of scenarios associated with the loss of a launch and an associated satellite and the benefits and costs that can result from the scenario. For example, the loss of a launch of a foreign satellite (i.e., a foreign satellite that could have been launched by a U.S. transportation service) would result in both benefits and costs. The benefits include an increase in launchpad availability for government launches, freeing up funds for possibly higher ROI endeavors, and increasing the profitability of the transportation service if the provision of the service had caused the provider to operate at a loss. The costs associated with the lost launch are those referred to as "lost launch revenue" as just stated and an increased likelihood of losing other launches. When the lost launch is associated with the placement of a U.S. satellite for providing domestic communication services (i.e., the second indicated scenario), the costs remain the same, but there are additional benefits. Presumably, launching the satellite via a foreign launch service was of higher value to the provider of the communication services. This increase in value (for example, the result of a lower priced foreign launch service) will increase the communications carrier profits and/or will lead to a decrease in end-use fees. In addition, there will be an increase in tax revenue from these higher profits. These would tend to partially offset the costs associated with the lost launch. Finally, the loss of a U.S. launch and associated U.S.-manufactured satellite (for non-U.S. services) results in additional costs associated with "lost satellite revenue" and increased likelihood of losing other satellites to foreign competition.

In the following paragraphs a number of these costs and benefits are quantified. Emphasis is placed upon the quantification of short-term losses. This quantification is in terms of the loss of a single launch. Transportation services based upon the use of the Atlas and Delta launch vehicles are considered in combination with the situations just identified.

The implication of reductions in R&D expenditures is discussed in terms of medium- and long-term losses. The reduction in R&D spending from both a lost launch and a lost satellite are considered. In addition, the difference in the degree to which U.S. technology is utilized in U.S. and foreign satellites is considered.

B. Mathematical Models

The obtained results are based upon the mathematical models presented in the following pages and discussed in the following paragraphs. The basic models are first described followed by the data utilized and associated rationale, and thence the obtained results.

Table 2.4 Lost launch[a] scenarios: benefits and costs

Scenario	Benefit[b]	Cost[c]
Loss of a launch of a foreign satellite	Increased launchpad availability for government launches Freed up funds for higher ROI endeavors Increased profitability when operating at a loss	Lost launch revenue Increased likelihood of losing other launches[d]
Loss of a launch of a U.S. communication satellite for providing domestic service	Increase in carrier profits and/or decrease in end-user fees[e] Increased tax revenue from increased carrier profits Increased launchpad availability for government launches Freed up funds for higher ROI endeavors Increased profitability when operating at a loss	Lost launch revenue Increased likelihood of losing other launches[d]
Loss of a launch of U.S.-provided satellite for non-U.S. services	Increased launchpad availability for government launches Freed up funds for higher ROI endeavors Increased profitability when operating at a loss	Lost launch revenue Increased likelihood of losing other launches[d]
Loss of a launch and associated satellite for non-U.S. services	Increased launchpad availability for government launches Freed up funds for LV provider for higher ROI endeavors Increased profitability of LV provider when operating at a loss	Lost launch revenue Increased likelihood of losing other launches[d] Lost satellite revenue Increased likelihood of losing other satellites

[a]A launch lost by a U.S. company to foreign providers of space transportation services as a result of an internationally competed launch services procurement.
[b]An improvement that results from the loss of a launch competition.
[c]Costs are relative to winning the launch competition.
[d]Except when there are government imposed launch constraints (i.e., quantity restrictions).
[e]Assuming that the winning foreign launch service offered a lower price than the U.S. launch service.

1. Loss of a Launch of a Foreign Satellite (refer to Fig. 2.41)

The reduction in direct income from a launch lost to foreign competition is equal to the launch price plus foreign expenditures made in the U.S. for satellite test and integration. This amount is reduced by payments made for foreign launch vehicle hardware that is incorporated into the U.S. launch vehicle and reduced corporate taxes. Foreign expenditures in the U.S. are related to the number of foreign visitors participating in launch and related operations, their average duration of stay, and their average daily expenditures. Because the provision of launch services is normally but one of many ongoing corporate activities and it is not possible to obtain meaningful data relating to product line profitability, the reduction in corporate taxes is approximated as the launch price multiplied by the aggregated industry corporate return on sales and by the corporate tax rate.

The reduction in indirect income is accounted for by multiplying the reduction in direct income plus the reduction in corporate taxes by the indirect income multiplier. (The indirect income multiplier is the income multiplier less 1.0.) The

Reduction in Direct Income:	Launch Price - Payment for Foreign LV Hardware + Foreign Expend. in U.S. for Satellite Test & Integration - Reduction in Corp. Taxes
Payment for Foreign LV Hardware:	Launch Price * LV Hardware Fraction * Foreign Fraction of LV Hardware
Foreign Expenditures. in U.S:	No. of Visitors to U.S. * Average Duration of Stay * Average Daily Expenditures
Reduction in Corp. Taxes:	Launch Price * Return on Sales * Corp. Tax Rate / 100
Reduction in Indirect Income:	[Reduction in Direct Income + Reduction in Corp. Taxes] * Indirect Income Multiplier
Reduction in Direct Employment:	Reduction in Direct Income * [1 - Benefit Rate / 100] / Average Aerospace Wage Rate + [Reduction in Direct Income * Benefit Rate / 100 + Reduction in Corp. Taxes] / Average Non-Aerospace Wage Rate
Reduction in Indirect Employ:	Reduction in Indirect Income / Average Wage Rate of Non-Aerospace Sector
Reduction in Federal Revenue:	Reduction in Corp. Taxes + [Reduction in Direct Income + Reduction in Corp. Taxes + Reduction in Indirect Income] * Income Tax Rate / 100 + [Reduction in Direct Income + Reduction in Corp. Taxes + Reduction in Indirect Income] * Unemployment Compensation Rate * Average Duration of Unemployment / 100
Reduction in R&D Expenditures:	LV Price * LV R&D Percentage / 100
Reduction in Productivity:	Reduction in R&D Expenditures * Productivity Multiplier

Fig. 2.41 Mathematical model: loss of a U.S. launch of a foreign satellite.

indirect multiplier accounts for the income generated (or lost) in the community as a result of expenditures by employees of the firms (including suppliers) that sold the basic product or services. Corporate tax revenues are also considered because they are used to finance a broad range of activities that then generate the indirect income.

The determination of the reduction in employment is predicated upon the assumption that all income is eventually converted into labor.

The aerospace industry wage rate is assumed to be the wage rate for those working directly for the aerospace industry as well as its suppliers of goods and services (for example, components incorporated into the launch vehicle). The nonaerospace wage rate is assumed to be the wage rate paid to those who provide benefits to the aerospace industry (for example, the health industry), as well as those who are recipients of the so-called indirect income.

To obtain an estimate of direct employment, the price of the launch is divided into two parts: one being the portion that goes to pay benefits (and this is divided by the nonaerospace wage rate) and the other the portion that pays everything else (and this is divided by the aerospace industry wage rate). To obtain an estimate of indirect employment, indirect income is divided by the nonaerospace industry wage rate.

The reduction in federal revenue consists of the reduction in corporate taxes, the reduction in direct and indirect income multiplied by the income tax rate, and the increase in unemployment compensation payments. The unemployment compensation payments take into account the average compensation rate and the average duration of unemployment. It is assumed that there is a growing economy that will eventually reemploy those who become unemployed.

Finally, the medium- and long-term effects of the reduction in R&D expenditures need to be mentioned. Lower R&D expenditures can result in lower productivity than would otherwise be the case. The effect on productivity can be estimated by multiplying the reduction in R&D expenditures by the productivity multiplier. The reduction in R&D expenditures is the launch price multiplied by the R&D investment rate (expressed as a percentage of sales). As indicated in Refs. 32–35, the productivity multiplier is rather difficult to measure. Estimates have been made that range from slightly less than three to slightly more than 14. These multipliers are based upon a broad-based technology program and utilize historical results based upon the technology program considered. None of the studies addressed the marginal problem, that is, the change in productivity which might result from a small change in the level of R&D. For this reason no further consideration is given to the effect of R&D expenditures on the U.S. economy other than to note that it is likely that productivity increases will probably be lower with lower R&D spending.

2. *Loss of a Launch of a Communication Satellite for Providing Domestic Communication Services (refer to Fig. 2.42)*

In the following it is presumed that the loss of a launch to foreign competition will result in an increase in the profit of the communication carrier that has procured the launch services. This increase in profit can be maintained in the corporation or distributed in a number of ways including a reduction in end-

Reduction in Direct Income:	. Launch Price - Payment for Foreign LV Hardware + Expenditures for Sat. Test & Integration - Increase in Carrier Profit - Reduction in Corp. Taxes
Payment for Foreign LV Hardware:	Launch Price * LV Hardware Fraction * Foreign Fraction of LV Hardware
Expenditures for Sat. Test & Integ:	No. of U.S. Participants * Average Duration of Stay * Average Daily Expenditures
Reduction in Corp. Taxes:	[Launch Price * Return on Sales - Increase in Carrier Profit] * Corp. Tax Rate / 100
Reduction in Indirect Income:	[Reduction in Direct Income + Reduction in Corp. Taxes] * Indirect Income Multiplier
Reduction in Direct Employment:	Reduction in Direct Income * [1 - Benefit Rate / 100] / Average Aerospace Wage Rate + [Reduction in Direct Income * Benefit Rate / 100 + Reduction in Corp. Taxes] / Average Non-Aerospace Wage Rate
Reduction in Indirect Employ:	Reduction in Indirect Income / Average Wage Rate of Non-Aerospace Sector
Reduction in Federal Revenue:	Reduction in Corp. Taxes + [Reduction in Direct Income + Reduction in Corp. Taxes + Reduction in Indirect Income] * Income Tax Rate / 100 + [Reduction in Direct Income + Reduction in Corp. Taxes + Reduction in Indirect Income] * Unemployment Compensation Rate * Average Duration of Unemployment / 100
Reduction in R&D Expenditures:	LV Price * LV R&D Percentage / 100
Reduction in Productivity:	Reduction in R&D Expenditures * Productivity Multiplier
Increase in Carrier Profit (Gross):	Launch Price * [1 - Foreign LV Price Fraction]

Fig. 2.42 Mathematical model: loss of a U.S. launch of a communication satellite (for providing domestic service).

user fees paid for the provided communication services. The reduction in direct income from a launch lost to foreign competition is therefore equal to the launch price less a compensating increase in carrier profit, which results from the use of the foreign launch service, payment for foreign launch vehicle hardware that would normally be incorporated into the U.S. launch vehicle, and the reduction in corporate taxes (the net of launch provider and communications carrier). [Because the change in profit is less than the change in consumer surplus that will result from a price change (a result of price elasticity), the former has been used in order to be conservative.] The increase in carrier gross profit is the launch price associated with the U.S. launch service less the foreign launch service price. Because the provision of launch services is normally but one of many ongoing corporate activities and it is not possible to obtain meaningful data relating to product line profitability, the reduction in corporate taxes is therefore approximated as the launch price multiplied by the corporate return on sales and the corporate tax rate.

The reduction in indirect income is accounted for by multiplying the reduction in direct income by the indirect income multiplier. The indirect multiplier accounts for the income generated (or lost) in the community as a result of expenditures by employees of the firm that sold the basic product or services. Because the expenditures for satellite test and integration services are provided outside of the U.S., there is a reduction in the indirect income in the U.S. Corporate tax revenues are also considered because they are used to finance a broad range of activities that then generate the indirect income.

All of the other terms and considerations are similar to those described for the consideration of the loss of a launch of a foreign satellite (as per Fig. 2.41).

3. *Loss of a U.S. Launch of a U.S.-Provided Satellite for Non-U.S. Services (refer to Fig. 2.43)*

The reduction in direct income from a launch lost to foreign competition is equal to the launch price less payments that would have been made for foreign launch vehicle hardware that is incorporated into the U.S. launch vehicle plus spending in the country providing the launch by the U.S. team concerned with payload test and integration and less corporate taxes. Because the provision of launch services is normally but one of many ongoing corporate activities and it is not possible to obtain meaningful data relating to product line profitability, the reduction in corporate taxes is approximated as the launch price multiplied by the corporate return on sales and by the corporate tax rate.

The reduction in indirect income is accounted for by multiplying the reduction in direct income plus corporate taxes by the indirect income multiplier. The indirect multiplier accounts for the income generated (or lost) in the community as a result of expenditures by employees of the firm that sold the basic product or services. Because the expenditures by the U.S. satellite test and integration team pay for services that are provided outside of the U.S., there is a reduction in the indirect income in the U.S.

Corporate tax revenues are also considered because they are used by the government to pay salaries of those providing goods and services to the government that are then spent in the community.

All of the other terms and considerations are similar to those described for the consideration of the loss of a launch of a foreign satellite (as per Fig. 2.41).

4. *Loss of a U.S. Launch and Associated Satellite for Non-U.S. Services (refer to Fig. 2.44)*

The reduction in direct income from a launch and an associated satellite lost to foreign competition is equal to the launch price less payments that would have been made for foreign launch vehicle hardware which is incorporated into the U.S. launch vehicle, satellite price less payments that would have been made for foreign satellite hardware which is incorporated into the U.S. satellite but offset by the amount of U.S. hardware included in the foreign satellite, and less corporate taxes. Because the provision of launch services and manufacture of a satellite are

Reduction in Direct Income:	Launch Price - Payment for Foreign LV Hardware + Expenditures for Sat. Test & Integration - Reduction in Corp. Taxes
Payment for Foreign LV Hardware:	Launch Price * LV Hardware Fraction * Foreign Fraction of LV Hardware
Reduction in Corp. Taxes:	Launch Price * Return on Sales * Corp. Tax Rate / 100
Expenditures for Sat. Test & Integ:	No. of U.S. Participants * Average Duration of Stay * Average Daily Expenditures
Reduction in Indirect Income:	[Reduction in Direct Income + Reduction in Corp. Taxes] * Indirect Income Multiplier
Reduction in Direct Employment:	Reduction in Direct Income * [1 - Benefit Rate / 100] / Average Aerospace Wage Rate + [Reduction in Direct Income * Benefit Rate / 100 + Reduction in Corp. Taxes] / Average Non-Aerospace Wage Rate
Reduction in Indirect Employ:	Reduction in Indirect Income / Average Wage Rate of Non-Aerospace Sector
Reduction in Federal Revenue:	Reduction in Corp. Taxes + [Reduction in Direct Income + Reduction in Corp. Taxes + Reduction in Indirect Income] * Income Tax Rate / 100 + [Reduction in Direct Income + Reduction in Corp. Taxes + Reduction in Indirect Income] * Unemployment Compensation Rate * Average Duration of Unemployment / 100
Reduction in R&D Expenditures:	LV Price * LV R&D Percentage / 100
Reduction in Productivity:	Reduction in R&D Expenditures * Productivity Multiplier

Fig. 2.43 Mathematical model: loss of a U.S. launch of a U.S. provided satellite for non-U.S. services.

normally but one of many ongoing corporate activities and it is not possible to obtain meaningful data relating to product line profitability, the reduction in corporate taxes is approximated as the launch price plus the satellite price multiplied by the corporate return on sales and by the corporate tax rate. [The return on sales (ROS) and the tax rate used are for the industry sector "Guided Missiles and Space Vehicles."]

The determination of the reduction in employment is predicated upon the assumption that all income is eventually converted into labor. Because the average wage rates of the aerospace and nonaerospace industries differ, the reduction in direct employment is considered in two parts with the former being aerospace employees and the latter being nonaerospace employees. The reduction of indirect employment is the result of the reduction in indirect income.

The data utilized for the analyses and associated data sources are indicated in Tables 2.5 and 2.6 with obtained results for the loss of Atlas and Delta launches presented in Tables 2.7 and 2.8, respectively. Aerospace industry statistics on R&D aerospace industry wage rates, percent of company funds devoted to R&D,

Table 2.5 Data utilized for evaluating the impacts of the loss of an Atlas launch

Input variable	Launch of foreign satellite	Launch of Comsat (domestic)	Launch of U.S. satellite for non-U.S. service	Launch and U.S. satellite for non-U.S. service	Source[a,38]
Launch price, M$	100	100	100	100	Refs. 38, 39
LV hardware fraction, %	80	80	80	80	[b]
Satellite price, M$	N/A	N/A	N/A	85	——[b]
Satellite hardware fraction	N/A	N/A	N/A	1	——
Fraction of foreign satellite that is U.S. hardware, %	N/A	N/A	N/A	20	——
Fraction of U.S. satellite that is foreign hardware, %	N/A	N/A	N/A	5	——[b]
Foreign fraction of LV hardware, %	0	0	0	0	——[b]
Number of participants	40	40	40	40	——[c]
Average duration of stay, days	60	60	60	60	——[c]
Average daily expenditures, $/day	132	132	132	132	——[c]
Return on sales, %	6	6	6	6	Ref. 40
Corporate tax rate, %	35	35	35	35	Ref. 40
Income multiplier	2	2	2	2	Refs. 32, 41, 42
Average aerospace weekly wage rate 1994, $	766	766	766	766	Ref. 43
Average nonaerospace hourly wage rate 1994, $	11	11	11	11	Ref. 44
Average nonaerospace hours/week	40	40	40	40	——
Benefit rate, %	33	33	33	33	Ref. 45

(Continued)

[a]Ref. 42 found an income multiplier of 2.1 associated with the NASA 1990 procurement expenditures. Ref. 32 used an income multiplier of 1.96, whereas Ref. 41 cites other studies that derived income multipliers ranging from 2.0 to 2.95.
[b]Industry sources.
[c]Based on estimates provided by industry.
[d]Internal Revenue Service.

Table 2.5 Data utilized for evaluating the impacts of the loss of an Atlas launch (continued)

Input variable	Launch of foreign satellite	Launch of Comsat (domestic)	Launch of U.S. satellite for non-U.S. service	Launch and U.S. satellite for non-U.S. service	Source[a,38]
Income tax rate, %	20	20	20	20	——[d]
Unemployment compensation rate, %	36	36	36	36	Ref. 46
Duration of unemployment, weeks	17.9	17.9	17.9	17.9	Ref. 46
LV R&D percentage, %	3.6	3.6	3.6	3.6	Ref. 47
Foreign LV price fraction	N/A	0.83	N/A	N/A	Ref. 39
Satellite R&D percentage	N/A	N/A	N/A	3.6	Ref. 47

[a]Ref. 42 found an income multiplier of 2.1 associated with the NASA 1990 procurement expenditures. Ref. 32 used an income multiplier of 1.96, whereas Ref. 41 cites other studies that derived income multipliers ranging from 2.0 to 2.95.
[b]Industry sources.
[c]Based on estimates provided by industry.
[d]Internal Revenue Service.

Table 2.6 Data utilized for evaluating the impacts of the loss of a Delta launch

Input variable	Launch of foreign satellite	Launch of Comsat (domestic)	Launch of U.S. satellite for non-U.S. service	Launch and U.S. satellite for non-U.S. service	Source[38]
Launch price, M$	45	45	45	45	Refs. 38, 39
LV hardware fraction, %	80	80	80	80	——[a]
Satellite price, M$	N/A	N/A	N/A	65	——[a]
Satellite hardware fraction	N/A	N/A	N/A	1	——
Fraction of foreign satellite that is U.S. hardware, %	N/A	N/A	N/A	20	——
Fraction of U.S. satellite that is foreign hardware, %	N/A	N/A	N/A	5	——[a]
Foreign fraction of LV hardware, %	0	0	0	0	——[a]
Number of participants	40	40	40	40	——[b]
Average duration of stay, days	60	60	60	60	——[b]
Average daily expenditures, $/day	132	132	132	132	——[b]
Return on sales, %	6	6	6	6	Ref. 40
Corporate tax rate, %	35	35	35	35	Ref. 40
Income multiplier[c]	2	2	2	2	Refs. 32, 41, 42
Average aerospace weekly wage rate 1994, $	766	766	766	766	Ref. 43
Average nonaerospace hourly wage rate 1994, $	11	11	11	11	Ref. 44
Average nonaerospace hours/week	40	40	40	40	——
Benefit rate, %	33	33	33	33	Ref. 45

[a]Industry sources.
[b]Based on estimates provided by industry.
[c]Ref. 42 found an income multiplier of 2.1 associated with the NASA 1990 procurement expenditures. Ref. 32 used an income multiplier of 1.96, whereas Ref. 41 cites other studies that derived income multipliers ranging from 2.0 to 2.95.
[d]Internal Revenue Service.

(Continued)

Table 2.6 Data utilized for evaluating the impacts of the loss of a Delta launch (continued)

Input variable	Launch of foreign satellite	Launch of Comsat (domestic)	Launch of U.S. satellite for non-U.S. service	Launch and U.S. satellite for non-U.S. service	Source[38]
Income tax rate, %	20	20	20	20	——[d]
Unemployment compensation rate, %	36	36	36	36	Ref. 46
Duration of unemployment, weeks	17.9	17.9	17.9	17.9	Ref. 46
LV R&D percentage, %	3.6	3.6	3.6	3.6	Ref. 47
Foreign LV price fraction	N/A	0.83	N/A	N/A	Ref. 39
Satellite R&D percentage	N/A	N/A	N/A	3.6	Ref. 47

[a]Industry sources.
[b]Based on estimates provided by industry.
[c]Ref. 42 found an income multiplier of 2.1 associated with the NASA 1990 procurement expenditures. Ref. 32 used an income multiplier of 1.96, whereas Ref. 41 cites other studies that derived income multipliers ranging from 2.0 to 2.95.
[d]Internal Revenue Service.

Reduction in Direct Income:	Launch Price - Payment for Foreign LV Hardware + Satellite Price - Payment for Foreign Satellite Hardware - Payment for U.S. Satellite Hardware Included in Foreign Satellite - Reduction in Corp. Taxes
Payment for Foreign LV Hardware:	Launch Price * LV Hardware Fraction * Foreign Fraction of LV Hardware
Payment for U.S. Sat. Hardware:	Satellite Price * Satellite Hardware Fraction * Fraction of Foreign Satellite that is U.S. Hardware
Payment for Foreign Sat. Hardware:	Satellite Price * Sat. Hardware Fraction * Foreign Fraction of Sat. Hardware
Reduction in Corp. Taxes:	[Launch Price + Satellite Price] * Return on Sales * Corp. Tax Rate / 100
Reduction in Indirect Income:	[Reduction in Direct Income + Reduction in Corp. Taxes] * Indirect Income Multiplier
Reduction in Direct Employment:	Reduction in Direct Income * [1 - Benefit Rate / 100] / Average Aerospace Wage Rate + [Reduction in Direct Income * Benefit Rate / 100 + Reduction in Corp. Taxes] / Average Non-Aerospace Wage Rate
Reduction in Indirect Employ:	Reduction in Indirect Income / Average Wage Rate of Non-Aerospace Sector
Reduction in Federal Revenue:	Reduction in Corp. Taxes + [Reduction in Direct Income + Reduction in Corp. Taxes + Reduction in Indirect Income] * Income Tax Rate / 100 + [Reduction in Direct Income + Reduction in Corp. Taxes + Reduction in Indirect Income] * Unemployment Compensation Rate * Average Duration of Unemployment / 100
Reduction in R&D Expenditures:	[LV Price * LV R&D Percentage + Satellite Price * Satellite R&D Percentage] / 100
Reduction in Productivity:	Reduction in R&D Expenditures * Productivity Multiplier

Fig. 2.44 Mathematical model: loss of a U.S. launch and associated satellite for non-U.S. services.

corporate tax rates, and return on sales and benefit rates were largely supplied by the Aerospace Industry Association, and where sufficient detail was available data representative of the complete guided missiles and space vehicles (SIC 3761) industry were used. Employment information, such as unemployment compensation and duration and the wage rate (which was used as the nonaerospace sector wage rate), was obtained for the overall economy from the Bureau of Labor Statistics. Industry contacts provided estimates on the number of people involved in satellite testing and integration, the length of time they might be needed, their daily expenditures, and average number of accompanying persons. The DoT's FAA STAR Database (developed by the FAA's Office of Commercial Space transportation) and the International Reference Guide to Space Launch Systems

Table 2.7 Impacts of a loss of an Atlas launch

Impact area	Launch of foreign satellite	Launch of comsat (domestic)	Launch of U.S. satellite for non-U.S. service	Launch and U.S. satellite for non-U.S. service
Payment for foreign LV hardware, M$	0.0	0.0	0.0	0.0
Foreign expenditure in U.S., M$	0.3	—	—	—
Increase in carrier gross profit, M$	—	16.7	—	—
Payment for U.S. satellite hardware, M$	—	—	—	17.0
Payment for foreign satellite hardware, M$	—	—	—	4.3
Satellite test and integration, M$	—	0.3	0.3	—
Reduction in corporate taxes, M$	2.1	(3.7)	2.1	3.9
Reduction in indirect income, M$	100.3	83.7	100.3	163.8
Reduction in R&D expenditures, M$	3.6	3.6	3.6	3.6
Reduction in exports, M$	100.3	100.3	100.3	163.8
Direct employment, person-yrs	3133	2548	3133	5120
Indirect employment, person-yrs	4291	3578	4291	7004
Reduction in employment, person-yrs	*7424*	*6126*	*7424*	*12,124*
Corporate taxes, M$	2.1	(3.7)	2.1	3.9
Income taxes, M$	40.1	33.5	40.1	65.5
Unemployment compensation, M$	24.9	20.7	24.9	40.6
Reduction in federal revenue, M$	*67.1*	*50.5*	*67.1*	*110.0*

Table 2.8 Impacts of a loss of a Delta launch

Impact area	Launch of foreign satellite	Launch of comsat (domestic)	Launch of U.S. satellite for non-U.S. service	Launch and U.S. satellite for non-U.S. service
Payment for foreign LV hardware, M$	0.0	0.0	0.0	0.0
Foreign expenditure in U.S., M$	0.3	—	—	—
Increase in carrier gross profit, M$	—	10.0	—	—
Payment for U.S. satellite hardware, M$	—	—	—	13.0
Payment for foreign satellite hardware, M$	—	—	—	3.2
Satellite test and integration, M$	—	0.3	0.3	—
Reduction in corporate taxes, M$	0.9	(2.6)	0.9	2.3
Reduction in indirect income, M$	45.3	35.3	45.3	93.8
Reduction in R&D expenditures, M$	1.6	1.6	1.6	4.0
Reduction in exports, M$	45.3	45.3	45.3	93.8
Direct employment, person-yrs	1415	1064	1415	2932
Indirect employment, person-yrs	1938	1511	1938	4010
Reduction in employment, person-yrs	*3354*	*2575*	*3354*	*6942*
Corporate taxes, M$	0.9	(2.6)	0.9	2.3
Income taxes, M$	18.1	14.1	18.1	37.5
Unemployment compensation, M$	11.2	8.8	11.2	23.2
Reduction in federal revenue, M$	*30.2*	*20.3*	*30.3*	*63.0*

served as the source for launch prices. Other studies estimating the effects of either NASA expenditures on the economy or the restriction of military exports provided justification for selecting an income multiplier of two. Employment multipliers from these studies were not used, but the resulting estimates of indirect employment relative to direct employment were within the range of employment multipliers quoted in these studies.

C. Summary

When foreign suppliers capture launch vehicle business, they gain many advantages. To begin, just as loss of business harms the U.S. manufacturers, gains in foreign business enable these suppliers to become more competitive (i.e., gain market share) by expanding their base upon which to write off fixed costs. To the U.S., there is a net loss of employment (i.e., jobs, or more appropriately, person-years of employment), with its ripple effects in the economy. There is a loss in revenue base that could, in part, form the basis for research and development that would improve the technology and make it more competitive in the future. And, more importantly, there is a loss of domestic access to space. In the near term use of foreign launch vehicles to place U.S.-manufactured satellites into orbit might increase the competitiveness of U.S.-manufactured satellites and reduce end-user fees. However, in the longer term loss of access to domestic launch vehicles by U.S. suppliers of satellite services could eventually result in losses to domestic satellite manufacturers.

Losing a competitive launch procurement to foreign competition can have far-reaching effects. These effects can be observed as both costs and benefits, can linger over an extended period of time, and can differ over time. In addition, the magnitude of the effects will depend upon the specific U.S. launch vehicle involved in the competition and the implications for U.S. satellite procurements. In general, the costs resulting from a lost launch will outweigh the associated benefits particularly if the procurement of a U.S. satellite is also lost in the process.

A launch lost to foreign competition impacts direct and indirect income, direct and indirect employment, exports and federal revenues. "Direct income" is income received (or lost) by an aerospace firm or other organization from sales of services related to a specific product or service (i.e., space transportation). "Indirect income" is income generated (or lost) in the community (e.g., among restaurants, doctors, retailers, etc.) as a result of expenditures from the receipt of direct income. "Direct and indirect employments" are defined in a similar manner. Federal revenues include taxes that are levied on direct and indirect income as well as by transfer payments (unemployment compensation).

The impacts of a launch lost to foreign competition are summarized in Tables 2.7 and 2.8 for the Atlas and Delta, respectively. The impacts depend upon the specifics of the events surrounding the lost launch. A number of these events, or scenarios, were considered and the impacts presented in these tables. Specifically considered was the loss of a U.S. launch of a foreign satellite, loss of a U.S. launch of a U.S. communication satellite for providing domestic U.S. communication services, loss of a U.S. launch of a U.S.-manufactured satellite for providing non-U.S. services, and the loss of a U.S. launch and associated U.S.-manufactured satellite. The impacts are summarized in terms of the reduction in U.S. exports, reduction in U.S. employment (direct, indirect, and total), and reduction in federal revenues.

The preceding discussion describes the economic impacts of the loss of individual launch competitions (loss of a U.S. launch of a foreign satellite, loss of a U.S. launch of a U.S. communication satellite providing domestic communication services, etc.). Placing them into an historical context can place these impacts into perspective. For this purpose, the specific launch events in the fourth quarter of 1994 are considered and the impacts developed that consider the actual events that occurred. The total and maximum impacts associated with these events reflect the jobs that would have been created and the Treasury revenue that would have resulted if the U.S. launch providers obtained 100% market share. These fourth quarter 1994 results are summarized in Table 2.3 and are the result of four launches "lost" to foreign competition: Long March 3A (1), Ariane 44L (1), and Ariane 42P (2). An Indian remote sensing satellite was launched on Polar Satellite Launch Vehicle (PSLV) (Indian), and numerous Russian satellites were launched on Russian vehicles for Russian use. These were not considered in the impact assessment. The Ariane 44L launched two Hughes satellites, and each Ariane 42P launched a Hughes satellite, one of which was for a U.S. operator. It was assumed that these launches "replaced" a total of four Atlas II launches and one Delta launch, with a Delta launch and an Atlas launch being "equivalent" to the Ariane 44L launch. As can be seen, potential impacts can become quite large.

IX. Technology Transfer: Assessment and Evaluation

Government agencies, for example, NASA, frequently undertake activities that are aimed at transferring the results of their research and development activities to other organizations within the agency, to other government agencies, and to the private sector. The objective of these activities is to influence investment decisions so that the R&D results are utilized beneficially. If investment decisions are not changed, there is no value obtained from the investment in technology transfer programs.

Technology transfer activities can influence investment decisions by 1) creating new opportunities, 2) creating an awareness of opportunities and technology, 3) reducing perceived uncertainty and resulting risk, and 4) reducing capital investment requirements. A number of NASA programs are specifically aimed at influencing investment decisions by performing experiments that will demonstrate technology and the ability to provide new and/or improved services. Two recent examples are the NASA's flight experiments program (In-Space Technology Experiments Program) and the ACTS program.

The flight experiment program aimed to measure variables, thereby reducing uncertainty in the knowledge of their values and to provide a convincing demonstration of capabilities or system performance. In each case the underlying purpose of a flight experiment is to reduce uncertainty ostensibly in the eyes of the program engineers or other decision makers who would be responsible for implementing or utilizing a new technology and/or in the eyes of upper management or investors who would approve or fund a program that makes use of a new technology. ACTS was launched to provide a means for demonstrating the performance of a broad range of satellite technology as well as to demonstrate the ability to provide a broad range of communication services.

Both the flight experiment and ACTS programs seek to influence investment decisions. These two programs are used to illustrate assessments and evaluations of technology transfer activities. The flight experiment program is considered from the point of view of determining a priori which flight projects should be selected for funding when limited resources are available, specifically taking into account the potential impacts of the experiment in terms of the decision-making processes of the target audience. The ACTS program is considered as an example of an opportunity to measure the value of a demonstration program in terms of a priori perceptions and measured experiment results and likely impacts on investment decisions.

Technology transfer can be defined as the process by which technology, knowledge, and/or information developed in one organization, in one area, or for one purpose is applied and utilized in another organization, in another area, or for another purpose. The end user might be the public, industry, another federal laboratory, or any other technology developer.

The technology transfer process is aimed at increasing the likelihood that technology in the form of hardware, software, patents and licenses, or knowledge developed by one organization will be used by other organizations. Thus technology transfer is concerned with creating an awareness of what is available to be transferred and then with increasing the likelihood that the known technology will be utilized. Technology transfer and commercialization go hand in hand. In a broad sense government technology transfer and commercialization activities form an integrated process of creating new opportunities for investment, creating awareness, and providing incentives by reducing decision-maker-required investment and reducing perceived risk. The private sector generally refers to this as advertising and promotion. Advertising creates awareness and promotion, through incentives, alters purchase/investment decisions. Because of these similarities, it might be possible for government policy makers to draw upon a large body of knowledge developed by the private sector to move technology and products to market.

A. Scope of Technology Transfer and Commercialization Activities

To have technology transfer, there must be an awareness of the availability of the technology. To have the technology utilized, there must be a perception that the use of the technology will lead to adequate net benefits. Government agencies, backed by legislation such as the Stevenson–Wydler Technology Innovation Act of 1980 (PL96-480), which required federal laboratories to take an active role in technical cooperation, the Bayh–Dole Act of 1980 (PL96-517), which authorized granting exclusive and partially exclusive licenses, the Cooperative Research Act of 1984 (PL98-462), which permitted industry consortia such as Sematech, the Federal Technology Transfer Act of 1986 (PL99-502), which made technology transfer a responsibility of all federal laboratory scientists, and the National Competitiveness Technology Transfer Act of 1989 (PL101-189), which required agencies to permit contractor managers of federal laboratories to enter into Cooperative Research and Development Agreements (CRADA), have developed

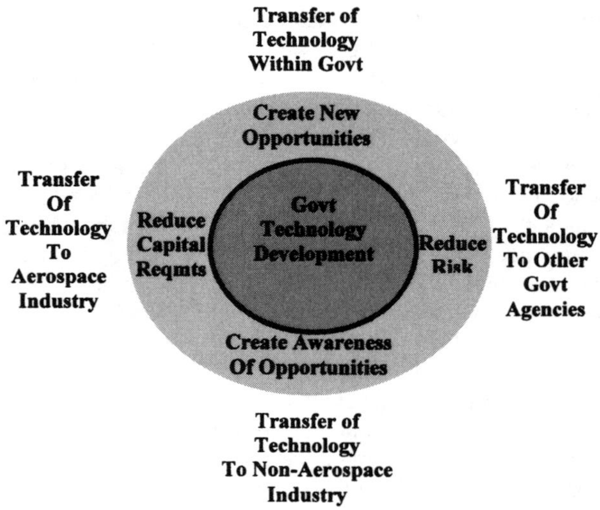

Fig. 2.45 Role of government in the technology transfer commercialization process. [Commercialization refers to a process that leads to a financing arrangement wherein the private sector has partial or long-term control over the design, development, and use of a space project or program. The private sector commits capital (at risk) to an endeavor and shares in the risk and return. The implication is that commercialized space projects or programs will be more market driven and consumer or end-user oriented than space endeavors under the government's control.]

aggressive policies with respect to the transfer of technology within a government agency, to other government agencies, and to various industry segments (refer to Fig. 2.45). Some of the developed technology transfer mechanisms are described in the following paragraphs.

Cooperative agreements are instruments entered into by the government, with industry, universities, and others, to support or stimulate research; agreements generally result in cost-shared research with the nonfederal participant.

Cost-shared contracts/subcontracts are a collaboration, through procurement, of mutual benefit to industry and the government. Often the government can agree not to disseminate commercially valuable data that are generated under a cost-shared contract for a limited period of time. Cost-shared contracts include in-cash and/or in-kind arrangements.

CRADAs are agreements between government laboratories and nonfederal parties in which both participants provide personnel, services, facilities, or equipment for the conduct of specified R&D. The nonfederal parties can also provide funds. (No direct funding is provided by the government laboratory.) Rights to inventions and other intellectual property are negotiated between the laboratory and participant, and certain data that are generated can be protected for up to five years.

R&D consortia are arrangements involving multiple federal and nonfederal parties working together for a common R&D objective. Funding for R&D consortia can be shared, but usually no funds are exchanged between participants.

Exchange programs are arrangements allowing government/federal laboratory staff to work in government laboratories to enhance technical capabilities and support research in specific areas. Costs are borne by the organization sending the personnel. Intellectual property arrangements can be addressed in exchange agreements.

Licensing is the transfer of less than ownership rights in intellectual property, such as a patent or a software copyright, to permit its use by the licensee. Licenses can be exclusive or nonexclusive, for a specific field of use or for a specific geographical area. The potential licensee must present plans for commercialization.

User facility agreements are arrangements permitting private parties to conduct research and development at a government laboratory. For proprietary R&D the laboratory is paid for the full cost of the activity. If the work will be published, cost can be adjusted. Intellectual property rights generally belong to the private party conducting the R&D.

Work for others are agreements whereby proprietary work for an industry can be done by technically qualified government laboratory staff using laboratory facilities with full costs charged to the industry. Title to intellectual property generally belongs to the sponsor. The government retains a nonexclusive, royalty-free license to such intellectual property.

In short, government agencies have entered into many different types of government-industry-academia cooperative arrangements. These include CRADAs, joint ventures, partnerships, consortia, cooperative agreements, and variants of these. In addition, activities such as the National Institute of Science and Technology's Advanced Technology Program, Advanced Research Project Agency's Technology Reinvestment Project (TRP), and similar programs in other agencies focus on the commercialization of advanced technology. The key element in most of these activities is that government agencies and industry attempt to identify common technology needs or requirements and create cooperative programs to develop the needed technology.

A NASA position paper stated:

> In order to ensure national economic security impacts of NASA programs, NASA will pursue a commercial technology mission concurrent to its aerospace mission. The commercial technology mission will require that each NASA program be carried out in a way that proactively involves the private sector from the onset, through a new way of doing business, to ensure that the technology developed will have maximum commercial potential.[48]

NASA has traditionally measured its progress in terms of technical performance, cost, and schedule. Now, in the post Cold War era there is another measure: contribution of technology to national economic security. In response, a strong emphasis on performing R&D partnerships with the private sector is the foundation of a new way of doing business with NASA. NASA's objective was to devote 10 to 20% of its budget to R&D partnerships with industry.

In addition to these mechanisms are a myriad of activities aimed at creating awareness. These include the production of publications such as NASA's "Tech Briefs" and "Spinoffs;" participation in workshops, seminars, and conferences; the preparation and distribution of strategic plans; electronic bulletin boards; etc. Other resources exist in the form of large databases that can put researchers in touch with the federal laboratory doing work in the field in which the researcher is interested or can allow those parties interested in seeking technology solutions to encountered problems to search, via keyboard, for applicable technologies and contacts. These resources include the Federal Laboratory Consortium, the Federal Laboratories Database that was developed by the Mid-Atlantic Technology Transfer Center, databases maintained by the National Technology Transfer Center such as Business Gold, a Directory of Federal Laboratory and Technology Resources published by the National Technical Information Service, and a private online technology service called Knowledge Express.

Other mechanisms can also have significant impacts on technology transfer. These include performing experiments (for example, the NASA flight experiments program) and demonstrations (for example, NASA's ACTS program), and the setting of standards (such as the setting of appliance efficiency and automobile mileage standards). Experiments and demonstrations aim to reduce uncertainty and risk that influence utilization and investment decisions. The setting of standards can (though not usually the major reason for imposing standards) encourage the use of new technology in order to meet the standards. Normally programs aimed at creating awareness are aimed at the providers of products and services. However, consumers through their purchase decisions can force manufacturers to provide improved products that require the use of advanced technology (for example, the Department of Energy's appliance efficiency labeling program influences consumer choices regarding the selection of higher energy-efficient products).

Thus the scope of technology transfer and commercialization activities is quite broad but with the common intent to influence investment/utilization decisions. In addition, the customer base for technology is diverse both with respect to business sector and level of acceptable risk. In NASA's case potential users of technology include NASA organizations, other government organizations, the aerospace industry, and nonaerospace industries. The role of the various activities in influencing decisions[4] is illustrated in Fig. 2.46.

B. Assessments and Evaluations

When the government provides incentives to achieve technology transfer through altering investment decisions, government is acting as an investment banker, that is, the government is allocating its scarce resources and should strive to maximize resulting benefits. It can be argued that government has a fiduciary responsibility to taxpayers in the same manner that the investment banker has responsibilities to shareholders. (Unfortunately, there is a significant difference in accountability.) Thus, there is a need to perform assessments and evaluations with the former being before the fact and the latter being after the fact. Assessments are performed to provide guidance for establishing the mix of activities that should/could be undertaken to affect technology transfer (for example, establish

Government program	Role of government: Reduce Risk	Ease Capital Form.	Create New Opportunities	Create Awareness
R&D	♠			
Demonstration	♠	♠		
Development of new infrastructure	♣	♣	♠	
Taxes				
Investment tax credit		♠		
R&D tax credit		♠		
Depreciation rules		♠		
R&D limited partnerships	♣	♠		
Cooperative Activities				
CRADA	♣	♠		
Joint endeavor agreements				
Cost reduction	♣	♠		
Market commitment	♠	♣		
Recoupment policies		♠		
Pricing policies	♣	♠		
Low-interest loans/bonds				
Insurance				
Regulation	♠			
Indemnification	♠			
Loan guarantees	♣	♠		
Standards	♠		♣	
Information dissemination				♠
Patents/proprietary rights	♠	♣		
Institutional arrangements	♠	♣		
Procurement policies	♠	♣	♣	
. .				

♠ = primary impact; ♣ = secondary impact.

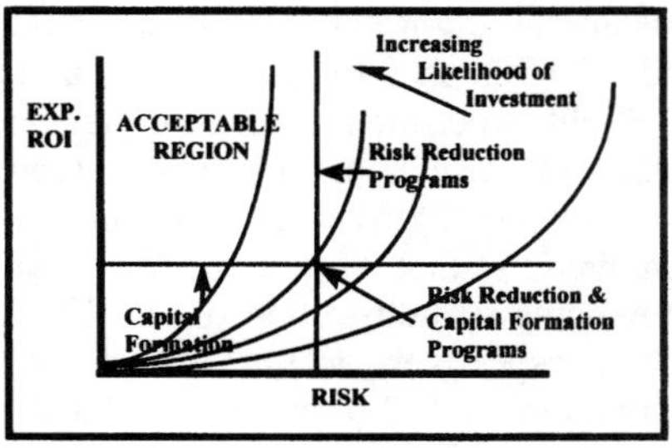

Fig. 2.46 Relationship of government programs to role of government.

the set of flight experiments that should be performed in order to reduce potential technology user risk perceptions), and evaluations are performed to measure actual performance, thus providing information for improving the overall technology transfer process. Therefore it is necessary to establish methods for ensuring that maximum value is obtained from limited budgets.

In general, assessments are performed to

1) provide estimates of project value according to selected metrics,
2) improve understanding of project potential impacts,
3) establish the need to terminate projects,
4) provide information for the efficient allocation of scarce resources,
5) serve as a basis for project prioritization and ranking aimed at maximizing program value, and
6) satisfy fiduciary responsibility to allocate scarce resources effectively.

Evaluations are performed to

1) ensure that project goals/objectives are clearly defined and
2) measure project effectiveness against stated goals/objectives, thereby
    (a) improving management efficiency by developing "lessons learned,"

(b) providing a basis for project termination,
(c) providing the basis for rewards, and
(d) justifying existence.

NASA's Flight Experiment and ACTS programs, though not normally considered as part of NASA's technology transfer program, were actually two very important and expensive components of technology transfer activities. For these reasons the flight experiments program is used to illustrate assessment methods, and ACTS is used to illustrate evaluation methods.

Flight experiments can be conducted for one of the following reasons: 1) to measure a variable or variables, thereby reducing uncertainty in the knowledge of their values; 2) to search for unknown phenomena, whether detrimental or beneficial, so that they can be further studied and planned for; and 3) to provide a convincing demonstration of capabilities or system performance. These reasons relate, respectively, to the following categories of experiments, which are summarized in Table 2.9; mission-oriented applied research, nondirected fundamental research, and engineering demonstrations. In each case the underlying purpose of a flight experiment is to reduce uncertainty, ostensibly in the eyes of the program engineers or other decision makers who would be responsible for implementing or utilizing a new technology for the first two of the preceding reasons and in the eyes of upper management or investors who would approve or fund a program that makes use of a new technology for the third reason.

In all cases the value of an experiment (and the associated resources utilized in/for the experiment) relates to how its outcome will affect decisions, for if decisions are not affected by the results of a planned experiment then the experiment has no value. The selection of experiments (i.e., the allocation of resources) aimed at technology transfer through their impact on decisions should of necessity consider both qualitative and quantitative factors. The current selection process relies primarily upon qualitative measures. To illustrate quantitative methods for assessing the value of experiments, two examples are presented, one dealing with mission-oriented research and one dealing with engineering demonstrations.

The purpose of mission-oriented experiments is to reduce uncertainty that engineers have regarding some aspect of a system. Without such experiments engineers would feel very uncertain whether a design would work, and they would take measures to guard against failure. A specific case is illustrated in Fig. 2.47, where prior to the performance of an experiment there is significant uncertainty as to the performance level x that can be achieved. The current state of knowledge (i.e., the a priori assessment as judged by one or more "experts") is that the true value of x will be in the range of A to B with the indicated likelihood. In the absence of better knowledge, the design engineer will normally be risk averse and select a design point close to A. If the outcome of the experiment is specified as a single value thereby implying no uncertainty, then there is no need for the experiment because the outcome is presumed to be known. Also associated with each design point (value of x) will be a different "economic" or other value. For example, if the variable were solar cell efficiency then different levels of efficiency will directly affect satellite design and cost and overall mission value.

Table 2.9 Categories of experiments

Concern	Nondirected fundamental	Mission-oriented applied research	Engineering demonstrations
Time	Before choosing applications or missions	Before choosing designs	Before flying
Purpose	Discover new options; strengthen research capabilities	Improve modeling capability; reduce uncertainty	Proof of design, ensure mission success
Important attributes	Technical quality Uniqueness Relevance	Benefit Cost Technical quality Relevance	Benefit Cost Technical quality Relevance
Evaluation measure	Qualitative	Quantitative plus qualitative	Quantitative plus qualitative
Type of review	Peer review Portfolio selection	Benefit/cost analysis Consideration of broader application Engineering-oriented peer review	Benefit/cost analysis Engineering oriented

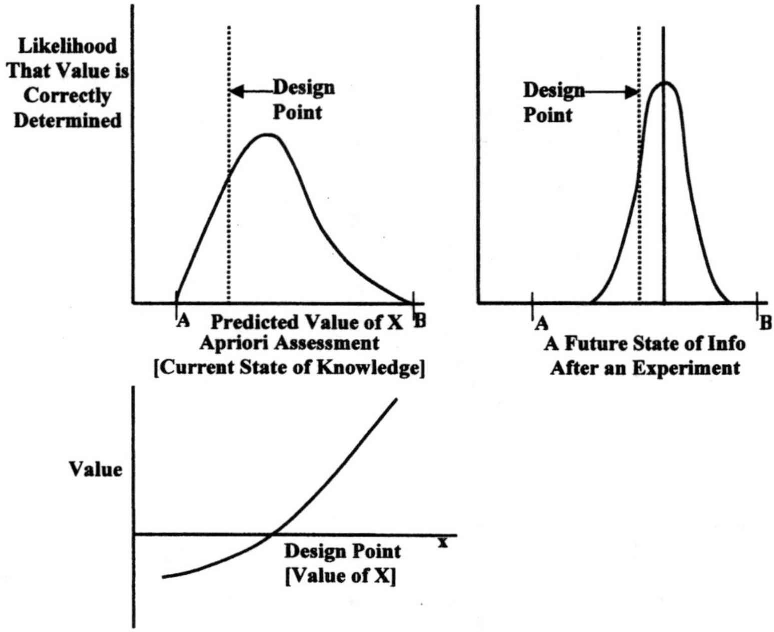

Fig. 2.47 Assessment of flight experiments: effect of design point.

The objective of performing an experiment is to alter the state of knowledge by achieving a measured result. The measured result of the experiment will lead to a new design point slightly lower than the measured outcome (because there will still be some uncertainty after an experiment). Because all outcomes in the range of A to B are possible as per the a priori likelihood function, the value of the experiment can be stated as the weighted or average value based upon the possible outcomes less the value based upon the use of the a priori design point:

$$\text{Expected Value} = \left[\int_A^B \text{Value}(X - \Delta x)P(X)\,\mathrm{d}x\right] - [\text{Value}(X = \text{A priori Design Point})]$$

Establishment of the value of a mission-oriented experiment requires expert opinion on the range and form of uncertainty with respect to the design variable in the absence of an experiment, the design point that would likely be selected if the experiment were not conducted, and the value associated with different design points. This approach not only adds significantly to the metrics currently used for selecting flight experiments that rely primarily upon nonquantitatively focused metrics, but also leads to the identification of specific variables that must be quantified (albeit subjectively) in order to establish metric values.

In the case of engineering demonstrations, which can be viewed as a go/no-go test of an engineering system or subsystem that is in the hardware phase, decision trees can be used to seek to capture the essence of the uncertainty that will be resolved. Figure 2.48 illustrates a simplified decision tree where the operation of the subsystem that is the subject of the demonstration is key to success of a mission. If the subsystem works, the mission is very likely to be a success; if it fails, the mission will be a total failure. Probability estimates are shown in the figure for the possible outcomes and would normally be based upon expert opinion.

At the right-hand side of the tree, the tips of the branches represent possible outcomes with probabilities of occurrence shown. If costs can be associated with each branch of the tree and benefits with each outcome, it is then possible to estimate the value of the demonstration as the expected value of a *yes* decision minus the expected value of a *no* decision. The benefit of the mission might be measured in terms of a revenue or cash-flow stream for a commercial system, a benefit to society for an applications mission, or as a cost-savings benefit for a science mission, where it is reasonable to assume that the mission would be redone in the event of a failure.

For the simplified example in Fig. 2.48,

$$\begin{aligned}\text{Expected Value} = {} & [(0.99 * \text{Net Mission Benefits} - 0.01 * \text{Mission Cost})] * 0.9 \\ & - \text{Experiment Cost}] \\ & - (0.9 * \text{Net Mission Benefits} - 0.1 * \text{Mission Cost})\end{aligned}$$

Whereas assessments are based upon a priori estimates, evaluations require a priori estimates as well as observed results. Evaluations reflect what was vs what might have been. Assessments are concerned with anticipating the value that might result from the performance of an experiment and are concerned with the efficient allocation of resources. Evaluations are concerned with estimating the

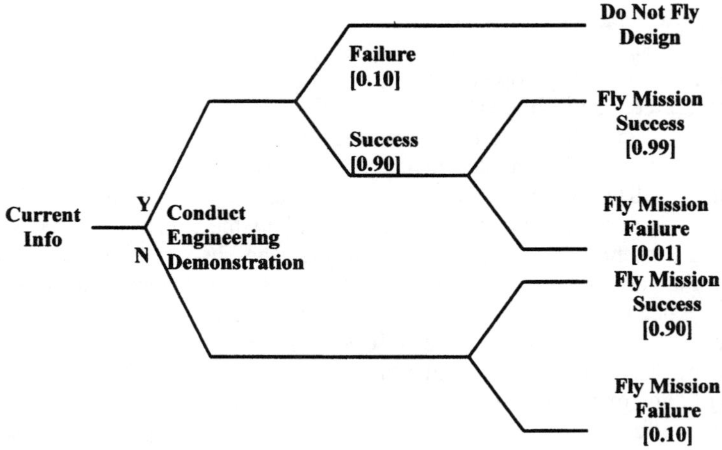

Fig. 2.48 Assessment of flight experiments: demonstration (revision of design prior to mission).

value that was obtained from the conduct of an experiment, with improving decision making, and demonstrating program efficacy.

ACTS was a demonstration program aimed at influencing both communication technology decisions as well as decisions concerning the provision of new communication services. Because of the high visibility and high cost of the ACTS program, it is anticipated that questions will be raised with respect to value received. The value of the ACTS flight demonstration program stems from both 1) the difference in the a priori and a posteriori design points, and 2) the difference in anticipated likelihood of utilization or observed utilization of the technology (or service) with the a priori estimates and the a posteriori experiment results. This last factor is extremely important because it is necessary to make judgments prior to performing the experiment (to establish the *what might have been* scenario) that will be compared with the measured results (which in the case of ACTS still require judgments of what is likely to happen given the measured experiment results).

The ACTS program encompassed experiments aimed at establishing ACTS technical performance (which would then serve as the basis for evaluating other experiments), demonstrating technology performance aimed at influencing technology utilization decisions, demonstrating service/performance aimed at influencing system and service decisions, and scientific knowledge. Because many of the demonstrations were undertaken for industry application and not specific firm application and because it is likely that in many cases delays will naturally occur prior to implementation based upon results, the value of the ACTS demonstration program as established shortly after program completion still requires subjective judgments to be made with respect to likely future decisions.

The basic evaluation procedure is to 1) establish experiment goals/objectives *prior* to the performance of experiments, 2) establish states of knowledge prior to and after the conduct of the experiment, and 3) estimate future courses of action in the absence of the experiment (*prior* to the conduct of the experiment) and likely actions given the results of the experiment.

When the experiment demonstrates technology performance aimed at influencing technology utilization decisions, the evaluation methodology is as described in Fig. 2.48. However, there is a difference. After the experiment has been performed, there is then a known measured result. The value of the measured result can be established in terms of resulting cash flows, benefits, etc. There is yet another difference. This relates to the need to estimate the likelihood that the technology will be utilized given the obtained and estimated results. This must then be compared to a priori estimates, including the a priori estimates of the likelihood that the technology will be used (in the absence of the experiment). Because the value of the experiment must be established relative to what would likely have been in the absence of the experiment, it is most desirable to make and document subjective estimates prior to obtaining results from the experiment.

A simplified example of an evaluation is illustrated in Fig. 2.49 and illustrates the use of the a priori and a posteriori subjective likelihood assessments. Because ACTS has already been launched without having made the necessary a priori judgments and experiments are well underway, and in most cases completed, it is no longer possible to estimate credibly what would have been, and a significant opportunity has been lost to perform a reasonable evaluation of one of NASA's

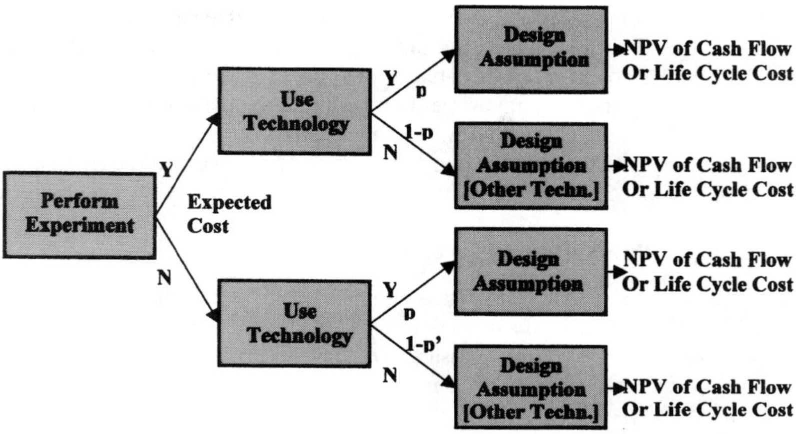

Fig. 2.49 Simplified experiment evaluation.

largest technology transfer activities. Upon completion of the ACTS demonstration program, evaluations must not only estimate what is in store for the future given the ACTS demonstrated results, but also the what would have been scenario will now have to be developed after the fact.

The ACTS program has already resulted in significant technology transfer with Motorola's Iridium LEO communication satellite system incorporating the ACTS beam-switching technology. This technology transfer resulted prior to the launching of ACTS. The ACTS flight demonstration therefore added no value in this particular arena except that other organizations' decisions might yet be affected (either by ACTS or the Iridium satellites).

C. Technology Transfer Metrics

Metrics are measures that are indicative of relative or absolute value, where "value" has meaning in terms of assessments and evaluations. In addition to their public relations value, metrics serve as management tools for assessing and evaluating performance. Because technology transfer involves numerous processes across multiple disciplines and organizations, appropriate metrics and methods for quantifying the metrics vary considerably. Also, the choice of appropriate metrics depends upon the availability of data and can change with time as new data and data collection processes are developed.

Metrics should provide answers to a very broad range of questions relating to the formulation, ongoing operation, and results of technology transfer activities. These questions include: How well are we likely to do? Are we likely to receive adequate value for our investment? How well have we performed? Is the program structured efficiently? How has performance changed? Have we received adequate value for our investment? Clearly the answers to these questions require the use of a broad range of metrics including those that might be described as activity and effectiveness related. Activity measures relate to anticipated or measured level of activity such as number of queries and papers presented or cited per unit time.

Effectiveness measures relate to anticipated or measured performance relative to one or more specified goals, for example, expected or measured industry investment per dollar of government investment or present value of savings where the goals are increasing industry investment and achieving cost savings from the use of specific technology, respectively. Because metrics are necessary for both program planning and program evaluation, past performance and judgments must be relied upon for the former and measurements combined with judgments for the latter.

According to NASA's July 1994 report,[48]

> The 1992 NASA Technology Transfer Report defined a minimum set of activity, efficiency and effectiveness metrics but stopped short of suggesting return on taxpayer investment measures. Database material and data collection tools already exist within NASA, and they are being used to begin the process of establishing an initial single, integrated set of metrics to support management's decision making.

The effectiveness measures include:

1) number of acknowledged uses (written statements from NASA's partners that they used a NASA technology in their commercial products, services or internal processes or technology from a NASA program that they used or adapted to commercial technologies),
2) number of NASA literature citations (the frequency of references to works of authors supported by NASA),
3) number of NASA patents generating license royalty,
4) size of the portfolio (total dollars spent on research partnerships by NASA and its partners),
5) types of products and services being produced,
6) when products and services would enter the marketplace,
7) effectiveness of the portfolio (ratio of activity generated per dollar spent),
8) frequency of use of various NASA business practices.

Activity-related metrics, such as number of queries per unit time, might prove useful for comparing alternatives that all aim to increase the number of queries and for indicating process change over time, but activity-related metrics are likely to have little or no value for influencing "portfolio" decisions in the absence of a relationship between the activity metric and technology transfer benefits. Activity-related metrics normally have little utility for answering such questions as "should x dollars be spent to increase the number of queries by y percent" unless there is an established relationship between the number of queries and the benefits from the queries (much in the same way that cost-estimating relationships are established).

Assessment metrics, expressed in terms of stated project goal/objective, include the following:

1) Create new identified opportunities: a) present value of net benefits divided by present value of cost and b) net jobs likely to be created.
2) Create awareness: a) audience size multiplied by probability of desired decision multiplied by estimated value divided by project cost and b)

reduced time to achieve benefits multiplied by net present value divided by present value of cost.

3) Reduce capital requirements: a) change in likelihood of investment multiplied by net present value divided by present value of cost and b) change in market penetration rate.
4) Reduce perceived risk: a) change in likelihood of investment multiplied by net present value divided by present value of cost and b) change in market penetration rate.

Similarly, evaluation metrics, expressed in terms of stated project goal/ objective, include the following:

1) Create new identified opportunities: a) number of new businesses divided by present value of cost and b) net present value divided by present value of cost.
2) Create awareness: a) number of requests divided by present value of cost and b) actual sales divided by anticipated sales.
3) Reduce capital requirements: a) private-sector funding attributable to project divided by present value of cost and b) present value of benefits divided by present value of cost.
4) Reduce perceived risk: a) private-sector funding attributable to project divided by present value of cost and b) present value of benefits divided by present value of cost.

Because industry advertising and promotion programs are quite similar to government information dissemination and incentive programs, an understanding of the response to advertising and promotion can prove helpful when planning and evaluating technology transfer programs. The following is an example.

Advertising and promotion expenditures are usually aimed at increasing sales and market share. Evaluations are performed to determine how well these objectives have been achieved. These evaluations are likely to be concerned with determining the sales response to advertising under both steady state and transient conditions. These evaluations are also concerned with how fast sales respond when advertising is increasing and how fast sales decay when advertising is decreased. A typical sales response curve to an advertising program conducted at a constant level for a finite period of time is illustrated in Fig. 2.50. The impact on sales is at a diminishing rate. Upon removal of the advertising program, sales start to decline.

The importance of the sales response curve to advertising is in the planning of evaluation programs. Response time is important because data collection must be accomplished with this curve in mind. It might or might not be appropriate to collect data during the growth portion of the curve. Saturation level data can be of major interest. This will dictate the duration of advertising or the use of measurement techniques that will provide for sufficient accuracy of growth to predict or forecast saturation credibly.

Measurement of the impact of a technology transfer activity can be very challenging and will depend frequently on the availability of data in related, or

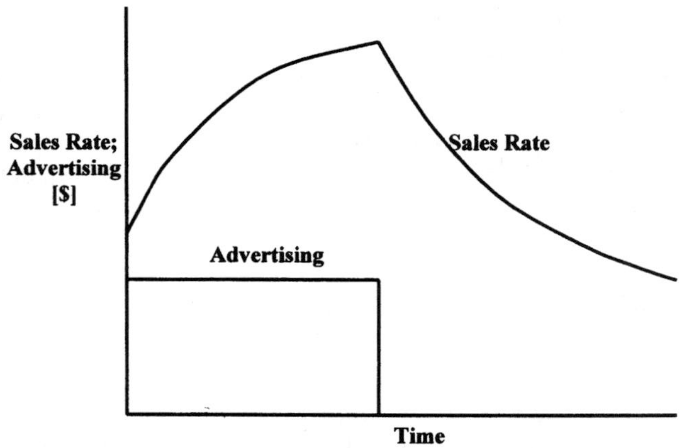

Fig. 2.50 Typical effect of advertising promotion (durable goods).

sometimes, apparently unrelated areas. For example, consider an energy information dissemination project whose goal is to reduce energy consumption through improved purchase and utilization decisions with a possible metric being energy savings per dollar of project investment.

Figure 2.51 presents a hypothetical illustration of energy consumption, budget, and temperature. It is obvious that an evaluation performed by measuring energy

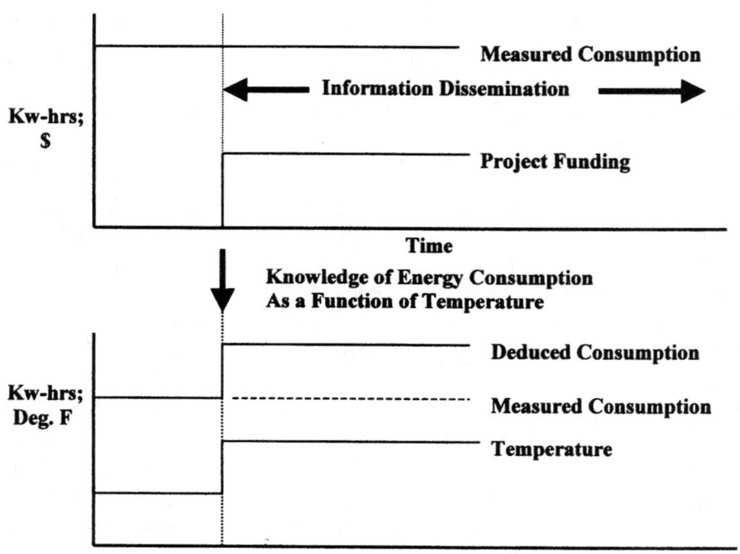

Fig. 2.51 Example of measuring the impact of an information dissemination program.

consumption, both before and during the information dissemination project, will result in a zero value for the selected metric. If temperature were also measured and a relationship developed between energy consumption and temperature, then the confounding influence of temperature can be taken into account and an adjustment made to the measured consumption (indicated as "deduced" consumption and indicates the likely energy consumption if the information dissemination project were not undertaken). The result is a nonzero value of the selected metric where savings are now in terms of deduced consumption less average prior (to the project) energy consumption. The point of the example is that it is necessary to consider all of the important factors that are likely to influence the chosen metric and establish data collection procedures for, and relationships between, the factors and the metric.

D. Summary

Technology transfer is a process by which technology, knowledge, and/or information developed by one organization, in one area, or for one purpose is applied and utilized in another organization, in another area, or for another purpose. The technology transfer process is aimed at increasing the likelihood that technology transfer and commercialization will occur. Technology transfer and commercialization go hand in hand with the objective of influencing investment and utilization decisions. In a broad sense government technology transfer and commercialization activities form an integrated process of creating new opportunities for investment, creating awareness, and providing incentives by reducing decision maker required investment, and reducing perceived risk. If decisions are not changed, there are no benefits obtained from the technology transfer program.

When the government provides incentives to achieve technology transfer through altering investment decisions, government is acting as an investment banker, that is, the government is allocating scarce resources and should strive to maximize resulting benefits. Because government should be accountable for its actions, metrics become important.

Assessments are performed to provide guidance for establishing the mix of activities that should/could be undertaken to effect technology transfer, and evaluations are performed to measure actual performance thus providing information for improving the overall technology transfer process. Because experiments and demonstrations are an important and expensive component of technology transfer activities, both qualitative and quantitative metrics need to be utilized for their assessment and evaluation.

Evaluations require that subjective judgments be made prior to the initiation of the experiments/demonstrations concerning likely decisions that will be made in the absence of the experiment/demonstration so that this can be compared with measured results on decisions or a posteriori judgments as to likely impact on decisions. The planning of the evaluation process, including the determination of appropriate metrics and associated data collection, should be an integral part of the assessment process. (Unfortunately this has not been the norm for NASA's flight experiment program and was not a part of the ACTS Program.)

References

[1]Greenberg, J.S., "Consideration of Uncertainty in Design," National Science Foundation, Final Rept. under Grant No. DMI-9900798, May 1999.

[2]Greenberg, J.S., "Space System Life Cycle Cost and Availability Analysis," *Space Economics*, edited by J.S. Greenberg and H.R., Hertzfeld, Progress in Astronautics and Aeronautics, Vol. 144, AIAA, Washington, DC, 1992, pp. 117–147.

[3]Greenberg, J.S., "Results of an Independent Economic and Market Analysis of Space Solar Power (SSP)," International Astronautical Federation, Paper 99-R.1.06, Oct. 1999.

[4]Greenberg, J.S., *Investment Decisions: The Influence of Risk and Other Factors*, American Management Association, New York, 1982.

[5]Greenberg, J.S., "Space Transportation Architecture Comparisons Explicitly Considering Uncertainty and Risk," International Academy of Astronautics, Paper 96-IAA.1.1.05, Oct. 1997.

[6]Morgenstern, O., et. al., *Economic Analysis of the Space Shuttle System*, Mathematica, Inc., Princeton, NJ, Jan. 1972.

[7]Hazelrigg, G.A., and Greenberg, J.S., "Cost Estimating for Technology Programs," International Academy of Astronautics, Paper 91-638, Oct. 1991.

[8]Bernstein, H., and Abbott, A.D., "Space Transportation Architecture Resiliency," Aerospace Corp., Los Angeles, March 1987, revised Aug. 1987.

[9]"Space Transportation Analysis," U.S. Air Force, Space Div., Contract No. F0470/-85-C-0055, Los Angeles, 2 March 1987.

[10]Greenberg, J.S., "Pricing and the Competitiveness of Space Transportation Services," International Academy of Astronautics, Paper 1.1-93-643, Oct. 1993.

[11]Bredley, R.A., and Myers, S.C., *Principles of Corporate Finance*, McGraw-Hill, New York, 1988.

[12]"Discount Rates to Be Used in Evaluating Time Distributed Costs and Benefits," Office of Management and Budget, OMB Circular A-94, Washington, DC, 27 March 1972.

[13]Markowitz, H.M., *Portfolio Selection*, Wiley, New York, 1959.

[14]Thierauf, R.J., and Klekamp, R.C., *Decision Making Through Operations Research*, Wiley, New York, 1975.

[15]Greenberg, J.S., "A Financial Risk Analysis of a District Heating Business Venture," *Proceedings of the 15th Intersociety Energy Conversion Engineering Conference*, AIAA in cooperation with the IECEC Steering Committee, Aug. 1980, pp. 600–608.

[16]*Financial Planning Model: PSI-RISK*, Princeton Synergetics, Inc., Princeton, NJ, March 1997.

[17]Greenberg, J.S., "Risk Analysis Results (A Space Processing Business Venture to Produce Monodispersed Polystyrene Latex Spheres)," NASA, July 1979.

[18]Greenberg, J.S., and Edelman, F., "Venture Analysis: The Assessment of Uncertainty and Risk," *Management: A Book of Readings*, edited by H. Koontz and C. O'Donnell, McGraw-Hill, New York, 1972, pp. 200–207.

[19]Greenberg, J.S., and Ross, D., "Manufacturer Impact Analysis of National Appliance Efficiency Standards," Operations Research Society of America/TIMS, Nov. 1985.

[20]Greenberg, J.S., and Gaelick, C., "Introduction of Fuel Efficient Technology into the Auto Industry: Financial Implications," Oak Ridge National Labs., Final Rept., Oak Ridge, TN, Sept. 1986.

[21]Hazelrigg, G.A., "Cost Estimating for Technology Programs," *Space Economics*, edited by J.S. Greenberg and H.R. Hertzfeld, Progress in Astronautics and Aeronautics, Vol. 144, AIAA, Washington, DC, 1992, pp. 97–115.
[22]Hertz, D., "Investment Policies That Pay Off," *Harward Business Review*, Jan.–Feb. 1968.
[23]Greenberg, J.S., and Edelman, F., "An Assessment of Uncertainty and Risk," *Financial Executive*, Vol. 37, No. 8, 1969, pp. 56–62.
[24]Greenberg, J.S., "Commentary: 'Guaranteeing' the Space Business," *Aerospace America*, June 1999, p. 3.
[25]Greenberg, J.S., "Tax Holidays vs. Improved Access to Space," *Aerospace America*, Jan. 1996, p. B44.
[26]Shaw, E.J., Taylor, D.T., and Hanamaker, J.W. "RLV Economics: Fiscal Evaluation of NASA's Reusable Launch Vehicle Efforts," *Space Policy*, Elsevier, May 1997, pp. 109–124.
[27]Hazelrigg, G.A., and Huband, F.L., "RADSIM—A Methodology for Large-Scale R&D Program Assessment," *IEEE Transactions on Engineering Management*, Vol. EM-32, No. 3, 1985.
[28]Greenberg, J.S., "NASA Aeronautics Investments: The Development of an Investment Portfolio," Princeton Synergetics, Inc., Princeton, NJ, March 1994.
[29]Greenberg, J.S., "Selecting a Launch Vehicle for GEO Missions," AIAA Paper 96-1112, Feb. 1996.
[30]Teets, P.B., "Guiding Transportation Policy," *Space News*, Vol. 5, No. 41, 24–30 Oct. 1994.
[31]"Agreement Between the Government of the United States of America and the Government of the Russian Federation Regarding International Trade in Commercial Space Launch Services," 2 Sept. 1993.
[32]Lousher, D.J., "Domestic Economic Impact of Russian and Chinese Launches of American Satellites on the Atlas Commercial Launch Program, 1993–2002," Foresight International, Victoria, Australia, 4 Dec. 1992.
[33]Hertzfeld, H.R., "Measuring Returns to Space Research and Development," *Space Economics*, edited by J.S. Greenberg and H.R. Hertzfeld, Vol. 144, AIAA, Washington, DC, 1992, pp. 151–169.
[34]Midwest Research Inst., "Economic Impact and Technological Progress of NASA Research and Development Expenditures," National Academy of Public Administration, Kansas City, MO, 20 Sept. 1988.
[35]Evans, M.K., "The Economic Impact of NASA R&D Spending," NASA NASW-2741, April 1976.
[36]Mathematica, Inc., "Quantifying the Benefits to the National Economy from Secondary Applications of NASA Technology," NASA CR-2674, March 1976.
[37]BETA, "Study of the Economic Effects of European Space Expenditure," Faculte des Sciences Economiques, Univ. Louis Pasteur, Strasbourg, Oct. 1988.
[38]Office of Commercial Space Transportation, U.S. Dept. of Transportation, Washington, DC, Bulletin 1194, Delta, Oct. 1994.
[39]Isakowitz, S.J., *International Reference Guide to Space Launch Systems*, AIAA, Washington, DC, 1991.
[40]Bureau of the Census, "Quarterly Financial Report for Manufacturing, Mining, and Trade Corps," *Aerospace Industries Association 1994 Year-End Review and Forecast*, Washington, DC, 1994.

[41]Bajusz, W.D., and Louscher, D.J., *Arms Sales and the U.S. Economy, The Impact of Restricting Military Exports*, Westview Press, Boulder, CO, 1988.
[42]"The Private Sector Economic and Employment Benefits to the Nation and to Each State of Proposed FY 1990 NASA Procurement Expenditures," Management Information Services, Inc., Washington, DC, April 1989.
[43]"Average Hourly Earnings in the Aerospace Industry, Guided Missiles and Space Vehicles," Aerospace Industries Association, Washington, DC, 1994.
[44]"Non-Agricultural Hourly Wage Rates," U.S. Bureau of Labor Statistics, Washington, DC, Nov. 1994.
[45]"Employer Costs for Employee Compensation in the Aircraft Manufacturing Industry," Aerospace Industries Association, 1 July 1994.
[46]"Employment Situation," U.S. Bureau of Labor Statistics, Washington, DC, Nov. 1994.
[47]"Aerospace R&D Company Funds as a Percentage of Net Sales (Averaged over the years 1987–1991)," *Aerospace Facts & Figures, 1993–94*, Washington, DC, 1994.
[48]"A New Way of Doing Business: Agenda for Change," NASA, July 1994.

Chapter 3

RLV Economics

I. Policy and Economic Analysis

WITH the aging of the space shuttle fleet and the overall desire to reduce significantly the cost of access to space, NASA, with the approval of Congress, initiated the Reusable Launch Vehicle (RLV) program. Upon its demise, it was followed by the Space Launch Initiative (SLI). The former program aimed to provide information on the technical and commercial feasibility of a single-stage-to-orbit (SSTO) fully reusable space transportation system and emphasized the role of industry with commercial funding the objective. The SLI program is more oriented toward technology development. NASA made the case that because RLVs would not depend on expendable hardware to achieve orbit they could take better advantage of economies of scale than expendable launch vehicles that discard costly hardware. It is hoped that if RLVs achieve prices to low Earth orbit (LEO) of less than $1000 per pound (current expendable launch vehicles and the space shuttle are in the range of $5000 to $10,000 per pound to LEO) they hold promise for eliciting an elastic response from the launch services market. As opposed to the capture of the existing market, this elastic market would represent new space-based industry businesses. New business creation contributes benefits to the economy by increasing U.S. Treasury revenue generation and increasing employment.

In short, the RLV program endeavored to open a new era in space flight by developing technology that supports commercial development of a next-generation heavy-lift launch vehicle, with the goal of lowering space transportation costs by an order of magnitude from the space transportation system. At this reduced cost NASA could maintain its current level of effort in vital space exploration and Earth science missions within budget constraints that, unfortunately, appear to be getting tighter. Perhaps, more importantly, the lower cost could encourage new and/or smaller customers to purchase space transportation services, thus encouraging new markets and allowing new applications in space to develop, along with the associated benefits.

The RLV program set a unique course aimed at developing a new reusable space transportation system that would sharply reduce space transportation costs for both commercial and government customers and to do so with funds provided, to a large extent, by the private sector. Because the goal was to have the private sector develop, build, and operate a fleet of RLVs and to provide space transportation services for commercial and government users, private investment and financing was critical to the program's success. The technological risks inherent in the development of a new reusable launch vehicle were addressed through a joint NASA–industry-funded program. The government-led partnership aimed to reduce risk perceptions by building and flying the X-33 SSTO demonstrator. However, developing and building the operational RLV fleet would require a substantial and costly effort beyond the X-33 program.

Prior to the private sector investing significant resources in a business venture based upon the development and use of an RLV fleet aimed at providing space transportation services to government and industry users of space transportation services, a number of financial hurdles must be surmounted. These hurdles relate to demonstrating that sufficient returns are likely to be achievable at a reasonable level of risk to warrant the anticipated sizable investment taking into account such factors as the uncertainty of future demand, the high level of initial investments, potential competition from existing and future expendable and reusable launch vehicles, and the difficulties of adjusting current policy mechanisms to new government–private cooperative arrangements.

Business planning is currently underway by the private sector, but because of long development schedules, the unpredictable outcome of the current technology development activity (both from the point of view of performance and cost), and competitive business pressures it is unlikely that definitive business plans and requirements will be finalized for several years. It is likely that the ultimate goal of achieving a commercial space transportation industry that provides launch services utilizing a fleet of RLVs might prove to be an elusive goal because of the combination of large required investment, inadequate rate of return, and perceived high risk.

It is likely that government assistance will be required in order to encourage and achieve private sector financing for a private sector space transportation business based upon a fleet of RLVs. Various forms of government "participation" or initiatives have been discussed with a number being analyzed in a preliminary fashion.[1] These initiatives include design, development, test, and engineering (RDT&E) participation by NASA, government in-kind contributions, anchor tenancy agreements, provision of termination liability, tax holidays, research and development (R&D) tax credits, subsidized loans, and loan guarantees and were described in Sec. V of Chapter 2. The various possible government initiatives affect private sector investment decisions in different ways, for example, some directly affect the magnitude of the private sector investment, some affect long-term rates of return, some affect risk perceptions, and some affect several of the private sector decision concerns. Because of perceived high risk, large capital requirements, and long payback periods, it is unlikely that the private sector will make the necessary investments without supportive government actions.

But what are appropriate government supportive actions and are they really necessary? NASA's Marshall Space Flight Center (MSFC) has been analyzing the need for and the form of government actions that might be required in order to achieve the goal of low-cost space transportation provided by a commercial venture. In spite of the misguided belief of many that the government should not "pick business winners and losers," it is clearly necessary for the government to perform an independent assessment of the role of, and need for, government actions that are likely to be needed to foster a commercial space transportation business venture based upon a fleet of RLVs. Not doing so would be financially irresponsible.

II. Lessons Learned (from Space Shuttle Economic Analysis)

The last major space transportation program undertaken in the United States was the Space Shuttle Transportation System (STS). It is instructive to look back at the environment within which program and policy decisions were made and to consider the economic and related analyses performed for and the resulting program and policy decisions made for the Space Shuttle Transportation System. Hopefully by identifying shortcomings and failures of the recent past, some lessons can be learned and pitfalls avoided for the development of new space transportation systems based upon RLVs. The intent of the following discussion is therefore to identify shortcomings of the STS analyses and resulting program and policy decisions in order to improve the program and policy decision process relating to the current reusable launch vehicle and related activities.

The following paragraphs identify a number of issues and associated lessons to be learned. These range from underlying Executive Branch and Office of Management and Budget (OMB) policy decisions within which the STS program was forced to function to limitations of analyses performed by NASA and its contractors. The following paragraphs are purposely kept brief so that the important issues and results can be quickly gleaned. The interested reader can delve into considerably more depth by reviewing the considerable amount of literature on this subject.[2–5]

"*The development of a space shuttle system is economically feasible assuming a level of space activity equal to the average of the United States* ***unmanned*** *program of the last 8 years.*"[6]

A. Background

The concept of developing a new space transportation system became a major issue in the late 1960s with detailed technical and economic studies performed through the early 1970s. The executive and legislative branches of government supported the development of the space shuttle. President Nixon said, on 5 January 1972, "The United States should proceed at once" to build this "entirely new type of space transportation system." This position was publicly supported by Presidential Science Advisor Dr. Edward E. David, Jr.; Casper W. Weinberger, then director of the OMB; the Secretary of the Air Force, Dr. Robert C. Seamans, Jr.; and by Dr. James C. Fletcher, Administrator of NASA. In the Congress, the shuttle program received continuing substantial support by both Houses. A proposed Senate amendment to kill the shuttle was defeated 64 to 22 during debate over the FY 1972 budget; a similar amendment was again defeated by a vote of 61 to 21 during the FY 1973 debates.

In addition, the shuttle received the endorsement of the Aeronautics and Space Engineering Board of the National Academy of Science's Advisory Committee to the Air Force Systems Command, and Dr. Charles Townes, Chairman of the National Academy of Sciences Space Science Board. Outside of government the most significant approval of the shuttle was given by the Executive Council of the AFL-CIO in a statement agreed upon at the union's annual meeting on 17 February 1972.

> The main reasons cited for these endorsements were that it [would] reduce the cost of space flight over a prolonged period and that it [would] give the U.S. new capability to accomplish practical work in space and continue scientific exploration. Because the Shuttle is essentially a transportation system, it [would] affects the entire space program of the nation, manned and unmanned. It [would] increase the national prestige as well as national defense while advancing science and technology, assisting in environmental surveys and control, promoting international cooperation, and developing and insuring a growing economy. Some of these benefits [would] not accrue for many years. In the short run, however, the Shuttle obtain[ed] its support because, as the AFL-CIO statement mention[ed], it [would] provide 50,000 jobs for people who have been displaced by prior government cuts in spending on the technical programs of the DoD and NASA. These considerations had a significant effect upon the decisions of the Executive and Legislative branches to back the Shuttle program.[2]

On 5 January 1972 President Nixon approved NASA's plan for the development of a space shuttle system. This commitment of $5.15 billion for development and flight test of two complete shuttle systems, with a $1 billion contingency reserve, was a momentous decision for the country and, in particular, for the aerospace industry. The results of economic analyses that influenced this decision have been the source of considerable public debate.

NASA funded extensive phase A and B design studies of reusable space shuttle concepts as well as studies of the economic justification of a reusable space transportation system in comparison with current and new expendable launch vehicles, including payload costs for various mission models that represented total U.S. space activity. Of primary importance in the latter category were the coordinated analyses performed by Lockheed Missiles and Space Company (LMSC),[7] the Aerospace Corporation,[8] and Mathematica, Inc.[6] LMSC was responsible for a detailed investigation of payload effects [i.e., the effects that a reusable space transportation system, as compared to an expandable launch vehicle (ELV) system, would have on the expected payload costs]. It was concluded that payload costs could be significantly reduced as a result of the relatively benign environment that would result from the use of a man-rated system, the large payload bay, the large mass delivery capability, and the ability to fly back payloads that failed to meet on-orbit tests. These attributes would allow high reliability to be achieved through increased redundancy using less costly, more massive lower reliability components. The Aerospace Corporation investigated the extent to which the payload effects identified by LMSC for specific individual spacecraft could be generalized across the mission models and also analyzed the cost of flying payloads with different launch vehicles. Mathematica then used the LMSC and Aerospace analyses in the performance of economic analyses.

As a result of these studies, it became clear that the economic benefits of the space shuttle depended primarily upon the ability to achieve the payload savings. The studies also indicated that even without the payload savings the "expected" space shuttle transportation cost would be on a par with those required by the continued use of the then current expendable launch vehicles. Thus it was generally concluded that the new prospects for payload savings, plus the introduction of versatility and flexibility in the space program resulting from the space shuttle characteristics, would make the development of a reusable shuttle system a worthwhile venture.

B. Hindsight

1. Budget Constraints

The economics of the shuttle was significantly affected by political budgetary decisions. In a memorandum to the NASA leadership in January 1970, George Low, NASA's deputy administrator said, "I think there is really only one objective for the Space Shuttle program, and that is to provide a low-cost, economical space transportation system. To meet this objective, one has to concentrate both on low development costs and on low operational costs."[9]

Low's objective was not a major consideration, however, because budgets for the STS were established in a political environment with OMB playing an active role in budget determinations. NASA had developed a budget based upon an all-liquid configuration and presented this budget to the President. There was apparently some confusion over what was approved with OMB asserting that the White House had agreed to a $5.1 billion shuttle with no contingencies. NASA thought it was $5.5 billion with contingencies. As a result, NASA agreed to the OMB figure and had to revise the shuttle configuration. In fact, NASA had to select a configuration that would both meet a total spending cap and a spending profile. The result was the use of the cheaper (nonrecurring cost) solid rockets. This was an important change because it significantly affected the nonrecurring and recurring cost tradeoff.

2. Utilization (Payload Budgets)

The economic justification of the space shuttle was predicated upon the ability to fly a large number of missions annually. Mission models were created to demonstrate the need for a large number of annual flights. Little or no serious consideration was given to the budgets that would be required (by both government and industry) to support the payload nonrecurring and recurring costs associated with the mission models. (Some felt that there was an effort to select a mission model with sufficient flights to justify the shuttle program.) Had this been seriously considered together with payload effects considerations (the possible lack, thereof), a lesser mission model than one requiring a launch rate of 60 flights per year (with 65,000 pounds of payload per flight) might have been considered as reasonable.

3. Payload Effects

The STS economic analysis concluded that the economic benefits of the space shuttle depended primarily upon the ability to achieve payload savings. It was assumed that these savings would result from the redesign of payloads to utilize the large benign environment of the shuttle cargo bay and to take advantage of the on-orbit operations made possible by the reusability of the space shuttle. The payload savings never materialized because of several factors. To achieve payload savings meant designing satellites that could only be launched by the shuttle.

History has shown that very few payloads have been designed (with the possible exception of government payloads having final orbits near that of the space shuttle) specifically for launch on the shuttle. To a large extent, this has been because of the desire to maintain a dual launch capability (for example, communication satellites designed for compatibility with Ariane and the shuttle). In addition, the envisioned reusable Space Tug was never developed, which would have significantly affected on-orbit operations, including service and repair. The space shuttle, per se, had little effect on the ability to achieve significant payload savings. The inability to achieve payload savings was as a result of the tenor of the times. (Because most technical studies showed that these savings were achievable and because it was mandated that all payloads would fly on the space shuttle, arguing that payload savings might not be achieved had little payoff.)

4. Contingency Planning for Failures

Contingency planning was not seriously addressed nor taken into account by the shuttle economic studies. The full range of development contingencies and operational contingencies was not considered. Most likely values were considered for schedules, performance, and costs associated with the RDT&E phase of the shuttle program. Because most likely and expected values can differ significantly and because only expected values are additive, the use of most likely values was likely to have contributed to the underestimation of costs and the overestimation of performance. [Because it is mathematically incorrect to add most likely values, except for the special case when expected values and most likely values are the same (for example, variability can be approximated by normal distributions), significant errors can result.] For example, this author participated in a meeting with NASA and an engine manufacturer where a development schedule was presented for the shuttle main engine. It was stated that the schedule would be very difficult to meet. When questioned, it became clear that there was little or no chance of reducing the development schedule, an almost certain guarantee of having delays and a significant budget overrun. In spite of this, the most likely estimates were used in the economic analysis.

The economic analysis did not consider the implications of shuttle failures on overall operations and on the cost of replacement. Government funding for production of ELVs was terminated after the shuttle program became operational. This was done without consideration of what the consequences might be if there were a shuttle failure. A policy was undertaken where all payloads were to fly on the shuttle. This policy helped the Ariane program to become a success (particularly after the shuttle failure) because, as stated earlier, most commercial payloads were designed for a dual launch capability and it was important to maintain a second source.

5. Policy Analysis (Need and Cost for Alternatives)

The shuttle economic analysis did not consider the need for and the cost of maintaining alternatives, that is, the ELVs. This would have reduced the economic advantage of the shuttle program and might have had significant impacts on the funding of the shuttle program. In other words, as in a number of other instances, the economic analysis did not question the guidance and assumptions that were presented for use in the economic analysis. Independent thought was not encouraged.

6. Maximization of Benefits vs Mission Satisfaction

The space shuttle configuration was determined by the need to satisfy all missions contained within the mission model. Of particular importance was the need to satisfy U.S. Department of Defense (DoD) missions; this proved to be a major factor in the sizing of the payload cargo bay. Ironically, it was these very same missions that DoD argued (successfully) required an alternative launch capability (the Titan) and therefore predetermined the unreasonableness of the "payload effects" savings. The shuttle analysis and the overall program planning and justification attempted to minimize the cost of flying the mission model with a shuttle configuration and not maximizing the benefits from a government investment. An approach aimed at maximizing net benefits would likely have resulted in a mixed fleet capability with the shuttle being configured to launch smaller payloads. This, however, is conjecture because behind the scenes was the presumed need to eventually support a space station program.

7. Government/Industry Relationship

NASA originally conducted in-house economic studies to justify the shuttle program. OMB desired an independent verification of the NASA findings, thus, the Mathematica economic study supported by LMSC and the Aerospace Corporation. LMSC and the Aerospace Corporation provided mission model, performance, and cost information utilized in the Mathematica economic analysis. No attempt was made by Mathematica to independently validate or question the data provided for the analysis. Presumably this decision was reached because the goal of the study was to assist with the justification of the shuttle program. In other words, the OMB desire to have an independent assessment was not really achieved.

8. Government Ownership vs Industry Ownership

The shuttle policy and economic analyses were set in a framework of government ownership and operation. Little or no consideration was given to the possibility of private ownership. Thus the government program was concerned with all of the costs associated with development through operation, and the government therefore continued its role as space transportation provider for both government and commercial launches. (Prior to the STS, the government placed commercial payloads into orbit utilizing ELVs.) The demand for government launches was overstated and for commercial launches was understated.

C. Lessons to Be Learned

As the U.S. embarks upon the development of the next major space transportation system, it is instructive to look back at the space shuttle program and identify some of the mistakes of the past that should not be made again. The preceding paragraphs have attempted to identify a number of the policy and economic considerations that played a major role in the structuring and conduct of the space shuttle program. Significant learning has already been accomplished and incorporated into the RLV program. The following lessons are those that should be noted, some of which have already been incorporated into the RLV program.

Development of an affordable RLV requires significant technological and programmatic advances. This implies that there are performance, cost, and schedule uncertainties. At best, it is possible to fix two of these factors. Initially, the shuttle program constrained both cost and schedule (i.e., through the establishment of a funding cap and spending profile), yet did not face up to the fact that in so doing it was significantly affecting the performance risk. When it was realized that performance was likely to suffer, cost constraints were relaxed, leading to large budget overruns. It appears that space station might have taken the same course. This lesson seems to have been learned by the RLV program because NASA is participating in a technology development and demonstration program aimed at reducing performance risk, and the ultimate goal is to achieve commercial financing for the development of the operational system.

Cost-estimating relationships (CER) have played a major role in the estimation of shuttle, space station, and RLV program costs. The appropriate role of CERs must always be reexamined in terms of the type of activity that is to be "costed." The use of CERs is inappropriate in those situations that require significant technology development in order to meet performance objectives. The development of CERs presumes that the task being costed is doable, but in situations that require significant technology development this is not the case; therefore, the CER method of estimating cost is not always appropriate, and alternatives such as RADSIM[10] must be utilized.

Forecasting the demand for space transportation must consider both government and commercial payloads. The forecasted demand for government launches should consider the budgets necessary for the development of the forecasted payloads. A new transportation system must also face up to the reality that there will be competition in the marketplace and that reasonable market shares should be considered (both caused by competition and the need for an alternative).

Contingency planning must be an integral part of the overall program planning. With the objective of achieving commercial space transportation with commercial financing, it is likely that contingency planning for development and operations will become an integral part of the overall program plan. However, government must exercise care in the development of alternatives. Because the market for space launches will be limited (and probably will not expand significantly without very significant price changes), increasing the number of viable alternatives might increase the difficulty of any one alternative obtaining commercial financing.

To ensure that there is an appropriate balance between available alternatives (i.e., viable transportation options), the need for alternative transportation services, the desire for establishing a commercial provider of transportation services utilizing RLVs, and the need for adequate transportation for the space station and other government programs, there should be an overall review of space transportation policy and related goals and objectives.

Maximization of benefits, financial viability of a commercial business venture providing space transportation services with a fleet of RLVs, and overall mission model satisfaction must be addressed. The same issues that faced the space shuttle relative to designing a launch vehicle to satisfy all mission requirements need to be considered by NASA prior to imposing requirements on a joint venture partnership. In other words, government objectives are not likely to coincide with private sector objectives.

Because it is likely that government support will continue to be necessary for commercial financing to become a reality, it is important that NASA establish a realistic approach for evaluating the financial and other requirements of industry in obtaining commercial financing. This should include being able to evaluate the financial viability of business plans, pricing policies, market forecasts, cost and schedule estimates, and other aspects of the business plan that affect the requirements for government participation. NASA has a fiduciary responsibility to spend taxpayers' money wisely; this implies that the government not be placed in the position of accepting information provided by industry (as appears to have been the case in the space shuttle program) without having the capability to provide knowledgeable independent assessments.

Government should continue its policy of seeking a solution for providing space transportation that is based upon private ownership and operation.

III. RLV Financial Modeling

The RLV program has set a unique course by aiming to develop a new reusable space transportation system that will sharply reduce space transportation costs for both commercial and government customers. Because the goal is to have the private sector develop, build, and operate a fleet of RLVs and to provide space transportation services for commercial and government users, private investment and financing is critical. The technological risks inherent in the development of a new reusable launch vehicle are currently being addressed and hopefully will be reduced through the government-led partnership that centered on building and flying the X-33 SSTO demonstrator and the follow-on space launch initiative program. However, developing and building an operational RLV fleet will require a substantial and costly effort beyond the X-33 and successor programs.

Private development of a business venture based upon the development and use of an RLV fleet must surmount both business and technical obstacles. These obstacles must be identified and approaches developed for their removal. The investment community has indicated that it is unlikely that significant funding from the capital markets will become available for an RLV business venture until technology risk is all but eliminated. The capital markets are attuned to normal business risk, not technology risk. Business obstacles that need to be considered relate to financial, insurance, liability, licensing, regulatory, legal, and policy issues. Financial hurdles include the uncertainty of future demand, the high level of initial investment, potential competition from existing and future expendable launch vehicles, and the difficulties of adjusting current policy mechanisms to new government–private cooperative arrangements.

Business planning is currently underway by the private sector, but because of long development schedules, the unpredictable outcome of the current technology development activity (both from the point of view of performance and cost), competitive business pressures, and the likelihood that today's business planner will not be tomorrow's implementer, it is likely that definitive business plans and requirements will not be finalized for several years.

In the meantime it is likely that a demonstration of government assistance will be required in the near term to encourage continuing private sector investment that will lead to a private sector space transportation business based upon a fleet of RLVs. Various forms of government participation or initiatives have been discussed with a number being analyzed in a preliminary fashion.[1] These initiatives include RDT&E participation by NASA, government in-kind contributions, anchor tenancy agreements, provision of termination liability, tax holidays, R&D tax credits, subsidized loans, and loan guarantees.[11] The various considered government initiatives affect investment decisions in different ways; for example, some directly affect the magnitude of the private sector investment, some affect long-term rates of return, some affect risk perceptions, and some affect several of the private-sector decision concerns.

Because it is likely that the private sector will look to NASA and to Congress for continuing assistance, it is important to have an understanding of the implications of differing government programs and policies on promoting RLV financing. This understanding must be developed prior to private-sector investment decisions and requires the government to have insight into the likelihood of different policies and programs affecting investment decisions and the reasonableness of alternative business scenarios. Safety concerns, liability determinations, insurance availability and cost, and licensing, regulatory, legal, and other related policy issues will affect the reasonableness of business scenarios.

Government RLV financial/economic analysis has been centered at the MSFC. The analysis utilizes financial models developed by MSFC and data provided by both government and industry. The MSFC RLV financial modeling has been conducted primarily within a spreadsheet environment with the resulting spreadsheet models incorporated within existing commercial risk analysis software (i.e., At-Risk) so as to provide a risk-analysis capability. This approach has several significant limitations that can become serious as attempts are made to expand upon the existing capabilities. These limitations are discussed in following paragraphs. The current spreadsheet modeling considers (interactively) both the government and the private sector. For the government the financial/economic analysis is concerned with developing both the government's cash-flow stream, net present value of the cash-flow stream, and internal rate of return. In addition, economic impacts are developed (discussed in following paragraphs) in terms of change in Treasury revenue and job creation. These benefits and costs take into account government expenditures, resulting cost savings (both caused by reduced space transportation costs and to the phase out of space shuttle operations), and other costs and benefits that might result from government policies and programs specifically undertaken to encourage and obtain private sector investment that results in the desired commercial RLV space transportation business venture.

A. Development of Financial Documentation

The results of the business planning and financial analyses performed need to be communicated to the financial community in order to assess the implications of obtained results on the likelihood that the private sector could obtaining financing given the government policies and programs under consideration. To accomplish this, it is necessary to present analyses results in a format familiar to and readily understandable by the financial community. Figures 3.1–3.3 depict a set of documentation formats (i.e., RLV transportation business proforma income statement, cash-flow projection, and balance-sheet projection) that are capable of translating the results of government RLV financial analyses into reports that the financial community is accustomed to dealing with.

Figure 3.1, the format of the **proforma income statement**, disaggregates total revenue into revenue from government and commercial services (with price per flight and number of flights indicated for each). Launch operations include payload integration and other operations, launch and liability insurance (provisions for the recovery of these expenses are assumed included in the pricing), launch fees, and maintenance/repair costs. Total annual operations expenses include the preceding items plus annual RDT&E expenses. General and administrative expenses are indicated as a single line item, whereas depreciation for facilities and the fleet of RLVs are considered separately. To the preceding items are added debt service expenses, allowances for uninsured losses, and continuing engineering/development expenses. The sum of all of these is subtracted from annual revenue to obtain before-tax profit. Carry-forward losses are considered in arriving at after-tax profit also taking into account the possibility of R&D tax credits, investment tax credits, and tax-holiday credits. These credits can be considered as government initiatives aimed at encouraging a commercial RLV business venture.

It should be noted that (in Figs. 3.1–3.3) risk measures have been omitted because the current spreadsheet approach is basically deterministic and does not generate measures of risk. (Hopefully, expected value results are generated.)

Figure 3.2 presents the format for the **cash-flow projection**. Total cash inflow is indicated as the sum of after-tax profit, decreases in receivables, increases in payables, decreases in cash, depreciation, debt financing receipts, and paid in capital. Total cash outflow is indicated as the sum of losses, increases in receivables, decreases in payables, increases in cash, capital expenditures (considering separately expenditures for facilities, original fleet vehicles, and replacement fleet vehicles), dividends and loan repayments, and stock buy-backs. The difference between total annual cash inflow and cash outflow is the total annual cash flow. Indebtedness is the negative of the cumulative cash flow over time. Finally, the net present value of the business is indicated at a number of different discount rates with the net present value (NPV) calculated separately for the contribution during the business-planning horizon and for the contribution over an infinite horizon. The total NPV is the sum of the two contributions. (In the calculation of NPV, care must be taken to subtract financing costs so that double counting does not occur.)

Figure 3.3 presents the format for the **balance sheet projection**. The balance sheet indicates the total assets as the sum of cash, accounts receivable, prepaid expenses, book value of facilities (facilities investment less accumulated depreciation), and book value of the RLV fleet (RLV fleet investments less accumulated depreciation). Total liabilities are indicated as the sum of accounts payable, prepayments, and long-term debt. Finally, total stockholder equity is

	Year					
	'02	'03	'04	'05	'06	'07 ········
No. of Flights						
Price						
Revenue [Government]						
No. of Flights						
Price						
Revenue [Commercial]						
Total Revenue						
P/L Integration						
Other Operations						
Launch Insurance						
Liability Insurance						
Launch Fees						
Maintenance/Repair						
Launch Operations						
DDT&E						
Total Operations Expenses						
General & Administrative Expense						
Facilities						
Fleet						
Depreciation Expense						
Debt Service [Interest]						
Allowance for Uninsured Loss						
Continuing Engineering/Development						
Before Tax Profit						
Carry-Forward Loss						
Gross Income Tax						
R&D Tax Credit						
Investment Tax Credit						
Tax-Holiday Credit						
After Tax Profit						
Return on Sales [%]						
Return on Assets [%]						
Debt Coverage						

Fig. 3.1 RLV transportation business proforma income statement.

indicated as the sum of paid in capital (indicating separately corporate funding and equity purchase) and retained earnings. The discounted rates of return on corporate investment and on equity investment are also indicated as a function of the cost of capital (i.e., the discount rate).

B. RLV Pricing Strategies

Goals have been set for significantly reducing the cost of placing payloads into orbit. But if costs are reduced, will prices also be reduced? Depending on demand

	Year					
	'02	**'03**	**'04**	**'05**	**'06**	**'07 ....**
After Tax Profit						
Decrease in Receivables						
Increase in Payables						
Decrease in Cash						
Depreciation						
Debt Financing						
Paid-In Capital [Equity]						
Total Cash Inflow						
Loss						
Increase in Receivables						
Decrease in Payables						
Increase in Cash						
Facilities						
Fleet [Original]						
Fleet [Replacement]						
Capital Expenditures						
Dividends						
Loan Repayment						
Stock Buy-Back						
Total Cash Outflow						
Annual Cash Flow						
Indebtedness						

	Discount Rate [%]										
	0	**5**	**10**	**15**	**20**	**25**	**30**	**35**	**40**	**45**	**≥50**
NPV Contribution [within planning horizon]											
NPV Contribution [infinite planning horizon]											
Total Net Present Value											

Fig. 3.2 RLV transportation business cash-flow projection.

elasticity, there might be little or no incentive for a commercial space transportation provider to reduce costs other than competitive pressures. Both RLV business viability and government achievement of transportation cost savings will depend upon pricing policies that are likely to be adopted by the provider of commercial space transportation services.

The following pricing schemes warrant consideration during the RLV business planning and evaluation process:

1) pricing to maximize profit but with prices constrained by competitive pricing;
2) two-tier pricing (government price for a specified duration with commercial pricing to maximize profits);

	Year						
	'02	'03	'04	'05	'06	'07	....
Cash							
Accounts Receivable							
Prepaid Expenses							
Facilities Investment							
Less Accumulated Depreciation							
Book Value of Facilities							
Fleet Investment							
Less Accumulated Depreciation							
Book Value of Fleet							
Total Assets							
Accounts Payable							
Prepayments [Deferred Income]							
Long Term Debt							
Total Liabilities							
Corporate Funding [% Ownership]							
Equity Purchase [% Ownership]							
Paid in Capital							
Retained Earnings							
Total Stockholder Equity							

	Discount Rate [%]										
	0	5	10	15	20	25	30	35	40	45	≥50
ROI [Discounted Rate of Return on Corporate Investment]											
ROE [Discounted Rate of Return on Equity Investment]											

Fig. 3.3 RLV transportation business balance-sheet projection.

3) pricing such that the RLV price is a specified fraction of the anticipated competitive price;
4) pricing in a competitive market so as to achieve a specified market share;
5) set price at a specified fraction of anticipated competitive price, then to maximize profit; and
6) set price as a constant times operating cost.

The mathematical formulations of these pricing schemes are developed in the following paragraphs so that they can be considered for inclusion in financial planning models that are likely to be or are currently under development. Of particular interest are the schema referred to as "pricing to maximize profit but with prices constrained by competitive pricing," and "two-tier pricing (a government price for a specified duration with commercial pricing to maximize profits)." The first is of concern because it is similar to current practice, and the latter is of extreme interest because it is a way of providing an assurance that the government can receive cost savings as a result of its investment in RLV technology and the provision of incentives.

Many of the variables described in the following paragraphs, including competitive pricing, can be considered as uncertainty variables and the related computations included within a Monte Carlo simulation model. The result is the establishment of probability distributions of financial performance measures taking into account pricing that includes uncertainty considerations.

1. Pricing to Maximize Profit but with Prices Constrained by Competitive Pricing

This pricing strategy attempts to establish RLV pricing such that the RLV business venture profit is maximized, subject, of course, to competitive (for example, ELV) pricing. The RLV business revenue is as follows:

$$\text{REVENUE}(T) = \sum_{\beta} U(\beta, T) * S(\beta, T) * P(T) * 0.01$$

where

β = market segment
$P(T)$ = price (\$/#) in year T
$S(\beta, T)$ = market share (%) in year T
$U(\beta, T)$ = units (#) sold in year T

The resulting profit is as follows:

$$\text{PROFIT}(T) = \sum_{\beta} U(\beta, T) * S(\beta, T) * P(T) * 0.01 - FC - VC * \sum_{\beta} U(\beta, T) * S(\beta, T) * 0.01$$

where

FC = fixed cost (\$)
VC = variable cost (\$/#)

It is assumed that demand is a function of price and has an exponential form. (This both simplifies the mathematics but is also reasonable in many real-world situations and is likely to be the case with those using space transportation services; more will be said about this in following paragraphs.) Demand is expressed as follows:

$$U(\beta, T) = U0(\beta) * \exp\left[-\lambda(\beta) * P(T)\right]$$

where

$U0(\beta)$ = maximum steady-state demand at a low price (selected via input data to provide a reasonable curve fit to demand elasticity data)
$\lambda(\beta)$ = curve-fit parameter that, together with $U0(\beta)$, relates steady-state demand to price

The profit equation can be written as follows:

$$\text{PROFIT}(T) = \sum_{\beta} U0(\beta) * S(\beta, T) * P(T) * 0.01 * \exp[-\lambda(\beta) * P(T)]$$
$$- FC - VC * \sum_{\beta} U0(\beta) * S(\beta,T) * 0.01 * \exp[-\lambda(\beta) * P(T)]$$

To maximize profit, the derivative of profit with respect to price is first established followed by the determination of the price that makes the derivative equal to zero as follows:

$$\frac{\text{dPROFIT}(T)}{\text{d}P} = 0.01 * \sum_{\beta} U0(\beta) * S(\beta, T) * \{-P(T) * \lambda(\beta)$$
$$* \exp[-\lambda(\beta) * P(T)] + \exp[-\lambda(\beta) * P(T)]\} + 0.01VC$$
$$* \sum_{\beta} U0(\beta) * S(\beta, T) * \lambda(\beta) * \exp[-\lambda(\beta) * P(T)] = 0$$

This can be simplified as follows and solved for the price $P(T)$:

$$\sum_{\beta} U0(\beta) * S(\beta, T) * \{\exp[-\lambda(\beta) * P(T)]\} * \{1 - \lambda(\beta) * P(T)\}$$
$$+ VC * \sum_{\beta} U0(\beta) * S(\beta, T) * \exp[-\lambda(\beta) * P(T)] = 0$$

subject to the constraint $P(T) \leq PC(T)$, where $PC(T)$ = competitive price (i.e., the price of a competing transportation system). An iterative process can be used to establish a reasonable estimate of $P(T)$.

To solve for $P(T)$, it is necessary to have a relationship between market share and RLV and competitive pricing. The value of market share $S(\beta, T)$ can be established as follows assuming that a reasonable relationship can be established between relative (i.e., RLV relative to competition) prices and market share. It is assumed that the current year's market share is a function of both the previous year's RLV price and the competitive transportation system price. Thus,

$$S(\beta, T) = f\{P(T-1)/PC(T-1)\}$$

where $f\{\ \}$ = functional market share relationship.

It is assumed that this functional relationship can be approximated by a linear relationship (see Sec. V in Chapter 2 for a discussion of the estimation of market share when multiple attributes are considered), such that

$$S(\beta, T) = 10 * CS(\beta) * [1 - P(T-1)/PC(T-1)] + SEP(\beta)$$

Subject to the following:

If $S(\beta, T) > 100$	then $S(\beta, T) = 100$
If $S(\beta, T) < 0$	then $S(\beta, T) = 0$

In the preceding equations,

$SEP(\beta)$ = market share (%) of RLV when the RLV price is equal to the competitive transportation system price

$CS(\beta)$ = percent change in RLV market share for a 10% reduction (from the equal price) in RLV price (e.g., a 10% reduction in price leads to a 20% point increase in market share: from 50 to 70%)

In keeping with the preceding discussion, Fig. 3.4 illustrates a method for inputting price elasticity data for each of several market segments to a financial planning model. Given a set of price elasticity data, the method illustrated in Fig. 3.5 can be used for rapidly modifying the price elasticity data through the use of an exponential relationship. It is assumed that the results of the exponential relationship would be displayed along with the database data so as to provide a basis for comparisons. Appropriate values of λ can be selected so as to establish a reasonable fit of the exponential to the data in the region of interest.

DATABASE: PRICE ELASTICITY DATA

MARKET SEGMENT: **Space Tourism** **3** β

ELASTICITY DATA:

α	PRICE [$/#]	DEMAND [$/YR]
1	100	50000
2	200	35000
3	300	25000
4	500	15000
5	700	10000
6	1000	5000
7	2000	1000
8	3000	500
9	5000	100
10	7000	0
α	PDB(α,β)	DDB(α,β)

Fig. 3.4 Inputting price elasticity data by market segment.

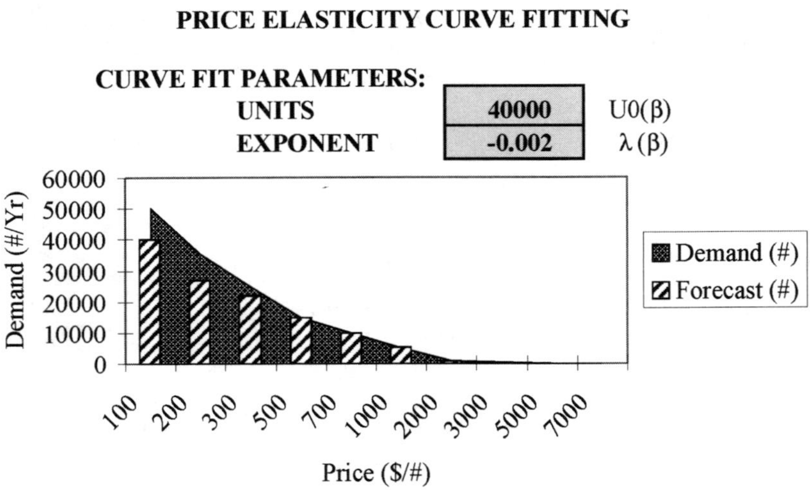

Fig. 3.5 Price elasticity curve fitting.

2. *Two-Tier Pricing (Government Price for Specified Duration with Commercial Pricing to Maximize Profits)*

It is assumed that as a result of government support provided during the development and/or startup years of the RLV business venture a special or preferred price will be offered for government payloads. This preferential pricing will be provided for a specified period of time. Thus a two-tier pricing policy will be maintained (i.e., government and nongovernment) for a period of time after which a single pricing policy will be utilized for both government and nongovernment customers.

Using the same notation as used in the preceding section, it is assumed that a government price (i.e., the government is considered as one of the multiple market segments) is set as a multiplier to annual operating cost. However, this price cannot exceed the price offered for nongovernment or commercial markets. The commercial price is assumed set to maximize annual profit.

In the following TST is the time at which the switch will be made to single-tier pricing, *PG*(*T*) is the government price and *OCM* is the operating cost multiplier (i.e., 1.5 implies multiplying operating cost by 1.5). Therefore,

when $T <$ TST and $\beta = 1$ (i.e., $\beta = 1$ represents the government sector)
then

$$PG(T) = OCM * VC \text{ subject to the constraint that } PG(T) \leq P(T)$$

when $T <$ TST and $\beta > 1$ (i.e., $\beta = 1$ represents the government sector)

Then the value of market share $S(\beta, T)$ can be established as noted before and the pricing can be established as follows:

$$\begin{aligned}\text{PROFIT}(T) = {} & \sum_{\beta=2}^{\beta} U0(\beta) * S(\beta, T) * P(T) * 0.01 * \exp[-\lambda(\beta) * P(T)] \\ & - FC - VC * \sum_{\beta=2}^{\beta} U0(\beta) * S(\beta, T) * 0.01 \\ & * \exp[-\lambda(\beta) * P(T)] + U0(\beta = 1) * S(\beta = 1, T) * PG(T) \\ & * 0.01 * \exp[-\lambda(\beta = 1) * PG(T)] - VC * U0(\beta = 1) \\ & * S(\beta = 1, T) * 0.01 * \exp[-\lambda(\beta = 1) * PG(T)]\end{aligned}$$

when $T \geq$ TST

then

$$\text{REVENUE}(T) = \sum_{\beta} U(\beta, T) * S(\beta, T) * P(T) * 0.01$$

and, as in the pricing scenario already described,

$$\begin{aligned}\text{PROFIT}(T) = {} & \sum_{\beta} U(\beta, T) * S(\beta, T) * P(T) * 0.01 \\ & - FC - VC * \sum_{\beta} U(\beta, T) * S(\beta, T) * 0.01\end{aligned}$$

Making the same assumptions as in the preceding pricing scenario with respect to the form of the demand–price exponential functional relationship, establishing the derivative of profit with respect to price, and equating the obtained derivative of profit with respect to price equal to zero results in the following equations, which can be solved for $P(T)$ utilizing an iterative process:

when $T <$ TST

then

$$\begin{aligned}& \sum_{\beta=2}^{\beta} U0(\beta) * S(\beta, T) * \{\exp[-\lambda(\beta) * P(T)]\} * [1 - \lambda(\beta) * P(T)\} \\ & \quad + VC * \sum_{\beta=2}^{\beta} U0(\beta) * S(\beta, T) * \exp[-\lambda(\beta) * P(T)] = 0\end{aligned}$$

when $T \geq$ TST

then

$$\begin{aligned}& \sum_{\beta} U0(\beta) * S(\beta, T) * \{\exp[-\lambda(\beta) * P(T)]\} * \{1 - \lambda(\beta) * P(T)\} \\ & \quad + VC * \sum_{\beta} U0(\beta) * S(\beta, T) * \exp[-\lambda(\beta) * P(T)] = 0\end{aligned}$$

with both of the preceding subject to the $P(T) \leq PC(T)$ constraint where $PC(T)$ is the competitive price (i.e., the price of a competitive transportation system). It is assumed that the value of market share is established in the same manner as that indicated in the pricing scenario already described.

3. Pricing Such That RLV Price is Specified Fraction of Anticipated Competitive Price

The objective of this pricing strategy is to establish an RLV price that is a specified fraction of the anticipated competitive price. In so doing it is realized that profit maximization will likely not be achieved, but it is likely that near-term profit maximization will be achieved. The RLV price is stated as follows:

$$P(T) = \delta(T) * PC(T)$$

and, as in the preceding pricing scenarios, the resulting value of market share $S(\beta, T)$ can be established as follows:

$$S(\beta, T) = f\{P(T-1)/PC(T-1)\}$$

where $f\{\ \}$ = functional market share relationship.

As just stated, it is assumed that this functional relationship can be approximated by a linear relationship such that

$$S(\beta, T) = 10 * CS(\beta) * [1 - P(T-1)/PC(T-1)] + SEP(\beta)$$

subject to the following:

$$\begin{array}{ll} \text{If } S(\beta, T) > 100 & \text{then } S(\beta, T) = 100 \\ \text{If } S(\beta, T) \leq 0 & \text{then } S(\beta, T) = 0 \end{array}$$

In the preceding equations,

δ = the fraction of the competitive price at which RLV pricing is set
$SEP(\beta)$ = market share (%) of RLV when the RLV price is equal to the competitive transportation system price
$CS(\beta)$ = percent change in RLV market share for a 10% reduction (from the equal price) in RLV price (e.g., a 10% reduction in price leads to a 20% point increase in market share: from 50 to 70%)

The profitability of the business venture with this pricing policy can be obtained from the following:

$$\text{PROFIT}(T) = \sum_{\beta} U0(\beta) * S(\beta, T) * P(T) * 0.01 * \exp[-\lambda(\beta) * P(T)] - FC$$
$$- VC * \sum_{\beta} U0(\beta) * S(\beta, T) * 0.01 * \exp[-\lambda(\beta) * P(T)]$$

4. Set Price at Specified Fraction of Anticipated Competitive Price, Then to Maximize Profit

This pricing strategy is similar to the preceding strategy with the exception that at a specified time TS a switch is made from a specified pricing to pricing to maximize annual profit. Thus,

when $T < TS$

then

$$P(T) = \delta(T) * PC(T)$$

and the value of market share $S(\beta, T)$ can be established as follows:

$$S(\beta, T) = f\{P(T-1)/PC(T-1)\}$$

where $f\{\ \}$ = functional market share relationship, and, as before, it is assumed that this functional relationship can be approximated by a linear relationship such that

$$S(\beta, T) = 10 * CS(\beta) * [1 - P(T-1)/PC(T-1)] + SEP(\beta)$$

Subject to the following:

$$\begin{array}{ll} \text{If } S(\beta, T) > 100 & \text{then } S(\beta, T) = 100 \\ \text{If } S(\beta, T) \leq 0 & \text{then } S(\beta, T) = 0 \end{array}$$

and δ, $SEP(\beta)$, and $CS(\beta)$ are as already defined. The resulting value of profit is then given by

$$\text{PROFIT}(T) = \sum_{\beta} U0(\beta) * S(\beta, T) * P(T) * 0.01 * \exp[-\lambda(\beta) * P(T)] - FC$$
$$- VC * \sum_{\beta} U0(\beta) * S(\beta, T) * 0.01 * \exp[-\lambda(\beta) * P(T)]$$

when $T \geq TS$

then revenue and profit are obtained, respectively, from the following equations and all terms are as already defined:

$$\text{REVENUE}(T) = \sum_{\beta} U(\beta, T) * S(\beta, T) * P(T) * 0.01$$
$$\text{PROFIT}(T) = \sum_{\beta} U(\beta, T) * S(\beta, T) * P(T) * 0.01$$
$$- FC - VC * \sum_{\beta} U(\beta, T) * S(\beta, T) * 0.01$$

As was done in the preceding pricing scenarios, it is assumed that demand can be expressed as a function of price with the following functional form assumed:

$$U(\beta, T) = U0(\beta) * \exp[-\lambda(\beta) * P(T)]$$

Taking the derivative of PROFIT with respect to $P(T)$, there results

$$\sum_{\beta} U0(\beta) * S(\beta, T) * \{\exp[-\lambda(\beta) * P(T)]\} * \{1 - \lambda(\beta) * P(T)\}$$

$$+ VC * \sum_{\beta} U0(\beta) * S(\beta, T) * \exp[-\lambda(\beta) * P(T)] = 0$$

which can be solved for $P(T)$, subject to the constraint $P(T) \leq PC(T)$, utilizing an iterative process.

5. *Pricing in Competitive Market so as to Achieve Specified Market Share*

The objective of this pricing strategy is to establish an RLV price that is likely to result in the achievement of a desired market share. This is accomplished in a competitive market. As in the pricing strategies already developed, it is assumed that a linear relationship exists between relative price and RLV market share; therefore,

$$S(\beta, T) = 10 * CS(\beta) * [1.0 - P(T-1)/PC(T-1)] + SEP(\beta)$$

such that

$$\begin{array}{ll} \text{If } S(\beta, T) > 100 & \text{Then } S(\beta, T) = 100 \\ \text{If } S(\beta, T) \leq 0 & \text{Then } S(\beta, T) = 0 \end{array}$$

Because different market shares can result for each of the market segments, it is necessary to select that market segment SFP which will be used for the price determination. It is also necessary to specify the desired market share in this market segment $DS(\beta = SFP, T)$. The desired price can then be obtained by solving the preceding market share equation and is as follows:

$$\begin{aligned} &P(T-1)/PC(T-1) \\ &\quad = 1.0 - [S(\beta = SFP, T) - SEP(\beta = SFP)]/[10 * S(\beta = SFP)] \end{aligned}$$

This value of relative price can then be used to establish market share in other market sectors and to establish annual profit.

6. *Set Price as a Constant Times Operating Cost*

The objective of this pricing scenario is to simply set the RLV price as a constant times the annual operating cost

where OCM = operating cost multiplier (i.e., 1.5 implies multiplying operating cost by 1.5).

$$P(T) = OCM * VC$$

Subject to the constraint:

$$P(T) \leq PC(T)$$

where $PC(T)$ = competitive price (i.e., the price of a competing transportation system) and the value of market share $S(\beta, T)$ can be established, as in the preceding pricing scenarios, as follows:

$$S(\beta, T) = f\{P(T-1)/PC(T-1)\}$$

and the functional relationship is approximated by a linear relationship such that

$$S(\beta, T) = 10 * CS(\beta) * [1 - P(T-1)/PC(T-1)] + SEP(\beta)$$

As before, the profit equation can be written as

$$\text{PROFIT}(T) = \sum_{\beta} U0(\beta) * S(\beta, T) * P(T) * 0.01 * \exp[-\lambda(\beta) * P(T)]$$

$$- FC - VC * \sum_{\beta} U0(\beta) * S(\beta, T) * 0.01 * \exp[-\lambda(\beta) * P(T)]$$

C. Economic Implications of RLV Pricing

A general methodology for assessing the economic impacts of policy changes was described in Sec. VIII in Chapter 2 with an example indicating the consequences of launch fee policy described in Sec. IV of Chapter 4. In the following paragraphs this methodology is further elaborated upon so as to indicate the economic impacts of RLV pricing policies.

If an RLV leads to lower space transportation prices, the demand for space transportation, in general, can increase, and the U.S. market share for internationally competed space launches can also increase. Both the increase in demand and increase in market share will impact the U.S. economy through increased job creation (both direct and indirect) and increased federal revenue. ("Direct" "indirect" employment is defined as employment that is a direct consequence of procurement of products and/or services relating to the construction of space hardware and the provision of space transportation services, and employment that is the consequence of money spent in a community by the direct employees, respectively. "Direct" and "indirect" income are similarly defined.)

In the following the indices T and m indicate time and mission, respectively. Mission might or might not be synonymous with market segment β (i.e., mission can be a subset of market segment or it can be the market segment). It is assumed that a price/quantity relationship can be estimated as indicated in Fig. 3.6 for different mission types. The following additional definitions also apply:

$P_{\text{To}}(T)$ = price for transportation to LEO (\$/kg) in the absence of an RLV

$Q_o(T, m)$ = payload quantity (kg) to be delivered to LEO given a price $P_o(T)$. The payload mass includes the mass of the satellite and the transfer stage (both including propellants).

$P_T(T)$ = price for transportation to LEO (\$/kg) when utilizing an RLV

$Q(T, m)$ = payload quantity (kg) to be delivered to LEO given a price $P(T)$. The payload mass includes the mass of the satellite and the transfer stage (both including propellants).

T = index referring to time

m = index referring to mission type

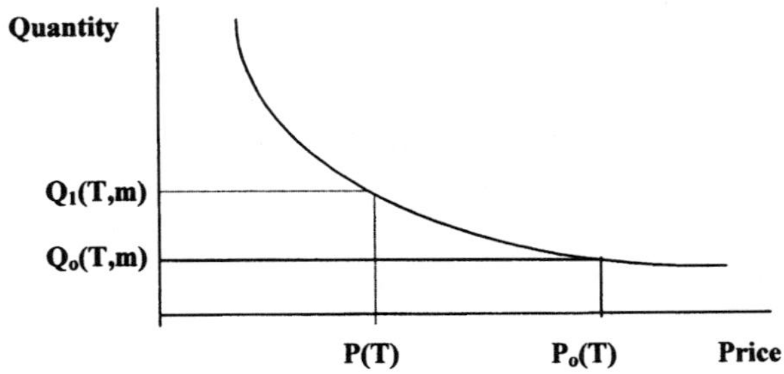

Fig. 3.6 Price/quantity relationship.

The increase in demand, resulting from a decrease in transportation price, will result in a change in transportation revenue, satellite revenue, and transfer stage revenue. It is assumed that the change in revenue (resulting from the use of transportation services, purchase of satellites, and transfer stages) is used to pay for additional labor hours that are required to support the increased demand and pay taxes.[12] It is assumed that profit margins (for example, return on sales) will remain relatively constant. Because certain missions will go beyond LEO, it is important to consider the cost of transfer stages when developing the price/quantity relationships for different classes of missions.

The change in transportation revenue (constant market share) is

$$\begin{aligned}&\Delta\text{Trans. Revenue}(1, T, m)\\&\quad = [Q(T, m) * P_T(T) - Q_o(T, m) * P_{\text{To}}(T)] * MS_T * 0.01\end{aligned}$$

where MS_T is the U.S. transportation market share (%) measured in terms of delivered mass to LEO.

There is also the possibility that transportation revenue will change as a result of a change in market share caused by a reduction in price that is not matched by foreign competition. This change in transportation revenue (variable market share) is approximated by

$$\begin{aligned}\Delta\text{Trans. Revenue}(2, T, m) &= \Delta\text{Trans. Revenue}(1, T, m) * \Delta MS_T * 0.01\\&\quad + Q_o(T, m) * P_{\text{To}}(T) * MS_T * \Delta MS_T * 0.0001\end{aligned}$$

where ΔMS_T is the percent change (If market share increases from 40 to 50%, then $\Delta MS_T = 10\%$.) in U.S. transportation market share that results from a reduction in transportation price. (The change in market share must take into account foreign transportation price changes that might be a reaction to the U.S. price changes.) The effect of relative pricing on launch vehicle market share is discussed in Sec. VII of Chapter 2 and Sec. II of Chapter 4.

The change in U.S. transportation revenue $\Delta\text{Revenue}_T(T, m)$ is

$$\Delta\text{Revenue}_T(T, m) = \Delta\text{Trans. Revenue}(1, T, m) + \Delta\text{Trans. Revenue}(2, T, m)$$

As a result of the change in transportation revenue, there is a change in corporate taxes $\Delta\text{Corp. Taxes}_T(T, m)$, a change in direct income ΔDirect

$\text{Income}_T(T, m)$, a change in indirect income $\Delta\text{Indirect Income}_T(T, m)$, a change in direct employment $\Delta\text{Direct Employ}_T(T, m,)$, a change in indirect employment $\Delta\text{Indir. Employ}_T(T, m)$, and a change in federal revenue $\Delta\text{Fed Rev}_T(T, m)$. These impacts that result from a transportation service price change are as follows:

$$\Delta\text{Corp. Taxes}_T(T, m) = \Delta\text{Revenue}_T(T, m) * ROS_T * CTR * 0.0001$$

where ROS_T is the corporate transportation related return on sales (%) and CTR is the corporate tax rate (%). These and other indicated variables are defined in Sec. VIII of Chapter 2.

$$\Delta\text{Direct Income}_T(T, m) = \Delta\text{Revenue}_T(T, m) - \text{Corp. Taxes}_T(T, m)$$

$$\Delta\text{Indirect Income}_T(T, m) = \Delta\text{Revenue}_T(T, m) * \text{Indirect Income Multiplier}$$

$$\begin{aligned}
&\Delta\text{Direct Employment}_T(T, m)\\
&\quad = \Delta\text{Direct Income}_T(T, m)\\
&\quad\quad \times [1.0 - \text{Benefit rate} \times 0.01]/\text{Ave Aerospace Wage Rate}\\
&\quad\quad + [\Delta\text{Direct Income}_T(T, m) * \text{Benefit Rate} * 0.01\\
&\quad\quad + \Delta\text{Corp. Taxes}_T(T, m)]/\text{Ave. Nonaerospace Wage Rate}\\
&\Delta\text{Indirect Employment}_T(T, m)\\
&\quad = \Delta\text{Indirect Income}_T(T, m)/\text{Ave. Wage Rate of Nonaerospace Sector}\\
&\Delta\text{Federal Revenue}_T(T, m)\\
&\quad = \Delta\text{Corp. Taxes}_T(T, m) + [\Delta\text{Direct Income}_T(T, m)\\
&\quad\quad + \Delta\text{Corp. Taxes}_T(T, m) + \Delta\text{Indirect Income}_T(T, m)]\\
&\quad\quad * \text{Income Tax Rate} * 0.01 + [\Delta\text{Direct Income}_T(T, m)\\
&\quad\quad + \Delta\text{Corp. Taxes}_T(T, m) + \Delta\text{Indirect Income}_T(T, m)]\\
&\quad\quad * \text{Unemployment Compensation Rate}\\
&\quad\quad * \text{Ave. Duration of Unemployment} * 0.01
\end{aligned}$$

Similar impacts will result from changes in revenue from the use of transfer stages. The change in transfer stage revenue (constant market share) is

$$\begin{aligned}
&\Delta\text{Trans. Stage Revenue}(1, T, m)\\
&\quad = [Q(T, m) - Q_o(T, m)] * PTS(T) * MS_T * MS_{TS} * 0.0001
\end{aligned}$$

where $PTS(T)$ is the price for transportation from LEO to final orbit location (\$/kg) and MS_{TS} is the U.S. transfer stage market share (%). The transfer stage revenue resulting from a change in U.S. transportation (to LEO) market share is

$$\begin{aligned}
&\Delta\text{Trans. Stage Revenue}(2, T, m)\\
&\quad = [Q(T, m) - Q_o(T, m)] * PTS(T) * \Delta MS_T * MS_{TS} * 0.0001\\
&\quad\quad + Q_o(T, m) * PTS(T) * \Delta MS_T * MS_{TS} * 0.0001
\end{aligned}$$

The change in U.S. transfer stage revenue $\Delta\text{Revenue}_{\text{TS}}(T, m)$, is

$$\Delta\text{Revenue}_{\text{TS}}(T, m) = \Delta\text{Trans. Stage Revenue}(1, T, m) + \Delta\text{Trans. Stage Revenue}(2, T, m)$$

As a result of the change in transfer stage revenue, there is a change in corporate taxes $\Delta\text{Corp. Taxes}_{\text{TS}}(T, m)$, a change in direct income $\Delta\text{Direct Income}_{\text{TS}}(T, m)$, a change in indirect income $\Delta\text{Indirect Income}_{\text{TS}}(T, m)$, a change in direct employment $\Delta\text{Direct Employ}_{\text{TS}}(T, m)$, a change in indirect employment $\Delta\text{Indir. Employ}_{\text{TS}}(T, m)$, and a change in federal revenue $\Delta\text{Fed Rev}_{\text{TS}}(T, m)$. These impacts can be computed in the same manner as just done for the transportation impacts with the exception that ROS_T is replaced by ROS_{TS} and $\Delta\text{Revenue}_T(T, m)$ is replaced by $\Delta\text{Revenue}_{\text{TS}}(T, m)$. ROS_{TS} is the return on sales (%) experienced by firms providing transfer stages.

Finally, similar impacts will result from changes in revenue from changes in satellite mass delivered to orbit. Only the recurring cost effects are considered because it is assumed that nonrecurring costs per unit mass will be small (amortized over multiple satellites in a mission) compared to the recurring costs per unit of satellite mass. The change in satellite revenue (constant market share) is

$$\Delta\text{Satellite Revenue}(1, T, m) = [Q(T, m) - Q_o(T, m)] * PS(T) * MS_T * MS_S * 0.0001$$

where $PS(T)$ is the price per unit mass of a satellite (\$/kg) and MS_S is the U.S. satellite market share (%). The satellite revenue resulting from a change in U.S. transportation (to LEO) market share is

$$\Delta\text{Satellite Revenue}(2, T, m) = [Q(T, m) - Q_o(T, m)] * PS(T) * \Delta MS_T * MS_S * 0.0001 + Q_o(T, m) * PS(T) * \Delta MS_T * MS_S * 0.0001$$

The change in U.S. satellite revenue $\Delta\text{Revenue}_S(T, m)$ is

$$\Delta\text{Revenue}_S(T, m) = \Delta\text{Satellite Revenue}(1, T, m) + \Delta\text{Satellite Revenue}(2, T, m)$$

As a result of the change in satellite revenue, there is a change in corporate taxes $\Delta\text{Corp. Taxes}_S(T, m)$, a change in direct income $\Delta\text{Direct Income}_S(T, m)$, a change in indirect income $\Delta\text{Indirect Income}_S(T, m)$, a change in direct employment $\Delta\text{Direct Employ}_S(T, m)$, a change in indirect employment $\Delta\text{Indir. Employ}_S(T, m)$, and a change in federal revenue $\Delta\text{Fed Rev}_S(T, m)$. These impacts can be computed in the same manner as just done for the transportation impacts with the exception that ROS_T is replaced by ROS_S and $\Delta\text{Revenue}_T(T, m)$ is replaced by $\Delta\text{Revenue}_S(T, m)$. ROS_S is the return on sales (%) experienced by firms, manufacturing satellites.

The total revenue, corporate tax, direct employment, indirect employment, and federal revenue impacts are as follows [These impacts can be summed over the multiple considered missions (m) to obtain total impacts as a function of time.]:

$$\Delta\text{Revenue}(T, m) = \Delta\text{Revenue}_T(T, m) + \Delta\text{Revenue}_{\text{TS}}(T, m) + \Delta\text{Revenue}_S(T, m) \quad (\$/\text{Yr})$$

$$\Delta\text{Corp. Taxes}(T, m) = \Delta\text{Corp. Taxes}_T(T, m) + \Delta\text{Corp. Taxes}_{\text{TS}}(T, m) + \Delta\text{Corp. Taxes}_S(T, m) \quad (\$/\text{Yr})$$

$$\Delta\text{Direct Employment}(T, m) = \Delta\text{Direct Employment}_T(T, m) + \Delta\text{Direct Employment}_{\text{TS}}(T, m) + \Delta\text{Direct Employment}_S(T, m) \quad (\text{person years/yr})$$

$$\Delta\text{Indirect Employment}(T, m) = \Delta\text{Indirect Employment}_T(T, m) + \Delta\text{Indirect Employment}_{\text{TS}}(T, m) + \Delta\text{Indirect Employment}_S(T, m) \quad (\text{person years/yr})$$

$$\Delta\text{Federal Revenue}(T, m) = \Delta\text{Federal Revenue}_T(T, m) + \Delta\text{Federal Revenue}_{\text{TS}}(T, m) + \Delta\text{Federal Revenue}_S(T, m) \quad (\$/\text{yr})$$

In the preceding calculations the following typical values can be used:

$ROS = 6\%$ (Ref. 13) (ROS_T, ROS_{TS}, and ROS_S)
$CTR = 35\%$ (Ref. 13)
Indirect income multiplier = 2.0 (Refs. 14–16)
Average aerospace wage rate = \$39,832/year (Ref. 17)
Average nonaerospace wage rate = \$22,880/year (Ref. 18)
Benefit rate = 33% (Ref. 19)
Income tax rate = 20%
Unemployment compensation rate = 36% (Ref. 20)
Duration of unemployment = 0.344 years (Ref. 20)

D. Generalized RLV Financing Scenario

The establishment of a business venture that provides space transportation utilizing a fleet of reusable launch vehicles is likely to require an investment measured in terms of billions of dollars. Preceding paragraphs discussed briefly a financial planning model developed by NASA to analyze and evaluate a wide range of technology and policy alternatives and their likely impact on the financial performance and ability to obtain financing for RLV and other space transportation business ventures. In particular, the developed capability is aimed at analyzing the likely performance of business ventures that will provide space transportation services utilizing a fleet of reusable launch vehicles and the implications of such business ventures on the need for government support and resulting expenditures and cost savings. The NASA-developed capability incorporates a financial planning model that allows for the consideration of a broad range of business concepts and related government programs and develops pro forma financial statements for the business and government cash flow projections and related information. The financial planning model allows for the explicit consideration of a broad range of government initiatives and allows their explicit impacts on both the business and government to be evaluated. Currently, consideration is being given to the inclusion of uncertainty and resulting risk.

Currently, however, the developed financial planning model does not consider RLV pricing policies nor does it consider business-financing scenarios. The former was previously addressed,[21] and the latter is the subject of the following paragraphs.

E. General RLV Financing Scenario (Description)

Financing arrangements for capital-intensive, risky space-related business ventures having relatively long payback periods are tailored to meet the specific needs of the venture seeking financing and the requirements of the financing institutions. History has shown the diversity of undertakings and associated financing schemes encompassing start-up situations building upon government contracts, start-up situations placing reliance on private financing, large company projects with corporate financing, large company projects with corporate financing supported by external (both private and public) financing, and large multi-organization projects. With the diversity of potential financing mechanisms for providing the financial resources required by a range of possible RLV business ventures, it is not possible to anticipate nor model all possible situations. However, it is anticipated that reasonable insights can be gained relating to the financial requirements and financial viability of RLV business ventures and the potential need for, and form of, government support, by considering the described generalized RLV financing scenarios

The generalized large company RLV financing scenario is summarized in Fig. 3.7. (The described scenario is based upon discussions with those knowledgeable and skilled in financing risky endeavors and has sufficient structure so that the implications of a broad range of government policies can be evaluated.) It considers a period of corporate investment followed by a period of debt financing and a period of equity financing. In theory, all financing decisions are based upon projections made at the point in time when the financing decision is to be made. This adds another dimension to the simulation modeling of the RLV business venture and is discussed further in the following section. In addition, the mathematical model should have sufficient flexibility so as to consider any combination of these various sources of funding.

Figure 3.7 illustrates a hypothetical cash-flow stream of an RLV business venture together with the indebtedness of the business venture. Indebtedness is defined as the negative of the cumulative cash flow measured at any point in time with the peak of the indebtedness curve indicating the total amount of the required funding. Neither the cash flow nor the indebtedness includes funds provided through financing. They do however take into account expenses associated with the funding (for example, interest payments). At any point in time, it is necessary that the cumulative provided financing be equal to or exceed the indebtedness. The large company-oriented financing scenario assumes that during the period T_0 through T_1 one or more organizations provides the necessary financing to a point in time [T_1] when it becomes possible to obtain debt financing at a reasonable cost.

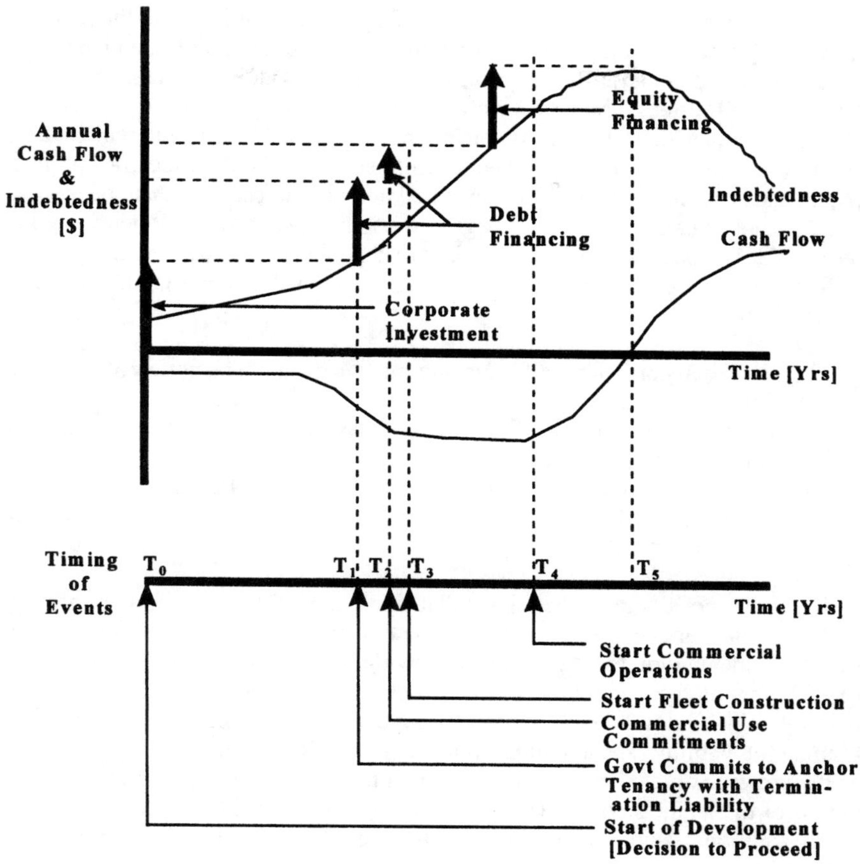

Fig. 3.7 Generalized RLV financing scenario (large company oriented).

One or more organizations can provide the necessary financing (up until T_1) with ownership being in direct proportion to the investment. For simplicity, it can be assumed that the financing proportions remain constant during this time period.

At time T_1 an initial commitment is made to utilize the RLV with additional commitments made at T_2. (The specific order shown in Fig. 3.7 for the events is for illustrative purposes only, and flexibility is necessary in the modeling to allow for other reasonable orderings.) It is assumed that these commitments will allow some level of debt financing based upon the perceived value of these commitments. The specific timing of these events should be coordinated with specific risk reduction demonstrations to minimize financing costs. When debt-financing possibilities have been exhausted, equity financing will be required to make up the shortfall between the maximum indebtedness and debt financing.

To achieve reasonably priced debt financing, it will be necessary for the RLV venture to reduce risk perceptions through technology development and flight demonstrations (of an RLV prototype) and/or receive commitments for a reasonable number of RLV flights at a specified price. It is likely that the financial community will consider this commitment with some degree of skepticism if it does not also include an assumption of termination liability. The cost of debt financing will differ without and with termination liability.

It is assumed that at some point in time the U.S. government will make a commitment to utilize the RLV. This commitment might be for a specified number of payloads or flights per year for a specified number of years. This can be thought of as anchor tenancy, to a certain degree. Various types of commitments need to be considered because each will affect risk perceptions, and financing costs, differently. At least the following need to be considered individually and in combination:

1) an initial commitment for placing mass in orbit over a period of time *with* termination liability,
2) an initial commitment for placing mass in orbit over a period of time *without* termination liability, and
3) annual commitments for placing mass in orbit. (Because this is on an annual basis it, can be assumed that this represents a "firm" contract and termination liability need not be considered.)

It should be assumed that the collateral value of the launch commitment (present value of the commitment payments discounted, at the lenders cost of capital, to the point in time when the borrowing is to take place) will be a function of the form of the commitment; for example, a commitment with termination liability will have a higher collateral value than the same commitment sans termination liability. The amount of financing provided will be the collateral value of the present value of the commitment with the cost of financing being a rate in excess of the lender's cost of capital so as to reflect the lender's risk perceptions.

As additional launch commitments are received, additional debt financing is assumed to occur. Because debt financing is assumed to be based upon launch commitments, as the launches occur debt repayment must be made. It is possible that zero coupon bonds will be utilized because at least in the early years revenue will not be available to cover interest payments. It is assumed that until an operational RLV fleet exists and its performance demonstrated significant borrowing will not be possible utilizing the fleet as collateral because the RLV, if there are problems, will have very little asset value. This will change as additional vehicles are added to the fleet.

At some point in time, debt financing will be accomplished utilizing all assets of value (from launch commitments to fleet RLVs). If it is found that the maximum indebtedness exceeds this amount, then it is necessary to establish equity financing. Thus the amount of equity financing required is the difference between the forecasted maximum indebtedness and the amount of corporate investment plus received debt financing. In actual fact, it is likely that the equity financing will occur as a number of small financings rather than one large financing. However, for the sake of simplicity and without compromising the intent of the analysis it is recommended that initially only a single equity financing be considered.

Equity ownership can be specified or it can be computed. A simple algorithm for computing ownership would establish equity ownership (investor) as a ratio of equity amount required and present value of the future cash-flow stream of the RLV business venture and utilizing a cost of capital that takes into account the variability of the present value (i.e., perceived risk).

F. General RLV Financing Scenario Modeling Considerations

The following considerations need to be taken into account in the development of the financing model:

1) From a modeling perspective launch commitments can be considered in terms of commitments to deliver mass to orbit. This can be converted to flights assuming some utilization factor or directly to future revenue given a pricing policy.
2) When considering uncertainty, it is necessary to develop at least expected values and standard deviations for the amount of corporate financing, debt financing, and equity financing. The problem is that these will be highly correlated, and great care must be taken in drawing conclusions without taking into account this correlation.
3) To simplify the initial modeling, consideration should be given to assuming that both principal and interest payments will be made when mass is delivered to orbit. This is equivalent to considering zero coupon bonds with balloon payments made at the time of mass placement into orbit.
4) It has been suggested that the RLV fleet would have little asset value from a collateral point of view because the market for RLVs, in the event of a decision to terminate the RLV launch business, would be very uncertain. However, it is likely that without too much difficulty a borrowing option against fleet RLVs could be incorporated into the financing model with the collateral value being considered as a variable.
5) When performing a Monte Carlo simulation (risk) analysis, each Monte Carlo run needs to establish the timing of borrowing actions (as keyed to specific events such as commitment timing), the amount of required debt financing, debt principal and interest payments, and equity timing (both amount and ownership).
6) Whereas debt financing is tied directly to asset value, which is known (at the time of the debt financing decision) and not dependent upon future perceptions, the amount of equity financing that is required must take into account perceptions of the future RLV business at the point in time that the equity financing decision is to be made. To sidestep this problem, each Monte Carlo run should establish the maximum value of indebtedness and the necessary equity financing. The necessary financing and ownership can then be established and all appropriate years adjusted to account for the financing and ownership.

G. Mathematical Formulation

The following mathematical formulation of the Generalized RLV Financing Scenario (Large Company Oriented) focuses on the presentation of the highlights of the mathematical model (of the financing scenario), leaving the details of implementation to the implementer. The mathematical formulation is based upon schema and definitions presented in Fig. 3.7. Additional variables and definitions are presented in the following paragraphs, which consider a single government commitment made at $t = T_1$ to launch payload mass in following years and a single commercial commitment made at $t = T_2$ to also launch payload mass in future years. There might be other launch commitments made after $t = T_1$ and $t = T_2$. In that case the values of T_1 and T_2 should be replaced by the appropriate value of T_i.

When $t < T$, where T is the lesser of T_1 and T_2, financing is achieved by corporate investment. Therefore,

$$CI(t) = \text{Corp. Investment}(t) = \sum_{t=0}^{T}[\text{Annual Cash Flow}(t)]$$

When $t \geq T_1$,
Debt financing is possible based upon the asset value of a continuing government commitment [assuming a single government commitment at T_1 to launch $MG(t = T_1)$ kg per year with T_1 and $MG(t = T_1)$ specified via input data] for use of the RLV. Therefore, the revenue $RGC(t)$ resulting from the government commitment is

$$RGC(t) = MG(t = T_1) * PG(t)$$

where (underlining indicates an input variable)

$\underline{MG(t)}$ = government commitment to launch mass (kg/yr) in year t
$\underline{PG(t)}$ = price (\$/kg) charged [This can either be input or can be a computed quantity (i.e., the output of a pricing algorithm). For the present it is assumed to be an input quantity.] to government for payload delivery in year t

The present value of revenue resulting from the government commitment (PVGC), is given by

$$PVGC(t = T_1) = \sum_{\mathrm{A}}^{\mathrm{B}} \{RGC(t)/[1.0 + 0.01 * CC(t = T_1)]^{t=T_1}\}$$

where

$$A = t = T_1 + STARTG(t = T_1)$$
$$B = T_1 + STARTG[t = T_1 + ENDG(t = T_1)]$$

$\underline{STARTG(t = T_1)}$ = time (year) of start of launching government mass as committed in year T_1 and $STARTG > T_1$
$\underline{ENDG(t = T_1)}$ = number of years covered by the government commitment that is made at time $t = T_1$
$\underline{CC(t = T_1)}$ = cost of capital (debt financing) at the time of obtaining financing

The amount of debt financing available as a result of the government commitment $DFGC(t = T_1)$ made at $t = T_1$ can be obtained from the following:

$$DFGC(t = T_1) = PVGC(t = T_1) * \delta G(t = T_1) * 0.01$$

where

$\underline{\delta G(t = T_1)}$ = percent of present value of the government commitment that will be lent (debt) and will depend upon the form of the government commitment (i.e., with/without termination liability)

In addition to an initial government commitment made at $t = T_1$, there might be other launch commitments made after $t = T_1$. In that case, the form of the preceding equations is appropriate, but the time frame must be suitably modified (i.e., replace all T_1 with the appropriate values of T_i).

When $t \geq T_2$,

Debt financing is also possible based upon the asset value of a continuing commercial commitment [The following equations assume a single commercial commitment at T_2 to launch $MC(t = T_2)$ kg per year with T_2 and $MC(t = T_2)$ specified via input data.] for use of the RLV. Therefore the revenue $RCC(t)$ resulting from the commercial commitment is

$$RCC(t) = MC(t = T_2) * PC(t)$$

where

$\underline{MC(t)}$ = commercial commitment to launch mass (kg/yr) in year t

The present value of revenue resulting from the commercial commitment (PVCC) is given by

$$PVCC(t = T_2) = \sum_{A}^{B} \{RCC(t)/[1.0 + 0.01 * CC(t = T_2)]^{t=T_2}\}$$

where

$A = t = T_2 + STARTC(t = T_2)$

$B = T_1 + STARTC(t = T_2) + ENDC(t = T_2)$

$\underline{STARTC(t = T_2)}$ = time (year) of start of launching commercial mass as committed in year T_2

$\underline{ENDC(t = T_2)}$ = number of years covered by the commercial commitment that is made at time $t = T_2$

The amount of debt financing available as a result of commercial commitments to launch payloads on the RLV, $DFCC(t = T_2)$, made at $t = T_2$ can be obtained from the following:

$$DFCC(t = T_2) = PVCC(t = T_2) * \delta C(t = T_2) * 0.01$$

where

$\underline{\delta C(t = T_2)}$ = percent of present value of the commercial commitment that will be lent (debt) and will depend upon the form of the commercial commitment (i.e.,with/without termination liability)

In addition to an initial commercial commitment for use of the RLV, there might be other launch commitments made after $t = T_2$. In that case T_2 in the preceding equations is replaced by the appropriate values of T_i.

It is assumed that prior to T_1, financing is through corporate investment. After T_1 it is assumed that debt financing will take place as long as the sum of available debt financing plus corporate investment (made prior to $t = T_1$) exceeds the indebtedness. When this is no longer possible, either additional corporate investment must be made or equity financing is necessary. Therefore, at time t if the sum of the corporate investment plus the debt resulting from both a government and commercial commitment to utilize the RLV exceeds the cumulative cash flow

$$\text{If } \sum_{x=1}^{x=t-1} CI(x) + \sum_{x=1}^{x=t} \{DFGC(x) + DFCC(x)\} \geq -\sum_{x=1}^{x=t} CF(x)$$

Then

$$CI(t) = 0$$
$$EF(t) = 0$$

$$\text{If } \sum_{x=1}^{x=t-1} CI(x) + \sum_{x=1}^{x=t} \{DFGC(x) + DFCC(x)\}$$
$$< -\sum_{x=1}^{x=t} CF(x) \textbf{ and } \text{EQUITY} = \text{“Yes”}$$

Then

$$CI(t) = 0$$

$$EF(t) = \sum_{x=1}^{x=t-1} CI(x) + \sum_{x=1}^{x=t} \{DFGC(x) + DFCC(x)\} + \sum_{x=1}^{x=t} CF(x)$$

$$EQSI(t) = \text{Equity Share of Investors}(t)$$

$$= EQA(t) * EF(t) / \sum_{x=1}^{x=t-1} \{CI(x) + EF(x)\}$$

[The calculation of equity share has been kept relatively simple but tries to capture the essence of the problem while providing insights into the implications of financing alternatives on financial performance. Ultimately it might be possible to relate equity share to perceived (future) financial performance and the rate of return required by investors when faced with the perceived performance.]

$$CEQSI(t) = \sum_{x=1}^{x=t} EQSI(t) * 100.0$$

$$CEQSC(t) = 100.0 - CEQSI(t)$$

The following definitions apply for the preceding equations and following paragraphs:

$EF(t)$ = equity financing amount ($) in year t
$CF(t)$ = annual cash flow in year t prior to the financing provided in year t
EQUITY = switch that allows for equity financing when equal to "Yes"
$EQA(t)$ = constant in the range of 0 to 1.0, which adjusts the equity share provided in return for financing
$CEQSI(t)$ = cumulative equity share obtained by investor in exchange for equity financing
$CEQSC(t)$ = cumulative equity share maintained by corporate investor

$$\text{If } \sum_{x=1}^{x=t-1} CI(x) + \sum_{x=1}^{x=1} \{DFGC(x) + DFCC(x)\} < -\sum_{x=1}^{x=t} CF(x) \text{ and EQUITY} \neq \text{“Yes”}$$

Then

$$CI(t) = \sum_{x=1}^{x=t-1} CI(x) + \sum_{x=1}^{x=t} \{DFGC(x) + DFCC(x)\} + \sum_{x=1}^{x=t} CF(x)$$

$$EF(t) = 0$$

$$EQSI(t) = 0$$

$$CEQSI(t) = 0$$

$$CEQSC(t) = 100.0$$

The preceding paragraphs consider the financing or funding side of the equation. The following paragraphs consider adjustments to the financial analysis to reflect debt repayment and the development of financial performance measures that take into account the debt and equity financing. It is assumed that debt repayment is proportional to launched government mass and launched commercial mass. It is necessary to treat these separately because different loan percentages [i.e., $\delta G(t)$ and $\delta C(t)$] are likely.

$MGF(t)$ = government mass delivered to orbit (flown) in year t
$MCF(t)$ = commercial mass delivered to orbit (flown) in year t
$DRG(t)$ = debt repayment for government mass delivery in year t
$DRC(t)$ = debt repayment for commercial mass delivery in year t

$$MGF(t) = \sum_{x=1}^{t} MG(x) * KG(x)$$

$$MCF(t) = \sum_{x=1}^{t} MC(x) * KC(x)$$

where

$$KG(x) = 1 \text{ \underline{if} } STARTG(x) \leq x \leq ENDG(x) + STARTG(x)$$
$$KG(x) = 0 \text{ \underline{if} } STARTG(x) > x > ENDG(x) + STARTG(x)$$
$$KC(x) = 1 \text{ \underline{if} } STARTC(x) \leq x \leq ENDC(x) + STARTC(x)$$
$$KC(x) = 0 \text{ \underline{if} } STARTC(x) > x > ENDC(x) + STARTC(x)$$

$$DRG(t) = PG(t) * \sum_{x=1}^{t} MG(x) * KG(x) * \delta G(x) * 0.01$$

$$DRC(t) = PC(t) * \sum_{x=1}^{t} MC(x) * KC(x) * \delta C(x) * 0.01$$

The preceding equations establish the debt repayment for the financing obtained as a result of both government and commercial commitments to have mass placed into orbit via the RLV. It is assumed that this debt will be repaid with interest paid at the time of the repayment (i.e., a balloon payment) but not at other times while the loans are outstanding.

$IRG(t)$ = interest payment on debt for government mass delivery in year t
$IRC(t)$ = interest payment on debt for commercial mass delivery in year t

$$IRG(t) = PG(t) * \sum_{x=1}^{t} [MG(x) * KG(x) * \delta G(x) * 0.01 * [1.0 + 0.01 * CC(x)]^{x+ENDG(x)-1} - DRG(t)]$$

$$IRC(t) = PC(t) * \sum_{x=1}^{t} [MC(x) * KC(x) * \delta C(x) * 0.01 * [1.0 + 0.01 * CC(x)]^{x+ENDC(x)-1} - DRC(t)]$$

The preceding equations can be used for establishing the adjustments to annual cash flow and profit computations to account for the debt and equity financing and the repayment of debt. A further adjustment must be made in the calculation of present value of cash flow and rate of return to account for equity financing. This can be accomplished as follows where the present value of the cash flow of the RLV business as seen by the corporate investor (PVCFCI) and by the equity investor (PVCFEI) are given by

$$PVCFCI = \sum_{x=1}^{t=\text{YEARS}} CF(x) * [1.0 + 0.01 * CC]^{-x} * CEQSC(x)$$

$$PVCFEI = \sum_{x=1}^{t=\text{YEARS}} [CF(x) + EF(x)] * [1.0 + 0.01 * CC]^{-x} * CEQSC(x)$$

where

YEARS = business planning horizon (years)
CC = corporate investor's cost of capital (This cost of capital is the overall cost of the corporate investor including the cost of the corporate investment and the debt financing.)

The internal rates of return of the business (corporate investor) and the equity partner can be determined from the preceding PVCFCI and PVCFEI equations.

H. Postscript: RLV Financial Modeling Concerns

The RLV financial simulation modeling performed by the government (i.e., MSFC)[22] has been conducted primarily within a spreadsheet environment with the resulting spreadsheet models incorporated within existing commercial risk analysis software (i.e., at risk) so as to provide a risk analysis capability. This approach has several significant limitations that can become serious as attempts are made to expand upon the existing capabilities. These limitations include the following:

1) The use of spreadsheet models implies a "deterministic mentality." That is, uncertainty and unreliability, and their consequences, are not normally considered in the development of spreadsheet models and, as a consequence, the necessary mathematical structure that is required for a valid risk analysis is frequently not developed. Wrapping a risk model around a spreadsheet does not add the structure with the result that performed risk analyses might be significantly limited. For example, the logic structure for simulating launch success/failure/recovery operations has not been incorporated; thus, the reliability characteristics of the RLV vs other launch vehicles cannot be considered, nor can the effect of RLV failures and recovery be taken into account in establishing business risk perspectives and required fleet size.
2) As mathematical model complexity increases, the ability to utilize spreadsheets quickly diminishes. (Spreadsheets were not developed to efficiently cope with complex mathematical formulations.) The MSFC-developed RLV financial spreadsheet model does not currently model pricing policies and financial investment scenarios. (These were elaborated upon in the preceding paragraphs.) As the need arises to

include these and other considerations (for example, uncertainty in the timing of major events such as the completion of first article, financing, delays that might result from failures or partial failures, and scheduling or throughput uncertainties), nonspreadsheet models will have to be considered with the use of spreadsheets perhaps limited to performing user-friendly input/output functions.

3) The inability to consider the implications of government programs and policies on business risk eliminates the possibility of comparing government programs and policies that impact business risk with those that primarily affect expected rates of return. For example, loan guarantees shift technology risk from industry to government, and there is no mechanism in the current model framework to account for this. Another example is the government entering into an anchor tenancy agreement. This reduces market uncertainty (hence reduces business risk) while having the ability to affect the overall financial structure. (Ability to obtain debt financing is increased.) The current spreadsheet modeling approach is not well suited to handling this situation. In other words, again the mathematical structure has not been provided wherein many areas involving uncertainty and risk can be investigated via the current risk modeling approach.

The inclusion of the preceding type of considerations would imply a different form of simulation model wherein launches are simulated with all failure/recovery/delay situations considered as they develop as a consequence of attempting to fly off manifests, which in themselves can be characterized as uncertain. These operational models would of necessity be imbedded within the financial simulation model. Typical examples of this type of modeling are *DOMSAT* and *SATCAV* and are described in the Sec. I of Chapter 4. In short, because of the increasing complexity of the financial planning model that is required to accommodate pricing policies, financing schema, and the need for flexibility to consider a broad range of government actions, it is necessary to seriously consider the development of an RLV financial planning model in a more formal mathematical setting (for example, FORTRAN or BASIC) rather than in a spreadsheet format.

IV. RLV Insurance Considerations[23]

Future deliveries of payloads to orbit are likely to utilize space transportation services that employ a fleet of reusable launch vehicles. To achieve this capability will require the investment of billions of dollars. Such sizable investments will require the formulation of detailed business plans that demonstrate the financial viability of the proposed business and that the business can function within an acceptable level of risk. Besides coping with the normal business risks associated with achievable level of technical performance, demand for launch services, competition, etc., the business must also cope with the risks associated with transportation system failures and delays. This requires the development of a viable risk management plan. Because risk management involves insurance (including the self-insurance option), the risk management plan must demonstrate that required levels of insurance cover (for both launch failures and third-party damage) will be available at affordable prices. This task is complicated by the relatively long delay between investment decisions and first operational flights and uncertainty with respect to RLV reliability. In addition, the level of damage that might be caused to third parties during both launch and recovery operations is uncertain as are the potential financial responsibility requirements that might be imposed by the FAA and the possibility and magnitude of government third-party loss indemnification. Financial responsibility and government indemnification are discussed in Sec. II of Chapter 5. It is unlikely that the current insurance industry will provide the necessary assurances with respect to insurance availability and pricing that will be required for RLV investment decisions.

The following discussion is concerned with describing a risk management scenario that can demonstrate, at the time that investment decisions must be made, that the necessary insurance cover at affordable prices will be available, over the long term, for operational missions. This discussion does not consider the provision of insurance cover for R&D flights required to demonstrate the achievement of an operational capability. Levels of insurance cover must be sufficient to recover from the loss of a billion-dollar asset with insurance priced so as to be but a small part of the price of a launch where the price of a launch is likely to be in the 50 to 100 million-dollar range. In addition, sufficient third-party coverage must be available to consider "maximum probable" third-party damage. (The concept of maximum probable loss, (MPL) and its use are described in Sec. II of Chapter 5.) Thus, the cost of insurance for an RLV must be comparable to or less than the cost of insuring an ELV to low Earth orbit. The considered risk management business structure involves three parties: the transportation provider, the insurance business, and the government, with risk sharing between all parties and appropriate incentives such that financial performance will be related to actual achieved transportation system performance.

A. Background

The space insurance industry has had its ups and downs. (This is discussed in Sec. IV of Chapter 5.) In the late 1970s insurance rates were less than 10% of the launch and the payload cost. During the 1980s, these rates sometimes significantly exceeded 20% (on occasion approaching 30%) of the launch and payload cost. Unrealistically low premiums were in part to blame for substantial space insurance industry losses in the early 1980s, and the resulting extremely high premiums coupled with a lack of long-term commitments created an insurance crisis in the mid-1980s. Some firms at this time (including Intelsat and RCA) elected to self-insure, rather than pay insurance premiums in excess of one-quarter of the value of their payloads and launch cost. Current launch insurance premiums (for geosynchronous communication satellites) are in the range of 13 to 18% and vary from launch vehicle to launch vehicle. The insurance capacity (the amount that insurers are willing to place at risk) is on the order of $500 to $800 million for launch (including payload) insurance with similar amounts being available for third-party coverage. The combined effect of the premium charge policy, relatively small number of launch events, difficulty of reliability assessments, and the competitiveness of the insurance business, have contributed to rather lackluster performance of the space insurance industry.

The space insurance industry is comprised of insurance brokers and insurance underwriters. Brokers in response to the demands of the space industry aggregate funds from underwriters and bids are made on a case-by-case basis. There has been a trend to reduce the risks associated with insuring small numbers to insuring groups of launches and has been extended to providing coverage for a group of launches provided by several different launch vehicles. The package arrangements have been for providing insurance at a specified premium for each individual launch of a group of insured launches as well as for providing deductible arrangements (i.e., if more than a specified number of failures occur then payouts are provided).

The pricing of insurance is directly related to launch-vehicle reliability perceptions. (Historically, because of frequent modifications to launch vehicles coupled with relatively low flight rates and concern with recent flight results calculated reliability measures lose significance, and perceptions of reliability are used in lieu of analytically derived measures. This is likely to continue with the introduction of RLVs, which will take time to establish a reasonable basis upon which to base reliability calculations.) A major problem of the insurance industry thus relates to the assessment of launch vehicle reliability. Estimating reliability higher than actual can result in significant financial loss, whereas estimating reliability lower that actual can result in significant financial gain. The latter situation, although desirable from the insurance industry's point of view, cannot be maintained for a significant period of time because of the competitive nature of the industry. In theory, in the long run insurance prices (expressed as a percentage of insured asset value) should be equal to failure rate plus overhead rate plus fee. Because statistics are limited, estimation of a failure rate for pricing purposes is difficult. For launch vehicles that have a significant history, the recent success rate (as well as the overall success rate) is considered together with engineering reviews of test data. This is complicated when a failure occurs and then considers the details of the cause of failure reviews and engineering review of corrective actions. Depending upon the form of the insurance contract, there might or may not be an increase in premiums. In the case of a new launch vehicle, considerable reliance must be placed on engineering review and analysis considering test data and inheritance from previous launch vehicles. In the past, for expendable launch vehicles it was necessary to develop confidence in the reliability regime of 0.9 to 0.99. For the reusable launch vehicle it will be necessary to develop confidence in the reliability regime of 0.99 to 0.999.

B. RLV Insurance Considerations

The space insurance industry normally will not make availability and premium commitments more than two or three years in advance of a launch. It is likely that the planning of an RLV business venture requires that there be an ensured mechanism for satisfying the RLV business venture insurance needs over the long term at the time of making commitments to invest in the development of a transportation business based upon the use of a fleet of RLVs. This might be on the order of 5 to 10 years (or possibly more) before the initiation of operational launch services. It is unlikely that the current space insurance industry will be willing to make the availability and pricing commitments that are required to satisfy the risk averseness of the business planners and subsequent investors. The necessary insurance might be available through self-insurance. However, the loss of an RLV can result in a replacement cost of \$500 to \$1000 million. This is comparable to the level of third-party indemnification currently provided (for ELVs) by the U.S. Government. [For licensed launches the U.S. Government will (if Congress authorizes the funds) cover third-party losses that are in excess of financial responsibility (insurance) requirements set by the Federal Aviation Administration (FAA) up to an amount of \$1.5 billion above this level. This is discussed in detail in Sec. II of Chapter 5.] Because third-party risk (the probability of third-parties sustaining loss) is much less than the risk of loosing a launch vehicle, willingness to self-insure against RLV loss could be construed as not really needing the third-party indemnification for providing space transportation services. Thus, self-insurance might not be a viable option.

RLV replacement cost can be in the range of \$500 to \$1000 million (per vehicle) with a goal of achieving a price (to LEO) of \$50 to \$100 million. Currently, insurance rates for placing payloads into LEO are about 6% (for a Delta ELV). At 6%, insurance for an RLV would be on the order of \$30 to \$60 million—clearly not affordable if prices are to be in the \$50 to \$100 million range. To be comparable to ELV "affordability," insurance rates must be in the neighborhood of 1 to 2%. This may ultimately come about when launch reliability is demonstrated and believed to be in excess of 0.99. But, if insurers are left to their own devices what will be the insurance rates until 0.99+ becomes believable?

The failure/recovery paths of an ELV and an RLV are significantly different and add to the complexity of establishing RLV insurance rates. This is illustrated in Figs. 3.8 and 3.9. (Third-party damage is not explicitly considered but can be considered as adding additional branches to the tree structure, for example, different payouts with different probabilities for each indicated "failure.") Whereas an ELV launch can succeed or fail, an RLV launch has other possible outcomes between complete success and complete failure with each outcome having a

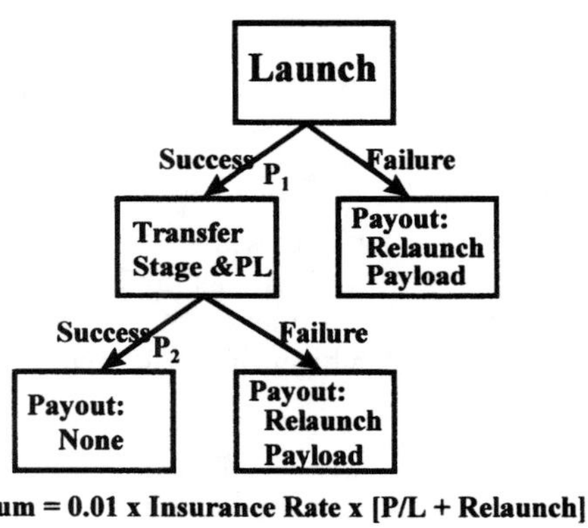

Fig. 3.8 ELV failure/recovery paths.

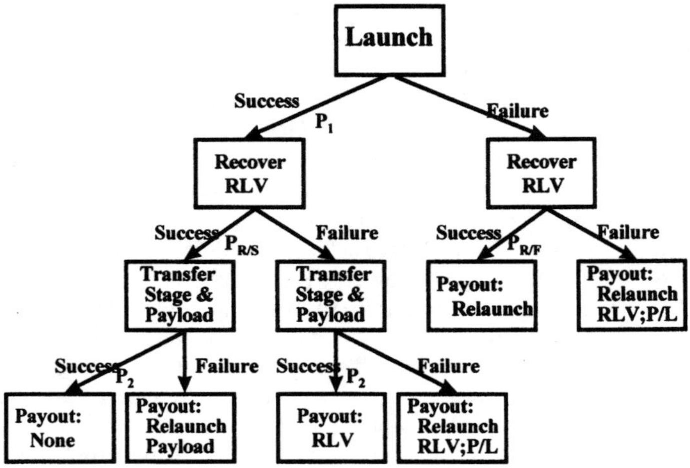

Fig. 3.9 RLV failure/recovery paths.

different probability of occurrence and different cost to recover from the failure. The probability distributions of ELV and RLV payouts are illustrated in Fig. 3.10 with the expected insurance payouts established as follows:

$$\text{ELV Expected Payout} = [1 - P_1 \times P_2] \times [\text{Payload Cost} + \text{Relaunch Cost}]$$

$$\begin{aligned}\text{RLV Expected Payout} = {} & P_1 \times P_2 \times [1.0 - P_{R/S}] \times [\text{RLV Replacement Cost}] \\ & + [1.0 - P_1] \times P_{R/F} \times [\text{RLV Relaunch Cost}] \\ & + P_1 \times [1.0 - P_2] \times P_{R/S} \times [\text{Payload Cost} \\ & + \text{RLV Replacement}] + [1.0 - P_1 \times [P_{R/S} \\ & + P_2 \times [1.0 + P_{R/S}]] - P_{R/F} \times [1.0 - P_1]] \\ & \times [\text{Payload Cost} + \text{RLV Relaunch Cost} \\ & + \text{RLV Replacement}]\end{aligned}$$

In the preceding equations all of the P terms are as defined in Figs. 3.8 and 3.9.

C. Business Concept

The ability to currently establish, with a high degree of confidence, that affordable insurance will be available as and when required requires the consideration of a rather specific business structure. This business structure and related business concerns is the subject of the following paragraphs.

1. Business Considerations

Risk is related to the chance that the value of a metric will exceed specified values, for example, the probability that return on investment (ROI), will exceed specified values. In many cases it is reasonable to characterize risk in terms of the

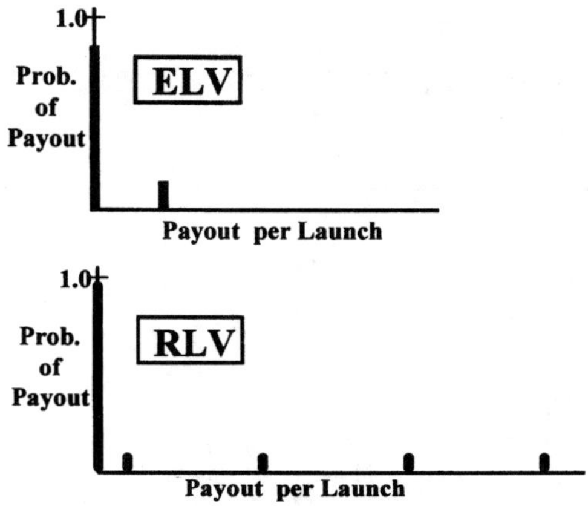

Fig. 3.10 Probability distribution of ELV and RLV payouts.

standard deviation of the metric under consideration where the smaller the standard deviation the lower the risk (i.e., the less variability there is relative to the expected value of the metric). Table 3.1, utilizing typical ELV numbers, illustrates the need for insuring large number of events. It can be seen that the risk, as indicated by the standard deviation of profit, is large relative to the expected profit until hundreds of launches are considered. Because an RLV must achieve reliability well in excess of 0.9, the situation will not be as bad as indicated in Table 3.1 but will still require that large numbers of launches be insured in order to achieve reasonable risk levels.

The problem is thus to be able to insure a large number of launches with the knowledge that the insurance cover will be available at an affordable price. To accomplish this requires the creation of a business venture for insuring RLV transportation services over an extended period of time where the 1) RLV transportation provider commits (i.e., guarantees) to insuring a large number of launches, 2) insurance venture commits to an insurance price or algorithm for

Table 3.1 Need for insuring large numbers of events[a]

Number of insured launches	Revenue, M$	Expected payout, M$	Profit, M$	
			Expected	Std. Dev.
1	15	10	5	30
5	75	50	25	67
10	150	100	50	95
100	1500	1000	500	300
200	3000	2000	1000	424
400	6000	4000	2000	600

[a]Reliability = 0.90. Insurance coverage = 100 M$. Premium = 15%.

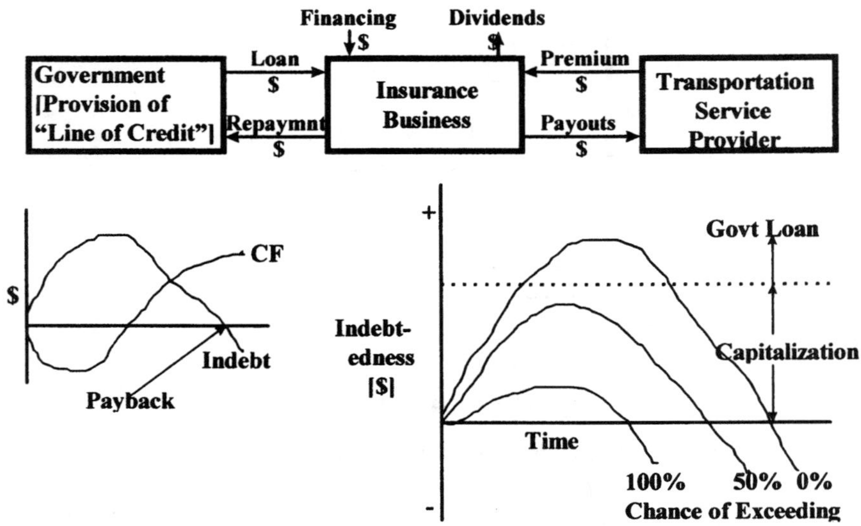

Fig. 3.11 Generic business structure.

establishing insurance price, and 3) U.S. Government (or other organization) acts as a guarantor covering (via loan) excessive levels of risk.

In other words, the three parties, the RLV transportation provider, the insurance venture, and the government, all share in the risk. The basic concept of a viable approach for providing long-term affordable insurance cover for RLVs is illustrated in Fig. 3.11. The transportation service provider guarantees, in essence, that it intends to stay in the RLV launch service business and will insure a specified large number of launches over an extended period of time according to an agreed upon price or pricing algorithm based upon actual flight experience. The insurance venture obtains financing and pays dividends with the dividends providing investors with a reasonable return on their investment. In the event that failures occur early prior to premium receipts and initial capitalization being adequate to provide the necessary payouts, the government would provide either a line of credit or guarantee loans to be repaid by future premium receipts. The level of capitalization will depend on a number of factors including the level of risk acceptable by the government. This concept is also illustrated in Fig. 3.11. The analysis of such a business venture would most likely utilize Monte Carlo simulation techniques[24] that would develop information such as that illustrated in Fig. 3.12. The simulation analysis would be required to establish a detailed understanding of this insurance business structure and the ramifications upon the transportation provider and the government.

2. *Insurance Product Pricing*

A pricing structure or pricing algorithm must be agreed to for all of the guaranteed space launches. Because, as indicated in Fig. 3.9, there are multiple possible outcomes involving different probabilities of occurrence, multiple

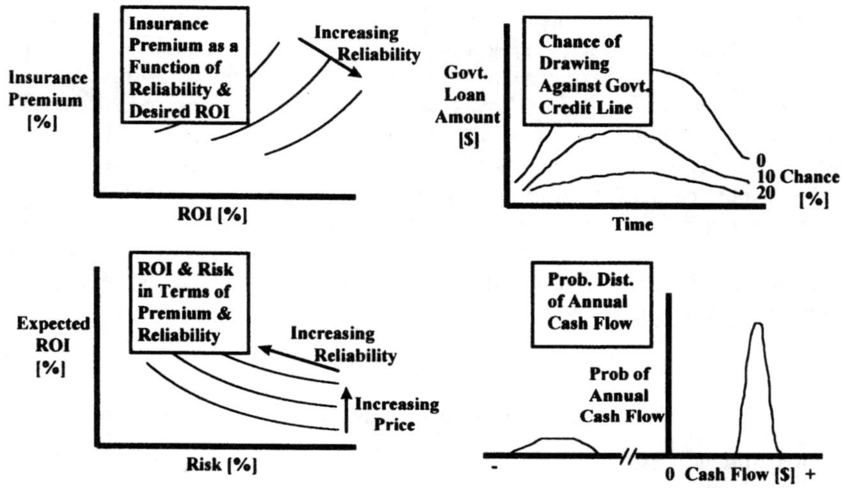

Fig. 3.12 Representative results of financial simulation analysis.

probability of success (reliabilities) must be considered. It is probably not adequate to agree to a single price for all launches because this does not provide an incentive for the RLV partner to continuously seek to improve reliability. A more appropriate approach would be to relate pricing to actual events, i.e., a Bayesian approach. The Bayesian approach (described in more detail in the Appendix to this chapter) considers that the reliability of a launch vehicle is not known with certainty, and each launch can be thought of as an experiment with measured outcomes (i.e., success or failure). The experiment result provides new information, that is, information that was not available at the time that the prior reliability estimate was made and that this new information can be used to update the prior estimate. Thus, a pricing algorithm (expressed as a mathematical equation) would require agreement relating to the initial estimate of the reliability and the form of updating given success or failure outcomes. Once this is established, the insurance price would be a function of actual observed performance with the current price equal to the prior price as modified (as per the algorithm) by the prior launch outcome. Poor performance leads to higher insurance rates, and performance improvements lead to lower insurance rates.

D. Summary

Investment decisions relating to the development and provision of space transportation services based upon a fleet of RLVs must be made in advance of assurances by the insurance industry that required insurance cover will be available at affordable rates. Because it is unlikely that the required investments will be made without an assurance that affordable insurance will be available, an insurance business scenario has been developed, which is likely to lead to the conclusion that affordable insurance will be available as and when required. The scenario is based upon the transportation provider guaranteeing a large number of insured flights, the insurance business agreeing to a specified insurance rate or pricing algorithm (based upon a Bayesian approach), and the government (or other organization) providing loans (or loan guarantees) when payouts exceed resources. Before drawing firm conclusions with respect to the viability of this concept, analyses must be performed to understand the relationships of flight guarantees, reliability perceptions, pricing policies, and other factors on insurance rates and the viability of the insurance business. It is likely that the analyses would utilize financial risk analysis and Monte Carlo simulation modeling techniques as described in *Space Economics.*[25]

V. Appendix: Algorithm for Dynamic Insurance Pricing

This Appendix is based upon notes provided by Dr. G.A. Hazelrigg, Jr., of the National Science Foundation.

The following paragraphs describe a Bayesian approach for dynamic insurance pricing. It is assumed that insurance price is directly related to an estimation of the probability of success. It is also assumed that the insurance rate would be set initially based upon an expectation of the probability of success and subsequently updated by evidence generated by launch experience. For example, a long string of successful launches would provide confidence in a high probability of success, whereas a failure or two would greatly lower the confidence in future successes. The mathematics given in the following paragraphs presents an approach, based on the theory of probability, to the computation of updated expectations.

In the phrase "the concept of updating prior estimates," the word "prior" is used to denote an estimate that is made before the conduct of an "experiment," and an experiment is anything for which there is a measurable outcome that cannot be known in advance of the experiment. For example, a launch can be thought of as an experiment with measured outcomes being success or failure. It is intuitive that the results of the experiment provide new information, that is, information that was not available at the time that the prior estimate was made and that this new information should be useful in updating the prior estimate. The updating of a prior estimate with the measured results of an experiment is a problem in probability theory that is solved using Bayes formula. Bayes formula is written as:

$$P(S_i|A) = \frac{P(S_i)P(A|S_i)}{\sum_{j=1}^{K} P(S_j)P(A|S_j)}$$

where S is a state of nature for which an estimate is desired. For example, S could be an estimate of the probability of success on any given launch. In the language of probability theory, the state of nature can be thought of as the actual probability of success. This probability cannot be known precisely. Rather it can only be estimated through subjective judgment, rational computation, and experimentation. The S_i are different possible values of S. The probabilities $P(S_i)$ comprise the prior on S. That is, they comprise the prior estimate of all possible values of S, namely all values that have nonzero probability, combined with the probabilities that each value is the actual state of nature. This equation says that the probability of any state of nature S_i updated by an experiment whose outcome is A is the product of the prior probability of S_i and the probability that A occurs given that the state of nature is S_i, divided by the total probability that A occurs.

Notice from Bayes formula that a prior estimate of a state of nature updates with the result of an experiment to a different estimate only if the original estimate is not known with certainty. For example, suppose that S is the probability that a coin will land heads on any given toss and, from years of experience and other evidence, there is a convincing argument that S is 0.5. That is, it is believed that $P(S = 0.5) = 1$ and $P(S \neq 0.5) = 0$. Now, regardless of evidence supplied by any additional number of flips the estimate of S remains 0.5, and the interpretation of a string of say 50 successive heads would be simply, "that's life." Given that $P(S = 0.5) = 1$, it is certainly possible, although highly unlikely, that 50 successive flips could be heads. Because it is "known" that $P(S = 0.5) = 1$, the string of heads is interpreted as simply an unlikely event that actually happened. Likewise, if it were asserted that the probability of a successful launch is 0.99 with probability 1, then a string of five failures would, of necessity, be interpreted merely as unlikely bad luck.

It will be helpful to generalize the Bayes formula just given. The probability of success for a launch vehicle on a given launch is a continuous variable. That is, it can take on any value between 0 and 1. In accordance with this, discrete distributions, of the form $P(S_i)$ are considered as continuous functions, that is, of the probability density function $f(S)$, and accordingly Bayes formula is written as

$$f(S|A) = \frac{f(S)P(A|S)}{\int_0^1 f(s)P(A|s)\,\mathrm{d}s}$$

A. Initial Representation of the Probability of Success

The initial representation of the probability of success should be reflective of the subjective opinion (albeit based on all available evidence) of the relevant decision maker. Note that it would be perfectly natural and not in violation of any mathematic for estimates of success made by the launch vehicle and the insurer to differ. It is convenient to represent the initial probability of success in the form of an equation. A useful form for this purpose is the beta distribution. The beta distribution is given by the formula

$$f(S) = \frac{\Gamma(\alpha + \beta)}{\Gamma(\alpha)\Gamma(\beta)} S^{(\alpha-1)}(1 - S)^{(\beta-1)}$$

where α and β are arbitrary positive numbers. $\Gamma(i)$ is called the gamma function, and the integer i is given by

$$\Gamma(i) = (i - 1)!$$

Figures 3.13 and 3.14 illustrate typical beta functions. In the following paragraphs it is assumed that the decision maker finds the beta distribution acceptable and chooses values of α and β to satisfy his/her subjective opinion.

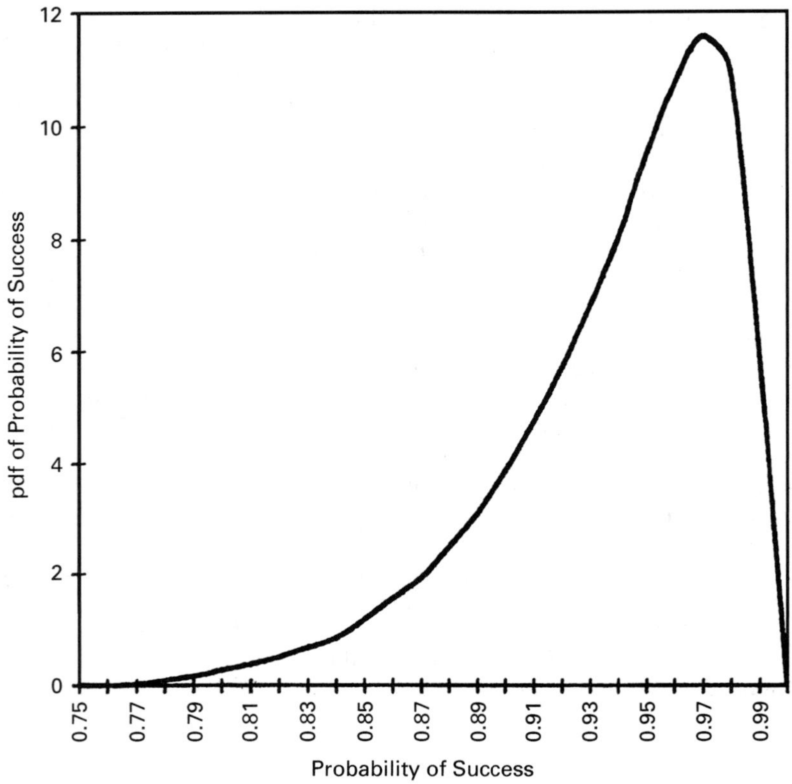

Fig. 3.13 Beta distribution with most likely probability of success equal to 0.968.

B. Updating a Prior Estimate of the Probability of Launch Success

The starting point is with a prior estimate such as that given by the preceding beta distribution. Note that, given this probability distribution of the probability of launch success, the expected loss is given by

$$E\{\text{Loss}\} = \int_0^1 V[1 - P(S)]\, f(S)\, \mathrm{d}S$$

where V is the insured value (the payout given a failure) and $[1 - P(S)]$ is the probability of a failure as a function of S. The expected loss can be considerably higher than the insured value times the most likely probability of failure. An initial insurance premium might be on the order of twice the expected loss as just given.

Each launch provides additional experience upon which an updated estimate of the probability of launch success can be based. In the following it is assumed that there are only two possible outcomes of a launch, success and failure, and that an updated probability of launch success is based on these outcomes.

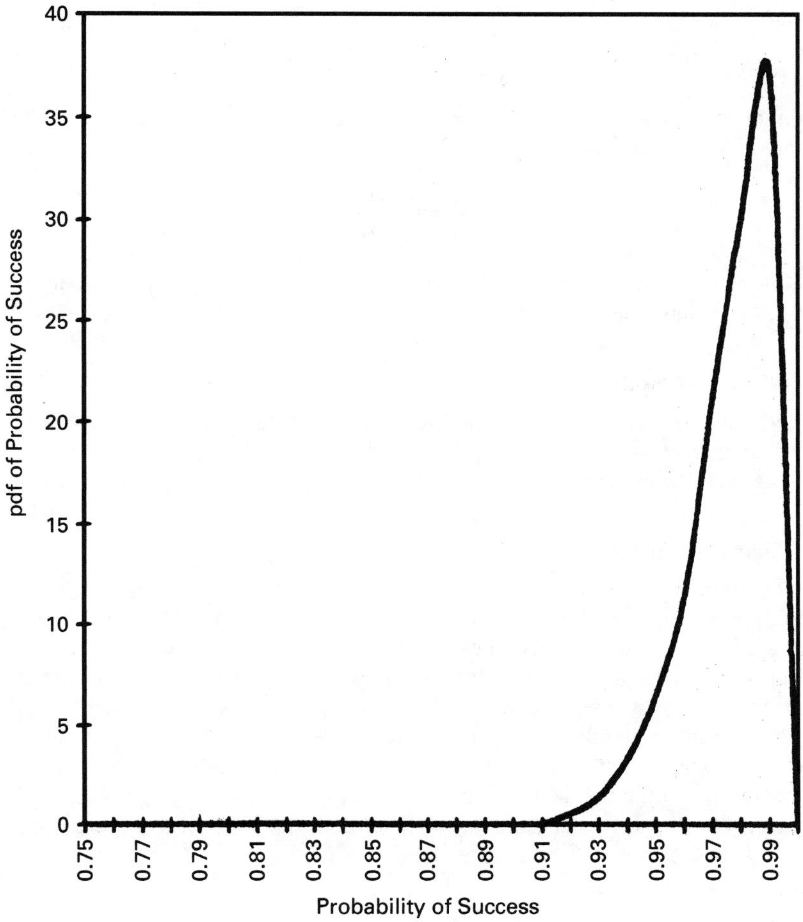

Fig. 3.14 Beta distribution with most likely probability of success equal to 0.988.

1. Success

Given that a launch is successful, the probability-of-success estimate can be updated by the following:

$$f(S|A) = \frac{f(S)}{\int_0^1 f(s)s\,\mathrm{d}s}$$

Note that $P(A|S) = S$, that is, the probability of a successful launch given that the probability of success is S is simply S.

2. Failure

Given that a launch is a failure, the probability-of-success estimate can be updated by the following:

$$f(S|A) = \frac{f(S)(1 - S)}{\int_0^1 f(s)s\,\mathrm{d}s}$$

Note that $P(A|S) = 1 - S$ is the probability of a failure given a probability of success is S and is simply $1 - S$.

C. Repeated Updates

Each update obtained by the preceding formula then becomes the new prior (prior to the next experiment) $f(S)$, and the preceding equations can be used recursively to continue to update $f(S)$ upon the outcome of each successive launch.

D. Word of Caution

If a Bayesian algorithm, such as that just given, is used as the basis for computing an updated probability of launch success and subsequent insurance rates are based upon the updated estimates, there is no guarantee that the insurers will ever achieve a positive expected cash flow. Suppose that the initial estimate of the probability of success is considerably optimistic, that is, the true state of nature is such that the probability of success is considerably lower than the estimate, then Bayesian updates might never reach an estimated probability of success that is representative of the true state of nature. The most that can be said is that it can be expected that the Bayesian estimate will tend, in the limit of a large number of launches, to the true state of nature.

References

[1]Hamaker, J., Taylor, D., and Shaw, E., "RLV Business Planning," NASA, George Washington RLV Financing Workshop, 1996.

[2]Layton, J.P., and Grey, J., *New Space Transportation Systems: An AIAA Assessment*, AIAA, New York, 9 Jan. 1973.

[3]Launius, R.D., "NASA and the Decision to Build the Space Shuttle, 1969–72," *The Historian*, Spring 1994.

[4]Trento, J.J., *Prescription for Disaster*, Crown Publishers, Inc., New York, 1987.

[5]Logsdon, J.M., "The Decision to Develop the Space Shuttle," *Space Policy*, Vol. 2, No. 2, May 1986, pp. 103–119.

[6]"Economic Analysis of the Space Shuttle System," Mathematica, 31 Jan. 1972.

[7]"Payload Effects Analysis—Final Reports Summary," Lockheed Missiles and Space Co., Rept. LMSC-A990594, June 1971. *Also* "Design Guide for Space Shuttle Low-Cost Payloads," Lockheed Missiles and Space Co., Rept. LMSC-A990558, June 1971.

[8]"Integrated Operations/Payload/Fleet Analysis Final Report," NASA, Aug. 1971.

[9]Low, G. to Myers, D., "Space Shuttle Objectives," George M. Low Collection, NASA Hq., 27 Jan. 1970.

[10]Hazelrigg, G.A., "Engineering Design and Decision Making," *Space Economics*, edited by J.S. Greenberg and H. Hertzfeld, Progress in Astronautics and Aeronautics, Vol. 144, AIAA, Washington, DC, 1992, pp. 381–402.
[11]Greenberg, J.S., "Guaranteeing the Space Business," *Aerospace America*, 1999.
[12]Greenberg, J.S., and Gaelick, C., "When a Launch Goes to the Competition," *Aerospace America*, Aug. 1995, pp. 36–39.
[13]Bureau of the Census, "Quarterly Financial Report for Manufacturing, Mining, and Trade Corporation," *Aerospace Industries Association 1994 Year-End Review and Forecast*, Washington, DC, 1994.
[14]Louscher, D.L., "Domestic Economic Impact of Russian and Chinese Launches of American Satellites on the Atlas Commercial Launch Program, 1993–2002," *Foresight International*, 4 Dec. 1992.
[15]Bajusz, W.D., *Arms Sales and the U.S. Economy, the Impact of Restricting Military Exports*, Westview Press, Boulder, CO, 1988.
[16]"The Private Sector Economic and Employment Benefits to the Nation and to Each State of Proposed FY 1990 NASA Procurement Expenditures," Management Information Services, Inc., Washington, DC, April 1989.
[17]"Average Hourly Earnings in the Aerospace Industry, Guided Missiles and Space Vehicles," Aerospace Industries Association, Washington, DC.
[18]"Non-Agricultural Hourly Wage Rate," U.S. Bureau of Labor Statistics, Washington, DC, Nov. 1994.
[19]"Employer Costs for Employee Compensation in the Aircraft Manufacturing Industry," Aerospace Industries Association, Washington, DC, 1 July 1994.
[20]"Unemployment Situation," U.S. Bureau of Labor Statistics, Washington, DC, Nov. 1994.
[21]Greenberg, J.S., "RLV Pricing Strategies," NASA, July 1997.
[22]Shaw, E.J., Taylor, D.T., and Hamaker, J.W., "RLV Economics: Fiscal Evaluation of NASA's Reusable Launch Vehicle Effort," *Space Policy*, May 1997, pp. 109–124.
[23]Greenberg, J.S., "Insuring RLV Transportation Services," International Academy of Astronautics, Paper 98-IAA.1.2.02, Sept.–Oct. 1998.
[24]Greenberg, J.S., "Financial/Investment Analysis," *Space Economics*, edited by J.S. Greenberg and H.R. Hertzfeld, Progress in Astronautics and Aeronautics, Vol. 144, AIAA, Washington, DC, 1992, pp. 3–33.

Chapter 4

Space Operations

SPACE operations involve the placement of satellites into orbit and the maintenance of services that they provide over a period of time. This has cost, schedule, availability, benefit, profit, cash flow, as well as other ramifications. Economic analysis of space operations is concerned with understanding the implications of both government and industry policy and investment decisions upon the cost, benefit, profit, and other aspects of space operations. The objective of the economic analysis is to provide improved information for making informed policy and investment decisions by understanding the cost, benefit, profit, and other implications of such decisions on resulting space operations.

A number of topics are discussed in the following pages, including a risk-based approach for comparing advanced space transportation architectures, the selection of a launch vehicle for geosynchronous orbit (GEO) communication satellite missions, the selection of launch vehicles for low Earth orbit (LEO) missions (both initial placement and replacement), determination of launch and satellite sparing strategies that involves the consideration of cost risk, life-cycle cost and availability analyses, implications of achieving a launch-on-demand capability on satellite sparing strategies and space transportation supply/demand (resiliency/operability) considerations, on-orbit service and repair, economic implications of orbital debris and remediation policies, and the implications of various government actions on international competitiveness of U.S. space transportation services.

Underlying the economic analysis of all space activities is the consideration of reliability and cost, performance, and market uncertainties. Thus, before proceeding with a discussion of space operations economic analyses two risk-based simulation models that have served as the basis for understanding a broad range of space operations economic issues are described. These are referred to as *DOMSAT* and *SATCAV*. A number of analyses based upon the use of these models are described. Several other models are also described together with their application.

I. Simulation Modeling

A. *SATCAV* Life Cycle Cost and Availability Model

This section discusses the problem of evaluating situations where there is no revenue nor profit, for example, the government need to maintain a system of global positioning or weather satellites. (A detailed discussion of the *SATCAV* Model is presented in Ref. 1.) The basic problem is similar to that encountered when evaluating business investment decisions with the exception that there are no revenues nor taxes. Thus, cash flow has only a negative component, that is, cash outflow or costs, and the objective is thus the minimization of life-cycle costs within performance constraints. Frequently the performance constraints can be simply expressed in terms of desired availability of sensors or transponders where availability represents the chance, or fraction of the time, that at least N out of M sensors or transponders are operational when required. The objective is to select the minimum life-cycle cost approach that meets the availability constraints.

Life-cycle cost is defined as the present value (i.e., discounted to the present) of all current and future costs associated with the mission or system being considered. This encompasses all nonrecurring and recurring costs and their specific timing. For a space mission nonrecurring costs include those costs incurred in satellite research, development, testing, and engineering (RDT&E), specific modifications that must be made to launch vehicles in order to provide compatibility, and the purchase of capital assets such as buildings and other special facilities. Again, for a space mission recurring costs include the cost of all satellite acquisitions (including initial, replacement, and spares), launch costs, engineering, and research and development (R&D) costs incurred in continuing support of the mission, mission control and operations costs, and related administrative and support costs. When there are impacts of the mission in other nonmission areas, the resulting cost impacts must also be considered. For example, if additional costs will be incurred to maintain compatibility of other system/missions then these costs need to be considered. If, on the other hand, other system or mission costs will be eliminated or foregone, then these cost reduction streams must also be considered (as a contra-expense or the equivalent of a revenue stream in the mission cash flow). These other costs can include both nonrecurring and recurring costs.

Life-cycle cost analysis is usually an important part of the process associated with the justification of investments in new technology, missions, and systems. It is used frequently to demonstrate the magnitude of the economic improvement that will result from the investment, taking into account the often faced problem of trading off increased near-term nonrecurring cost with longer-term reductions in annual operating costs. Life-cycle cost analysis frequently plays an important role in the evaluation and comparison of alternative courses of actions aimed at selecting the minimum cost approach for achieving a set of objectives. It is this latter area that is of current concern.

In the space business annual costs, and therefore the present value of life-cycle costs, must be considered as probability distributions. The reasons for this are many but are primarily as a result of uncertainties associated with both nonrecurring and unit recurring cost (in no small part resulting from the use of new technology); less than perfect reliability of launch vehicles,[2,3] failures of which can significantly effect costs and schedules; schedule uncertainty resulting from both failures and necessary rescheduling to recover from failures and to respond to uncontrollable delays (i.e., weather and standdown time required to correct previous failures); and random and wear out failures of operational satellites, which require replacement and therefore affect the number of cost incurring events and their timing.

As will be seen, the analysis of life-cycle costs is further complicated by a myriad of operational alternatives involving sparing and maintenance/repair strategies. In addition to satellite design considerations relating to sparing and redundancy, there are operational considerations such as the use of on-orbit active or dormant spare satellites, ground spares, and the consideration of on-orbit servicing or repair at a transportation node. Because sensor/transponder availability is a function of overall satellite configuration (including redundancy), sparing and maintenance strategies, and transportation system reliability and availability, it is necessary to consider the many complex interrelationships that exist when configuring a satellite and estimating mission life-cycle cost.

In the following paragraphs a number of factors are discussed that require consideration when estimating life-cycle costs. These include random and wear out failures, cost spreading, learning effects, operational analysis, and the explicit consideration of uncertainty and resulting risk. All are concerned with establishing the number of cost-incurring events and their timing and the magnitude of the associated cost. Each of these topics is discussed briefly. These are then followed by a discussion of a life-cycle cost and availability model, *SATCAV*, which builds on these concepts and adds a formal and consistent structure to life-cycle cost analysis. Finally, the general flexibility and applicability of such a model are described together with typical results.

1. Operational Analysis

Space system operational analysis is the quantification of the probability distributions of the number of recurring cost-associated events required to establish and maintain a space mission. These events can include the number of satellites purchased, the number of launches, the number of satellite retrievals or refurbishments, and so on. Each of these events has associated with it a cost that, at least in the planning phase, includes some degree of uncertainty. The random nature of the number of events can be caused by hardware failures, failure to accomplish certain prescribed events, for example, a rendezvous and docking maneuver, or variability in certain wear out phenomena such as running out of station-keeping propellant. Thus, in addition to analyzing the number of events operational analysis must also be concerned with their time of occurrence, which will ultimately affect system availability.

Operational analysis explicitly considers the possibility of failures both in the transportation system and the satellites, the chance of these failures occurring, and the consequences if the failures do indeed occur. More specifically the analysis considers: 1) the number, sequence, and complexity of operations to be performed; 2) the recovery modes, that is, given that a failure has occurred, the possible resulting sequence of events; 3) the probability of successfully performing each of the required operations, both in the success and failure recovery sequences; and 4) scheduling (or lack thereof) of events. The results of an operational analysis in total are referred to as the operational risk.

The transportation system operational analysis includes all of those events associated with the orbital placement, refurbishment, and retrieval of satellites, including the satellite deployment and initial operation. These last two events are included as transportation system events because if failures occur during these operations the recovery modes sometimes involve components of the transportation system.

The first step in performing an operational analysis of a given mission (for example, a satellite placement, a satellite placement and retrieval, etc.) is to establish the success-oriented mission profile. The success-oriented mission profile consists of the nominal mission timeline or sequence of events where each event can be successfully accomplished or unsuccessful. If the event is accomplished successfully, the mission proceeds on to the next event. If the event is unsuccessful, a failure recovery mode must be adopted. Sometimes the failure recovery mode can return the mission to the normal timeline, and sometimes major modifications are necessary in the mission. Some failures, such as the loss of a redundant subsystem, cause mainly inconvenience and possibly a minor cost item; others, such as the loss of a transfer stage and satellite, result in major cost items. It is generally possible and helpful to sketch the resultant success-oriented mission profile in cartoon form as shown by the example of Fig. 4.1. (Note that the reusable system shown can be made into an expendable system by simply setting the recovery probabilities to zero.)

The second step of an operational analysis requires an explicit definition of what constitutes completion of a mission. In a service-type space program, if a satellite is required to provide the service and the mission is to place that satellite in orbit then the satellite must be successfully placed and operational for the mission to be complete. If a particular flight fails to accomplish this, it must be repeated. On the other hand, if a satellite is to be retrieved from space and the retrieval attempt fails one can elect not to make a second retrieval attempt, but instead to augment the ground-based inventory via the purchase of a new satellite. Here it becomes apparent that one of a variety of failure recovery modes must be chosen in order to proceed. Ultimately, each of the alternatives should be investigated and the choice made on an economic basis. [See Ref. 4, which describes a model

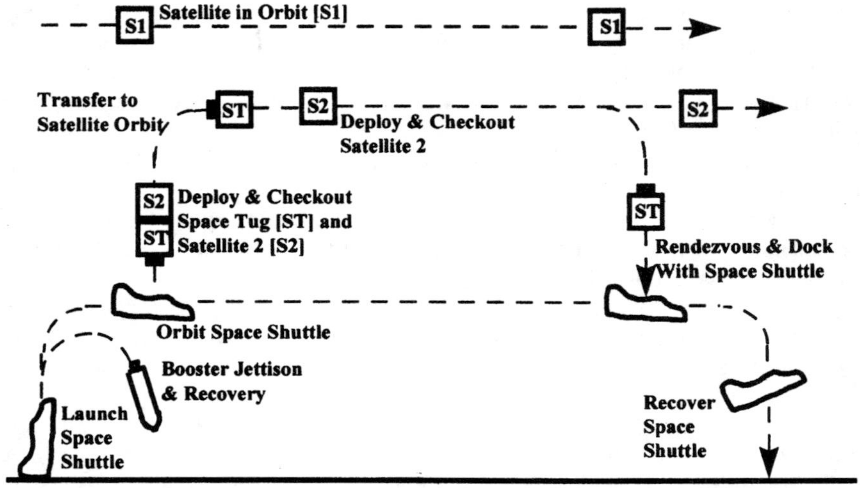

Fig. 4.1 Generalized satellite placement mission profile.[3]

that incorporates maintenance/repair decisions that depend upon the type and timing of failures (i.e., a random failure near the end of wear out life will have different consequences than the same random failure occurring at the beginning of life).]

The third step involves establishing the mission scenario. The mission scenario is defined as the timeline sequence of all possible events (within the desired level of detail) that can occur from start to completion of a mission. The success-oriented path through the mission scenario is the mission profile; however, the mission scenario includes all of the pertinent failure recovery paths as well. The mission scenario can be thought of as a series of nodes connected by branches. Each node is a decision point representing a group of events. Emanating from each node are branches for the success and failure recovery paths. The probability of departing the node on one branch or another depends upon the probability of success (or failure) of the events represented by the node. The failure recovery paths must ultimately provide a route to mission completion as defined in the preceding step two. In any event mission completion requires a proper restoration of the inventory to its premission level. For example, if the mission is to place a satellite and retrieve a satellite, and if there is a failure to retrieve the satellite, then the purchase of a new satellite is necessary to restore the inventory to its proper level.

The mission scenario can also be shown as a logic flow diagram as in Fig. 4.2 for the example mission shown in Fig. 4.1. In Fig. 4.2 the nodes are represented by the diamond-shaped boxes (representing probabilities of success or reliability) and the branches as lines with major cost-associated events given in the rectangular boxes.

When the mission scenario is established and the corresponding probability data are available, the operational analysis can be programmed for computer analysis. One method of computer analysis is by Monte Carlo techniques wherein many random walks (typically 1000) are made through the mission scenario and events counted. Decisions are taken at each diamond-shaped box depending upon the value of a computer-generated random number relative to the associated probability of success. (If the computer-generated number is equal to or less than the probability of success, the success path is chosen; otherwise, the failure path is chosen.) Because of the extremely low probability associated with certain of the paths, the finite Monte Carlo analysis can yield only approximate results.

The space transportation system operational analysis results in both the determination of the number of events (and their timing) that will be used in the life-cycle cost analysis (for each of the Monte Carlo runs) and histograms or probability distributions (across all of the Monte Carlo runs) of the number of occurrences of each major cost-associated event required to perform a given mission. (Costs must be determined prior to performing the next Monte Carlo run; otherwise, incorrect costs will be established because of the loss of event correlation effects.)

2. *Satellite Operational Analysis*

A service mission typically consists of two phases, namely, the establishment of a desired level of service and the maintenance of that level of service. The

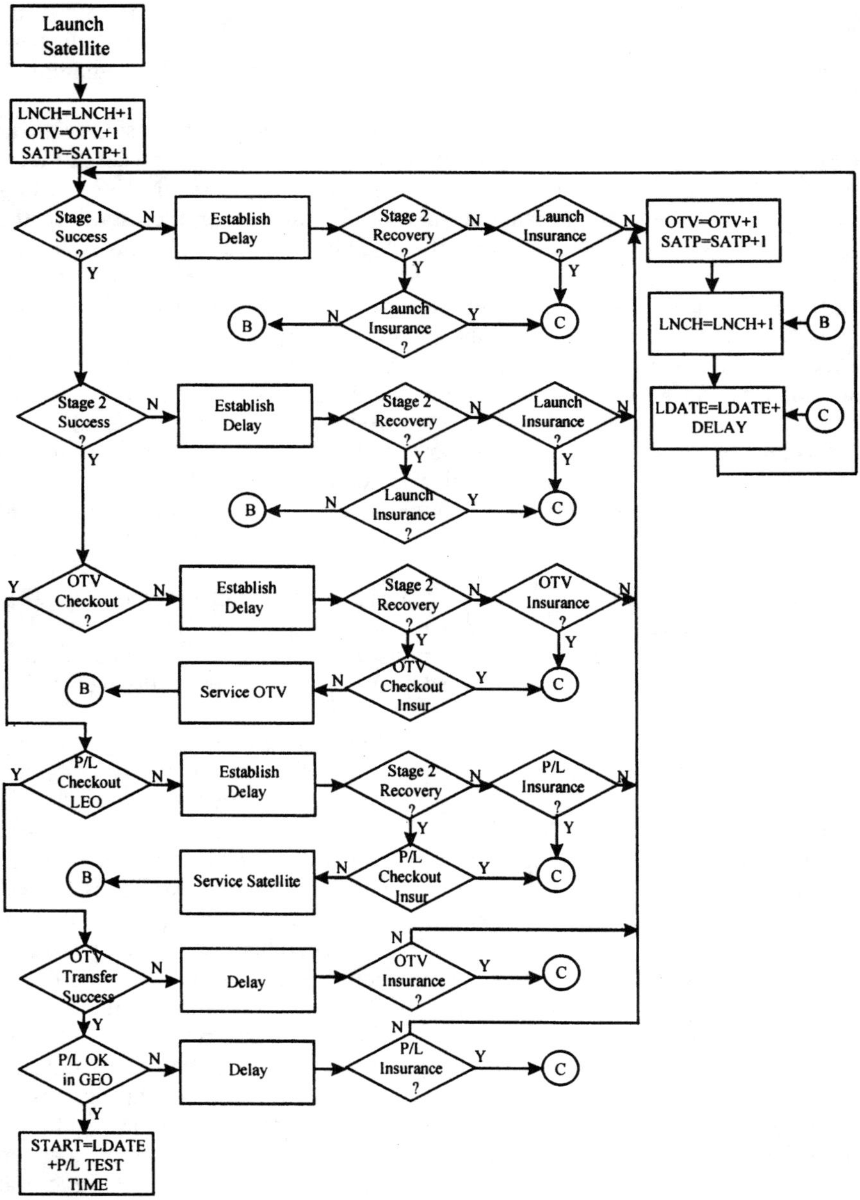

Fig. 4.2 Satellite placement scenario.

establishment of the service is concerned with successfully placing a number of satellites in orbit over a period of time as determined by service needs. If it is assumed that both the need and the capability of the satellites are known, it follows that the number of satellites required to establish the service is deterministic. The maintenance of the service is concerned with maintaining the desired number of operational satellites in service. As a result of less-than-perfect reliability and variability associated with wear out phenomena, such as attitude-control gas depletion, satellite failures will occur in a random manner with the result that the traffic associated with the maintenance of the service will be known only in a probabilistic sense. The probabilistic nature of the traffic as a result of satellite failures and replacements adds an important degree of uncertainty and risk to the service mission. It is, therefore, necessary to consider the impact of satellite reliability on the overall mission.

Satellite failures occur as the result of four causes: improper design or manufacture generally caused by an incomplete understanding of all related physical principles, imperfect quality control during manufacture, uncertainties in the environment during storage, transportation and operation, and because of design and technology limitations generally associated with lifetime. These failure causes generally manifest themselves in three different types of failures: early, chance, and wear out. In this book early satellite failures are considered in the space transportation system operational analysis. Their consequences are dependent upon the space transportation system and the operational modes.

The general satellite subsystem failure model, considering both random and wear out failures, is given by

$$R(n) = \frac{\exp(-\lambda n)}{\sigma(2\pi)^{1/2}} \int_n^\infty \exp\left[\frac{-(a-M)^2}{2\sigma^2}\right] \mathrm{d}a$$

This expresses the combined effects of the random and wear out phenomena as the probability of surviving through the nth time period (years) and λ is the failure rate (failures per year) or reciprocal of the mean-time between failures (MTBF). M is the mean or expected wear out life (years) and σ is the standard deviation of wear out life (years). The general shape of the reliability function is illustrated in Fig. 4.3, and the specifics are established by specifying λ, M, and σ. The probability of a satellite subsystem failing in time period n, $F(n)$, is thus

$$F(n) = R(n-1) - R(n)$$

The reliability model, taking into account the arrangement of subsystems (i.e., their series/parallel relationships and the criticality of sensors, individually and in combination), is used to establish the number and timing of satellite failures, based on the number in orbit and operational, and thus need to be replaced. This, together with the space transportation system operational analysis, establishes the probability distributions of all pertinent cost-associated events as a function of time.

It is necessary to define a satellite failure, that is, to specify the combination of sensor/transponder failures that will result in a cost-incurring event (for example, a satellite replacement or servicing). Because satellites can be comprised of multiple sensors or transponders, it is necessary to specify critical failures or

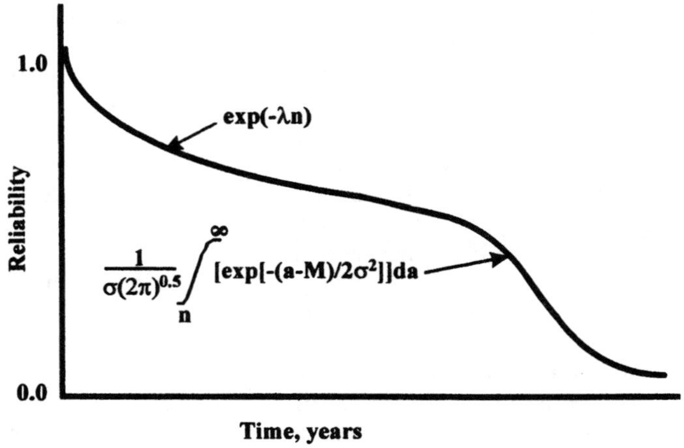

Fig. 4.3 Satellite subsystem reliability function.

combinations of failures either in terms of specific failures or the loss of value, where value is related to the relative importance of the information obtained from the sensors/transponders.[4] Thus, when the total satellite value is reduced below a specified threshold a cost-incurring event is required.

3. *Cost Analysis*

Costs can generally be categorized as either nonrecurring, that is, one-time costs such as RDT&E, or recurring, namely those costs that are activity related. Nonrecurring cost-related decisions can impact recurring costs (with the space shuttle decisions being a case in point). Costs frequently are categorized within the framework of a work breakdown structure as described in Sec. IX. During the planning phase of a program, both nonrecurring and recurring costs are nondeterministic quantities. Nonrecurring cost uncertainties arise from all aspects of a program with which there can be any amount of uncertainty: the outcome of a test, the purchase of various equipments, manpower requirements, etc. Recurring cost uncertainties arise from the difficulties of predicting the cost of producing an item before it has been designed. (This is particularly true when dealing with products that use new technology.) A life-cycle cost analysis must be concerned with the uncertainties in both nonrecurring and recurring costs.

The problem addressed is how to quantify uncertainty. This requires that informed estimates be made of the ranges of uncertainty of key cost variables and their probability distributions within the range. The estimates of uncertainty might be made, for example, at the cost estimating relationship (CER) level, or they might be made at the unit cost (transportation, satellite, etc.) level. The uncertainty assessments can be made by individuals with the assistance of an experienced analyst, or they can be made by an experienced group of individuals using Delphi-type techniques.[5,6] The estimates are very subjective in nature and quantitatively express the attitudes regarding the uncertainties. The estimates reflect past experience with similar efforts, problems that have been encountered in the past, insights into problems that might develop, etc.

A methodology for establishing the shape of the uncertainty profiles, illustrated in Fig. 4.4, has been employed in numerous analyses.[7] The first step is to establish the range of uncertainty based on knowledgeable persons assessing what can go right and what can go wrong. The range is then divided into five equal intervals, and a relative ranking of the likelihood of the cost variable falling into each of the intervals is performed. The general shape (skewed left, skewed right, central) of the uncertainty profile is thus established. (Many analysts prefer to utilize triangular distributions to characterize uncertainty. The triangular distribution is determined by specifying the minimum, maximum, and most likely values with the expected value being equal to their sum divided by three. The use of triangular distributions is appealing because of the ease of communication and simple calculations for establishing expected and standard deviation values. It is often overlooked that along with simplicity comes an approximation that can lead to significant errors in the establishment of expected values. For example, there is a 30% error in the calculation of expected value when the minimum, most likely, and

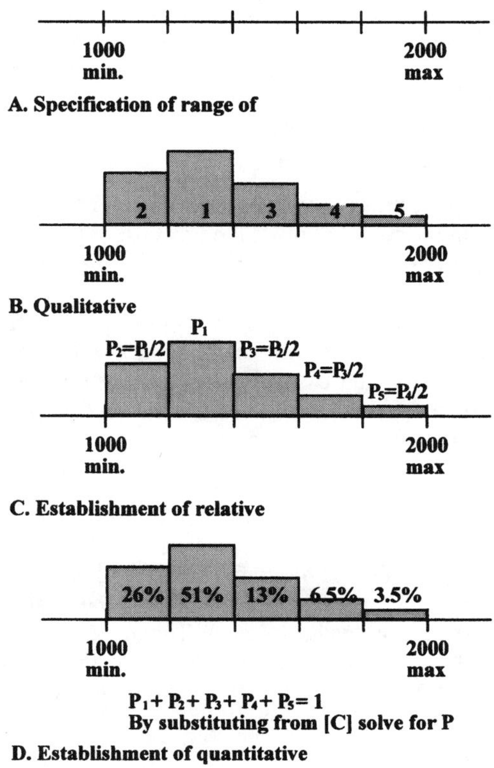

Fig. 4.4 Methodology for establishing shape of cost uncertainty profile.

expected values are specified as 0, 3, and 10 vs the specification of a five-point distribution having 10, 75, 10, 3, and 2% chances of being in the ranges of 0–2, 2–4, 4–6, 6–8, and 8–10, respectively. Errors in the determination of expected values are particularly onerous because the expected values can be added and/or subtracted.) The next step is to establish relative values of the chance of falling into each of the intervals. For example, in the illustration, the chance of falling into the first interval is estimated to be half as likely as falling into the second interval. This is repeated for each interval relative to the interval already considered. The last step is to solve the illustrated equation for the quantitative values by substituting the data from the preceding step.

To simplify this procedure, a large number of typical uncertainty profiles can be stored in the computer. The evaluator can thus simply specify the range of uncertainty (minimum and maximum values) and the name of the uncertainty profile that reasonably represents his/her feelings. If none of the stored profiles is suitable, then the procedure just described can be followed and the appropriate uncertainty profile data provided as part of the input data.

In some cases, such as satellite unit recurring costs, it can be assumed that future procurement costs will be highly correlated with first unit costs. Historically, the cost of the *n*th satellite of a procurement will be less than that of the *n*-1 satellite. As additional satellites are manufactured, costs decrease as a result of a "learning" process. Typically, the learning process is such that every time the quantity produced doubles the cost is reduced by a specified percentage, that is, one minus the learning rate. Therefore,

$$\frac{N^{th}\text{ Unit Cost}}{\text{First Unit Cost}} = N^{[\mathrm{Log}_{10}L - 2.0]/0.301}$$

where N is the cumulative number of units produced and L is the learning rate (%). Thus, if a 90% learning curve is followed the unit cost of the second unit is 90% of the first unit cost; the cost of the fourth unit is 81% of the first unit cost; and so forth.

Normally the cost of a satellite or its launch does not all occur in a single year. These costs are spread over several years, typically two to four. Thus, the computation of annual cost and cash flow must allow for this "cost spreading." Typically the time of an event is established and the cost spread backward in time according to an anticipated spending profile.

4. *Economic Analysis*

The economic analysis of space programs involves cash flow patterns that occur over several (n) years; thus, it is desirable to present the results of the economic analysis in terms of the present value of costs. The present value, which explicitly takes into account the magnitude and the timing of the cash-flow patterns, is defined as the summation of future annual costs discounted to the present and is given by

$$PV = \sum_{i=1}^{n} C_i/[1 + r/100]^i$$

where PV is the present value of cost, C_i is the cost in the ith year, and r is the discount rate (%) or the cost of capital. When working with constant dollars, the real cost of capital should be used.

The costs entering into the preceding equation, however, are not deterministic quantities. Variations in the yearly costs C_i occur because of the uncertainties in predicting future item (or per event) costs caused by the uncertainty both in the number of events necessary to perform the desired program and the time of occurrence of these events. Thus the present value of costs must also be characterized by a probability distribution.

Because the objective is to achieve a specified level of availability of sensors or transponders, comparisons of alternatives must be made at the desired level of availability. Because each present value of life-cycle cost is characterized by both an expected value and a standard deviation (i.e., the risk dimension), a tradeoff between expected cost and risk must be made to select the best alternative. This is illustrated in Fig. 4.5, where all of the points represent alternatives that will result in the required level of availability. Here alternatives 1 and 2 have the same level of risk (i.e., $\sigma_1 = \sigma_2$), but the expected PV of cost of alternative 2 is greater than that of alternative 1. Therefore, alternative 1 is preferable to alternative 2. In a similar manner, it can be argued that alternative 3 is preferable to alternative 4. Also in a similar manner, alternative 3 is preferable to alternative 1 because both have the same expected PV, but alternative 1 is riskier. This process can be continued with all alternatives being considered. In the limit it can be seen that a frontier of "best" alternatives can be established. Each of the points or alternatives represented by the frontier is different in the respect that the risk and expected PV are different. The class of best alternatives has thus been obtained, and the best alternative can be selected based on the decision maker's risk judgment. That is, the decision maker must decide what the tradeoff is between a reduction in expected PV of cost and an accompanying increase in risk.

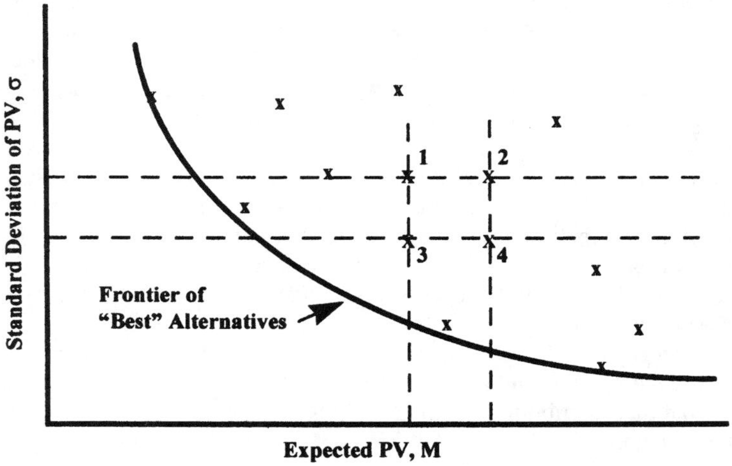

Fig. 4.5 General problem of decision making under uncertainty.

5. SATCAV Life-Cycle Cost and Availability Model

Utilizing the general considerations discussed in the preceding paragraphs, a generalized space mission life-cycle cost and availability model was developed[8–10] and is discussed in the following paragraphs. Typical results are then described.

Because mission requirements are frequently specified in terms of required sensor availability, the objective then is to select that system which minimizes the present value of life-cycle cost (and risk) and satisfies mission requirements, including sensor availability. Sensor availability is a function of overall satellite configuration (including redundancy), sparing and maintenance strategies, and transportation system reliability and availability. It is therefore necessary to consider the many complex interrelationships that exist in order to select the system and associated operational approach that minimizes life-cycle cost and achieves availability objectives. The following paragraphs describe methods and procedures that are required and have been developed for considering this complete system framework. This is accomplished by describing a developed simulation model (*SATCAV*) and presenting typical results that consider alternative maintenance and sparing strategies.

SATCAV is a dynamic stochastic life-cycle cost and availability model that simulates the launch and on-orbit operations associated with the initiation and continuing operation of a generalized space mission comprising multiple satellites with multiple "sensors." A user-friendly menu-driven input/output system is used to create a data file that is accessed by a compiled FORTRAN Monte Carlo program that performs the life-cycle and availability computations.

SATCAV simulates satellite launch operations of a multistage vehicle by considering the reliability associated with the performance of major operations (including payload checkout and testing) and the consequences of a priori specified types of failures in terms of cost-incurring events and time delays. Delays that can result from different types of failures are taken into account in the simulations. Both expendable and recoverable launch vehicles and upper stages can be considered. *SATCAV* simulates the random and wear out characteristics of a multi-"sensor" satellite determining when specific failures occur and when maintenance actions are required to respond to critical failures. ("Sensor" is used broadly to either represent actual sensors or communication subsystems or antennas.)

SATCAV encompasses alternative maintenance scenarios that include both ground and on-orbit active or dormant spares. In addition, both launch on failure and launch in anticipation of wear out failure alternatives are available. The model can also be used to analyze scheduled replacement options. The model develops cost, event, availability, and cause of failure statistics reports. It also develops a typical event timeline report.

The consideration of subjective uncertainties, transportation system reliability, and satellite subsystem random and wear out failures results in the establishment of the probability distribution of annual events and the probability distributions of annual costs and present value of life-cycle costs. The annual events report includes the probability distribution of launch attempts and transfer vehicles utilized. The cost report summarizes the annual recurring and nonrecurring costs and the present value of costs (at up to five discount rates) in terms of expected values and associated standard deviations.

SATCAV develops sensor availability statistics that are presented as the probability density function of the number of sensors available (for each sensor type) annually as well as over the mission duration. The statistics take into account sensor and subsystem random and wear out failure characteristics, sparing strategy, transportation scenarios, delays associated with different types of failures, and duration of on-orbit testing. Associated with this report is a cause of failure statistical report that indicates the probability that different subsystem failures were the cause of the sensor failures. Also provided is a timeline report that indicates (for a single Monte Carlo simulation run) the status of satellites and subsystems, the timing of satellite placements and maintenance actions, and the subsystem failures that were the cause of sensor failures.

Important attributes of the described version of *SATCAV* include the following:

1) time horizon up to 25 years;
2) consideration of multiple operational satellites;
3) sparing strategies that include
    a) multiple ground and on-orbit active or dormant spares,
    b) launch on failure of critical sensors, and
    c) launch on random failure but in anticipation of wear out failure of critical sensors (Additional strategies, such as scheduled replacement, can be considered through appropriate structuring of input data.);
4) identification of critical sensors (individually and in combination);
5) specification of desired launch dates for initial operational satellites and on-orbit spares (actual launch dates take into account failures and associated delays);
6) specification of completion date of ground spares;
7) specification of the probability that a dormant satellite will operate successfully when activated and the time delay to move the dormant satellite into replacement position;
8) consideration of satellite cost learning effects;
9) transportation scenario described in terms of the probability of successfully performing each major operation in the launch/recovery sequence taking into account the consequences of failures in terms of cost-incurring events and time delays;
10) a priori defined uncertainty variables (i.e., subjectively specified as ranges of uncertainty and the form of uncertainty) include
    a) delays (function of failure type),
    b) transportation cost from Earth to LEO,
    c) transportation cost from LEO to Earth,

d) transportation cost from LEO to P/L orbit.
e) satellite unit recurring cost and nonrecurring cost (Satellite recurring and nonrecurring cost are specified in terms of expected value and standard deviation.),
f) transfer vehicle repair cost,
g) capital expenditures (up to 10),
h) satellite control operations cost, and
i) engineering expenses;

11) satellite configuration can consist of multiple sensors and multiple satellite bus subsystems. Each sensor and subsystem is described in terms of its random and wear out failure characteristics. (A fictitious sensor is included with a random failure rate that can be specified as a function of time. This allows the effects of orbital debris to be considered. This is discussed in detail in Sec. XII of this Chapter.);
12) cost spreading for nonrecurring cost of capital items and satellite and launch recurring cost; and
13) develops cost, event, availability, and cause of failure statistics, and satellite expected life and associated standard deviation.

The *SATCAV* model allows the effects of 1) utilizing different spacecraft bus and sensor technologies, 2) launch and procurement schedules, 3) achieving different subsystem reliabilities (random and wear out), 4) alternative sparing strategies, 5) utilizing alternative launch vehicles, 6) utilizing expendable or reusable launch vehicles, and 7) launch vehicle reliability and cost to be evaluated in terms of 1) annual cost (expected value and standard deviation), 2) present value of annual cost and risk, and 3) sensor availability.

The *SATCAV* model creates the following reports that are available upon request:

1) Annual cost projection report summarizes recurring and nonrecurring cost statistics, present value of life-cycle cost statistics, and quantity of satellite-produced statistics.
2) Probability of indicated events report presents the probability distributions of annual placement and repair launches and payloads procured and serviced.
3) Percentage of time that sensors are available over planning horizon report presents the percentage of time that a specific number of sensors of each of five (5) types are available over the planning horizon.
4) Percentage of time that sensors are available each year report presents the percentage of time that a specific number of each of five types are available each year.
5) Probability that indicated subsystem will cause sensor failure report presents the probability that each subsystem will cause a sensor failure (for each of five sensor types) and also indicates expected value and standard deviation of satellite life.

6) Event timeline report presents detailed timing and status information for all satellites for a single Monte Carlo simulation run and also indicates the subsystem failure that was the cause of each sensor failure on each satellite.

The *SATCAV* Model, as indicated in Fig. 4.6, has been integrated with a Phase-A satellite design model, SMALLSAT,[11,12] which is a computer-aided Phase-A design and technology evaluation tool for small satellites. It enables satellite designers, mission planners, and technology program managers to observe the likely consequences of their decisions in terms of satellite configuration, nonrecurring and recurring cost, launch vehicle selection and, through the *SATCAV* Model, mission life-cycle costs and availability statistics.

In SMALLSAT a satellite is configured in terms of sensor attributes, such as dynamic mass, dimensions, data and power profiles, and pointing and stabilization requirements, and bus subsystem technology. SMALLSAT models the satellite in terms of nine interrelated subsystems: attitude reference and control; power; thermal; orbital maintenance; propulsion; data handling; onboard computer; tracking, telemetry, and command (TT&C); and structure. The design of these subsystems is iterated so as to take into account their mass and power interactions.

SMALLSAT contains sensor, launch vehicle, and bus subsystem technology databases. The user can select input from these databases during satellite configuration. The user can also selectively override this input to model new

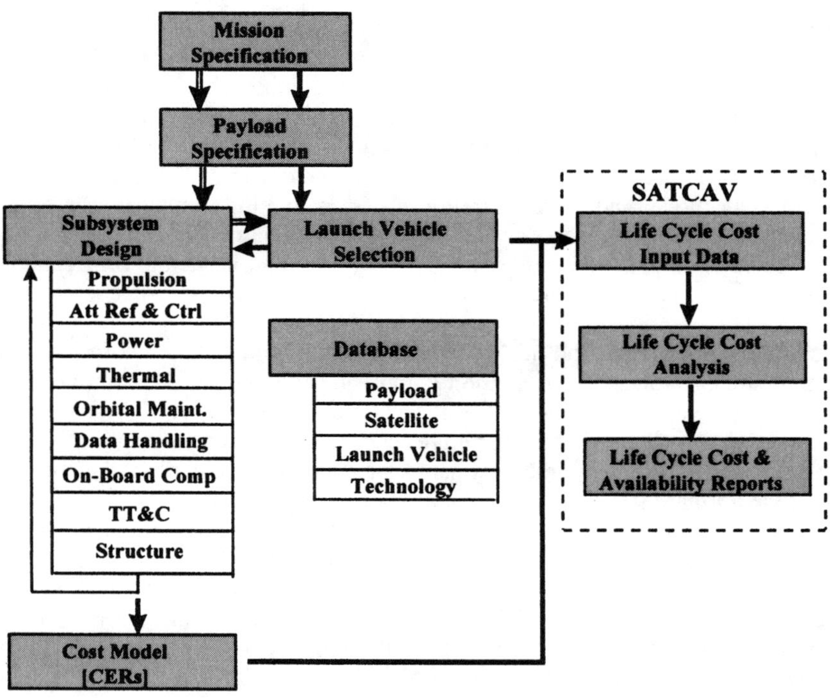

Fig. 4.6 Overview of the SMALLSAT/*SATCAV* model.

technology developments. This enables analysis of technology impacts on satellite configuration and costing. For example, launch vehicles can either be selected from the database, in which case they place constraints on satellite configuration, or they can be selected based upon the resulting satellite configuration. Modifying launch vehicle attributes allows the effect of new launch capability to be ascertained in terms of satellite configuration and cost and mission life-cycle cost. In addition, bus subsystem technology levels, such as solar cell efficiency, can be selectively altered to reflect likely technology advances or the technology advance that is sought through a specific technology program. This will allow the likely effects of technology programs to be observed in terms of satellite configuration, cost, and the ability to switch launch vehicles.

The impact of the resulting satellite configuration on mission life-cycle cost and sensor availability statistics can also be determined using SMALLSAT. A set of CERs (specifically developed for small satellites by the Aerospace Corporation) contained within SMALLSAT is utilized to provide estimates of both nonrecurring and unit recurring costs at the subsystem level. The estimated costs can be selectively overridden by specifying expected values and/or standard deviations. This allows the uncertainty of costing associated with the introduction of new technology to be explicitly and quantitatively considered. The integrated Monte Carlo simulation model, *SATCAV*, then determines the resulting life-cycle cost and availability statistics based on user-specified mission parameters such as number of satellites required, their desired launch dates, satellite sparing and replacement strategies, cost-spreading functions, and probability of orbital debris impact per year per unit cross-section area as a function of time.

The SMALLSAT/*SATCAV* model is an integration of Microsoft Excel and FORTRAN (compiled code) software and operates within the Microsoft Windows environment on personal computers.

A number of diverse analyses made possible through the use of the *SATCAV* model are described in following sections. However, to complete the discussion of the *SATCAV* model and place its utility in perspective the results of a life-cycle cost and availability analysis are discussed. Of particular concern is the determination of an appropriate sparing strategy for a multisatellite constellation. This is discussed briefly in the following paragraphs and then revisited in detail in Sec. IX.

Figure 4.7 illustrates the life-cycle cost and availability tradeoffs in terms of sparing strategy for a typical mission comprising five operational satellites plus spares with the expected wear out life of each satellite being 7 years. Availability is measured as the chance that four or more sensors (one sensor per satellite) will be operational at any point in time. Dormant spares are assumed not to fail while they are in the dormant state, but the probability of startup of a dormant spare may be less than 1.0 (as indicated by the dormant spares curve). Three sparing alternatives are indicated: active spares (with 0, 1, 2, or 3 spares), 1 or 2 dormant spares (with different probability or turn-on), and launch in anticipation (from 0 to 2 years) of expected wear out failure but launch on random failure. It can be seen that the minimum cost approach is a function of the required level of availability. For example, at low level of required availability (i.e., 85–95%) the launch in anticipation of wear out failure strategy is best offering lower expected values of life-cycle cost than other alternatives to achieve the same level of availability. At

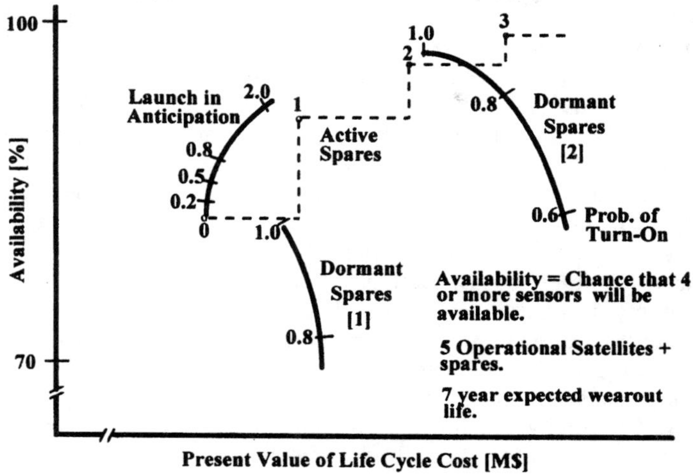

Fig. 4.7 Life-cycle cost and availability tradeoff in terms of sparing strategy.

higher required levels of availability, the active spares option is best. For a very limited set of conditions, dormant spares (2) might be preferred, but at high levels of availability active sparing is the only viable alternative.

The preceding information is presented for illustrative purposes only and should not be taken as a general rule. The minimum life-cycle cost approach within an availability constraint is a function of anticipated stand-down time (given a failure), launch vehicle reliability, and other factors. The preceding approach demonstrates an important role of life-cycle cost analysis relating to the evaluation and comparison of alternatives.

6. Summary

In the space business annual costs, and therefore the present value of life-cycle costs, must be considered as probability distributions. The reasons for this are many but are primarily because of uncertainties associated with both nonrecurring and unit recurring cost (in no small part resulting from the use of new technology); less than perfect reliability of launch vehicles, failures of which can significantly effect costs and schedules; schedule uncertainty resulting from both failures and necessary rescheduling to recover from failures and to respond to uncontrollable delays, that is, weather and stand-down time required to correct previous failures); random and wear out failures of operational satellites that require replacement.

The analysis of life-cycle costs is further complicated by a myriad of operational alternatives involving sparing and maintenance/repair strategies. In addition to satellite design considerations relating to sparing and redundancy, there are operational considerations such as the use of on-orbit active or dormant spare satellites, ground spares, and the consideration of on-orbit servicing or repair at a transportation node. Because sensor/transponder availability is a function of overall satellite configuration (including redundancy), sparing and maintenance strategies, and transportation system reliability, it is necessary to consider the many complex interrelationships that exist when configuring a satellite and estimating mission life-cycle cost.

The life-cycle cost analysis, in addition to the explicit and quantitative consideration of cost and delay uncertainties, requires the consideration of cost-learning effects and cost spreading. Because launch and space operations play a large part in both life-cycle cost and availability, the specific launch scenarios must be defined for both initial placement and maintenance flights including reliability or probability of success per major operation, cost (including uncertainties), and delays that can result from different types of failures.

The net result of these considerations is that life-cycle cost must be established as a probability distribution and decisions made taking into account quantitative risk measures: the chance that annual costs will exceed annual budgets and the chance that present value of life-cycle costs will exceed specified levels. The net result of these considerations is also the determination of quantitative measures of availability. Availability is a function of launch reliability and delays caused by failures, satellite subsystem reliability characteristics, and sparing/maintenance strategies. Different system parameters (ranging from launch system reliability to sparing strategy) will result in different sets of life-cycle cost (expected value and standard deviation) data as well as availability data. Thus, it is necessary to compare life-cycle costs of alternatives at the same level of availability. In other words, the costs associated with equal capabilities must be compared.

B. *DOMSAT* Communication Satellite Financial Planning Model

Chapter 2 considered the financial analysis of a profit-oriented business where a typical objective was the maximization of net present value of cash flow or return on investment. It was pointed out that, because of uncertainty and resulting risk, it is most desirable to describe the net present value and return on investment as probability distributions. The selection from among a number of alternatives should thus consider differences in both the expected value and risk dimensions. Fortunately, in most cases the probability distributions of important financial performance measures tend to be normally distributed; thus, the necessary comparisons are between different sets of expected values and standard deviations. The decision maker's risk-avoidance preferences can be exercised to choose the best alternative from among the set of alternatives lying upon or near the frontier of best alternatives.

The *DOMSAT* communication satellite business financial planning model embellishes upon these concepts and extends them to a specific form of business, a GEO communication satellite business venture. It encompasses many of the factors contained within the *SATCAV* model (for example, the same reliability and cost-spreading functions are included and the same approach is taken with respect to the consideration of uncertainty and the development of risk profiles). *DOMSAT* was developed to transform technology, design, and policy options into impacts on financial performance. It was originally developed at Princeton University[13] in support of NASA's Lewis Research Center's activities to assess the impact of several satellite technology programs on the financial performance of satellite business ventures. Since its initial development, it has been significantly modified and upgraded so as to keep current with the evolving space technology and operations.[9,14,15] As a result, the model has also been used, as will be discussed in following sections, to analyze the possibility of predatory pricing by foreign space transportation service providers and the effects of pricing on international competitiveness. The following discussion presents the general characteristics of the model and its application. The application demonstrates the model's use in assessing the likely impacts of satellite on-orbit propulsion and power technology programs on fixed satellite service (FSS) business ventures. The impacts are evaluated by comparing the financial performance based upon a satellite configured with existing technology and then reconfiguring and recosting the satellite taking into account the satellite design choices and anticipated results of the technology programs. These estimates are then combined with a business scenario (i.e., number of satellites as a function of time, number and type of transponders, demand for transponders by type of service provided, pricing, price elasticity, launch system scenario as a function of time, scheduling and related uncertainties, transfer time from LEO to GEO, cost of insurance, and other expense items) to establish a range of financial performance measures such as annual profit (loss), annual and cumulative cash flow, payback period, and return on investment (ROI). The financial performance measures are all described by probability distributions (risk profiles) because cost, scheduling, and demand uncertainties and launch vehicle and satellite subsystem reliability are considered.[14] The impacts of the technology programs and related policies are reflected in satellite design changes, which then alter overall business financial performance measures that are indicative of the overall impacts and the value of the programs and/or policies.

The impact of satellite design, technology programs, and related policies can be assessed in terms of the resulting changes in financial performance, cost and service attributes that are the result of the design, technology program, and policy decisions. Two analyses are necessary: one to establish a base case from which changes are measured and one taking into account the design change, technology program, etc. The difference between the two is assumed to be directly attributable to the decision. In most instances it is necessary to consider uncertainty and risk explicitly so that comparisons can consider both the expected and risk dimensions.

The *DOMSAT* model allows a business scenario to be specified in terms of 1) number of years in the business plan, 2) maximum number of operational satellites, 3) desired launch schedule, 4) time to transfer from LEO to GEO, 5) number of narrow- and wide-band transponder groups per satellite, 6) number of transponders per narrow- and wide-band groups, 7) number of spare transponders per narrow- and wide-band groups, 8) transponder reliability characteristics [mean time to failure, expected wear out life, and variability of wear out life (This type of reliability characterization was discussed in the preceding section, which described the *SATCAV* life-cycle cost and availability simulation model.)], 9) satellite support system reliability characteristics (mean time to failure, expected wear out life, and variability of wear out life), 10) types of communication services provided (protected, protected/preemtable, unprotected, preemptable), 11) rates per narrow- and wide-band transponders for each type of service, 12) annual demand for narrow- and wide-band transponders in terms of type of service, 13) relaunch threshold in terms of number of operational transponders, 14) annual cost of satellite operations, 15) annual General and Administrative (G&A) expense (fixed and variable components), 16) annual R&D expense (fixed and variable components), 17) other annual expenses, 18) insurance cost and cost spreading, 19) satellite unit recurring cost and cost spreading, 20) satellite nonrecurring cost and cost spreading, 21) satellite unit recurring cost-learning rate, 22) launch cost and cost spreading, 23) launch scenario as a function of time (described in terms of the probability of success of each major step in the launch sequence[2]), 24) depreciation lives, 25) interest rate, 26) tax-related data, 27) discount rates, and 28) balance-sheet related data.

Many of the preceding variables (those in italics) are considered as uncertainty variables, requiring the range and form of uncertainty be specified. In addition, unreliability and its effects are explicitly considered. The explicit and quantitative consideration of uncertainty and unreliability is absolutely necessary when considering design and program choices that are specifically aimed at reducing uncertainty and altering reliability, both of which affect perceived risk and hence investment decisions (for example, alternative transponder sparing concepts primarily affect the risk dimension). To establish the quantitative measures of risk, the model utilizes Monte Carlo techniques, where the complete business scenario is simulated a large number of times (typically 1000 or more), each time randomly sampling from the uncertainty profiles and the reliability characteristics that are specified.

The model develops many financial performance measures including annual after-tax profit, annual cash flow, cumulative cash flow, return on sales, return on assets, payback period, and net present value. The probability distribution of return on investment (internal rate of return) is developed from the developed probability distribution of net present value (at different rates of return). (This avoids the potential multiple-root problem when attempting to establish the internal rate of return at the completion of each Monte Carlo simulation run.)

The *DOMSAT* model consists basically of two interactive parts. The first, utilizing the desired schedule of events, demand for communication services, the satellite configuration, specified launch scenario and reliability characteristics, establishes the specific timing and number of events and their costs. The availability of transponders (taking into account failures, sparing concepts, and services offered) is matched against demand to establish the schedule for replacement launches and the timing of additional capital expenditures for replacement satellites and launches. The timing and cost information is then passed to the second part of the model, which performs the financial computations and establishes values of the financial performance measures.

A typical but hypothetical FSS business venture was used as a base case for a number of technology program assessments. The postulated venture represented a carrier that launched and operated satellites with the objective of generating revenue through the leasing of transponders. The base-case satellite was spin-stabilized and transmitted in the Ku frequency band and was based upon the Hughes H376 model. The satellite had a 20 for 16 redundancy, with 16 active transponders and four spares. This particular business was based upon placing and maintaining three operational satellites in orbit; the first launched midway through the fourth year, and the others launched midway through years five and seven. Satellites that failed or worn out were replaced subject to a launch delay (between 0.5 and 0.8 years) and three months delay to allow for transit from LEO to GEO and on-orbit testing and checkout. On each satellite up to 14 transponders were available for lease as protected service and up to two as unprotected, at a price less than half that of the protected transponders.

Figure 4.8 illustrates typical results obtained from the model in the form of risk profiles of present value of cash flow. The risk profiles indicate the chance that present value exceeds different levels. The figure illustrates the present value risk profiles for the same business scenario (with unreliability considered but with all uncertainties set to zero) but with on-orbit propulsion system life of 8 and 12 years. Increasing the wear out life from 8 to 12 years (all other factors remaining constant) increased the expected net present value of the business by $27 million. Thus the value of extending the satellite propulsion system life by four years is worth $27 million to the venture.

The *DOMSAT* model was used to investigate the implications of the development of advanced technology on the already described FSS business venture. The considered technology programs [ion thrusters for on-orbit propulsion and gallium arsenide (GaAs) solar cells] have the ability of reducing overall satellite mass without altering overall performance. The satellite can be designed with reduced mass with the result that transportation charges can be reduced (when charges are proportional to mass or a smaller and cheaper launch vehicle might suffice), leading to an increase in expected ROI with little or no change in risk. On the other hand, the mass can be put back in a number of ways, each of which alters satellite attributes, such as on-orbit propulsion system life, number of active transponders, number of spare transponders, and so on. This is illustrated hypothetically in Fig. 4.9, where the financial implications (relative to

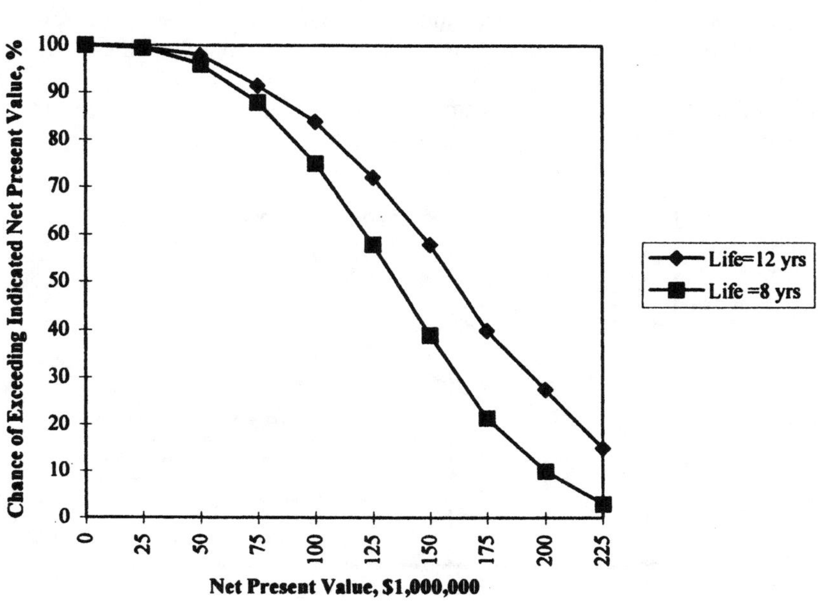

Fig. 4.8 Risk profiles of a typical FSS business venture indicating impact of on-orbit propulsion system life.

the base case) of the possible satellite configurations (i.e., use of mass savings resulting from the introduction of the new technology) are indicated in terms of impacts on expected ROI and risk as measured by the standard deviation of ROI, σ. For example, a considerable increase in expected ROI, with an accompanying increase in risk, can result from introducing an ion propulsion system with sufficient propellant to extend satellite wear out life but with a perceived reduction in mean time to failure.

Determination of the best use of the mass requires the establishment of the preference curve or risk-aversion attitudes. To illustrate, given the risk aversion attitudes indicated in Fig. 4.9, alternative D offers the best use of the mass savings and therefore represents the maximum value of the technology program when the results of the program are used in the postulated business scenario. When alternative technology programs are to be compared, the comparison must use the maximum values that result from the use of each of the technologies.

To evaluate and compare the ion-thruster and GaAs solar-cell technologies, the FSS satellite was reconfigured so as to make efficient use of the technologies. This resulted in mass savings on the order of 90 and 15 kg, respectively. These mass savings allowed four additional active transponders and two years of life to be redesigned into the ion-thruster satellite and two additional active transponders into the gallium arsenide satellite. In this way, added capability was designed into the satellites with the mass at liftoff remaining approximately constant. The business scenario data were adjusted for the new parameters: new satellite nonrecurring and recurring costs and cost uncertainties, the number of transponders, and on-orbit life. The type of obtained results are summarized in Fig. 4.10, which illustrates the risk profiles associated with the base case and with the results of successful ion thruster and GaAs solar-array programs incorporated into the satellite design. Both programs offer the possibility of significant savings to the FSS business with little change in the overall risk. (There is a slight difference in the risk associated with the incorporation of the ion thrusters and GaAs solar arrays with the latter offering a slight risk reduction.)

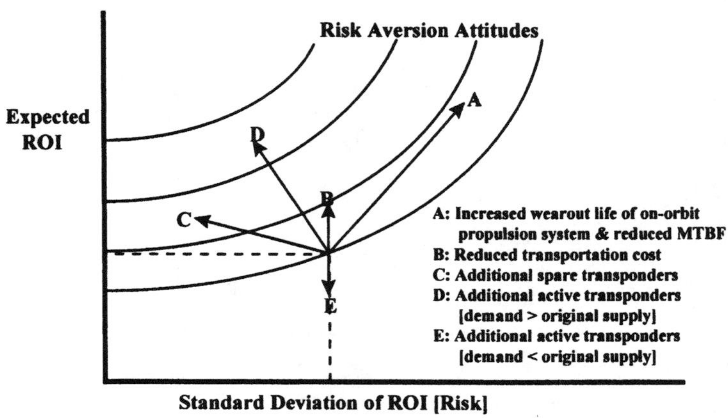

Fig. 4.9 Assessment of the best use of mass savings for a particular technology program.

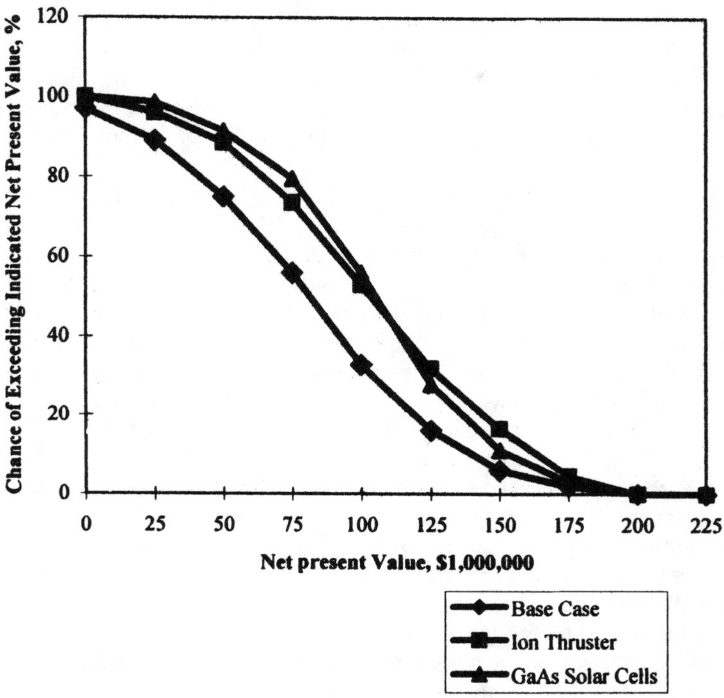

Fig. 4.10 Impacts of utilization of new technology (typical impacts as measured in terms of net present value).

To summarize, the *DOMSAT* model has been used to analyze and evaluate (in addition to those applications discussed in the following sections) the implications of the development of new technology on the design of communication satellites and the results of such design on the financial performance of typical communication satellite business ventures. The model allows the effects of launch vehicle and satellite attributes and the demand for communication services to be considered in the development of the financial performance of communication satellite business financial performance. The explicit and quantitative consideration of uncertainty and unreliability allows risk profiles to be developed that indicate the chance that selected financial performance measures will exceed different levels. This makes possible the comparison of alternatives in terms of both expected values and risk (as measured by the standard deviation of the performance measure).

II. Selecting a Launch Vehicle (Commercial Missions)

A. Introduction

The decision process concerned with the selection of a launch vehicle for a particular mission is a complex process because of the many attributes of launch vehicles that affect financial decision variables.[17] Launch vehicle attributes include price, payment schedule, reliability, lift capability, final orbit placement accuracy, insurance rate, scheduling (including possible delays), and other factors. These influence, either directly or indirectly, financial decision variables such as profit, payback period, ROI, and risk (the perceived variability of financial performance measures).

The *DOMSAT* communication satellite financial planning simulation model, described in Sec. I, includes models of launch and on-orbit operations, was developed to evaluate communication satellite business venture financial performance when the business utilizes different launch services having different "packages" of service attributes. Financial performance is developed in terms of these attributes so that the value of different launch services to the business can be evaluated in financial terms. Developed financial performance measures include net present value of cash flow, ROI, and risk, where risk is measured quantitatively in terms of the standard deviation of the developed financial performance measures. *DOMSAT* is thus used to transform the launch service attributes into financial performance measures. The result is that launch vehicles having multiple service attributes can be transformed into a set of financial performance measures having a combination of expected value and standard deviation (m, σ). Tradeoffs can then be made based upon the launch vehicle financial values taking into account both the expected value and risk dimensions.

The methodology for comparing the launch vehicle alternatives is developed in the following paragraphs, and this is then followed by several examples.

Because government actions such as investment in technology programs, procurement policies, and regulatory actions will affect launch service attributes, and because government actions are often undertaken with the expressed intent of improving competitiveness (i.e., influencing industry investment or launch vehicle selection decisions), the developed methodology is expanded in a predictive manner. A space transportation service user decision database has been developed, and is described, which relates the likelihood that different choices will be made (i.e., different launch vehicles will be selected) to financial performance measures (specifically, relative expected ROI and relative risk). Through the use of the *DOMSAT* model (used to develop the financial performance measures) in combination with the database, the likelihood of selecting different launch vehicles in a competitive environment can be established. This likelihood is equivalent to market share. Thus a methodology that allows the effect of government actions to be assessed in terms of likely user choices and impacts on launch vehicle market share is presented. (This is also discussed in Secs. VI and VIII in Chapter 2.)

B. Methodology

The determination of the financial impacts of alternative launch vehicle choices is accomplished by creating a business plan for the communication satellite business under consideration and utilizing the different launch vehicles to place the communication satellites into geostationary orbits. This is accomplished by specifying the detailed characteristics of the business together with pertinent attributes of the communication satellite and launch service and then simulating the operations of the business over a relatively long period of time. The *DOMSAT* communication satellite business venture financial planning model is used to simulate the business operations as well as satellite launch and on-orbit operations.

The *DOMSAT* model simulates the communication satellite business and utilizes a launch service/vehicle that is specified in terms of an identified set of attributes (i.e., price, reliability, insurance rate, stand-down time, payload delivery schedule, and expected on-orbit satellite life). Additional information relating to the *DOMSAT* model is presented in the Appendix to this section. This results in the determination of various financial performance measures, including the expected net present value (NPV) of cash flow and takes into account perceived uncertainties (for example, nonrecurring and recurring satellite cost). Because decision makers' perceptions of the values of a number of the attributes can differ and because certain attributes such as price and insurance rate might be open to negotiation, each of the pertinent launch vehicle attributes is varied one at a time in order to establish the functional relationship between NPV and the varied launch vehicle attribute. This, in essence, establishes sensitivity coefficients that are not necessarily constant over the range of potential variability of the attributes. The relationship between the change in NPV that results from a change in satellite life (i.e., wear out life of the station-keeping subsystem) is illustrated in Fig. 4.11, where the intersection of the x axis and the y axis represents the nominal or base-case scenario from which all changes are measured. The results indicated in Fig. 4.11 are based upon a satellite having a 14-year expected life of the station-keeping system.

A brief digression is necessary to further consider the value of satellite life. The loss of value caused by a reduction in satellite life depends upon perceptions of future transponder demand and the availability and capability of a next-generation satellite. The results shown in Fig. 4.11 are based upon a satellite having 24

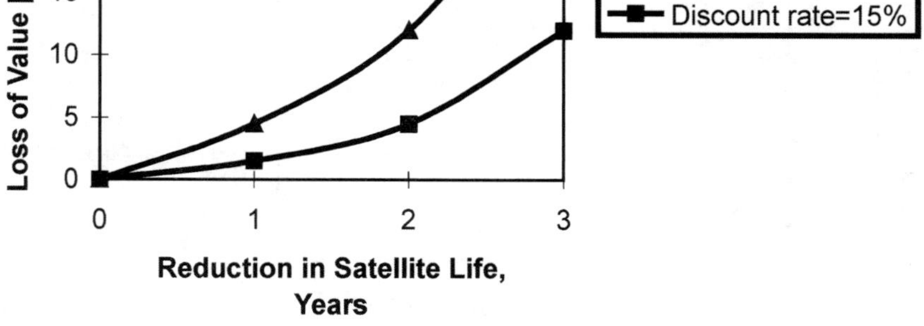

Fig. 4.11 Sensitivity of NPV to changes in satellite life (business based upon the use of an Intelsat VIIA class satellite launched on Long March).

C-band and 14 Ku-band transponders, an increase in demand of 7%/year and replacement as required (upon failure) by a similar capability satellite. If a second-generation capability satellite with 24 C-band and 28 Ku-band transponders were available, additional revenue could be generated when demand exceeds the original satellite capability. The implication of using a shorter life satellite (i.e., 12 years) in combination with a next-generation satellite brought online at different points in time is indicated in Fig. 4.12 in terms of transponder demand perceptions and internal rate of return. The results indicate that perceptions of relatively low-demand growth rates (i.e., 7%) coupled with high discount rates (i.e., 15%) lead to loss of value from reduced satellite life. On the other hand, perceptions of somewhat higher growth rates (i.e., 10%) coupled with lower discount rates (i.e., 10%) lead to substantial potential increased value from reduced satellite life. The conclusion is that reduced satellite life (caused by reduced launch vehicle lift capability) may or may not be a disadvantage depending upon demand perceptions and required rates of return.

Returning to the main topic, having established sensitivity coefficients for each of the launch service attributes, it is then possible to establish the expected change in NPV, ΔNPV, in terms of the combined effect of the possible changes in the multiple attributes. (This is an approximation because it assumes that all higher-order derivatives are negligible. Expected values should be utilized for the changes in each of the launch service attributes.) This can be stated as

$$\Delta NPV = \sum_{i=1}^{N} K_i(\Delta A_i) * \Delta A_i$$

where ΔA_i is the expected change in value of the ith attribute, K_i (ΔA_i) is the sensitivity coefficient of the ith attribute and can be a function of the magnitude of the attribute change, and N is the number of attributes considered.

A value of ΔNPV for the communication satellite business can thus be established for each of a number of launch vehicles and the relative value of one

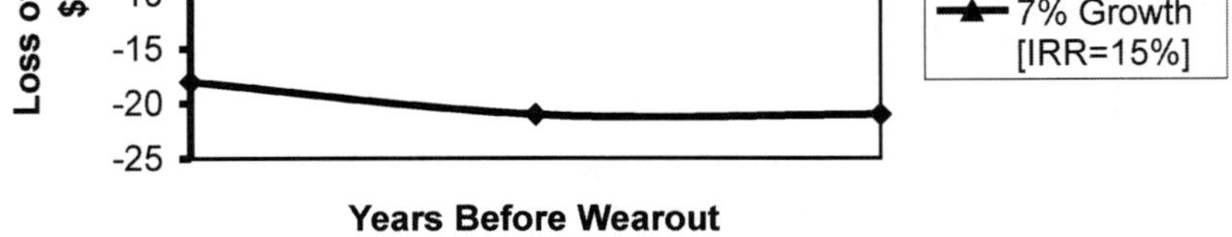

Fig. 4.12 Impact of introducing an increased capability next-generation satellite (loss in value caused by use of 12-year life satellite).

launch service vs another established for the considered business situation. Thus,

$$V = (\Delta NPV)_x - (\Delta NPV)_y + (NPV)_x - (NPV)_y$$

where V represents the value of launch service x relative to launch service y and $(\Delta NPV)_j$ represents the change in net present value of the jth launch service as measured relative to base cases $(NPV)_j$ using nominal values for the set of attributes. Thus, typical values can be specified for the attributes of each launch service to be considered and the effect on overall value of one launch service to another established in terms of possible variations from the nominal values. This is described in more detail in the following pages with typical results presented.

Suffice it to say at this point that a business scenario is developed around a specified satellite configuration. This entails specifying satellite recurring and nonrecurring cost, the number of transponders and their reliability characteristics, services provided, demand for and pricing of transponders in terms of service provided, and other satellite and business descriptive data. (See Sec. I for more detailed information on the *DOMSAT* model used for analyzing the business situation with different launch services.) Also provided are the number of satellites and their desired launch dates that are included in the business venture.

Each launch vehicle/service is explicitly described in terms of its reliability or probability of successfully placing the communication satellite into its geosynchronous orbit, transportation cost or pricing associated with each launch, and cost spreading. In addition, delays that might result from various types of failures can be described in terms of probability distributions that are conditional upon the type of failure. Anticipated scheduling delays and insurance rates can also be specified. The effects of launch vehicle payload delivery capability and final payload placement accuracy can also be considered indirectly through their effect on station-keeping subsystem useful life characteristics.

C. Choice of a Launch Vehicle

As already described, the *DOMSAT* model can be used to develop a financial evaluation of a contemplated communication satellite business venture. The results of the analyses include quantitative values of after-tax profit, cash flow, ROI, and NPV of cash flow. Each of these is described in terms of both expected values and risk, with the risk represented by the standard deviation of the performance measures. The values take into account launch vehicle reliability, satellite subsystem reliability, scheduling uncertainties, and their interrelationships and cost uncertainties.

The financial analysis, if performed using the various considered launch vehicles and their associated attributes as perceived by the user, results in the determination of the expected value and standard deviation of the net present value of the cash flow of the considered business venture. If the resulting risks from the use of the alternative launch vehicles are comparable, then the appropriate choice should be the launch vehicle that results in the largest value of NPV of cash flow. When expected values of NPV appear to be comparable, then the appropriate choice should be the launch vehicle that minimizes risk. Finally, when neither the expected values nor levels of risk are comparable and the NPV associated with the use of one launch vehicle is greater than that of another but the level of risk associated with the latter is less than that of the former, choices are more difficult and risk avoidance preferences, either stated explicitly or implied, must be considered.

Alternatively, as already described, the *DOMSAT* model can be used to establish sensitivity coefficients and the net present value of savings resulting from the use of alternative launch vehicles. To illustrate the type of analyses that can be performed, an example (hypothetical yet close to reality) is indicated based upon an Inmarsat-type satellite configuration. Nominal values of a subset of parameters used for the Atlas and Proton launch vehicles are indicated in Table 4.1. (The reason for using the indicated price of the Proton will become apparent in following paragraphs.) Sensitivity coefficients have been established for these factors as well as payment schedule and launch delay.

The sensitivity coefficients are established by varying the different attributes of the launch vehicles and performing the business simulation analysis. (The sensitivity analyses are based upon varying only one attribute at a time.) The results are indicated in Fig. 4.13. The data from Table 4.1, together with the satellite configuration data and general business scenario data, result in the expected difference, or savings, in NPV of $28.7 million. As indicated in Fig. 4.13, an equation for the expected NPV of savings for one launch vehicle vs another (for the indicated example, Proton vs Atlas for the Inmarsat-type business venture) has been developed. For simplicity, only the sensitivity coefficients associated with the variables indicated in Table 4.1 are presented. This then provides a reasonable indication of the relative value of the launch vehicles within the business framework and can be used to establish the NPV associated with different combinations of launch vehicle attributes, either real or perceived.

Additional sensitivity coefficients are provided in Fig. 4.14 in graphical form for two different missions using different launch vehicles.

To minimize the computation problem, the sensitivity computation results have been placed within a spreadsheet framework, as indicated in Fig. 4.15. The computations utilize the nominal values unless a non-zero value is placed in the "to be used" column. Thus a Proton price of $62.3 million is used in lieu of the nominal $35 million. The $62.3 million price is the "value in use" price because it yields a relative value of NPV of zero. (The value in use price is the launch service price of service B that would make the user indifferent between the selection of services A or B. In other words, it is the price that compensates for all

Table 4.1 Nominal attributes associated with Proton and Atlas launch services

	Launch vehicle	
Attribute	Atlas	Proton
Price, M$	62.5	35.0
Reliability, %	93.0	96.0
Insurance rate, %	17.0	20.0
Expected on-orbit satellite life, yrs	12.0	13.0

		Valid Range
PROTON =	- 0.75 * [Price(M$) – 35.0]	25.0 to 60.0
	+2.80 * [Reliability(%) – 96.0]	85.0 to 99.0
	1.00 * [Insurance Rate(%) – 20.0]	15.0 to 22.0
	+ K1 * 9.1 * [Sat On-Orbit Life (Yrs) –13.0]	K1=1 when $10.0 \le Life \le 13.5$
	+ [1 – K1] * [0.2 * Sat On-Orbit Life (Yrs) +1.9]	K1=0 when $13.5 < Life \le 16.0$
PROTON =	- 1.01 * [Price(M$) – 62.5]	50.0 to 75.0
	+2.00 * [Reliability(%) – 93.0]	85.0 to 99.0
	1.33 * [Insurance Rate(%) – 17.0]	15.0 to 22.0
	+ K2 * 13.32* [Sat On-Orbit Life (Yrs) –12.0]	K2=1 when $10.0 \le Life \le 12.9$
	+ [1 – K2] * [0.88 * Sat On-Orbit Life (Yrs)]	K2=0 when $12.9 < Life \le 16.0$

NPV OF SAVINGS [M$] = 28.69 + PROTON - ATLAS

Fig. 4.13 NPV functional relationship for proton relative to Atlas and sensitivity coefficients (Inmarsat-type mission).

other attribute differences.) Figure 4.15 also indicates the effect of changing the Proton payment schedule. (Payment schedule indicates the percent of the price paid two years prior to launch, one year prior to launch, and in the year of launch.) Thus, changing the indicated payment schedule increases the value to the user by $2.3 million.

To further illustrate the concept of value in use, if Proton reliability is perceived as less than the nominal value, then Proton's value in use price should be reduced. Similarly, if using Proton leads to a satellite with shortened life then its value in use should decrease. In other words, as the value of the service provided decreases an offsetting reduction in price is necessary if the user is to remain indifferent with respect to launch vehicle choice.

The use of the spreadsheet with the embedded sensitivity coefficients then allows the user to compare and evaluate received bids for the provision of launch services while interjecting user perceptions regarding the specific considered attributes as well as the possible variations in the values of the attributes.

D. Government Impacts on User Choices

Government actions (such as investing in launch vehicle technology development) and policies (such as the setting of insurance requirements, priorities of commercial launches from government facilities, and the setting of international agreements relating to launch constraints and pricing) can influence user choices by changing user perceptions relating to launch service attributes. These attributes, as already discussed, influence financial decision attributes that might alter user choices.

A somewhat different type of analysis is required to assess likely launch vehicle market shares and the effect that government policies might have on these shares with changes in market shares being a reflection of changes in user perceptions. To accomplish this requires 1) a method for establishing communication satellite business venture financial performance measures in terms of launch vehicle attributes and 2) an understanding of likely decisions or choices in terms of the developed financial performance measures. The former has already been described

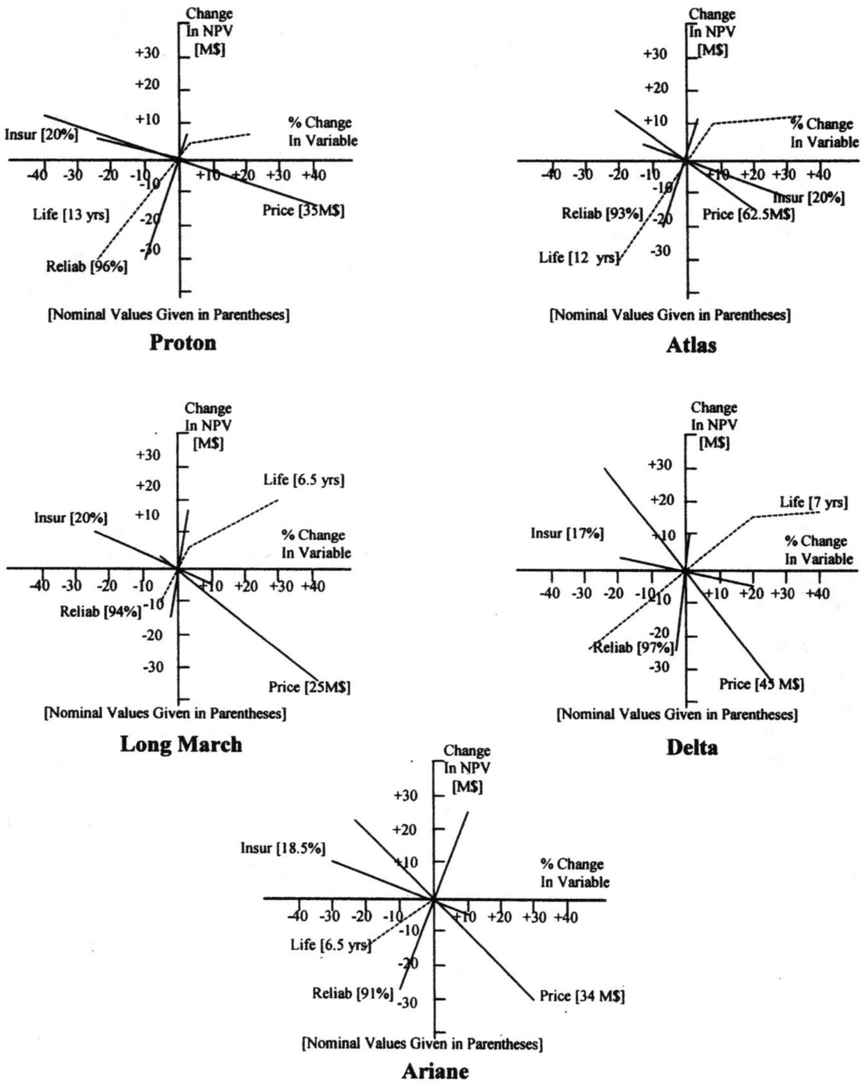

Fig. 4.14 NPV sensitivity coefficients (Inmarsat and Arabsat type of missions).

and is based upon the use of the *DOMSAT* model to develop expected NPV and return on investment and risk measures, where risk is measured as the standard deviation of NPV and return on investment. The latter is discussed in following paragraphs.

To understand the characteristics of the demand for space transportation services for launching communication satellites better, it is necessary to understand

PROTON vs. ATLAS for Inmarsat III Mission

Nominal Payment Scheme:

NPV of PROTON relative to ATLAS = 28.69[2] - 28.7 –0.00

NPV of PROTON relative to ATLAS = 0.00

	PROTON		***ATLAS***		***Change in NPV***	
LV Attributes	***Nominal***	***To Be Used***	***Nominal***	***To Be Used***	***PROTON***	***ATLAS***
Price [M$]	**[35.0]**	**[62.3]**	**[62.5]**	**[ ]**	**-28.7**	**0.0**
Payment Scheme	**[2]**	**[2]**	**[2]**	**[2]**	**0.0**	**0.0**
Reliability [%]	**[96.0]**	**[ ]**	**[93.0]**	**[ ]**	**0.0**	**0.0**
Insurance Rate [%]	**[20.0]**	**[ ]**	**[17.0]**	**[ ]**	**0.0**	**0.0**
Launch Delay [Yrs]	**[0.25]**	**[ ]**	**[0.25]**	**[ ]**	**0.0**	**0.0**
Sat On-Orbit Life [Yrs]	**[13.0]**	**[ ]**	**[12.0]**	**[ ]**	**0.0**	**0.0**
Other Cost elements [M$]						
Launch Support		**[0.0]**		**[0.0]**	**0.0**	**0.0**
Trans/Shipping		**[0.0]**		**[0.0]**	**0.0**	**0.0**
Payload Integration						
Initial Launch		**[0.0]**		**[0.0]**	**0.0**	**0.0**
Other Launches		**[0.0]**		**[0.0]**	**0.0**	**0.0**
Perigee Kick Motor		**[0.0]**		**[0.0]**	**0.0**	**0.0**
Political Risk Ins. Rate [%]		**[0.0]**		**[0.0]**	**0.0**	**0.0**

Alternative Payment Scheme:

NPV of PROTON relative to ATLAS = 28.69 – 26.4 –0.00

NPV of PROTON relative to ATLAS = 2.305

	PROTON		***ATLAS***		**Change in NPV**	
LV Attributes	***Nominal***	***To Be Used***	***Nominal***	***To Be Used***	***PROTON***	***ATLAS***
Price [M$]	**[35.0]**	**[62.3]**	**[62.5]**	**[ ]**	**-28.7**	**0.0**
Payment Scheme	**[2]**	**[5]**	**[2]**	**[2]**	**2.3**	**0.0**
Reliability [%]	**[96.0]**	**[ ]**	**[93.0]**	**[ ]**	**0.0**	**0.0**
Insurance Rate [%]	**[20.0]**	**[ ]**	**[17.0]**	**[ ]**	**0.0**	**0.0**
Launch Delay [Yrs]	**[0.25]**	**[ ]**	**[0.25]**	**[ ]**	**0.0**	**0.0**
Sat On-Orbit Life [Yrs]	**[13.0]**	**[ ]**	**[12.0]**	**[ ]**	**0.0**	**0.0**
Other Cost elements [M$]						
Launch Support		**[0.0]**		**[0.0]**	**0.0**	**0.0**
Trans/Shipping		**[0.0]**		**[0.0]**	**0.0**	**0.0**
Payload Integration						
Initial Launch		**[0.0]**		**[0.0]**	**0.0**	**0.0**
Other Launches		**[0.0]**		**[0.0]**	**0.0**	**0.0**
Perigee Kick Motor		**[0.0]**		**[0.0]**	**0.0**	**0.0**
Political Risk Ins. Rate [%]		**[0.0]**		**[0.0]**	**0.0**	**0.0**

Changing PROTON Payment Schedule				
From:	**20%**	**30%**	**50%**	**[2]**
To:	**0%**	**0%**	**100%**	**[5]**
Has a User Value of $2.3 million				

Fig. 4.15 Spreadsheet sensitivity model (illustration of the effect of payment schedule on relative value). Determined from *DOMSAT* analysis utilizing the indicated nominal values.

the implications of financial performance measures in selecting alternatives. The basic problem can be stated as follows: Given that a decision has been made to commit resources to a communication satellite business venture and that there exist multiple implementation alternatives (i.e., the selection of a specific space transportation service based upon vehicles such as Ariane, Atlas, Long March, Proton, and Delta), what is the chance that different alternatives will be selected in terms of perceived differences in resulting financial performance measures?

Clearly in a riskless environment and with all other factors being equal, the choice would be made to select that alternative which maximizes NPV of cash flow or return on investment. Enter risk, and the problem becomes more complex because it is necessary to understand the tradeoffs between perceptions of risk and other financial performance measures.

To understand these tradeoffs, a limited survey was conducted of decision makers (i.e., the financial community) within communication satellite businesses. The survey (described in the Appendix to this section, presents the instructions for responding to the survey, which was augmented with telephone discussions) sought to establish the likely choices between alternatives having different expected value of ROI and risk characteristics. (This survey is similar in many respects to that described in Ref. 7.) The result of this limited survey was the development of a family of curves that indicates the chance of selecting alternative x relative to alternative y. The already described decision survey (described in Sec. IV of Chapter 2) was concerned with establishing the likelihood of investment for independent investment decisions, whereas the communication satellite business survey was concerned with mutually exclusive decisions, that is, selecting either transportation service A or transportation service B where the selection of one excludes the selection of the other.

The general form of the obtained results is indicated in Fig. 4.16. The curves represent the chance of selecting x relative to y, $p_{x/y}$, in terms of the ratio of expected values of ROI, m_x/m_y, and the ratio of perceived risk, σ_x/σ_y. Thus if a business is defined and the *DOMSAT* model utilized to evaluate the financial performance measures with two different launch vehicles, their associated financial attributes m_x, m_y, σ_x, and σ_y can be obtained. Entering the set of curves in Fig. 4.16 with these values results in an estimate of the probability that x will be selected ($p_{x/y}$).

Thus, for business ventures that have been selected so as to represent market segmentation, the likelihood of utilizing different launch vehicles can be determined. If three different launch vehicles (a, b, c) are considered with values $\{m_a, \sigma_a\}$, $\{m_b, \sigma_b\}$, and $\{m_c, \sigma_c\}$ determined through the use of the *DOMSAT* model and resulting values of $p_{a/b}$, $p_{a/c}$, etc., as determined from Fig. 4.16, it is possible

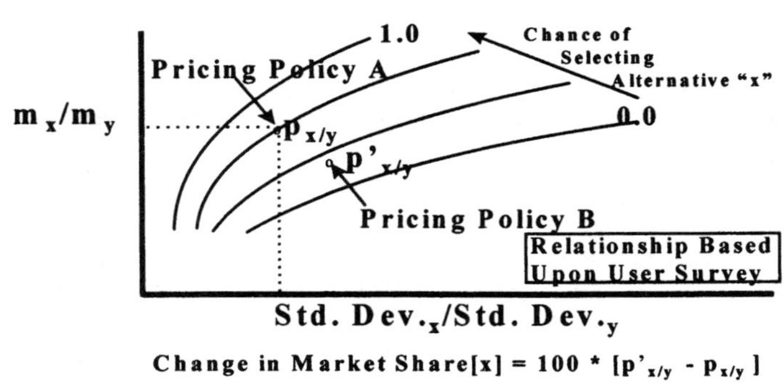

Fig. 4.16 Chance of selecting alternative x (typical survey results): impact of pricing policy on market share.

to establish market share for launch vehicle a, S_a, as

$$S_a(\%) = 100[p_{a/b} + p_{a/c}]/[p_{a/b} + p_{a/c} + p_{b/a} + p_{b/c} + p_{c/a} + p_{c/b}]$$

The effect of government policies and programs can be assessed in terms of impacts on launch vehicle attributes or business parameters that then result in changes in values of m and/or σ for one or more launch vehicles. For example, large government procurements of launch services will reduce costs through learning effects that can be passed on to commercial users of launch services in the form of lower prices. These lower prices will affect the m and σ values of the users of launch services that will affect the probability of selecting the launch vehicle. This then results in changes in market share through the use of the family of curves illustrated in Fig. 4.16.

Thus a mechanism has been developed that can be used to assess, in a rational and consistent analytic manner, the likely impacts of government policies and programs on launch vehicle market share as a result of effects on user perceptions.

E. Summary

A number of different parties are concerned with the provision of launch services. These include the users or customers for the launch services (i.e., the communication satellite business ventures), the providers of the launch service, and governments. Concerns range from business-related issues to international fair-trade issues. The following specific concerns were addressed:

1) Given a decision to establish a business to provide satellite communication services, which of a number of competing launch services should be utilized?
2) What effect will various government actions and policies have on user decisions?

To answer these and other related questions, it is necessary to plan and evaluate the financial performance of communication satellite business ventures utilizing each of the alternative launch services. The evaluation of financial performance requires the use of a simulation model that not only simulates the business but also simulates both launch and on-orbit operations. It is necessary to explicitly and simultaneously consider the multiple attributes of the launch services. These include price, reliability, insurance rate, availability/schedule and related uncertainties, and the effect of launch vehicle performance on satellite life. As a result of the financial analyses, financial performance measures such as net present value of cash flow and return on investment (both expected values and standard deviations, with the latter being the quantitative indicator of risk) can be determined and can be used to answer the preceding questions.

The choice of launch vehicle should be based upon user maximization of net present value of cash flow or ROI taking into account differences in risk associated with each of the considered launch vehicles and also taking into account risk aversion preferences.

The effect of government policies and programs can be judged in terms of impacts on launch vehicle market share. The determination of market share requires the analysis of typical communication satellite business ventures with each of the considered launch vehicles and the determination of the resulting expected return on investment and the standard deviation of ROI. This information, together with information available on attitudes relating to launch service choices in terms of financial performance measures, leads to estimates of launch service market share. The effect of government policies and programs can then be measured in terms of likely changes in these market shares.

F. Appendix: Financial Choices (Survey Instructions)

Suppose that your firm has decided to launch one or more communications satellites. Taking this decision as a given, you must choose between two possible launch vehicle/service alternatives. Although there are many attributes that are associated with a given system, suppose that these have been translated into financial performance measures, such as represented by a histogram describing a probability distribution of return on investment. This value is expressed as your normalized ROI, that is, ROI divided by your cost of capital. This normalized value is referred to as *R*. Your responsibility, then, is to choose the better of two systems by evaluating their respective *R* curves.

Through the tip of each curve, a dashed line has been drawn, indicating the expected, or most likely, normalized return on investment. The spread, or width, of the curve represents the associated risk. Thus if a curve is tall and narrow, it indicates a rather sure choice, with little chance of large deviations, for better or for worse, from the expected return. A shorter, wider curve, on the other hand, represents a riskier choice, with a greater chance that the realized *R* will be more or less than anticipated.

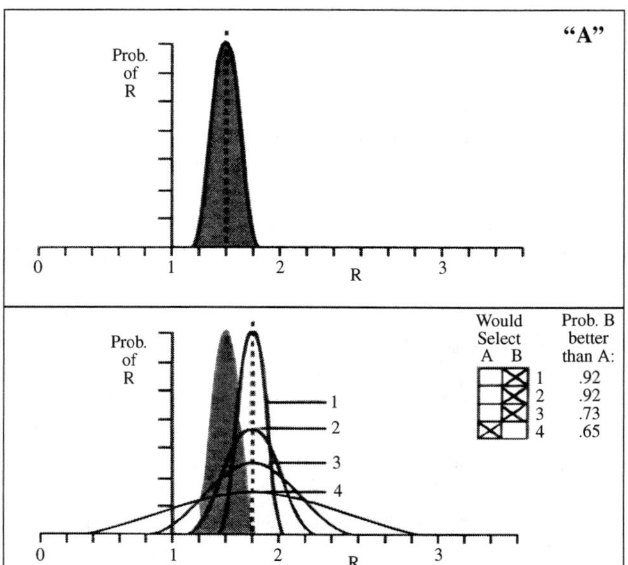

Survey Instructions: Example of choices

On the following pages are shown a series of pictures, each made up of histograms of ROI, normalized to your cost of capital. For each column, there is a top histogram, labeled "A," which will always serve as one of your choices. Listed next are five other pictures, each labeled "B." Within each of the B pictures are four curves that you will individually compare to the A curve. You might notice that each B picture contains the same four histograms of *R*, except that they have successively higher expected values. The A curve is drawn in gray on each picture to simplify comparison.

The horizontal axis measures *R*. At the point labeled 1 your return on investment is equal to your cost of capital. At the point labeled 2 your return on investment is double your cost of capital, and so on. Each tic mark in between thus represents a change of 0.2 in *R*.

Each set of graphs also has an associated list of probabilities. These indicate the chance that a particular curve B will produce a higher return than curve A. This is only one factor in your decision, and not the only criteria. Some other factors that you might wish to consider are as follows:

1) What is the relative difference in expected value between the two curves?
2) Does either curve extend beyond the 1 line? That is, could the firm lose money on either investment?

For each picture depicting the B alternatives, please indicate by placing an X in the appropriate boxes, whether you believe your firm would, based upon the *R* information shown, select alternative A or B. This should be done for each of the four curves within each or the five B alternatives relative to the A curve at the top of the column and for each of the columns on the following pages.

At the top of this page is an example to illustrate the procedure. Please make sure that only one X is placed in each of the four indicated rows (that is, the A and B alternatives are mutually exclusive).

III. On-Orbit Service/Repair (Commercial Missions)

During the 1980s, NASA conducted studies and undertook development activities to define possible uses of the space station by the communication industry. One such study[18] performed by Ford Aerospace for NASA's Lewis Research Center postulated a number of uses of the space station infrastructure by communication satellites to obtain cost and technical advantages over current practices. Preliminary results[19] indicated that retrieval operations had very high value for satellite failures that could be retrieved and fixed in space. In particular, the results can be summarized as follows:

1) The cost of making a communication satellite retrievable is likely to be relatively low—approximately 1% of the $130 million total capital expenditures (including insurance and launch costs but not the cost of making the satellite more easily repairable or the cost of the repair) for a typical geostationary communications satellite of the 1990s.
2) Probabilities of a retrievable failure (i.e., satellite is nonoperational but intact) are 8% of all launches based upon historical data.

3) Cost of the retrieval operation is dominated by transportation and insurance costs. Return of the entire satellite to Earth for repair is usually too expensive. Repair in space (if possible) has a potentially large economic value.
4) Retrieval costs are estimated as follows: For LEO retrieval-repair-relaunch scenarios, \$9 million is required for repair at the space station vs \$95 million for Earth repair. For GEO retrieval-repair-relaunch scenarios, \$85 million is required for station repair vs \$149 million for Earth repair.

The conclusion of this "static" analysis was that in-space repair could have high value (i.e., costs significantly less than the value of the satellite), whereas return-to-Earth for repair is not cost effective for GEO satellite failures and is marginally possible for failures in LEO. It was also concluded that a more sophisticated analysis of the retrieval problem was required because there is an interplay between probability of failure (which also determines insurance rates), satellite cost, launch cost, and retrieval mission costs, which can only be analyzed via a probabilistic model. For example, if there are few or no retrieval and repairable failures there is little value in retrieval. Another extreme case is when transportation costs are very high or very low, both of which favor retrieval—either to allow in-space repair and reduce launch costs or to save on replacement costs. Finally, because different launch operations, failures, and sparing methods affect the availability of satellites and the services that they provide, availability must be considered when performing comparisons of alternatives. To accomplish this requires the use of a probabilistic or Monte Carlo simulation model.

Since the study just referenced was completed, space station plans and actual capabilities have been scaled back. There are currently no known plans to utilize the space station as a transportation node or as an in-orbit inventory depot or repair facility. There are no known plans for the space basing of orbital transfer stages. But as space becomes more accessible and advances in related technologies and capabilities make possible more complex space operations, economic issues surrounding the placement and use of a space-based transportation node and in-orbit satellite retrieval and repair are likely to reemerge. The following discussion presents tools and techniques that have been developed and are likely to be useful if not required for an informed comparison of in-orbit satellite repair, satellite return to Earth for repair and replacement, and satellite return to a space-based transportation node for repair and replacement.

To evaluate the wide range of possible future satellite service (i.e., repair at the satellite location) and repair (i.e., repair at a transportation node) options, it is necessary to place these options within the context of business situations (private sector considerations) and government missions. Commercial communication satellites and service/repair options should be considered within the framework of a business model such as *DOMSAT*, and service/repair options associated with government missions should be considered within the framework of a life-cycle cost and availability model such as *SATCAV*. Both of these models are discussed briefly in Sec. I and in more detail in the indicated references. The modeling of a service/repair business option (i.e., a business devoted to providing transportation and related service/repair services to multiple satellite owners) is not discussed.

This may very well be a very viable business option requiring analysis but is likely to be considered only after individual satellite system business (i.e., a communication satellite business concerned with both its satellite placement and service/repair) options have been considered. Such an analysis would require a much more complex analysis and would draw upon *DOMSAT/SATCAV* techniques and pricing and investment modeling as described in Sec. IV of Chapter 3.

For discussion and illustration purposes the following is limited to the consideration of GEO communication satellite businesses where the *DOMSAT* model is an appropriate analysis tool. The following paragraphs describe a number of transportation scenarios that have been developed and incorporated into the *DOMSAT* model specifically so that service/repair options could be evaluated taking into account operations, reliability, cost, etc., and developing expected value and risk metrics so that the likely consequences of service/repair options could be judged both in terms of expected values and standard deviations.

Several terms that require clarification are used in the following discussions: OTV refers to a generic orbital transfer vehicle capable of transporting required system elements from one orbit to another orbit; OMV refers to a generic orbital maneuvering vehicle that is capable of docking with space assets; and space station is used to refer to a generic space asset that is an element of a space transportation node. In following figures a reusable launch vehicle is illustrated. As was already pointed out, the various components of a reusable system can be made expendable (for analysis purposes) by setting appropriate reliabilities to zero.

To review briefly, the *DOMSAT* financial model is a Monte Carlo model that runs many cases in order to take into account the probabilities of the various segments of a mission. The inputs to the model are as follows. The mission is broken down into segments with probability of success and cost (including cost uncertainties) for each segment. Insurance is based upon expected loss times a multiplier (i.e., 1.2) to allow for overhead and profit. (The expected loss is based on the input probability of success for each operation and the failure/recovery paths.)

The operation of the model is as follows. A mission model is selected (i.e., initial launch of three satellites and retrieval repair at the space station for repairable failures). One-thousand or more different cases of each scenario are run. An independent choice of success or failure for each mission segment is made based on the input probability of success of each operation. The result of each case is the rate of return on investment (and other financial metrics). The standard deviation of the rate of return around the mean rate of return for all cases is a measure of the financial risk. The mean rate of return (also referred to as the expected internal rate of return or IRR) must be high enough to justify the mission. If the standard deviation of the rate of return is high, the mission is risky; if it is low, there is low financial risk.

A number of nonretrieval and retrieval cases based on different transportation scenarios will likely require analysis and comparison. Typical transportation scenarios incorporated into the *DOMSAT* model and available for analyzing service/repair options are illustrated in Figs. 4.17–4.21. Two transportation scenarios must be specified for each case: one for initial payload (P/L) placement operations and the other for P/L maintenance (i.e., service/repair) operations.

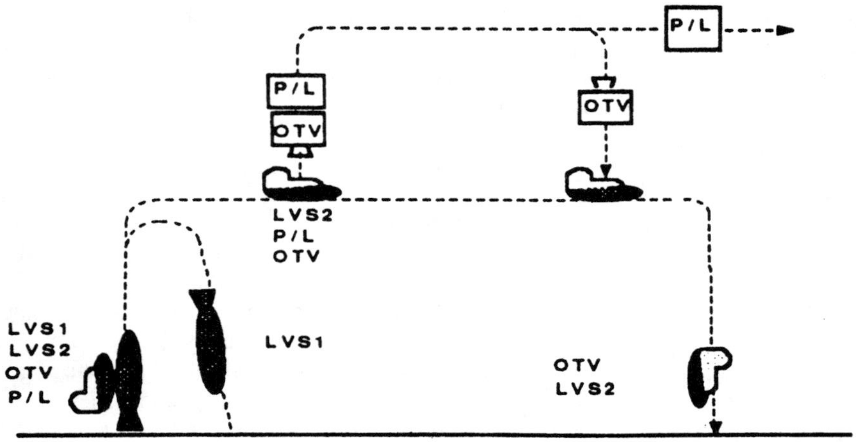

Fig. 4.17 Transportation scenario 1: direct placement using ground-based assets.

It is assumed that a transportation system consists of a generic launch vehicle (LV) that contains two stages (LVS1 and LVS2) and an OTV. Each of these stages can actually contain other stages but can be considered from a reliability, cost, and recovery point of view as being lumped into a single stage.

Transportation Scenario 1 consists of a reusable LV or LV and OTV or expendable LV and OTV for initial P/L placement and/or replacing failed P/L. (Stage 1, LVS1, can be reusable or expendable.) Scenario 1 can be used in conjunction with scenario 7 for the placement portion of the replace–return–repair mission. If reusable launch vehicles are considered, OTV and P/L checkout failures in LEO can be corrected when and if returned to Earth.

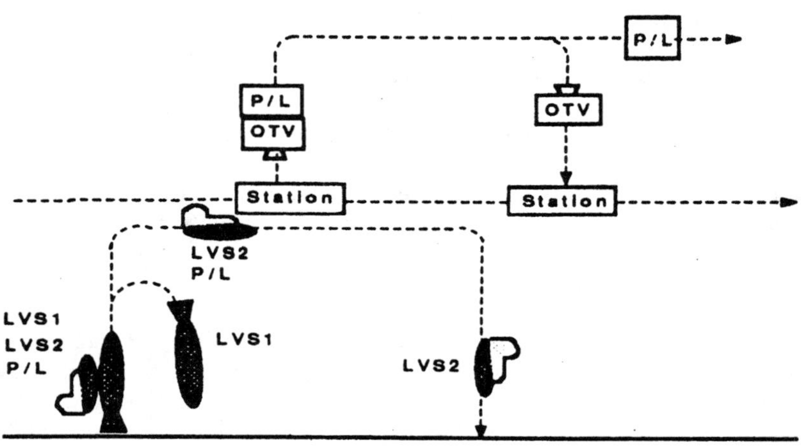

Fig. 4.18 Transportation scenario 5: direct placement using space-based assets.

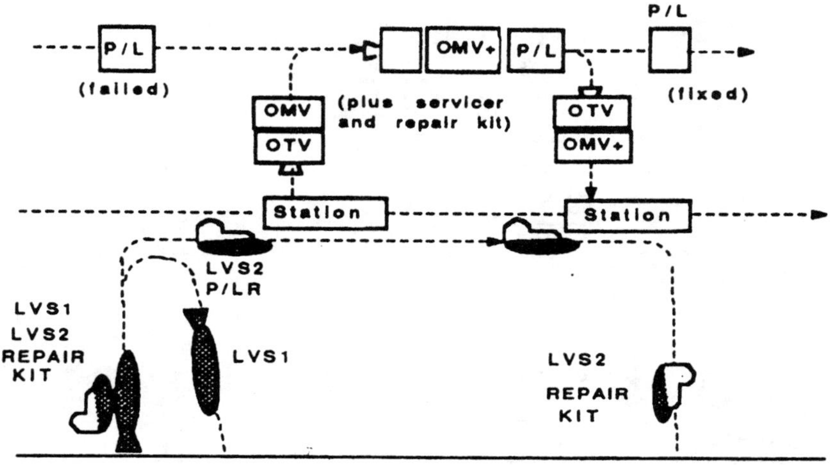

Fig. 4.19 Transportation scenario 7: on-orbit repair using space-based assets.

Transportation Scenario 5 consists of a reusable or expendable LV for transportation P/L to the space station. The OTV, based at the space station, is used to place the P/L into final orbit. The OTV can be reusable or expendable. It is assumed that P/L and OTV checkout failures can be corrected at the space station.

Transportation Scenario 7 consists of a reusable LV, LV, and OTV, or expendable LV and OTV for performing on-orbit P/L repair. The OTV and OMV are located at the space station. It is assumed that the OMV is capable of docking with specifically configured P/L. If repair cannot be accomplished, replacement

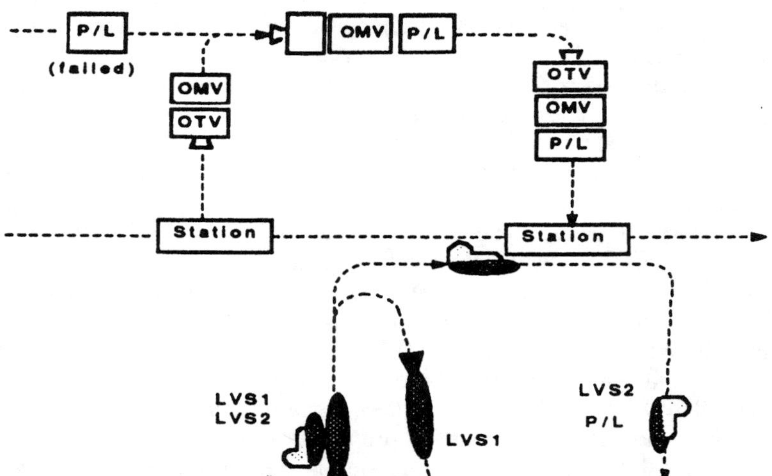

Fig. 4.20 Transportation scenario 8: return to earth for repair using space-based assets.

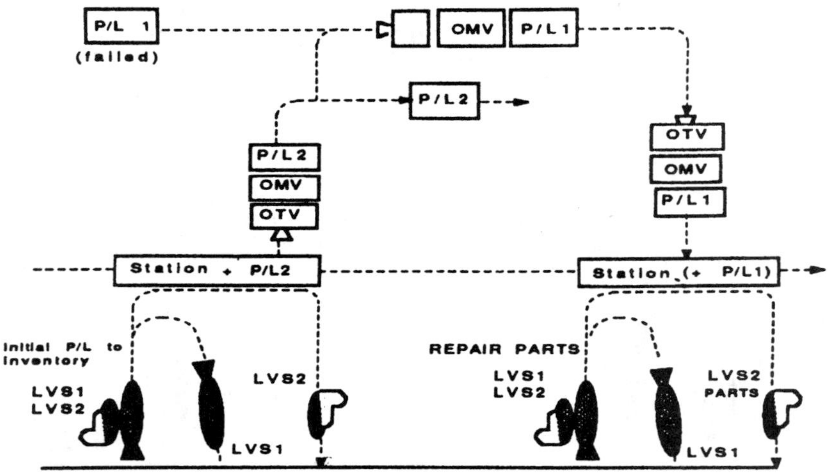

Fig. 4.21 Transportation scenario 9: replace, return to station for repair using space-based assets.

will be accomplished either via scenarios 1 or 5 (as specified). It is assumed that OTV checkout failures can be corrected at the space station. A repair kit is delivered from the Earth to the space station and, upon mission completion, returned to Earth.

Transportation Scenario 8 consists of a reusable OTV for acquiring and returning a failed P/L to the space station. The P/L is then returned to Earth for repair using a reusable LV. It is assumed that the OMV is capable of docking with specifically configured P/L. Replacement is performed (prior to returning the failed P/L) using an appropriate specified scenario. It is assumed that OTV checkout failures can be corrected at the space station.

Transportation Scenario 9 consists of a reusable LV and OTV transportation system for placing a P/L into orbit and returning a failed P/L in the same OTV flight. A P/L is stored in inventory on the space station, and repair is performed at the space station. An initial flight is required to place a P/L into inventory on the space station.

To compare alternative repair/service scenarios, these and other transportation scenarios have been converted into a series of nodes and branches that includes all success and pertinent failure recovery paths. Each node is a decision point representing a group of events. Emanating from each node are branches for the success and failure paths. The probability of departing the node on one branch or another depends upon the probability of success (or failure) of the events represented by the node. The failure recovery paths must ultimately provide a route to mission completion and requires a proper restoration of the inventory to its premission level. For example, if the mission is to place a satellite and retrieve a satellite, and if there is a failure to retrieve the satellite, then the purchase of a new satellite is necessary to restore the inventory to its proper level. The scenarios have been converted into detailed logic flows that include all cost-incurring events and

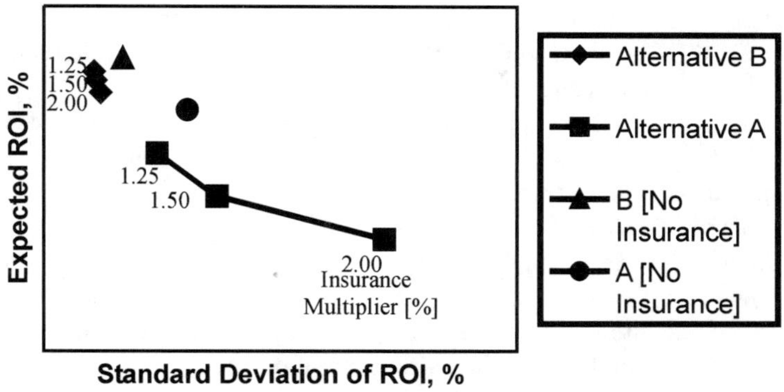

Fig. 4.22 Comparison of cases A and B indicating the sensitivity of results to insurance rate assumptions.

time delays. These logic flows have been included within the framework of the *DOMSAT* model. Thus different service/repair concepts can be evaluated by specifying one or more communication satellite missions, structuring these missions in terms of the alternative service/repair scenarios, and establishing the related mission metrics (i.e., net present value of mission life-cycle costs).

The results of an analysis of six nonretrieval and retrieval cases, based on different transportation scenarios, are described in the following paragraphs and summarized in Figs. 4.22 and 4.23. Specific quantitative results are not presented, but general trends are presented and their robustness described. The six cases are identified by letter as follows:

A: Launch is from Earth direct to GEO using transportation scenario 1 and a satellite of appropriate design. The space station is not utilized.

B: Launch is from Earth via the space station using transportation scenario 5 and a satellite of appropriate design.

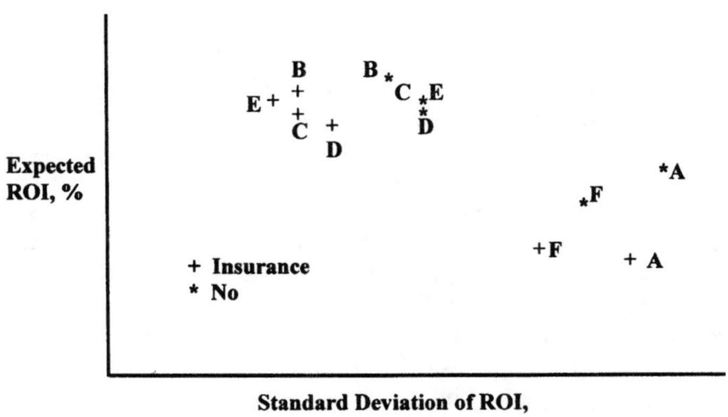

Fig. 4.23 Comparison of service/repair missions in terms of expected ROI and risk.

C: Launch is to GEO via the space station using transportation scenario 5. The OTV is used for in-orbit repair missions, as required, utilizing transportation scenario 7.

D: Launch is to GEO via the space station utilizing transportation scenario 5. The OTV is utilized to replace a failed satellite with a new satellite, and a failed satellite is returned to Earth for repair utilizing transportation scenario 8.

E: Launch is to GEO via the space station utilizing transportation scenario 5. The OTV is used to replace the failed satellite with a new satellite, and the failed satellite is repaired at the space station and placed into inventory. This is accomplished utilizing transportation scenario 9.

F: Launch is from Earth direct to GEO utilizing transportation scenario 1. The OTV is used for in-orbit repair missions utilizing transportation scenario 7.

A. Baseline Cases (A and B)

The methodology provides for all "new" retrieval-repair cases to be compared with a baseline nonretrieval case in order to assess accurately the difference in economic performance. Case A, the "business-as-usual" launch from Earth to GEO of an appropriate satellite design using transportation scenario 1, is the baseline for comparison with case F. Case B, the launch from Earth to GEO via the space station utilizing an appropriately designed satellite and transportation scenario 5, is the baseline for comparison with cases C, D, and E. The obtained results indicate that not only is case B likely to achieve a significantly higher expected IRR, but the variation (standard deviation) of the IRR is likely to be much less than that achieved through the use of case A. Typical results are summarized in Figs. 4.22 and 4.23, where each considered alternative (as well as sensitivities) is indicated in the expected value—standard deviation of IRR space. Specifics are discussed in following paragraphs.

B. Repair in Orbit

Two in-orbit repair cases, F and C, were considered. When satellites fail, they are repaired in orbit by utilizing the OTV to carry an OMV plus servicer and repair kit to GEO. [In all cases both nonrecurring and recurring costs of all items (such as repair kits, satellites, etc.) are taken into account.] The repair kit is brought up from Earth to the space station. Because not all satellites that fail are repairable, a "repairability" parameter (i.e., probability that a satellite that has failed in such a way that it is repairable) is considered that is less than 1.0. When satellites fail and are not repairable, new satellites are launched. Appropriate delays must be taken into account because the continuity of revenue streams and the timing of cash flows are important factors in the economic comparisons. An analysis of cases F and C shows no significant benefit for in-orbit repair. This is because of a combination of high reliability, few incidents of failure that can be successfully repaired in orbit, and the high cost of in-orbit repair. This conclusion will likely change as technology changes and estimates of cost and reliability are affected. The tools and methodology have been developed and can be utilized to reexamine the economics of service/repair as the need arises.

Case F includes the initial satellite placements in GEO utilizing transportation scenario 1 and the use of an expendable launch vehicle with an appropriately designed satellite. The OTV is used for an in-orbit repair mission as per transportation scenario 7. Results indicate that there is a slight improvement in internal rate of return (IRR) (relative to the baseline) with an accompanying decrease in the standard deviation of IRR. The difference appears to be small and insufficient to argue for a change from the business-as-usual case A. However, the financial risk (standard deviation of the IRR) is reduced and could lead to a preference for case F.

Case C includes initial placement of satellites based upon the utilization of transportation scenario 5 with launch to GEO via the space station of an appropriately designed satellite. The OTV is used for in-orbit repair missions utilizing transportation scenario 7. The nonrepair decision factor of 90% was used in the analyses and reflects the statistics that 90% of the in-orbit failures are judged to be unrepairable in orbit, and therefore no repair mission would be launched. This factor is high in part because initial launch scenario 5 uses deployment and checkout at the space station to minimize repairable problems. When case C is compared with case B, it is seen that there is a slight decrease in IRR with essentially no change in financial risk (i.e., standard deviation of IRR). The difference in IRR appears not to be large enough to argue for a change from the baseline case B. Future technology changes could alter this conclusion.

C. Replace, Retrieve, Repair

Two in-orbit replace–retrieve–repair cases, D and E, were considered. These cases are based upon the use of OTVs based at the space station. New satellites are delivered to the space station and placed into GEO via the space-based OTVs. When satellites fail, a spare satellite maintained in inventory on the space station is placed into GEO, and the failed satellite is returned via the OTV to the space station. Again, no significant change in return-on-investment or financial risk is noted, and there is no compelling argument for the retrieval, repair, and relaunch scenarios.

Case D returns the failed satellite via the space station to the Earth (utilizing transportation scenario 8) for repair and subsequent return to inventory. It is assumed that there is a low probability of nonrepairability (5%) and a high (92.5%) reliability of repair because repair is performed on Earth.

Case E utilizes transportation scenario 9 and repairs the failed satellite on the space station. The repaired satellite is then placed into inventory on the space station. The probability that the satellite cannot be repaired on the space station is 30% (considerably higher than the Earth repair of 5%). When satellite failures are not repairable, new satellites are placed into orbit via the use of the space station and the OTV. The reliability of repair was considered to be somewhat lower at 60% because the repair is carried out at the space station.

D. Summary and Conclusions

A number of service/repair scenarios were considered in order to demonstrate the type of modeling and analysis that is necessary in order to draw intelligent decisions with respect to the development of desirable options for improving overall space operations. As continually emphasized throughout this book, both expected value and risk information must be created for these decisions. The *DOMSAT* model, containing a range of transportation scenarios, has been used to evaluate typical GEO communication satellite business ventures with the different service/repair scenarios. The change in economic value of the business ventures is the maximum affordable investment that can be made to provide the various service/repair capabilities. The general form of obtained results is indicated in Figs. 4.22 and 4.23 with the former comparing only cases A and B and illustrating the type of sensitivity analyses that might be performed. The indicated sensitivities are with respect to insurance costs, where the insurance multiplier refers to an insurance rate that is the multiplier times the expected loss. (For example, a multiplier of 1.25 indicates that the cost of insurance is 1.25 times the expected loss. If the launch reliability is 0.9 and the total loss is $100 million, then the insurance cost is $12.5 million.) The implications of taking or not taking insurance are indicated.

Figure 4.23 compares the options of not taking insurance (i.e., self-insurance) vs taking insurance for each of the six considered cases. Only general trends are indicated with the specifics of the graph scales intentionally omitted because of the sensitivity of specific results to assumptions.

Several preliminary conclusions can be drawn at this point, but continued analysis is necessary as technology changes particularly with respect to transportation costs, telerobotics and transportation service, and satellite reliability.

Use of space-based assets is the first topic. The results indicate that there can be a substantial economic benefit to establishing a transportation node in space (i.e., the space station) for launch of commercial communication satellites. Not only is return-on-investment improved, but also financial risk is reduced (by using case B, the space station launch scenario rather than case A, a satellite launch direct to GEO). The reduced financial risk is because of an assumed increase in reliability of the space station and OTV operations vs the business-as-usual-launch.

In-orbit repair considers repair scenarios for the GEO communication satellites that appear to offer little economic value for the satellite scenarios analyzed. This appears to be because of a combination of high reliability, few incidents of failure that can be successfully repaired in orbit, and the high cost of space operations. Concern is with GEO communication satellites, and different conclusions have been reached with respect to other types of satellites (refer to Sec. XI).

Retrieve, repair, replace is the third topic under consideration. The considered repair scenarios also add little value of significance to communication satellite economics for both Earth-based and station-based repair scenarios. This is because of a combination of few repairable failures and the high cost of space transportation. Again, it must be realized that certain specific mission cases might gain greatly from retrieve, repair, and relaunch operations.

Insurance, in general, reduces risk with a slight reduction in ROI for the space station scenarios. For the less reliable non-station scenarios, risk remains relatively high with and without insurance—the main impact of insurance is to reduce return on investment. Insurance rate has little effect on ROI for the space station scenarios. The insurance rate has a great effect on financial risk for the non-station scenarios, with increased insurance rates causing increased risk.

Launch costs directly influence ROI for all cases. The station scenarios are more sensitive to upper stage (OTV) costs, whereas the non-station scenarios are more influenced by the initial stage launch costs. Non-station scenario risk is very sensitive to changes in launch costs.

IV. Analysis of User Fees

The federal government provides a range of services to the space industry. These include the provision of launch and related services, launch safety reviews, and licensing of launch activities. The government-incurred costs associated with the provision of these services, in some cases, are not reimbursed. To some, these nonreimbursed costs are considered as subsidies, whereas to others these nonreimbursed costs are necessary for an industry to compete with foreign competitors that are subsidized by their respective governments. The following paragraphs address the issue of government full cost recovery for the provision of government services. The full cost recovery of launch fees (for commercial launches from government facilities) is used as a proxy for this discussion.

The U.S. Department of Defense (DoD), and the U.S. Air Force in particular, owns and operates a significant portion of the U.S. space launch infrastructure. The Commercial Space Launch Act states that the DoD can be reimbursed by commercial users of the DoD facilities, but only for the incremental costs of using the facilities. DoD cannot recapture the investment in the facility that is, in many cases, the bulk or the cost of the launch, excluding the launch vehicle itself. When the Commercial Space Launch Act was enacted, commercial launches were but a small fraction of the total launch activity, and pricing based upon incremental cost seemed reasonable. In recent years the composition of launch activity has changed with commercial launches being a large part of the overall launch activity. In the current environment pricing based upon incremental cost no longer seems reasonable, particularly when budgets are tight and there is a need to update the launch infrastructure in order to accommodate both government and commercial users.

DoD is considering revising pricing policies for services it provides to launch companies that utilize DoD launch infrastructure. (DoD currently includes both direct costs and certain overhead items in its determination of recoverable costs.) DoD's ultimate goal is to achieve full cost recovery: "... [*set pricing so that*] industry pays the full freight of whatever actual [*cost*] is. What's the incremental cost? We don't know. The initial numbers look like maybe a few percent extra on each launch. What does that mean in terms of competitiveness? We don't know. . . ."[20] However, in trying to achieve this goal, it is a foregone conclusion that price increases will occur. In theory, full cost recovery is desirable, but in the real world where foreign providers of space transportation might be supported to varying degrees by their governments this might not be the case. There might also be a linkage between pricing of federal range services provided in support of commercial launches and services provided by nonfederal (commercial) launch sites. If the federal range does not price to achieve full cost recovery, it is likely that pricing for comparable services will be higher at commercial launch sites than at federal ranges. This can add additional weight to rethinking federal range pricing policies.

Because the U.S. commercial space transportation industry is in competition with foreign providers of space transportation, it is possible that price changes will affect the competitive position of the U.S. launch companies. Not increasing prices for DoD-provided services, on the other hand, will have a direct impact on DoD budgets. Not recovering the full cost of provided services is a form of subsidization. Negatives of subsidization include promoting inefficient systems and distorting the competitive environment, which ultimately leads to higher prices. However, this must be balanced against the influence and actions of foreign competition and international trade implications. But what are the overall economic impacts if pricing policy is changed or maintained?

The following pages describe an approach for establishing the likely impact of changes in launch fees on U.S. launch vehicle (space transportation service) market share and the resulting U.S. economic impacts in terms of job creation and U.S. Treasury revenue. Application of the methodology would result in establishing the general relationship between pricing of DoD launch services and economic impacts. No attempt is made to argue the merits of one pricing strategy vs another, but an attempt is made to show the implications of price changes in international competitions. The discussion is limited to the consideration of launching communication satellites, which are, by far, the largest share of the nongovernment launch market. In addition, there is a further focusing on the geosynchronous communication satellite market segment. (The narrowing of the discussion to GEO communication satellites does not imply that the described methodology is not applicable to other market segments; the narrowing has been done to simplify explanations.)

A. Implications of Launch Fee Changes

Competition between the providers of space transportation (i.e., the launch operators) has become intense with the competition being between both domestic and international launch operators. The different launch operators (for example, Lockheed Martin, Sea Launch, and ArianeSpace) offer services having a mix of attributes that can affect user choices. These attributes, including price, payment schedule, reliability, payload placement accuracy, delivery capability, insurance, and schedule/availability, can have significant impacts on user financial performance and, through user decisions that reflect financial performance perceptions, affect market share. Thus, changes in the federal range fee structure that affect the attribute mix offered by U.S. launch operators (that partake of federal range services) are likely to impact U.S. launch operator market share by influencing user choices for internationally competed launch services.

The following paragraphs first describe user decisions as they can be influenced by perceptions of financial performance and the resulting impact on market share. This is followed by a discussion of the development of financial performance measures and is based upon the use of the financial planning model *DOMSAT*. Finally, the economic impacts that are likely to result from a change in market share are developed.

1. Price Change Impacts on Market Share

To understand the characteristics of the demand (i.e., market share) for space transportation services for launching communication satellites better, it is necessary to understand the implications of financial performance measures in selecting among alternatives. The basic problem can be stated as follows: given that a decision has been made to commit resources to a communication satellite business venture and that there exist multiple implementation alternatives (i.e., the selection of a specific space transportation launch service such as Ariane, Atlas, Long March, Sea Launch, etc.), what is the chance that different alternatives will be selected in terms of perceived differences in resulting financial performance measures?

Clearly in a riskless environment and with all other factors being equal, the choice would be made to select that alternative which maximizes the net present value of cash flow or return on investment. Enter risk, and the problem becomes more complex because it is necessary to understand the tradeoffs between perceptions of risk and other financial performance measures.

To understand these tradeoffs, a limited survey was conducted of decision makers (i.e., the financial community) within communication satellite businesses. (A similar survey is described in Ref. 7.) The survey (Sec. IV of Chapter 2 presents a description of the survey technique) sought to establish the likely choices between alternatives having different expected value of return on investment and risk characteristics. The result of this limited survey was the development of the risk-expected return attitudes of those responsible for selecting launch services and is depicted by a family of curves that indicate the chance of selecting alternative x relative to alternative y. The general form of the obtained results is indicated in Fig. 4.16. The curves represent the chance of selecting x relative to y, $p_{x/y}$, in terms of the ratio of expected values of ROI m_x/m_y and the ratio of perceived risk σ_x/σ_y. Risk is defined as the variability of ROI as expressed by the standard deviation σ of the ROI probability distribution. The variable m represents the expected value of this distribution. Thus if a business is defined and a financial simulation model (such as the *DOMSAT* model) utilized to evaluate the financial performance measures with two different launch vehicles with each being described by a mix of service and performance attributes, their associated financial attributes m_x, m_y, σ_x, and σ_y can be obtained. Entering the set of curves in Fig. 4.16 with these values results in an estimate of the probability that x will be selected ($p_{x/y}$). If changes in the DoD pricing structure are passed on as changes in the price of launch services, a change will result in the values of m and σ, which will result in changes in the probability of selecting a particular launch service (i.e., a change will occur in market share). This is depicted in Fig. 4.16 as the probability $p_{x/y}$ with pricing policy A and $p'_{x/y}$ with pricing policy B. There is a higher launch cost associated with alternative x under policy B than under policy A.

Thus, the likelihood of utilizing different launch vehicles can be determined for different communication satellite business ventures. This can be accomplished for a number of typical business ventures that are representative of the market for launch services with obtained results scaled appropriately. If three different launch vehicles (a, b, c) are considered with values $\{m_a, \sigma_a\}$, $\{m_b, \sigma_b\}$, and $\{m_c, \sigma_c\}$ determined through the use of a financial simulation model such as *DOMSAT*, and resulting values of $p_{a/b}$, $p_{a/c}$, etc. as determined from Fig. 4.16, it is possible to establish market share for launch vehicle a, S_a, as

$$S_a(\%) = 100 * [p_{a/b} + p_{a/c}]/[p_{a/b} + p_{a/c} + p_{b/a} + p_{b/c} + p_{c/a} + p_{c/b}]$$

The effect of government policies and programs can be assessed in terms of impacts on launch vehicle attributes or business parameters that then result in changes in values of m and/or σ for one or more launch vehicles. This is not limited to the implications of DoD price changes but any action that can affect decision-maker perceptions of financial performance including risk. For example, government procurements of launch services will reduce costs through learning effects that can be passed on to commercial users of launch services in the form of lower prices. These lower prices will affect the m and σ values of the users of launch services that will affect the probability of selecting the launch vehicle. Launch successes or failures alter reliability perceptions that also affect pro forma financial performance including risk. Similarly, anchor tenancy arrangements will affect market uncertainty and resulting business risk (i.e., σ) perceptions. These actions result in changes in market share through the use of the family of curves illustrated in Fig. 4.16.

Thus a mechanism has been developed that can be used to assess, in a rational and consistent analytic manner, the likely impacts of government policies and programs on launch vehicle market share. But, alas, the methodology has not been applied, and decisions are being made without the benefit of crucial information that could readily be obtained.

2. *Development of Measures of Financial Performance (Selecting a Launch Vehicle)*[21]

Because risk is an important decision making attribute, it is necessary to quantify financial performance measures that include both expected values and risk. To accomplish this, the financial simulation model *DOMSAT*[22] was developed and utilized to evaluate the pro forma performance of communications satellite business ventures. As discussed in Sec. I, *DOMSAT* is a communication satellite business venture financial planning model specifically developed to provide a means for evaluating the impacts of a broad range of policies and decisions on the financial performance of communication satellite business ventures. *DOMSAT* provides the means for evaluating the impacts of both government and industry policies and perceptions and decisions in terms of financial performance measures such as after-tax profit, cash flow, present value of cash flow, and ROI. Both expected and risk measures are developed.

As already described, *DOMSAT* is a Monte Carlo simulation model that stochastically simulates the performance of a broad range of communication satellite business ventures, explicitly and quantitatively taking into account uncertainty, unreliability, and resulting risk. The model works with data often provided by a business planner describing the parameters of a communication satellite business. However, in addition to the expected value data, many variables can be described in terms of a range of uncertainty and the form of the uncertainty (i.e., a probability distribution). Included are data that describe the reliability (initial, random, and wear out failures) and performance of satellite and launch vehicle subsystems, desired launch schedule and possible launch delays, low-

Earth-orbit to geocentric-Earth-orbit transfer time, number of transponders, sparing and grouping of transponders, services to be provided and prices for each, annual demand and elasticity, recurring and nonrecurring satellite costs, launch costs, insurance costs, depreciation, and financial- and balance-sheet-related data. Many of these input parameters can be considered either as deterministic or specified as uncertainty variables requiring the specification of the range and form of uncertainty. In addition, the model is specifically configured so that situations involving replacement of operating satellites with satellites of different capabilities can be examined; two generations of satellites can be considered with data provided for both the current and next-generation satellites.

The use of Monte Carlo techniques allows the complete business scenario to be simulated a large number of times, each time randomly sampling from the uncertainty profiles and the reliability characteristics that are specified. The results of all of the business analyses are saved and histograms developed of the financial performance measures. These are summarized in terms of expected values and standard deviations and serve as input to the multiattribute decision database for estimating market share.

The model has been used to analyze the financial consequences of launching on different launch vehicles (such as Ariane vs Atlas), taking into account differences in reliability, pricing, insurance rates, final payload placement accuracy, schedules, and time delays that can result from failures. It can also be used to assess the impacts of changing range service pricing or to assess the impacts of requirements to change orbit at near end of life. This can be accomplished by establishing a base-case scenario from which variations can be measured.

DOMSAT can be used to help answer questions such as the following:

1) What are the financial impacts of launch delays? satellite life? insurance rates? pricing?
2) What is the value of space launch infrastructure investments?
3) Are foreign launch vehicles priced to account for their capability differences?
4) How will the availability (or lack of availability) of less costly foreign launch services affect the U.S. communication satellite business?
5) How sensitive are forecasts at required levels of satellite replacement to assumptions about the relationship between satellite design life and replacement schedule and the impact of specific technology developments?
6) What will be the financial impacts on communication satellite businesses of alternative strategies to reduce orbital debris (e.g., moving GEO satellites to higher altitudes prior to end of life and restricting transfer stage trajectories)?
7) Which launch vehicle is most cost effective?
8) What policies should be followed to improve transportation service market share?

DOMSAT is specifically configured to provide the expected value and risk data used in the analysis of market share and changes in market share that can result from changes in launch vehicle, satellite, or business attributes.

3. Economic Impacts of a Launch Lost to Foreign Competition[23]

As discussed in Sec. VIII of Chapter 2, when foreign suppliers capture launch vehicle business they gain many advantages. To begin, just as loss of business harms the U.S. manufacturers gains in foreign business enable these suppliers to become more competitive (i.e., gain market share) by expanding their base upon which to write off fixed costs. To the U.S., there is a net loss of employment (i.e., jobs, or more appropriately, person-years of employment) with its ripple effects in the economy. There is a loss in revenue base that could, in part, form the basis for research and development which would improve the technology and make it more competitive in the future. And, more importantly, there is a loss of domestic access to space. In the near term use of foreign launch vehicles to place U.S.-manufactured satellites into orbit can increase the competitiveness of U.S.-manufactured satellites and reduce end-user fees. However, in the longer term loss of access to domestic launch vehicles could eventually result in losses to domestic satellite manufacturers. At the least one could expect that foreign providers of launch services would give priority to the launching of their domestically manufactured satellites.

The economic impacts of a lost launch can be categorized as: 1) income opportunity, direct and indirect; 2) employment, direct and indirect; 3) federal revenue; and 4) other. "Direct income" is income received (or in the case under consideration, income lost as a result of competition) by an aerospace firm from sales of services related to an export item. "Indirect income" is income generated (or lost) in the community (e.g., among restaurants, doctors, service stations, auto dealers, retailers, etc.) as a result of expenditures by employees of the firm that has sold the basic product or service. In other words, the loss of a launch to foreign competition will reduce direct income that in turn will reduce purchases made in surrounding communities (reduction in indirect income).

"Direct" and "indirect" employment are defined in the same manner, where direct employment is the consequence of procurement of products and/or services relating to the construction and provision of space transportation services and indirect employment is the consequence of money being spent in a community by the direct employees. Because the impacts of a lost launch are discussed, direct and indirect income refers to a reduction in income (i.e., a loss) relative to having won the launch competition, and direct and indirect employment refers to employment (measured in person-years) associated with the launch, the satellite if applicable, and employment associated with the spending of income earned in the production of the launch (the indirect effects). These impacts are relative to that which would have been if the U.S. firm were the winner of the launch competition.

In addition to the preceding information, there are other impacts including increased costs as a result of reduced reliability caused by reduced flight experience, and loss of competitiveness capability resulting from increased unit costs (i.e., the result of a smaller production base). In addition, there will be long-term impacts associated with specific uses of revenue. A firm utilizes the revenue from the sale of a product or service for a number of different purposes: labor, material, overhead, and profit (ultimately all of these go to labor). However, the specific uses differ and have different consequences on the economy. For example, part of the firm's expenditures can be traced to R&D. The use of funds for R&D

has different long-term consequences than the use of funds for manufacturing. R&D expenditures tend to increase productivity in the long term (more than other types of expenditures) and therefore result in a long-term multiplier effect—in other words, dollars spent on R&D today will generate several times their amount in the long term. Thus R&D expenditures require separate considerations.

These short-, medium-, and long-term impacts of a competitively procured launch lost to foreign competition are summarized here. *The short-term impacts* are as follows: 1) direct (i.e., aerospace sector and its suppliers) income loss; 2) indirect (i.e., nonaerospace sectors) income loss; 3) reduced direct employment (i.e., aerospace sector and its suppliers); 4) reduced indirect employment (i.e., nonaerospace sector); 5) reduced exports; and 6) reduced tax revenues (from both direct and indirect income) and increased transfer payments (i.e., unemployment compensation). *The medium-term impacts* (in addition to those identified as short-term impacts) include the following: 1) loss of competitive capability caused by higher unit costs; 2) failure to provide a base for research and development to upgrade systems; 3) lower reliability from less flight experience; 4) potential loss of satellite procurements and technology; 5) potential failure to maintain world-class facilities if demand is reduced; and 6) reduced opportunity to increase productivity (i.e., caused by reduced investment in R&D). *The long-term impacts* (in addition to those identified as both short- and medium-term impacts) are as following: 1) failure to develop space activities to their fullest potential; 2) potential loss of satellite procurements, technology, and manufacturing base; 3) failure of the U.S. to remain a leader in the development of space; and 4) reduced productivity as a result of lower investments in R&D. The type and magnitude of the impacts are a function of the circumstances associated with the lost launch that is, loss of a U.S. launch of a foreign satellite, loss of a U.S. launch of a U.S.-manufactured communication satellite (for providing domestic service), loss of a U.S. launch of a U.S.-manufactured communication satellite (for non-U.S. service), and loss of a U.S. launch and U.S.-manufactured satellite. The costs associated with the lost launch include a reduction in direct and indirect income and employment, reduction in tax revenue and increase in transfer payments, reduction in R&D expenditures, loss of competitive capability resulting from increased unit costs, and lower reliability as a result of less flight experience.

To illustrate the short-term impacts of a launch lost to foreign competition, consider the loss of a communication satellite for providing domestic communication services, where it is presumed that such a loss will result in an increase in the profit of the communication carrier. This increase in profit can be maintained or distributed in a number of ways including a reduction in end-user fees paid for the provided communication services. The reduction in direct income is therefore equal to the launch price less a compensating increase in carrier profit that results from the use of the foreign launch service, payment for foreign hardware that would normally be incorporated into the U.S. launch vehicle, and the reduction in corporate taxes (net of launch provider and communications carrier). [Because the change in profit is likely to be less than the change in consumer surplus that will result from a price change (a result of price elasticity), using the former is conservative.] The increase in carrier gross profit is the launch price associated with the U.S. launch service less the foreign launch service price. Because the provision of launch services is normally but one of many ongoing corporate activities and it is not possible to obtain meaningful data relating to product line profitability, the reduction in corporate taxes can be approximated as the launch price multiplied by the corporate return on sales (typically about 6%) and the corporate tax rate.

The reduction in indirect income is accounted for by multiplying the reduction in direct income by the indirect income multiplier (about one dollar of indirect income is generated for every dollar of direct—a multiplier of two). Because the expenditures for satellite test and integration (about 40 people for 60 days) are provided outside of the U.S., there is a reduction in the indirect income in the U.S. Corporate tax revenues are also considered because they are used to finance a broad range of activities that then generate the indirect income.

The determination of employment reduction is predicated upon the assumption that all income is eventually converted into labor. To estimate direct employment, the price of the launch is divided into two parts: a portion that goes to pay benefits and is divided by the nonaerospace average wage rate, and a portion that pays everything else and is divided by the aerospace industry average wage rate. To obtain an estimate of indirect employment, indirect income is divided by the nonaerospace industry wage rate.

The reduction in federal revenue consists of the reduction in corporate taxes, the reduction in direct and indirect income taxes, and the increase in unemployment compensation payments. The unemployment compensation payments take into account the average compensation rate and the average duration of unemployment. It is assumed that there is a growing economy that will eventually reemploy those who become unemployed.

Finally, the medium- and long-term effects of the reduction in R&D expenditures need to be mentioned. Lower R&D expenditures can result in lower productivity than would otherwise be the case. The effect on productivity can be estimated by multiplying the reduction in R&D expenditures by the productivity multiplier. The reduction in R&D expenditures is the launch price multiplied by the R&D investment rate (expressed as a percentage of sales). However, because the productivity multiplier is rather difficult to measure, its estimation has been based upon broad-based technology programs that utilize historical results based upon the technology program considered rather than the "narrow" focused program in support of launch vehicles, and the change in productivity that might

Table 4.2 Economic impacts of a launch lost to foreign competition

	Typical per launch impacts	
Impact areas	Atlas	Delta
Reduction in exports, M$	100–165	45–94
Reduction in		
Direct, person-years	2,500–5,120	1,060–2,930
Indirect, person-years	3,580–7,000	1,510–4,010
Total, person-years	6,130–12,130	2,580–6,940
Reduction in federal revenue, M$	50–110	20–63

result from a small change in level of R&D has not been addressed. (Estimates have been made that range from slightly less than 3 to slightly more than 14.) No consideration has been given to the effect of R&D expenditures on the U.S. economy other than to note that it is likely that productivity increases will probably be lower with lower R&D spending.

Typical per launch impacts are indicated in Table 4.2 based upon the loss of an Atlas or Delta launch to foreign competition. Actual impacts depend upon the specific events, for example, a loss of a U.S. launch of a foreign satellite, loss of a U.S. launch of a U.S. communication satellite for providing domestic communication services, etc. The indicated ranges encompass the earlier identified situations with the higher bounds being associated with the loss of a launch and of a U.S.-manufactured satellite.

A change in federal range fees can effect the federal range, the launch operators (the commercial users of the range), and the customers of the launch operators (for example, a provider of communication satellite services), who make the decisions regarding the choice of a launch service and ultimately determine market share. The form of these impacts is summarized in Table 4.3.

B. Measuring the Economic Impacts of Launch Price Changes

The available transportation services, each with its mix of known and perceived attributes, have resulted in the market shares of today. The market shares will change over time as the perceptions of launch service attributes change. These changes in perceptions, imagined or real, will affect anticipated financial performance and will result in changes in decisions relating to the selection of launch services. If all of the launch service attributes, with the exception of price, are assumed to be fixed, it is possible to anticipate the likely impacts of changes in launch fees, first in terms of changes in market share, and then, as a result, changes in U.S. Treasury revenue and job creation. The method for accomplishing this is illustrated in Fig. 4.24 and is based upon the use of the *DOMSAT* model for transforming transportation service attributes into user financial performance, the use of the decision database described earlier for transforming financial performance measures into probability of selecting each of the various competitive launch services (i.e., determining market shares), and then transforming the changes in market share that are the direct result of attribute changes (in the case being considered, DoD infrastructure use fees) into economic impacts. (Actually, the same methodology is applicable for analyzing the impacts of changes in any one or more of the launch service attributes so long as changes in the attributes can be transformed into changes in the financial performance of the business ventures that are the customers for space transportation services.)

Referring to Fig. 4.24, the assessment of the economic impacts of launch service pricing starts with the specification of the communication satellite business ventures that are to be used as representative of the users of space transportation services. This requires that business data that are associated with each mission (i.e., launch competition) be specified together with the perceived attributes of the competitive launch services. The specification of the business ventures entails quantifying the various inputs for the *DOMSAT* model. These inputs include the range and form of uncertainties associated with a number of variables such as

Table 4.3 Summary of impacts of changes in federal range fees

Government price change	Impacts on: Range operator	Launch operator	U.S. economy
Increase	Increased cost recovery	*Maintain price*[a]	
	Reduced budget need	Reduced profit	Reduced U.S. treasury revenue
		Reduced taxes	(tax reduction)
		Increased price	
		Maintain profit[b]	Reduced U.S. treasury revenue
		Reduced competitiveness	Reduced jobs
		(reduced no. of launches)	
Decrease	Decreased cost recovery	*Maintain price*	
	Increased budget need	Increased profit	Increased U.S. treasury
		Increased taxes	revenue
		Decrease price	
		Maintain or increase profit[c]	Reduced U.S. treasury revenue
		Increased competitiveness[d]	Increased jobs
		(increased no. of launches)	

[a]In the absence of a government price change, this is the price that maximizes launch operator objectives.

[b]The objective is to maintain profit, but this might not be achieved because of reduced competitiveness (i.e., reduction in the number of launches).

[c]The objective is to maintain or increase profit; this can be achieved if competitiveness is increased sufficiently to at least offset the price reduction.

[d]Change in market share (assuming price change is insufficient to alter total demand for launches).

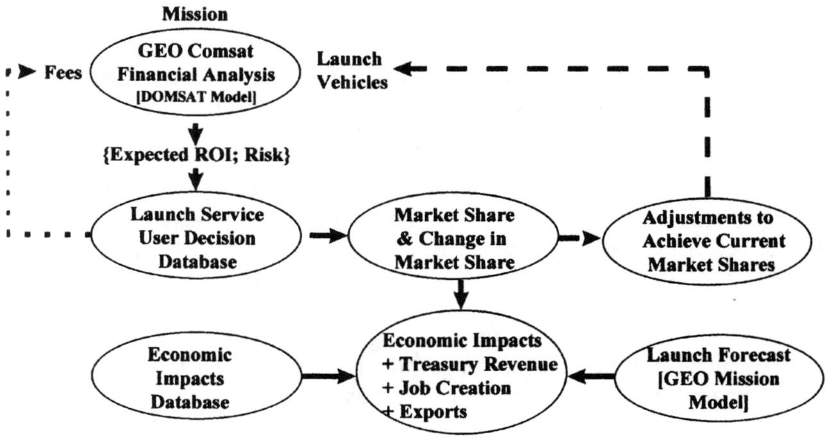

Fig. 4.24 Methodology for assessing economic impacts of launch service pricing.

demand, expenses, and capital expenditures. When considering market share, it can be assumed that the price of launch services is a known quantity (i.e., quotes have been obtained). To analyze the effect of price changes, all variables are assumed fixed, and launch service price is varied so that its effect on financial performance can be isolated. The *DOMSAT* model is then used to transform the business and launch service attributes into financial performance measures. Both expected ROI and risk (i.e., the standard deviation of ROI) are determined based upon the use of each of the competitive launch services. If three different launch vehicles (a, b, c) are considered, the result is the determination of $\{m_a, \sigma_a\}$, $\{m_b, \sigma_b\}$, and $\{m_c, \sigma_c\}$.

As already discussed, the ratio of expected values of ROI m_x/m_y and the ratio of perceived risk σ_x/σ_y are then used to enter the launch service user decision database (as illustrated in Fig. 4.16). This results in the determination of values of $p_{a/b}, p_{a/c}$, etc., which indicate the probability of selecting each service relative to each of the other launch services. From these results it is possible to establish market share for launch service "a," S_a, as

$$S_a(\%) = 100[p_{a/b} + p_{a/c}]/[p_{a/b} + p_{a/c} + p_{b/a} + p_{b/c} + p_{c/a} + p_{c/b}]$$

Changes in launch fees will be reflected in service prices that will change the values of m and/or σ for one or more launch vehicles. This in turn will cause changes in market share that will result in economic impacts. If the calculated market share differs from that which currently exists in the market place, adjustments can be made to the perceptions of launch vehicle and service attributes and the preceding steps repeated. This process can be repeated until current market shares result. This can then be considered as the base case from which changes are measured.

The final step is to multiply the determined changes in market shares by the number of launches anticipated for each launch service alternative given the current DoD pricing policy (i.e., the base case). This results in the change in the number of launches for each launch service. Using the data indicated in Table 4.2 together with the changes in number of launches, economic impacts can be developed in terms of change in exports, change in direct and indirect employment, and changes in federal revenue. The data indicated in Table 4.2 are based upon Atlas and Delta; similar data can be developed for other launch vehicles.

When current market shares have been obtained, changes can be made to variables such as launch price in order to determine the effect on market share—in essence, performing a sensitivity analysis around the base case. Varying launch service fee, and adjusting launch service price accordingly, results in the establishment of the relationship between the change in launch service fee and the resulting change in market share. Utilizing this relationship together with an economic impacts database consisting of the type of data illustrated in Table 4.2, the mathematical model for calculating the economic impacts and the specific data (such as aerospace industry average wage rate, average number of weeks per unemployment claim, etc.) required by the economic impacts model results in the economic impacts per lost launch (to foreign competition) in each of the typical business scenarios analyzed.

The final step is to multiply the economic impacts resulting from each launch competition by the data in the launch forecast model (the number of launches by mission type). This weighted average results in the total annual economic impacts.

Changes in DoD (U.S. Air Force) fees for launch service support will affect U.S. Air Force revenue and will have economic impacts on the U.S. economy. These changes are depicted conceptually in Fig. 4.25. The change in U.S. Air Force revenue is indicated as a function of the percent increase in price. It is shown as a linear function, although this is not likely to be the case (because of the likely

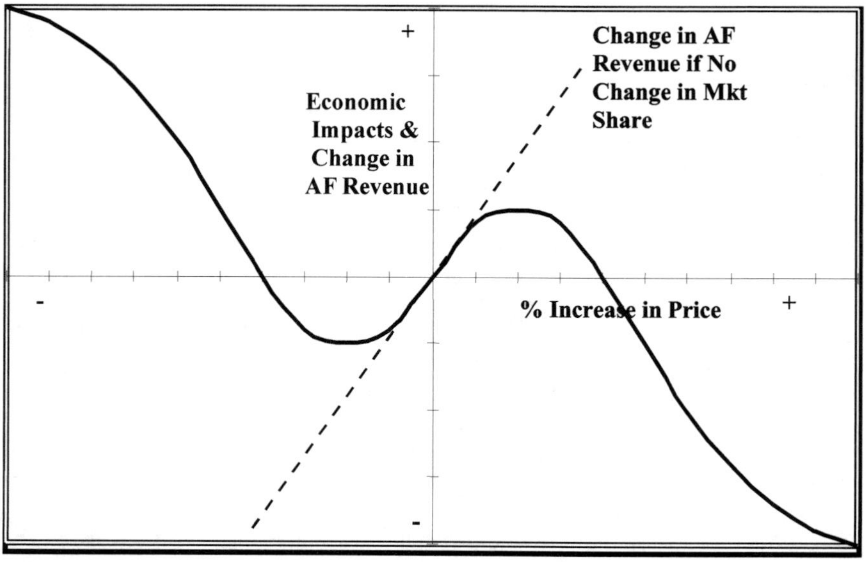

Fig. 4.25 Economic impacts as a function of launch service price change.

change in market share for transportation services that use the DoD launch infrastructure). Unfortunately, it is likely that the linear assumption will be used in evaluating government pricing policies. Also shown in Fig. 4.25 are the economic impacts as a function of the percent change in price. It is likely that economic impacts will be positive with small increases in price (i.e., little or no change in market share), but as price continues to increase it is anticipated that the economic impacts will become negative as more decisions are made to utilize non-U.S. launch services (i.e., as a result of decreased U.S. market share). It is anticipated that just the opposite will result from decreases in DoD fees.

C. Summary and Conclusions

DoD, and the Air Force in particular, own and operate a significant portion of the U.S. space launch infrastructure. The Commercial Space Launch Act states that DoD can be reimbursed by commercial users of the DoD facilities, but only for the incremental costs of using the facilities. When the Commercial Space Launch Act was enacted, commercial launches were but a small fraction of the total launch activity, and pricing based upon incremental cost seemed reasonable. In recent years the composition of launch activity has changed with commercial launches being a large part of the overall launch activity. In the current environment pricing based upon incremental cost may no longer be reasonable, particularly when budgets are tight and there is a need to update the launch infrastructure in order to accommodate both government and commercial users. DoD's ultimate goal is full cost recovery. But such a pricing policy is likely to result in higher U.S. launch costs that might affect international competitiveness. A methodology has been developed that will allow the effects of changes in launch fees for government-provided services to be observed in terms of changes in U.S. space transportation service market share. The changes in market share are then developed in terms of impacts on the U.S. economy as measured in terms of job creation and U.S. Treasury revenue. The methodology first converts changes in launch service attributes, or other factors that can have an effect on business financial performance, into changes in perceived (by those affected by the government programs and policies) financial performance. These changes in financial performance are then, through the use of a multiattribute decision database, transformed into likely changes in market share. Finally, these changes in market share are converted into impacts on the U.S. economy.

The developed methodology is extremely robust and will allow the effect of a broad range of government policies and programs, which can affect either or both decision-maker expected perceptions of financial performance or perceptions of business risk, to be assessed in terms of impacts on the U.S. economy.

V. Fair Pricing: Impacts on International Competition

Competition between the providers of space transportation has become intense with launch services being offered by organizations from both market and nonmarket economies. Primarily the U.S's Atlas and Delta, France's Ariane, Russia's Proton, and China's Long March provide GEO commercial launch services. Recently, Sea Launch, a U.S.–Russian joint venture, has entered the fray and rapidly become a major factor. This competition, combined with the possibility that governments provide both direct and indirect subsidies in order to influence transportation user choices, bodes well (at least in the near term) for the users of space transportation services but has serious ramifications with regard to the future makeup of the international space transportation industry.

In the recent past it was felt that the governments of nonmarket economy countries provided significant subsidies that resulted in the pricing of provided launch services to bear little or no relationship to actual costs. This, if allowed to occur, could have caused significant distortions in free market competitions. The U.S. government reaction was to place quantity restrictions on these launch services and to enter into agreements with nonmarket economy governments regarding international trade in commercial launch services. The quantity restrictions were in the form of placing limitations (i.e., quantity restrictions) on the number of U.S.-manufactured satellites and satellites utilizing U.S. technology that could be launched annually by these subsidized launch services.

The U.S. Government and the Government of the People's Republic of China (PRC) entered into a Memorandum of Agreement that addressed certain issues regarding international trade in commercial launch services including continued PRC participation in the international market for commercial launch services. The PRC agreed that its providers of commercial launch services would provide such services "at prices, terms, and conditions which are on a par with those prices, terms and conditions prevailing in the international market for comparable commercial launch services offered by commercial services provides from market economy countries, including the United States."[24] The agreement went on to state that when the differential between a bid, offer, or contract to provide commercial space launch services by a Chinese launch service provider and the bid, offer, or contract by a provider from a market economy country is greater than 15% (relative to the market economy provider) and after taking into consideration "comparability factors" the U.S. believes that China's launch service prices are not consistent with the foregoing, then the U.S. and China shall have special consultations so as to resolve the problem of pricing.

Another part of the agreement established PRC launch quantity constraints in terms of allowed average annual number of commercial launches. These constraints were tempered by anticipated market (demand) conditions and infrastructure availability.

Both parties to the Agreement acknowledge the fact that differences in launch service attributes (such as lift capability and its effect on satellite useful life, reliability, etc.) have an impact on pricing and should be taken into account in the assessment of the fairness of pricing. Both parties agreed to the following six "factors" for comparing or evaluating launch in the international market. It is understood that such factors can often explain legitimate distinctions in the price offered for the launch of a particular payload relative to market economy providers of commercial launch services. The agreed-to factors are discussed next together with the impact that the factor could have on the ultimate price to the customer when applied in a competition. These cost ranges represent the values associated with the factors for contracts involving PRC launch providers in the international commercial launch services market.

The factors considered include the following:

Intended orbit is the first factor. It was agreed that recent contracts involving the purchase of a perigee kick motor (PKM) by the customer have resulted in additional costs of $6 to $7 million (US dollars) for the purchase of the PKM alone. Integration and risk management costs for the payload/PKM will be included in this factor, when appropriate.

Risk management is the second factor. It was agreed that the basic risk management insurance rates for PRC vehicles could be 1 to 4% higher than the rates for market economy vehicles, depending upon the particular vehicles in question. It was also agreed that factors such as political risk can be considered in this calculation, if appropriate.

Additional costs is the next factor. It was agreed that additional costs, in the amount of $4 to $6 million, might be incurred including those relating to integration, and mission software/hardware modification.

Lift capability is another factor. It was agreed that lift capability, which can affect satellite on-orbit useful life, might sometimes be applied as a comparability factor because of differences in vehicle prices from one class of performance to the next. It was also agreed that the use of some PRC launch vehicles can result in satellite lifetimes that are 1 to 2.5 years less than launches on a market economy vehicle, although in some cases there might be no impact on satellite lifetime. Evaluation of this factor is complex and must be done on a case-by-case basis.

Payment conditions is the final factor. It was agreed that a launch provider's ability to offer favorable credit terms, which would produce a favorable payment schedule, can be important and should be considered when necessary.

It is apparent from the Agreement that it is understood that multiple attributes of a launch service affect the decisions of the potential user of the launch service. The problem is thus to identify when inappropriate pricing is occurring in the market place, and this to be done in light of multiple launch service attributes. This is the subject of the following paragraphs.

Different launch services offer users a mix of attributes that can affect user choices, the ultimate indicator of competitiveness. The metric for measuring competitiveness is market share, which reflects user choices as impacted by the suite of launch service attributes and government-imposed constraints. These attributes, including price and payment schedule, reliability, placement accuracy, delivery capability, and availability, can have significant impacts on user economics and can significantly affect the influence exerted by government support and policies.

A number of different parties are concerned with the provision of launch services. These include users or customers for the launch services (i.e., the communication satellite business ventures), providers of the launch services, and governments. Competitiveness concerns range from business-related issues to international fair trade issues. These concerns can be summarized in terms of a number of questions including the following:

1) Given a decision to establish a business to provide satellite communication services, which of a number of competing launch services should be utilized? This is an indicator of competitiveness from the provider's perspective based upon user perceptions and resulting decisions.

2) Is a launch provider utilizing unfair or predatory pricing in order to increase the likelihood of winning launch contracts and thereby increasing market share?

Because many intangible and often nonquantifiable factors bare consideration, answers to these questions cannot be developed in totally quantitative terms. (In at least one known launch competition, such factors as government stability, provision of nonrelated products and services, and human rights issues were taken into account in the launch competition and the final selection of a launch provider.) However, quantitative information will obviously be of import when answering the preceding questions. (If for no other reason than to have insight into what is being given up in order to make choices other than those indicated by the quantifiable factors.) To further complicate the problem, the decision process relating to the selection of a launch service for a particular mission usually considers many launch service attributes and the effect of these attributes and their interrelationships with financial decision variables. Answers to all of the preceding questions depend upon the relationships between the multiple attributes of launch vehicles and their effect on user choices. There is no one single launch service attribute that is indicative of competitiveness, not even price.

The former question was addressed in Sec. II; the latter question is addressed in the following paragraphs. Both rely on the use of the *DOMSAT* model or an equivalent capability.

The assessment of competitiveness of launch services is concerned with establishing the *value in use* of the various available launch services to the user and the launch service price that would result in the same value in use (i.e., the user would be indifferent to the choice of launch service). [Given a price for launch service B, price(B), there is a price for launch service A, price(A), that will make the user indifferent to the choice of A or B. These specific prices, denoted as price(A) and price(B), are referred to as the *value in use* or *on a par* prices. This implies that the probability of selecting A, P_A, is equal to 0.50.] This is sometimes also referred to as pricing on a par. The value in use will vary from one mission or business to another and from one satellite configuration to another.

The value in use is established as described in Sec. II using the *DOMSAT* model. The satellite configuration and the launch vehicle attributes and business-related data are specified to the *DOMSAT* model for a specific launch service competition. Results are obtained first with a U.S. launch service that is in the internationally competed procurement. This establishes the pertinent financial performance measures of the communication satellite business when utilizing the U.S. transportation service with its perceived attributes. Of particular importance is the determination of expected return on investment, that is, the ROI of the communication satellite business when utilizing the U.S. transportation service at the likely price that would be used in the competitive bidding together with other pertinent attributes (for example, payment schedule, reliability, insurance rate, etc.).

The next step is to replace the U.S. transportation service and its associated attributes with a competitive nonmarket economy transportation service and its associated attributes, including the quoted price and other specific attributes of the bid that will affect overall communication satellite business performance. With this information the *DOMSAT* model is again utilized to establish the resulting ROI. The objective is then to adjust the price of the competitive transportation service until the resulting ROI of the communication satellite business venture is comparable to that obtained when considering the U.S. transportation service. This then is the *on a par* price or the price that should be bid when taking into account all of the transportation service attributes and the U.S. transportation service attributes. Thus, differences in reliability, lift capability, payment schedules, insurance rates, etc, are automatically taken into account. [For example, if the U.S. launch vehicle (A) has a greater lift capability than a nonmarket economy launch vehicle (B), with the result that a satellite will have a greater on-orbit life potential when placed into orbit with A than when placed into orbit with B, then B should charge a lower price for its service than should A. U.S.–Chinese launch agreements reflect this by specifically including sensitivity coefficients that are used to establish allowed pricing in terms of differences in launch vehicle attributes. The *DOMSAT* model served as the basis for the determination of these coefficients.] Significant differences between bid price and on a par price should raise a red flag indicating that predatory pricing may be taking place. If this happens once, it might be the result of an introductory price or marketing strategy when a new service or launch vehicle is introduced. If this occurs a number of times, then government action might be required to stabilize the market.

VI. Risk-Based Approach for Comparing Advanced Transportation Architectures

A. Introduction

To make informed choices with respect to investments that will lead to lower cost and internationally competitive space transportation systems, it is necessary to evaluate and compare options that are in various stages of research, development, and operation and, therefore, are faced with different levels of risk. This implies that comparisons should be based upon metrics that explicitly and quantitatively include risk-related measures as well as expected value measures. In addition, because a major goal of technology programs is risk reduction it is necessary to measure the impacts of risk reduction on future decisions and costs. This requires metrics that include measures of risk. Convenient and informative metrics include the expected value and standard deviation of the present value of space transportation system life-cycle cost and expected value and standard deviation of savings resulting from a transportation system relative to a base case. (This begs the issue of how a new capability will effect the base case and therefore the effect on the selected metrics.) This implies that each considered space transportation system alternative or architecture will be described in terms of a pair of attributes $\{m, \sigma\}$ that relate to expected life-cycle cost (or savings) and the variability of life-cycle cost (or savings) in terms of its standard deviation, respectively.

There are a relatively large number of transportation architectures that require evaluation when considering long-term technology programs that can result in highly reusable space transportation systems with the potential to significantly reduce Earth- to low-Earth-orbit transportation costs. These include various single-stage-to-orbit (SSTO), two-stage-to-orbit (TSTO), and continued use of the current and enhanced expendable launch vehicles (ELV) and the space shuttle. Each of these architectures requires a different mix of R&D and capital investment and results in different performance, schedule, and operational cost uncertainties and associated levels of risk.[2,3] An approach is described in the following paragraphs for performing early comparisons of transportation architectures explicitly taking into account quantitative measures of uncertainty and risk. The approach considers the uncertainty associated with the achievement of technology goals, the effect that the achieved level of technology will have on transportation system performance and the relationship between transportation system performance/capability, and the ability to accommodate variations in payload mass. The consequences of system performance are developed in terms of expected values and associated standard deviations of nonrecurring, recurring, and the present value of transportation system life-cycle costs.

There are three degrees of freedom associated with the development of new architectures: performance, cost, and schedule. It is not possible to fix all three but might be possible to fix two. For example, if it is desired to achieve a specified level of performance, then cost (and possibly schedule) must be considered as an uncertainty variable. If cost and schedule are specified, then performance must be considered as uncertain. To simplify the initial analysis that explicitly and quantitatively considers risk, it is assumed that all schedules are known and that either or both performance and cost are specified with resulting development cost risk and/or performance risk.[25]

The objective is to establish the probability distribution of the present value of life-cycle costs of performing a mission model over an extended period of time and to accomplish this at a rather high level of abstraction so that attention can be focused on a small number of architectures that can then be the subject of more detailed analyses. Because not all highly reusable space transportation systems will satisfy all mission requirements, architectures can include both reusable and expendable launch systems. Life-cycle costs include nonrecurring cost associated with development of transportation architectures, capital costs associated with launch facilities and launch vehicle fleets, transportation recurring costs, payload nonrecurring costs and recurring costs (particularly as effected by different space transportation architectures), and relative timing of all costs. The payload considerations must take into account the effect of transportation margins on payload design margins and the ensuing impact on payload cost.

Because of the complexity of the analysis of transportation architectures, it is not reasonable to seek a closed mathematical solution for development of risk metrics. Therefore Monte Carlo simulation techniques are used for the development of both expected value and risk measures.[26] The Monte Carlo techniques make possible the determination of the probability distribution of the net present value of life-cycle cost and related savings associated with different space transportation architectures.

B. Analysis: SSTO

The basic approach, as depicted in Fig. 4.26, is to randomly sample a subjective assessment (provided in the form of a probability density function and referred to as an uncertainty profile) of achieved capability (for example, injected mass). (The

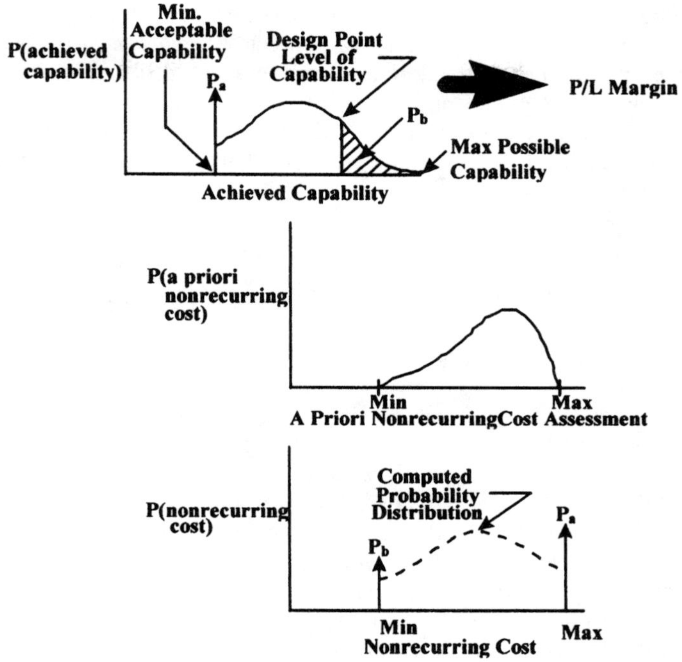

Fig. 4.26 Development of SSTO nonrecurring cost probability distribution.

detailed mathematical model for this SSTO analysis is provided in the Appendix to this section.) The random sampling of the injected mass probability density function results in the determination of the probability that the minimum acceptable capability will not be exceeded (P_a). However, an implied assumption is that funding will be provided so as to at least achieve the minimum acceptable level of capability. Thus, if a random sample of achieved capability is obtained that is less than the minimum acceptable level of capability the minimum acceptable level of capability is assumed to be achieved at the maximum value of the a priori estimate of nonrecurring cost. If the random sample of achieved capability results in a level of capability greater than the design point capability, the random sample is set equal to the design point. The assumption is that the design point will not be exceeded (funding will be adjusted accordingly). Thus there is a probability P_b that the minimum cost will be achieved (synonymous with the probability of exceeding the design point) and a probability P_a that the maximum cost will be achieved (synonymous with the probability of achieving the minimum acceptable level of capability).

If the random sample of achieved capability is less than the design point and greater than the minimum acceptable level of capability, the a priori assessment of nonrecurring cost is randomly sampled to establish a nonrecurring cost between the specified minimum and maximum values. When this process is performed a large number of times, the nonrecurring cost probability distribution is established, which takes into account the a priori assessment of nonrecurring cost uncertainty (at a specific design point) and the a priori assessment of performance or achieved capability uncertainty. Cost spreading of nonrecurring cost must be specified so that appropriate timing of annual nonrecurring cost is obtained.

In addition to using the random sample of achieved level of capability to develop the nonrecurring cost probability distribution, the random sample is used to establish available payload design margins. This a posteriori payload (P/L) design margin is defined as

$$\text{P/L Margin} = [\text{A Priori P/L Design Margin}] + [\text{Achieved Injected Mass Capability/Design Point Injected Mass Capability}] - 1.0$$

If payload design margin is affected by the achieved level of transportation capability, then what happens to cost when design margin changes? It is assumed that subjective estimates are made for P/L nonrecurring cost per unit mass, Fig. 4.27, at a specified or a priori P/L design margin, which is based upon achievement of the transportation system design point level of capability. It is assumed (Fig. 4.27) that as P/L margin decreases, then the minimum cost remains the same and the maximum (and most likely) cost increases. It is assumed that the uncertainty profile remains the same, but the range of uncertainty is changed. This range change requires the estimation of a sensitivity coefficient that relates nonrecurring cost increase to decrease in P/L margin. This sensitivity coefficient of P/L design margin is described in terms of a second-order polynomial. In a similar manner, when P/L margin increases then the maximum remains the same,

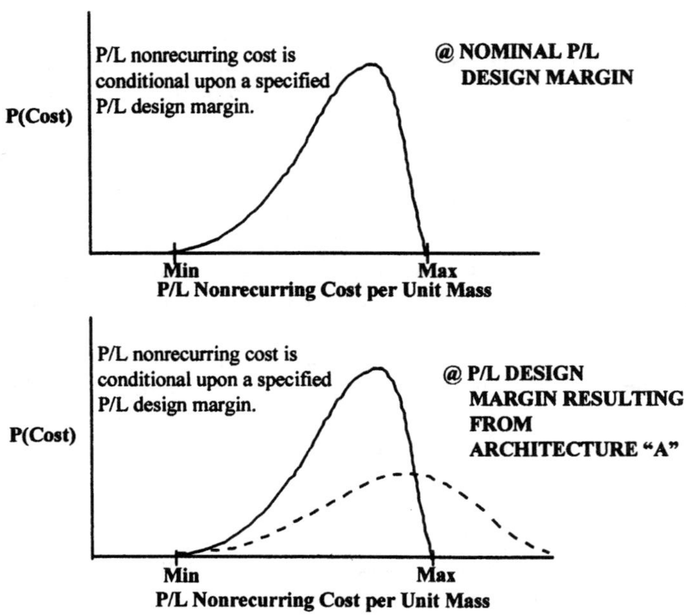

Fig. 4.27 Subjective assessment of P/L nonrecurring cost.

and the minimum (and most likely) cost decreases. It is anticipated that this sensitivity coefficient will require specification at a number of values of P/L design margin.

Single-stage-to-orbit fleet capital investment, launch facility investment, transportation system delivery cost per unit mass, and payload recurring cost per unit mass are also established using random sampling of specified ranges of uncertainty and weighted by a triangular probability distribution based upon the specification of maximum, minimum, and most likely values. (No consideration has been given to the establishment of fleet size and the associated implications of reliability resiliency, and system operability.) Because there might not be a one-for-one correspondence between the number of satellites developed and satellites flown, an estimate is required of the ratio of satellites flown to satellites developed. Finally, an estimate of delivered payload mass (at the a priori design margin) is required as a function of time. To maintain flexibility for comparison purposes, allowance is made for the use of ELVs during transition to the reusable system and for those payloads that for one reason or another will likely be launched by ELVs. Thus when comparing architectures, the annual payload mass delivered to LEO is maintained as constant across the architectures.

As a result, annual cost can be established as follows:

$$\begin{aligned}
\text{Annual Cost}(I, R) = {} & \text{Nonrecurring Transportation System Cost}(I, R) \\
& + \text{Fleet Capital Investment}(I, R) \\
& + \text{Launch Facility Capital Invest}(I, R) \\
& + [\text{RLV Transportation System Cost per Unit Mass}(R) \\
& + K * \text{Payload Nonrecurring Cost per Unit Mass}(I, R) \\
& + \text{Payload Recurring Cost per Unit Mass}(I, R)] \\
& * [\text{Payload Delivered Mass}(I) * F(I)] \\
& + [\text{ELV Transportation System Cost per Unit Mass}(R) \\
& + K * \text{Payload Nonrecurring Cost per Unit Mass}(I, R) \\
& + \text{Payload Recurring Cost per Unit Mass}(R)] \\
& * [\text{Payload Delivered Mass}(I) * (1 - F(I))]
\end{aligned}$$

where K is the ratio of satellites developed to satellites flown (or stated another way, the reciprocal of the average number of P/Ls per mission), I is an index that refers to time (i.e., years), R is the Monte Carlo run index, and $F(I)$ is the fraction of payload mass placed into orbit on a reusable launch vehicle. Thus each variable containing an R index will be dimensioned according to the specified number of simulation runs to be performed. The present value of life-cycle costs (PVLCC) is

$$PVLCC(R) = \sum_{I} [\text{Annual Cost}(I, R)]/[1 + d]^{I}$$

where d is the discount rate. The expected value m and standard deviation σ of the distribution of PVLCC are given by

$$m = \left[\sum_{R} PVLCC(R)\right] / MAXR$$

$$\sigma = \left[\left\{\sum_{R} PVLCC(R)^2\right\} / MAXR - m^2\right]^{0.5}$$

where *MAXR* is the number of Monte Carlo runs performed.

To establish the expected value and standard deviation (i.e., risk) of a SSTO architecture, the following types of data are required (Specific data requirements are presented in the Appendix to this section.): 1) design point and minimum acceptable injected (into LEO) mass capability; 2) minimum, maximum, and most likely achievable injected mass (into LEO) capability; 3) a priori minimum, maximum, and most likely SSTO nonrecurring cost; 4) a priori minimum, maximum, and most likely SSTO LEO delivery cost per unit payload mass; 5) a priori minimum, maximum, and most likely SSTO capital cost per vehicle; 6) a priori minimum, maximum, and most likely launch complex investment; and 7) number of vehicles in the SSTO fleet. In addition, data must be provided for the following types of variables that can be reasonably assumed to be independent of transportation system architecture: 1) minimum, maximum, and most likely payload nonrecurring cost per unit mass at the nominal payload design margin with the design point injected mass capability; 2) sensitivity of payload nonrecurring cost to decreasing payload design margin; 3) minimum, maximum, and most likely payload recurring cost per unit mass at the nominal payload design margin with the design-point injected mass capability; 4) sensitivity of payload recurring cost to decreasing payload design margin; 5) nominal payload design margin; 6) average number of payloads launched per mission; and 7) annual payload mass delivered to LEO at nominal payload design margin. Finally, cost-spreading functions need to be specified to establish reasonable timing of expenditures for present value computations.

C. Analysis: TSTO

The TSTO system is presumed to consist of an upper stage having both performance and cost uncertainty that are conditional upon the performance achieved by the lower stage. (The detailed mathematical model for the TSTO analysis is provided in the Appendix to this section.) The general approach for considering the effect of the first-stage upon the TSTO second-stage nonrecurring costs is illustrated in Fig. 4.28. It is also presumed that both the performance and cost of each stage must be described, a priori, in terms of probability distributions as in the case of the SSTO. The major difference between the TSTO and SSTO analyses is that the a priori probability distribution of second-stage performance is specified given the first-stage design point. The analysis proceeds (using Monte Carlo simulation techniques) by randomly sampling the first-stage achieved capability probability distribution. If the sample results in a capability less than the minimum acceptable value, the maximum nonrecurring cost is utilized; otherwise, the a priori probability distribution of nonrecurring cost is sampled. When this

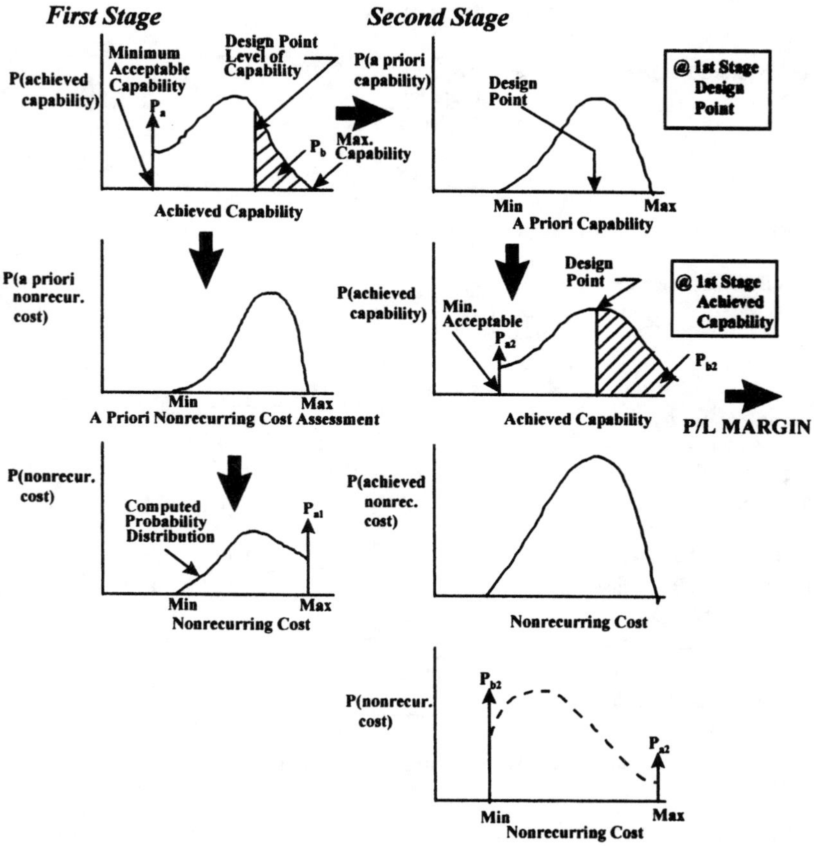

Fig. 4.28 Development of TSTO nonrecurring cost probability distribution.

process is repeated a large number of times, the a posteriori probability distribution of cost is established with the maximum cost occurring with a probability P_{a1}. The second-stage a priori probability distribution of capability is also randomly sampled with a new level of capability being established based upon the sampled capabilities of both stages. This implies that a functional relationship exists between first- and second-stage performances.

The a posteriori probability distributions of second-stage performance and nonrecurring cost are established by setting the achieved capability to the minimum acceptable value and the cost to the maximum value when the random sample of capability is less than the minimum acceptable capability level. Similarly, if the capability (as adjusted to reflect the first-stage capability) exceeds design point the a posteriori capability is set equal to the design point, and the cost is set equal to the minimum of the range of costs. If the random sample is between these two extremes, then the probability distribution of capability and cost are sampled. This results in the a posteriori nonrecurring cost distributions as

ARCHITECTURE SUMMARY

Architecture Description:	**Argus with Maglifter**			
	Typical Example: Georgia Tech - AE			
Reference Number:	**1003**	**Date:**	**8/24/99**	
Start of HRV Operations:	**2010**			
Can Exceed Design Point:				
Stage 1	**1**			
Stage 2	**1**			

	Expected	**Std. Dev.**	
Trans. Sys. Nonrec. Cost	**5,416**	**1661**	**(M\$)**
HRV Unit Cost	**704**	**242**	**(M\$)**
HRV Fleet Investment	**2,112**	**725**	**(M\$)**
Infrastructure Investment	**2,333**	**522**	**(M\$)**
Launch Cost/Unit PL Mass	**5**	**0.00**	**(K\$/kg)**
PL Nonrec. Cost/Unit Mass	**42**	**3.97**	**(K\$/kg)**
PL Rec. Cost/Unit Mass	**20**	**1.21**	**(K\$/kg)**
PL Design Margin	**5**	**4.23**	**(%)**
Infinite Horizon Discounting	**N**		
Present Value of Oper. Sys. LCC	**394661**	**11784**	**(M\$)**
Present Value of Tech. Prog.			**(M\$)**

Fig. 4.29 Architecture summary information.

indicated in Fig. 4.29 with the indicated probabilities P_{a2} and P_{b2}. In addition to using the random sample of achieved level of capability to develop the nonrecurring cost probability distribution, the random sample is used to establish available payload design margins. The P/L design margins take into account the capability established for both stages and are used to modify the a priori P/L nonrecurring cost per unit mass estimates.

As for the SSTO case, TSTO fleet capital investment, launch facility investment, transportation system delivery cost per unit mass, and payload recurring cost per unit mass are also established. Since there may not be a one for one correspondence between the number of satellites developed and satellites flown, an estimate is required of the ratio of satellites flown to satellites developed. Finally, an estimate of delivered payload mass (at the a priori design margin) is required as a function of time. Allowances can be required for first-stage reliability/reusability.

As a result, annual cost is established as described for the SSTO. The SSTO and TSTO life-cycle costs as computed by the preceding process are comparable because they include both the expected value and risk dimensions. Differences are reflected in the computed a posteriori probability distributions that then result in different expected values and risk.

The TSTO analysis requires the specification of at least the following key data items: 1) first-stage design point and minimum acceptable level of capability; 2) first-stage minimum, maximum, and most likely achievable capability; 3) first-stage a priori minimum, maximum, and most likely nonrecurring cost; 4) second-stage design point and minimum acceptable level of capability; 5) second-stage minimum, maximum, and most likely achievable capability at the first-stage design point; 6) second-stage a priori minimum, maximum, and most likely nonrecurring cost; 7) functional relationship between first- and second-stage capabilities; 8) a priori minimum, maximum, and most likely TSTO LEO delivery cost per unit payload mass; 9) a priori minimum, maximum, and most likely TSTO capital cost per vehicle; 10) a priori minimum, maximum, and most likely launch complex investment; and 11) number of vehicles in the TSTO fleet.

In addition, the same data must be provided as for the SSTO for the variables that can be reasonably assumed to be independent of transportation system architecture, and cost-spreading functions need to be specified to establish reasonable timing of expenditures for present value computations.

D. Space Transportation Architecture Risk System

The Space Transportation Architecture Risk System (STARS), is the implementation of the just-described methodology for the economic comparison of advanced transportation concepts explicitly considering uncertainty and risk. STARS is a Monte Carlo simulation model that operates totally within the Excel spreadsheet structure. Operating on a 133-MHz computer, 100 simulation runs takes several seconds. The model (details presented in the Appendix to this Section) operates at a relatively high level of abstraction, that is, detailed mathematical modeling is replaced by input data that are based upon results of other analyses and subjective judgments. Many of the input variables are considered as ranges of uncertainty, thus allowing uncertainties to be captured quantitatively.

The STARS was developed to allow rapid and early economic comparisons to be made of advanced transportation concepts explicitly considering uncertainty and risk. STARS allows for the consideration of SSTO and TSTO concepts augmented by expendable or other competitive transportation systems (i.e., market shares as a function of time may be considered). STARS allows for the consideration of stage performance uncertainties with interactions between stage performance and overall capability and payload design margins. Nonrecurring cost uncertainty is related to achieved performance, and payload nonrecurring and recurring cost are related to achieved design margins. The model uses Monte Carlo techniques to establish the probability distribution of the present value of life-cycle costs of performing a mission model over an extended period of time. The life-cycle cost takes into account the nonrecurring costs, fleet and launch facility capital investments, annual transportation cost, and payload nonrecurring and recurring cost as effected by achieved transportation capabilities. The result is the determination of expected values and standard deviations of life-cycle cost that take into account subjective estimates of performance and cost uncertainties and thereby allow alternatives to be compared taking into account both the expected value and risk dimensions. STARS is programmed in Excel for operation within the DOS environment and has a user-friendly menu-driven input/output system.

STARS utilizes a series of user-friendly menu-driven screens for inputting of the data described in preceding paragraphs, and results are available in both numerical and graphical formats.

E. Typical HRST Architecture

The following discussion is presented to clarify and expand upon the use of the STARS model. The discussion uses a magnetic levitation, maglev architecture as a typical example of a highly reusable space transportation architecture. The data used and the results obtained are presented for illustrative purposes only and should not be taken as a definitive analysis and/or evaluation of a maglev architecture.

Maglev architecture refers to an Earth-to-orbit (ETO) launch concept that employs a catapult which uses superconducting maglev to achieve dramatically augmented payload capacity in ETO transportation systems while reducing mission costs.[27,28] Unlike other "gun" concepts, the Maglev architecture does not require extremely high accelerations, does not involve radical changes in payload (i.e., spacecraft) design or components, and does not require very high launch rates to achieve economical operations.

Maglev is straightforward in its conception. The payload capacity of a wide range of vehicles—but especially SSTO vehicles—can be significantly increased with the provision of a relatively small "assist" during the first minute of the launch to LEO by means of a ground-based catapult system. Maglev would provide this assist via an advanced, high-speed maglev guideway and carrier vehicle(s). The system would consist of the following major elements:

1) The first element is a highly robust structural support system (to altitude), typically a tunnel inside a mountain (acceleration phase, having a length of several miles), and an external guideway support system on the mountain (deceleration phase, also having a length of several miles).
2) A long maglev guideway, including the accelerator system and the carrier decelerator, is another element. Typically, the accelerator system would be enclosed in a tunnel (or pressurized tube) that would be filled with a gas at partial pressure, i.e., helium, with a low density and a high speed of sound.
3) The third element is a local power supply system, such as a superconducting magnetic energy storage system, which can be charged from the local power grid and then discharged during a launch sequence. Other options exist (such as using a battery of gas turbines for direct power generation during launch). The final choice would be based on the results of studies of life-cycle costs, R&D investment values, etc.
4) A set of fully reusable maglev "accelerator-carriers" would provide the initial acceleration for the vehicles to be launched. These carriers, which might need to be "ganged" for launching larger vehicles, would accommodate "cradles" capable of structural support to vehicles during acceleration as well as rapid, controlled release at the appropriate point in

the catapult sequence. They would also provide any needed support for vehicles during the launch sequence (approximately 1 min in duration).

5) A launch/exit system would provide a managed transition from the environment inside the tunnel and on the guideway to free flight in the external environment. Active control of both the vehicle and the accelerator-carrier would be required during the transition.
6) The last element is a staging facility, including maglev carrier staging, vehicle-carrier integration, launch vehicle staging (specific to vehicles and payloads), servicing and maintenance facilities, and an operations control center.

Typically, the launched vehicle would emerge from the maglev system at an altitude of approximately 10,000 ft at velocities of about 600 mph. At exit from the launch system, the angle of the velocity vector of the center of gravity of the vehicle would be approximately 45 deg (measured from the local horizontal). The maximum acceleration during the acceleration phase of launch would be approximately three gravities. All of these parameters are subject to variation through design trades and analyses currently underway.

Typical STARS-developed results are presented in a series of reports as illustrated in Figs. 4.29–4.31. Figure 4.29 presents information that provides an overview of the architecture being analyzed. All computed results include both

	ARCHITECTURE ANNUAL COST (M$)					
Architecture Description:	Argus with Maglifter					
	Typical Example: Georgia Tech - AE					
Reference Number:	1003				Date:	8/24/99
Start of HRV Operations:	2010	Year				
	2009	2010	2011	2012	2013	2014
HRV Stage 1	39	0	0	0	0	0
HRV Stage 2	463	231	0	0	0	0
Infrastructure	0	0	0	0	0	0
HRV Payloads	0	0	3,723	10,110	19,642	22,258
ELV Payloads	15,651	17,203	14,553	10,015	2,229	1,160
Fleet Investment	422	211	0	0	0	0
Fleet Replacement	0	0	106	106	106	106
Technology Program	0	0	0	0	0	0
Total Nonrecurring Cost	16,575	17,645	18,382	20,230	21,977	23,523
* Std. Deviation *	1673	1832	1655	1354	1886	2061
HRV Launch Ops.	0	0	79	214	416	472
ELV Launch Ops.	5,670	6,124	5,262	3,572	771	414
HRV Payloads	0	0	5,212	14,154	27,499	31,161
ELV Payloads	22,326	24,112	20,718	14,065	3,036	1,630
Total Recurring Cost	27,996	30,235	31,271	32,006	31,723	33,677
* Std. Deviation *	1390	1501	1303	1179	1679	1902
Total Annual Cost	44,571	47,881	49,653	52,236	53,699	57,200
* Std. Deviation *	2175	2368	2106	1796	2525	2805

Fig. 4.30 Architecture annual cost summary.

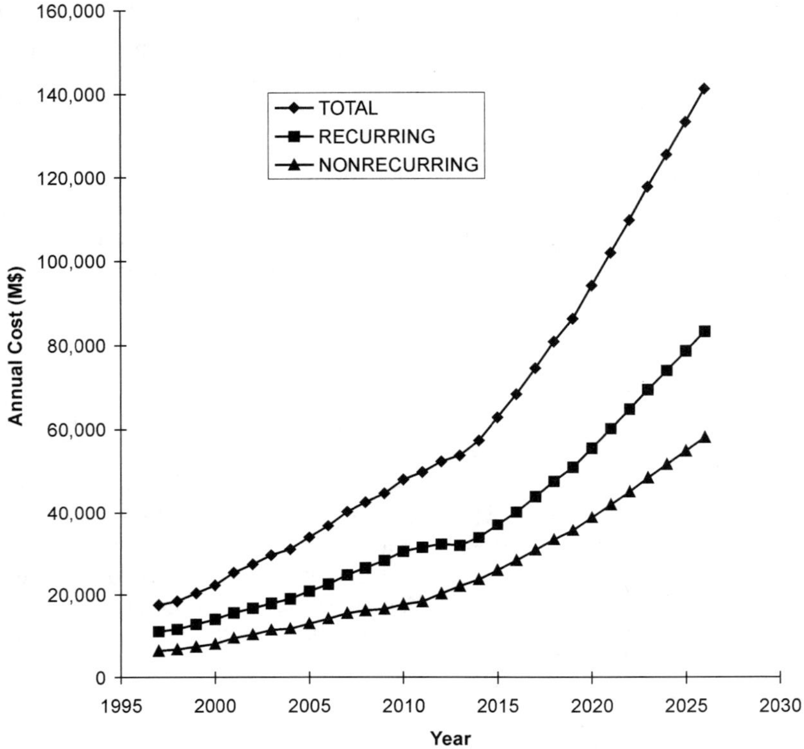

Fig. 4.31 Graphical representation of architecture expected annual cost (includes transportation and payload costs).

expected and standard deviation values. Of particular importance are the present value expected value and standard deviation data that provide the basic information for the comparison of alternative architectures. The "Infinite Horizon Discounting?" response indicates whether infinite horizon discounting is considered or discounting within the specified life-cycle cost time frame (finite) is considered. Figure 4.30 provides an annual cost summary of the considered architecture. Figure 4.30 indicates the annual nonrecurring and recurring costs. The nonrecurring cost includes HRV stage 1, HRV stage 2, infrastructure, HRV payload, ELV payload (when ELV flights are required in conjunction with HRV flights to place the mission model payloads into orbit), HRV fleet investment, HRV fleet replacement, and research and technology costs. The recurring cost includes HRV launch operations, ELV launch operations, and HRV and ELV payload recurring costs. Figure 4.31 presents the architecture annual costs in graphic form with expected nonrecurring, recurring, and total costs illustrated as a function of time.

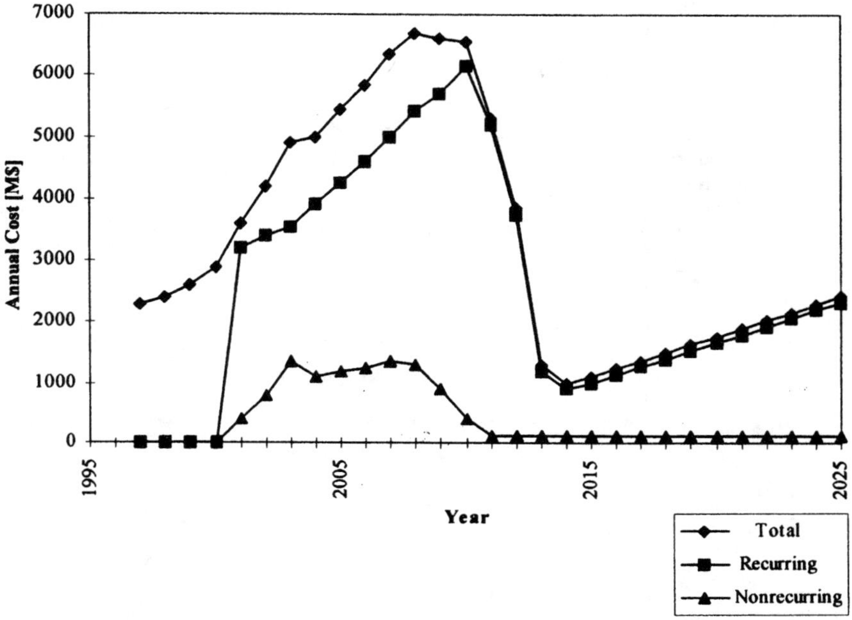

Fig. 4.32 Graphical representation of architecture expected annual cost with payload-related costs eliminated.

The dominance of payload-related costs can be observed by setting the payload recurring and nonrecurring costs per unit mass to zero. Obtained results are illustrated in Fig. 4.32 and indicate only transportation related costs.

F. Comparison of Architectures

The economic comparison of advanced space transportation architectures involves establishing cash-flow patterns that occur over many years; thus, it is desired to present the results of analyses in terms of the present value of costs. The present value takes into account the magnitude and the timing of cash-flow patterns. Also as already described, the annual costs entering into the present value computation are not deterministic quantities because of the uncertainties in predicting level of achieved performance and recurring and nonrecurring costs. Thus the present value of costs must also be characterized by a probability distribution that can be summarized in terms of its expected value m and standard deviation σ, with the latter indicating the risk dimension.

Because the present value of life-cycle cost of each transportation architecture is characterized by both an expected value and a standard deviation (i.e., the risk dimension), a tradeoff between expected cost and risk must be made to select a best or most desired alternative. This is illustrated in Fig. 4.33, where all of the points in the m-σ space represent alternatives that will result in satisfying a specified mission model. Indicated are alternatives 1 and 2 having the same level

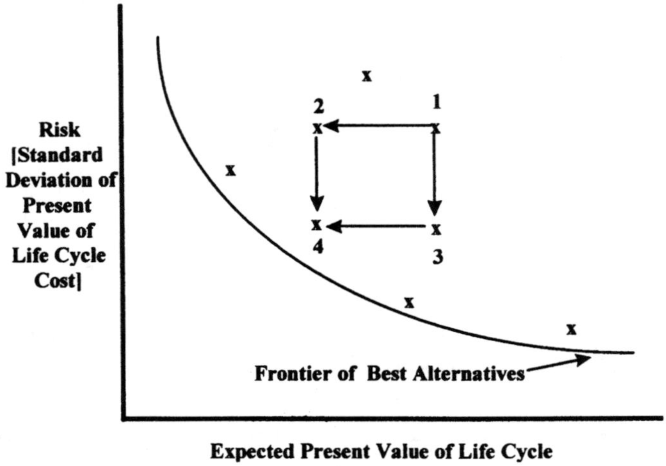

Fig. 4.33 Comparison of alternatives when considering risk.

of risk (i.e., $\sigma_1 = \sigma_2$), but the expected present value of cost of alternative 1 is greater than that of alternative 2. Therefore, alternative 2 is preferable to alternative 1. In a similar manner, it can be argued that alternative 4 is preferable to alternative 3. Also in a similar manner, alternative 3 is preferable to alternative 1 because both have the same expected present value of cost, but alternative 1 is riskier. This process can be continued with all alternatives being considered. In the limit it can be seen that a frontier of best alternatives can be established. Each of the points or alternatives represented by the frontier are different in the respect that the risk and expected present value are different. The class of best alternatives has thus been obtained, and the best alternative can be selected based on the decision-maker's risk judgment. That is, the decision maker must decide what the tradeoff is between a reduction in expected present value of cost and an accompanying increase in risk.

The analysis and the determination of expected present value and risk make it possible to identify and then focus attention on those alternatives that appear to be most attractive (i.e., lie on or are close to the frontier of best alternatives). In accomplishing this, it is necessary to compare alternatives on a common basis. To achieve this common basis, a demand scenario (i.e., mass delivered to orbit as a function of time) has been assumed. However, demand will most likely be related to transportation price (so far only cost has been considered), and architectures having significantly different costs (and prices) will ultimately have to be considered as well as their effect on payload demand through price elasticity considerations. All is not lost, however. Because architectures that are not significantly different in cost are likely to be compared, utilizing the same demand scenario can still give reasonably comparative results. In addition, utilizing a common demand scenario is likely to overstate the cost differences between architectures without changing the order of desirability. In any event, to totally avoid this problem would require the development of a pricing and demand component for STARS and the consideration of metrics other than present value of life-cycle costs.

G. Summary

A methodology for performing comparisons of transportation architectures that explicitly and quantitatively considers risk has been developed. The methodology, developed for screening comparisons between SSTO and TSTO architectures and having application to a broad range of transportation architectures, takes into account the uncertainty with respect to achievement of technology goals, the effect that the achieved level of technology will have on transportation system performance, and the relationship between transportation system performance capability and the ability to accommodate variations in payload mass. The consequences of system performance are developed in terms of nonrecurring, recurring and the present value of transportation system life-cycle costs. Transportation system impacts on payload costs are also taken into account in the development of life-cycle costs. The results are in a form that allows future space transportation system options to be compared explicitly taking into account expected present value of transportation system life-cycle cost and the associated level of risk as indicated by the standard deviation of the present value of life-cycle costs. The capability was developed specifically for the evaluation and comparison of a large number of architectures resulting from NASA studies aimed at establishing a technology program that would lead to significant reductions in future space transportation costs.

H. Appendix: STARS Model

1. Data Requirements (Definition of Input Variables)

The following is the set of input variables that is used in the STARS model. These variables are defined next and are organized for data entry into a series of user-friendly input data screens in EXCEL.

APLMARG	= a priori payload design margin, %
APLMASS(I)	= annual payload mass delivered to low Earth orbit, kg/yr
CSHRVFI(N)	= percent (%) of fleet investment spent in year N [Cost spreading is backward in time with $N = 1$ being the year of completion of fleet investment (CYFI).]
CSHRVS1(N)	= percent (%) of stage 1 nonrecurring cost spent in year N. [Cost spreading is backward in time with $N = 1$ being the year of start of HRV operations (CYSO).]
CSHRVS2(N)	= percent (%) of stage 2 nonrecurring cost spent in year N [Cost spreading is backward in time with $N = 1$ being the year of start of HRV operations (CYSO).]
CSIF(N)	= percent (%) of infrastructure investment spent in year N [Cost spreading is backward in time with $N = 1$ being the year of completion of the infrastructure investment (CYLF).]

CSTP(I) = percent (%) of technology program cost spent in year I starting in the year of the start of the analysis
CY = calendar year (for example, 1999) of start of analysis [The annual time index I is equal to 1 at calendar year (CY).]
CYFI = calendar year of completion of HRV fleet acquisition
CYLF = calendar year of completion of launch infrastructure
CYSO = calendar year (for example, 2005) of start of HRV operations
DR = discount rate (%) used in computation of present value of life-cycle cost
FIMLHRV = most likely cost (price) of an HRV incorporated into the HRV fleet, M$
FIMNHRV = minimum possible cost (price) of an HRV incorporated into the HRV fleet, M$
FIMXHRV = maximum possible cost (price) of an HRV incorporated into the HRV fleet, M$
FLTRP = percentage (%) of HRV fleet cost spent per year for replacements
IHD = when IHD = yes, infinite horizon discounting will be included, and when IHD ≠ yes, infinite horizon discounting will not be included
LFML = most likely launch infrastructure investment, M$
LFMN = minimum possible launch infrastructure investment, M$
LFMX = maximum possible launch infrastructure investment, M$
MAXR = number of Monte Carlo simulation runs to be performed
MAXYRS = maximum number of years to be considered in the analysis
MLTPC = most likely technology program cost, M$
MNTPC = minimum technology program cost, M$
MXNRCH = maximum possible reduction in nonrecurring cost caused by an increase in P/L margin, % reduction in cost
MXRCCH = maximum possible reduction in recurring cost caused by increase in P/L margin, % reduction in cost
MXTPC = maximum technology program cost, M$
NOHRV = number of HRVs purchased for the fleet
PLMLNRC = most likely P/L nonrecurring cost per unit mass at the nominal P/L design margin, K$/kg
PLMLRC = most likely P/L recurring cost per unit mass at the nominal P/L design margin, K$/kg
PLMNNRC = minimum possible P/L nonrecurring cost per unit mass at the nominal P/L design margin, K$/kg
PLMNRC = minimum possible P/L recurring cost per unit mass at the nominal P/L design margin, K$/kg
PLMXNRC = maximum possible P/L nonrecurring cost per unit mass at the nominal P/L design margin, K$/kg

PLMXRC = maximum possible P/L recurring cost per unit mass at the nominal P/L design margin, K$/kg
PLPM = average number of payloads launched per mission
PLSENNRC = sensitivity of P/L nonrecurring cost to increasing payload design margin, % decrease in cost/% increase in margin
PLSENRC = sensitivity of P/L recurring cost to increasing payload design margin, % decrease in cost/% increase in margin
PPL(I) = percent of payload mass delivered by ELVs as a function of time, %/yr
S1ALC = HRV stage 1 minimum acceptable level of capability relative to nominal design point, % of design point capability
S1DEC = decision to allow a stage 1 capability to exceed the design point level of capability (1 = can exceed, 0 = cannot exceed)
S1MLLC = most likely stage 1 level of capability relative to nominal design point, percent of design point capability
S1MLNRC = most likely stage 1 nonrecurring cost, M$
S1MNLC = HRV stage 1 minimum achievable level of capability relative to nominal design point, % of design point capability
S1MNNRC = stage 1 minimum possible a priori nonrecurring cost, M$
S1MXLC = HRV stage 1 maximum achievable level of capability relative to nominal design point, % of design point capability
S1MXNRC = stage 1 maximum possible a priori nonrecurring cost, M$
S2ALC = HRV stage 2 minimum acceptable level of capability relative to nominal design point, % of design point capability
S2DEC = decision to allow a stage 2 capability to exceed the design point level of capability (1 = can exceed, 0 = cannot exceed)
S2MLLC = most likely stage 2 level of capability relative to nominal design point, % of design point capability
S2MLNRC = most likely stage 2 nonrecurring cost, M$
S2MNLC = HRV stage 2 minimum achievable level of capability relative to nominal design point, % of design point capability
S2MNNRC = stage 2 minimum possible a priori nonrecurring cost, M$
S2MXLC = HRV stage 2 maximum achievable level of capability relative to nominal design point, % of design point capability
S2MXNRC = stage 2 maximum possible a priori nonrecurring cost, M$
S2S1A = percent change in stage 2 capability per percent change in stage 1 capability (at stage 1 and stage 2 design points)
S2S1B = percent change in stage 2 capability per percent change squared in stage 1 capability (at stage 1 and stage 2 design points)
TCMLELV = most likely ELV transportation cost per unit payload mass, K$/kg

TCMLHRV = most likely HRV transportation cost per unit payload mass, K$/kg

TCMNELV = minimum possible ELV transportation cost per unit payload mass, K$/kg

TCMNHRV = minimum possible HRV transportation cost per unit payload mass (operating cost does not consider purchase or replacement of fleet), K$/kg

TCMXELV = maximum possible ELV transportation cost per unit payload mass, K$/kg

TCMXHRV = maximum possible HRV transportation cost per unit payload mass (operating cost does not consider purchase or replacement of fleet), K$/kg

2. *STARS Mathematical Model*

The details of the STARS mathematical model are described in the following pages. In the mathematical relationships, *italics* are used to indicate input variables (i.e., input data are provided), and roman characters are used to refer to computed quantities. All input variables and computed quantities have mnemonic names (for example, *S2MLLC* indicates "stage 2 most likely level of capability").

1) Determination of stage 1 level of capability: (%)
S1CAP(R) $1 \leq \mathrm{R} \leq MAXR$

Generate random number, RN.
Using a random sampling algorithm, establish value of S1CAP(R):

$$[S1MXLC; S1MNLC; S1MLLC] \Longrightarrow \mathrm{S1CAP(R)}$$

2) Determination of stage 1 nonrecurring cost: (M$)
S1NRC(R) $1 \leq \mathrm{R} \leq MAXR$

Generate random number, RN.
Using a random sampling algorithm, establish value of S1NRC(R):

$$[S1MXNRC; S1MNNRC; S1MLNRC] \Longrightarrow \mathrm{S1NRC(R)}$$

3) Determination of stage 2 a priori level of capability: (M$)
S2CAP(R) $1 \leq \mathrm{R} \leq MAXR$

Generate random rumber, RN.
Using a random sampling algorithm, establish value of S2CAP(R):

$$[S2MXLC; S2MNLC; S2MLLC] \Longrightarrow \mathrm{S2CAP(R)}$$

4) Adjustment of a priori stage 2 capability to account for achieved stage 1 capability: (%)
S2CAP(R) $1 \leq R \leq MAXR$

$$S2CAP(R) = S2S1A * [S1CAP(R) - 100.0] + K1 * S2S1B * [S1CAP(R) - 100.0]^2 + S2CAP(R)$$

If S1CAP(R) ≥ 100.0
Then

$$K1 = 1.0$$

If S1CAP(R) < 100.0
Then

$$K1 = -1.0$$

5) Determination of stage 2 nonrecurring cost: (M$)
S2NRC(R) $1 \leq R \leq MAXR$

Generate random number, RN.
Using a random sampling algorithm, establish value of S2NRC(R):

[*S2MXNRC*; *S2MNNRC*; *S2MLNRC*]⟹S2NRC(R)

6) Determination of ELV payload recurring cost per unit mass: (K$/kg)
ELVPLRC(R) $1 \leq R \leq MAXR$

Generate random number, RN.
Using a random sampling algorithm, establish value of ELVPLRC(R):

[*PLMXRC*; *PLMNRC*; *PLMLRC*]⟹ELVPLRC(R)

7) Calculation of P/L design margins: (%)
PLMARG(R) $1 \leq R \leq MAXR$

$$PLMARG(R) = APLMARG + [S2CAP(R) - 100.00]$$

If PLMARG(R) < 0.0
Then

$$PLMARG(R) = 0.0$$

8) Determination of range of uncertainty of P/L nonrecurring cost per unit of P/L mass: (K$/kg)
PLMXNR(R); PLMNNR(R); PLMLNR(R) $1 \leq R \leq MAXR$

If PLMARG(R) ≤ *APLMARG*

Then

$$\text{PLMNNR(R)} = \mathit{PLMNNRC}$$
$$\text{PLMXNR(R)} = \mathit{PLMXNRC} * \{1.0 + \mathit{PLSENNRC} * [\text{PLMARG(R)} - \mathit{APLMARG}] * 0.0001\}$$
$$\text{PLMLNR(R)} = \mathit{PLMLNRC} * \{1.0 + \mathit{PLSENNRC} * [\text{PLMARG(R)} - \mathit{APLMARG}] * 0.0001\}$$

If PLMARG(R) $>$ *APLMARG*
Then

$$\text{PLMNNR(R)} = \mathit{PLMNNRC} * \{1.0 + \mathit{PLSENNRC} * [\text{PLMARG(R)} - \mathit{APLMARG}] * 0.0001\}$$
$$\text{PLMXNR(R)} = \mathit{PLMXNRC}$$
$$\text{PLMLNR(R)} = \mathit{PLMLNRC} * \{1.0 + \mathit{PLSENNRC} * [\text{PLMARG(R)} - \mathit{APLMARG}] * 0.0001\}$$

If PLMLNR(R) $<$ *MXNRCH* $*$ *PLMLNRC* $*$ 0.01
Then

$$\text{PLMLNR(R)} = \mathit{MXNRCH} * \mathit{PLMLNRC} * 0.01$$

If PLMNNR(R) $<$ *MXNRCH* $*$ *PLMLNRC* $*$ 0.01
Then

$$\text{PLMNNR(R)} = \mathit{MXNRCH} * \mathit{PLMNNRC} * 0.01$$

9) Determination of P/L nonrecurring cost per unit mass: (K$/kg)
PLNRC(R) $1 \le \text{R} \le \mathit{MAXR}$

Generate random number, RN.
Using a random sampling algorithm, establish value of PLNRC(R):

$$[\text{PLMXNR(R); PLMNNR(R); PLMLNR(R)}] \Longrightarrow \text{PLNRC(R)}$$

10) Determination of range of uncertainty of P/L recurring cost per unit of P/L mass: (K$/kg)
PLMXR(R); PLMNR(R); PLMLR(R) $1 \le \text{R} \le \mathit{MAXR}$

If PLMARG(R) $\le$ *APLMARG*

Then

$$PLMNR(R) = PLMNRC$$
$$PLMXR(R) = PLMXRC * \{1.0 + PLSENRC * [PLMARG(R) - APLMARG] * 0.0001\}$$
$$PLMLR(R) = PLMLRC * \{1.0 + PLSENRC * [PLMARG(R) - APLMARG] * 0.0001\}$$

If PLMARG(R) $>$ *APLMARG*
Then

$$PLMNR(R) = PLMNRC * \{1.0 + PLSENRC * [PLMARG(R) - APLMARG] * 0.0001\}$$
$$PLMXR(R) = PLMXRC$$
$$PLMLR(R) = PLMLRC * \{1.0 + PLSENRC * [PLMARG(R) - APLMARG] * 0.0001\}$$

If PLMLR(R) $<$ *MXRCCH* $*$ *PLMLRC* $*$ 0.01
Then

$$PLMLR(R) = MXRCCH * PLMLRC * 0.01$$

If PLMNR(R) $<$ *MXRCCH* $*$ *PLMLRC* $*$ 0.01
Then

$$PLMNR(R) = MXRCCH * PLMNRC * 0.01$$

11) Determination of P/L recurring cost per unit mass: (K$/kg)
PLRC(R) $1 \leq R \leq MAXR$

Generate random number, RN.
Using a random sampling algorithm establish value of PLRC(R):

$$[PLMXR(R); PLMNR(R); PLMLR(R)] \Longrightarrow PLRC(R)$$

12) Determination of HRV transportation cost per unit mass: (K$/kg)
TCUM(R) $1 \leq R \leq MAXR$

Generate random number, RN.
Using a random sampling algorithm, establish value of TCUM(R):

$$[TCMXHRV; TCMNHRV; TCMLHRV] \Longrightarrow TCUM(R)$$

13) Determination of ELV transportation cost per unit mass: (K$/kg)
ELVTCUM(R) $1 \leq R \leq MAXR$

Generate random number, RN.

Using a random sampling algorithm, establish value of ELVTCUM(R):

$$[TCMXELV; TCMNELV; TCMLELV] \Longrightarrow \text{ELVTCUM(R)}$$

14) Establish HRV unit cost (purchase price for inclusion into HRV fleet: (M$/unit)
HRVUC(R) $1 \leq \text{R} \leq MAXR$

Generate random number, RN.
Using a random sampling algorithm, establish value of HRVUC(R):

$$[FIMXHRV; FIMNHRV; FIMLHRV] \Longrightarrow \text{HRVUC(R)}$$

15) Establish launch infrastructure cost: (M$)
INFC(R) $1 \leq \text{R} \leq MAXR$

Generate random number, RN.
Using a random sampling algorithm, establish value of INFC(R):

$$[LFMX; LFMN; LFML] \Longrightarrow \text{INFC(R)}$$

16) Determination of total technology program cost: (M$)
TPC(R) $1 \leq \text{R} \leq MAXR$

Generate random number, RN.
Using random sampling algorithm, establish value of TPC(R):

$$[MXTPC; MNTPC; MLTPC] \Longrightarrow \text{TPC(R)}$$

17) Cost spreading of HRV stage 1 nonrecurring cost: (M$)
HRVNRC1(I,R) $1 \leq \text{R} \leq MAXR$
$1 \leq \text{N} \leq 10$

When $\text{I} = CYSO - CY - 10 + \text{N}$ **and** $\text{I} \geq 1$
Then

$$\text{HRVNRC1(I, R)} = CSHRVS1(N) * 0.01 * \text{S1NRC(R)}$$

18) Cost spreading of HRV stage 2 nonrecurring cost: (M$)
HRVNRC2(I,R) $1 \leq \text{R} \leq MAXR$
$1 \leq \text{N} \leq 10$

When $\text{I} = CYSO - CY - 10 + \text{N}$ **and** $I \geq 1$
Then

$$\text{HRVNRC2(I, R)} = CSHRVS2(N) * 0.01 * \text{S2NRC(R)}$$

19) Cost spreading of infrastructure expenditures: (M$)
INFNRC(I,R) $1 \leq \text{R} \leq MAXR$
$1 \leq \text{N} \leq 10$

When $\text{I} = CYLF - CY - 10 + \text{N}$ **and** $I \geq 1$

Then

$$INFNRC(I, R) = CSIF(N) * 0.01 * INFC(R)$$

20) Cost spreading of fleet investment: (M$)
FNRC(I,R) $1 \leq R \leq MAXR$
$1 \leq N \leq 10$

When $I = CYFI - CY - 10 + N$ **and** $I \geq 1$
Then

$$FNRC(I, R) = CSHRVFI(N) * 0.01 * HRVUC(R) * NOHRV$$

21) Determination of annual HRV payload nonrecurring cost: (M$)
HRVPLNRC(I,R) $1 \leq R \leq MAXR$

$$HRVPLNRC(I, R) = PLNRC(R) * APLMASS(I) * 0.001 * [100 - PPL(I)] * 0.01/PLPM$$

22) Determination of annual ELV payload nonrecurring cost: (M$)
ELVPLNRC(I,R) $1 \leq R \leq MAXR$

Generate random number, RN.
Using a random sampling algorithm, establish value of EXPLNRC(R):

$$[PLMXNRC; PLMNNRC; PLMLNRC] \Longrightarrow EXPLNRC(R)$$

$$ELVPLNRC(I, R) = EXPLNRC(R) * APLMASS(I) * 0.001 * PPL(I) * 0.01/PLPM$$

23) Determination of annual research and technology program cost: (M$)
TPNRC(I,R) $1 \leq R \leq MAXR$

$$TPNRC(I, R) = 0.01 * CSTP(I) * TPC(R)$$

24) Determination of expected value and standard deviation of transportation system nonrecurring cost: (M$)

$$ETSNRC = \left\{ \sum_{R=1}^{MAXR} [S1NRC(R) + S2NRC(R)] \right\} / MAXR$$

$$SDTSNRC = \left\{ \left\{ \sum_{R=1}^{MAXR} [S1NRC(R) + S2NRC(R)]^2 \right\} / MAXR - [ETSNRC]^2 \right\}^{0.5}$$

25) Determination of expected value and standard deviation of HRV unit cost: (M$)

$$\text{EHRVUC} = \left\{\sum_{\text{R}=1}^{MAXR} \text{HRVUC(R)}\right\}/MAXR$$

$$\text{SDHRVUC} = \left\{\left\{\sum_{\text{R}=1}^{MAXR} [\text{HRVUC(R)}]^2\right\}/MAXR - [\text{ELVUC}]^2\right\}^{0.5}$$

26) Determination of expected value and standard deviation of HRV fleet investment: (M$)

$$\text{EHRVFC} = NOHRV * \text{EHRVUC}$$
$$\text{SDHRVFC} = NOHRV * \text{SDHRVUC}$$

27) Determination of expected value and standard deviation of infrastructure investment: (M$)

$$\text{EINFC} = \left\{\sum_{\text{R}=1}^{MAXR} \text{INFC(R)}\right\}/MAXR$$

$$\text{SDINFC} = \left\{\left\{\sum_{\text{R}=1}^{MAXR} [\text{INFC(R)}]^2\right\}/MAXR - [\text{EINFC}]^2\right\}^{0.5}$$

28) Determination of expected value and standard deviation of launch cost per unit payload mass: (K$/kg)

$$\text{ELCUM} = \left\{\sum_{\text{R}=1}^{MAXR} \text{TCUM(R)}\right\}/MAXR$$

$$\text{SDLCUM} = \left\{\left\{\sum_{\text{R}=1}^{MAXR} [\text{TCUM(R)}]^2\right\}/MAXR - [\text{ELCUM}]^2\right\}^{0.5}$$

29) Determination of expected value and standard deviation of payload nonrecurring cost per unit payload mass: (K$/kg)

$$\text{EPLNRC} = \left\{\sum_{\text{R}=1}^{MAXR} \text{PLNRC(R)}\right\}/MAXR$$

$$\text{SDPLNRC} = \left\{\left\{\sum_{\text{R}=1}^{MAXR} [\text{PLNRC(R)}]^2\right\}/MAXR - [\text{EPLNRC}]^2\right\}^{0.5}$$

30) Determination of expected value and standard deviation of payload recurring cost per unit payload mass: (K$/kg)

$$\mathrm{EPLRC} = \left\{ \sum_{\mathrm{R}=1}^{MAXR} \mathrm{PLRC(R)} \right\} / MAXR$$

$$\mathrm{SDPLRC} = \left\{ \left\{ \sum_{\mathrm{R}=1}^{MAXR} [\mathrm{PLRC(R)}]^2 \right\} / MAXR - [\mathrm{EPLRC}]^2 \right\}^{0.5}$$

31) Determination of expected value and standard deviation of payload design margin: (%)

$$\mathrm{EPLDM} = \left\{ \sum_{\mathrm{R}=1}^{MAXR} \mathrm{PLMARG(R)} \right\} / MAXR$$

$$\mathrm{SDPLDM} = \left\{ \left\{ \sum_{\mathrm{R}=1}^{MAXR} [\mathrm{PLMARG(R)}]^2 \right\} / MAXR - [\mathrm{EPLDM}]^2 \right\}^{0.5}$$

32) Determination of expected value and standard deviation of present value of operational system life-cycle cost: (M$)

If $\mathrm{I} \leq CYSO - CY + 1$

Then

$$\mathrm{AFRNRC(I,R)} = 0$$

If $\mathrm{I} > CYSO - CY + 1$

Then

$$\mathrm{AFRNRC(I, R)} = 0.01 * FLTRP * NOHRV * \mathrm{HRVUC(R)}$$

$$\begin{aligned} \mathrm{TACA(I, R)} = \{&\mathrm{HRVNRC1(I, R)} + \mathrm{HRVNRC2(I, R)} + \mathrm{INFNRC(I, R)} \\ &+ \mathrm{HRVPLNRC(I, R)} + \mathrm{ELVPLNRC(I, R)} + \mathrm{FNRC(I, R)} \\ &+ \mathrm{AFRNRC(I, R)}\} + \{[\mathrm{TCUM(R)} + \mathrm{PLRC(R)}] \\ &* 0.001 * APLMASS(I) * [1 - 0.01 * PPL(I)] \\ &+ [\mathrm{ELVTCUM(R)} + \mathrm{ELVPLRC(R)}] * 0.001 \\ &* APLMASS(I) * 0.01 * PPL(I)\} \end{aligned}$$

When $IHD \neq$ yes

Then

$$\mathrm{EPVC} = \left\{ \sum_{\mathrm{R}=1}^{MAXR} \left\{ \sum_{\mathrm{I}=1}^{MAXYRS} \mathrm{TACA(I, R)}/[1 + 0.01 * DR]^{\mathrm{I}-1} \right\} \right\} / MAXR$$

$$\mathrm{SDPVC} = \left\{ \left\{ \sum_{\mathrm{R}=1}^{MAXR} \left\{ \sum_{\mathrm{I}=1}^{MAXYRS} \mathrm{TACA(I, R)}/[1 + 0.01 * DR]^{\mathrm{I}-1} \right\}^2 \right\} / MAXR - [\mathrm{EPVC}]^2 \right\}^{0.5}$$

When *IHD* = yes

Then

$$\mathrm{EPVC} = \left\{ \sum_{R=1}^{MAXR} \left\{ \sum_{I=1}^{MAXYRS} \mathrm{TACA(I, R)}/[1 + 0.01 * DR]^{I-1} \right\} + \mathrm{TACA}(I = MAXYRS, R)/[0.01 * DR] * [1 + 0.01 * DR]^{MAXYRS} \right\} / MAXR$$

$$\mathrm{SDPVC} = \left\{ \left\{ \left\{ \sum_{R=1}^{MAXR} \left\{ \sum_{I=1}^{MAXYRS} \mathrm{TACA(I, R)}/[1 + 0.01 * DR]^{I-1} \right\} + \mathrm{TACA}(I = MAXYRS, R)/[0.01 * DR] * [1 + 0.01 * DR]^{MAXYRS} \right\}^2 \right\} / MAXR - [\mathrm{EPVC}]^2 \right\}^{0.5}$$

33) Determination of HRV stage 1 expected annual nonrecurring cost: HRVS1NC(I) (M$)

$$\mathrm{HRVS1NC(I)} = \left\{ \sum_{R=1}^{MAXR} \mathrm{HRVNRC1(I, R)} \right\} / MAXR$$

34) Determination of HRV stage 2 expected annual nonrecurring cost: HRVS2NC(I) (M$)

$$\mathrm{HRVS2NC(I)} = \left\{ \sum_{R=1}^{MAXR} \mathrm{HRVNRC2(I, R)} \right\} / MAXR$$

35) Determination of infrastructure expected annual nonrecurring cost: INFNC(I) (M$)

$$\mathrm{INFNC(I)} = \left\{ \sum_{R=1}^{MAXR} \mathrm{INFNRC(I, R)} \right\} / MAXR$$

36) Determination of HRV payloads expected annual nonrecurring cost: HRVPLNC(I) (M$)

$$\mathrm{HRVPLNC(I)} = \left\{ \sum_{R=1}^{MAXR} \mathrm{HRVPLNRC(I, R)} \right\} / MAXR$$

37) Determination of ELV payloads expected annual nonrecurring cost: ELVPLNC(I) (M$)

$$\mathrm{ELVPLNC(I)} = \left\{ \sum_{R=1}^{MAXR} \mathrm{ELVPLNRC(I, R)} \right\} / MAXR$$

38) Determination of fleet investment expected annual nonrecurring cost: HRVFNC(I) (M$)

$$\mathrm{HRVFNC(I)} = \left\{\sum_{\mathrm{R}=1}^{MAXR} \mathrm{FNRC(I, R)}\right\}/MAXR$$

39) Determination of fleet replacement expected annual nonrecurring cost: HRVRNC(I) (M$)

If I $\leq CYSO - CY + 1$
Then

$$HRVRNC(I) = 0$$

If I $> CYSO - CY + 1$
Then

$$\mathrm{HRVRNC(I)} = 0.01 * FLTRP * \mathrm{EHRVFC}$$

40) Determination of technology program expected annual nonrecurring cost: TPNC(I) (M$)

$$\mathrm{TPNC(I)} = \left\{\sum_{\mathrm{R}=1}^{MAXR} \mathrm{TPNRC(I, R)}\right\}/MAXR$$

41) Determination of HRV expected annual launch operations recurring cost: HRVRC(I) (M$)

$$\mathrm{HRVRC(I)} = \left\{\sum_{\mathrm{R}=1}^{MAXR} \mathrm{TCUM(R)}\right\} * 0.001 * APLMASS(I) * [1 - 0.01 * PPL(I)]/MAXR$$

42) Determination of ELV expected annual launch operations recurring cost: ELVRC(I) (M$)

$$\mathrm{ELVRC(I)} = \left\{\sum_{\mathrm{R}=1}^{MAXR} \mathrm{ELVTCUM(R)}\right\} * 0.001 * APLMASS(I) * 0.01 * PPL(I)/MAXR$$

43) Determination of HRV payload expected annual recurring cost: HRVPLC(I) (M$)

$$\mathrm{HRVPLC(I)} = \left\{\sum_{\mathrm{R}=1}^{MAXR} \mathrm{PLRC(R)}\right\} * 0.001 * APLMASS(I) * [1 - 0.01 * PPL(I)]/MAXR$$

44) Determination of ELV payload expected annual recurring cost: ELVPLC(I) (M$)

$$ELVPLC(I) = \left\{ \sum_{R=1}^{MAXR} ELVPLRC(R) \right\} * 0.001 * APLMASS(I) * 0.01 * PPL(I)/MAXR$$

45) Calculation of total expected annual nonrecurring cost, total expected annual recurring cost, and total expected annual cost: TNRC(I); TRC(I); ETAC(I) (M$)

$$\begin{aligned} \mathrm{TNRC(I)} &= \mathrm{HRVS1NC(I) + HRVS2NC(I) + INFNC(I) + HRVPLNC(I)} \\ &\quad \mathrm{+ ELVPLNC(I) + HRVFNC(I) + HRVRNC(I) + TPNC(I)} \\ \mathrm{TRC(I)} &= \mathrm{HRVRC(I) + ELVRC(I) + HRVPLC(I) + ELVPLC(I)} \\ \mathrm{ETAC(I)} &= \mathrm{TNRC(I) + TRC(I)} \end{aligned}$$

46) Calculation of total expected cumulative nonrecurring cost, total expected cumulative recurring cost, and total expected cumulative cost: CTNRC(I); CTRC(I); CETAC(I) (M$)

$$\mathrm{CTNRC(I)} = \sum_{A=1}^{I} \mathrm{TNRC(A)}$$

$$\mathrm{CTRC(I)} = \sum_{A=1}^{I} \mathrm{TRC(A)}$$

$$\mathrm{CETAC(I)} = \mathrm{CTNRC(I) + CTRC(I)}$$

47) Calculation of standard deviation of total nonrecurring cost: SDNRC(I) (M$)

$$\mathrm{SDNRC(I)} = \left\{ \left\{ \sum_{R=1}^{MAXR} [\mathrm{HRVNRC1(I, R) + HRVNRC2(I, R) + INFNRC(I, R) + HRVPLNRC(I, R) + ELVPLNRC(I, R) + FNRC(I, R) + AFRNRC(I, R) + TPNRC(I, R)}]^2 \right\} / \mathrm{MAXR} - [\mathrm{TNRC(I)}]^2 \right\}^{0.5}$$

48) Calculation of standard deviation of total recurring cost: SDRC(I) (M$)

$$\mathrm{SDRC(I)} = \left\{\left\{\sum_{\mathrm{R}=1}^{MAXR} [\mathrm{TCUM(R)} * 0.001 * APLMASS(I) * [1 - 0.01 * PPL(I)] + \mathrm{ELVTCUM(R)} * 0.001 * APLMASS(I) * 0.01 * PPL(I) + \mathrm{PLRC(R)} * 0.001 * APLMASS(I) * [1 - 0.01] * PPL(I) + \mathrm{ELVPLRC(R)} * 0.001 * APLMASS(I) * 0.01 * PPL(I)]^2\right\} / MAXR - [\mathrm{TRC(I)}]^2\right\}^{0.5}$$

49) Calculation of standard deviation of total annual cost: SDTAC(I) (M$)

$$\mathrm{SDTAC(I)} = \left\{[\mathrm{SDNRC(I)}]^2 + [\mathrm{SDRC(I)}]^2\right\}^{0.5}$$

3. *Initialization List*

$$\mathrm{HRVNRC1(I, R)} = 0 \quad \text{for } 1 \leq \mathrm{I} \leq MAXYRS$$
$$\mathrm{HRVNRC2(I, R)} = 0 \quad \text{for } 1 \leq \mathrm{I} \leq MAXYRS$$
$$\mathrm{INFNRC(I, R)} = 0 \quad \text{for } 1 \leq \mathrm{I} \leq MAXYRS$$
$$\mathrm{FNRC(I, R)} = 0 \quad \text{for } 1 \leq \mathrm{I} \leq MAXYRS$$
$$\mathrm{TPNRC(I, R)} = 0 \quad \text{for } 1 \leq \mathrm{I} \leq MAXYRS$$

VII. Selecting a Launch Vehicle (Government Missions)

Selection of a launch vehicle to be utilized in commercial business ventures was discussed in Sec. II, where it was concluded that the selection process should rely heavily on a multiplicity of metrics such as expected return on investment, expected present value of cash flow, and risk measured in terms of the variability of ROI or present value of cash flow. The development of such metrics requires the use of a financial planning model such as the stochastic (Monte Carlo) *DOMSAT* model. This model, as described in Sec. I, converts market, design, and performance attributes, including in many instances their perceived uncertainty profiles, into the necessary ROI and net present value expected value and risk metrics. Thus, all of the pertinent launch service attributes, and their interrelationships, are taken into account in the determination of these metrics, and appropriate choices can be made having knowledge of both the resultant expected values and risk associated with the metrics utilized for making informed choices.

Government space missions are concerned with the establishment and maintenance of services that utilize individual satellites, or more frequently, constellations of satellites. The establishment and maintenance of services imply the placement, operation, and maintenance of individual satellites or constellations of satellites in orbit over extended periods of time. Satellites that have been placed successfully into LEO constellations have already failed and been replaced in order to maintain the desired continuity of service. As additional constellations are launched, competition for provided launch services will increase, and the importance of cost minimization within availability objectives will increase. The determination of satellite launch, sparing, and replacement strategies that will minimize the present value of mission life-cycle cost while achieving annual and long-term availability constraints will increase in importance. The analysis of these strategies and the determination of efficient (i.e., cost and risk minimization) strategies is the subject of Sec. IX. The determination of efficient strategies requires the consideration of cost and schedule uncertainties, launch and satellite (subsystem) reliability, and delays that might result from failures. Launch, sparing, and replacement strategies include launch (both for placement and replacement) according to a planned schedule, replacement on satellite failure, replacement in anticipation of wear out failures, on-orbit operating and dormant spares, and various combination strategies.

As is readily apparent, government missions are frequently concerned with achieving objectives that include both short- and long-term availability of service. When this is the case, the selection of a launch vehicle cannot, and should not, be considered without simultaneous consideration of sparing and maintenance strategies. Launch vehicle availability, reliability, and rescheduling delays given that a failure has occurred are interrelated with satellite design characteristics and sparing/maintenance strategies. For this reason it is necessary to consider the complete mission (i.e., constellation placement and maintenance) when selecting a launch vehicle.

The overall goal is to minimize the cost of achieving both short- and long-term availability requirements, where availability is expressed as the percentage of the time within a specified duration of time (for example, one year or over the mission duration of 10 years), so that at least a required number of sensors or transponders will be operational. Note that this is equivalent to considering the probability that at least a required number of sensors or transponders will be operational as and when required. The problem becomes more complicated when multiple (different) sensors and/or transponders are on-board each satellite and the value of each is different. This is illustrated in Sec. XI.

To establish the best choice of a launch vehicle for a mission, it is necessary to simulate the mission utilizing a stochastic simulation model such as *SATCAV*, as described in Sec. I, or the yet to be implemented launch campaign simulation model, as described in Sec. VIII. In either case the results are the probability distribution of the present value of life-cycle costs and annual and overall mission duration availability statistics for each sensor/transponder type. The results are totally dependent on the assumed attributes of the considered space transportation service in combination with selected sparing/maintenance strategies. Selecting different sets of space transportation service attributes is equivalent to considering different launch vehicles and in so doing allows the cost and availability results to be compared when the mission is performed utilizing the different launch vehicles. Again, the objective is to select that launch vehicle or transportation service which minimizes the present value of the life-cycle cost of achieving both short- and long-term availability requirements. The decision process is complicated by the fact that the present value of the mission life-cycle cost must be described as a probability distribution, that is, expected value and standard deviation with the latter being a measure of risk. Thus, in the true sense the selection of a launch vehicle or transportation service is a multiattribute decision problem having both cost and availability (of multiple sensors/transponders) statistics. The problem is indeed complex, but the decision maker should be provided with all possible information, particularly when making decisions that involve very significant resources.

VIII. Launch Campaign Modeling Considerations

The objective of the following paragraphs is to describe the requirements for the development of a computer simulation model [referred to as the **Launch Campaign Simulation (LCS) model** for assessing the annual and life-cycle (discounted) costs associated with alternative launch campaigns that might be considered for placing and maintaining constellations of communication or other satellites. The approach taken is to identify the requirements for a basic analytic tool, that is, a simulation model that considers scheduled placements and replacements of satellites, and then to describe a number of additional capabilities (for example, launch in anticipation of wear out failure, on-orbit operational spares, etc.) that could be added as and when required. The goal is to provide a powerful tool that would be useful for analyzing launch campaign strategies and alternatives as soon as possible and to then work toward expanding its range of application over a period of time into the areas relating to population of multiple orbital planes and the explicit consideration of sparing strategies.

A. Requirements

The envisioned simulation model would initially be aimed at analyzing launch campaign strategies and alternatives and would be expandable so as to analyze sparing strategies and alternatives. The model would establish satellite availability statistics and the probability distributions of annual and life-cycle cost (i.e., present value of life-cycle cost at a number of discount rates) associated with the placement and maintenance of constellations of communication or other satellites. The constellations can encompass multiple satellites operating in multiple planes, but the multiple planes initially would be considered as operating independently, at least from the points of view of satellite sparing (spares cannot move between planes), system availability (availability statistics reflect satellite operations within a plane), and satellite placements (i.e., only one plane can be populated as a result of any single launch). (Because of the realities associated with the development of complex models and associated software, a phased capability approach is considered. The complications that result from the consideration of multiple planes that are not independent would be developed following the development of an initial capability.) It is therefore envisioned that the simulation model would, at least initially, simulate a constellation contained within an orbital plane and launches, satellites, and associated costs summed appropriately outside of the simulation model when it is necessary to consider multiple plane scenarios.

A three-step process is envisioned and described in terms of requirements or capabilities to be implemented. Step 1 would be concerned with the development of a basic LCS model for analyzing launch campaign strategies and alternatives. This model would allow launch campaigns and strategies and satellite procurement and delivery schedules to be evaluated in a multilaunch vehicle environment considering system availability and cost statistics. Step 2 would be concerned with expanding the Step 1 capability so that populating multiple planes as a result of a single launch could be considered as a strategy alternative. Step 3 would be concerned with expanding the Step 1 and 2 capabilities so that other alternative sparing and maintenance strategies could be analyzed and evaluated.

The objective of the development of the LCS model is to provide a means for understanding the implications of alternative launch campaigns and satellite sparing strategies on annual and life-cycle costs and system availability. It is envisioned that the simulation model would utilize Monte Carlo techniques as described in Sec. I.

The following general capabilities should be included in the Step 1 Basic LCS model:

1) Launches should be allowed to take place according to a specified desired schedule for both the placement and planned replacement of satellites with launches for recovery from launch failures taking into account availability and scheduling of replacement vehicles and satellites. The actual launch schedule differs from the desired schedule as a result of failures and related rescheduling delays. Both the desired launch schedule and replacement schedule should be specified via input data.
2) The LCS model should consider launch vehicle failures and should reschedule launches among a specified family of launch vehicles taking into account schedule delays, launch vehicle availability, delivery capability (i.e., number of satellites), and other factors.
3) Initially (Step 1), a constellation of satellites would be considered as multiple satellites within a single orbital plane. It should be assumed that if multiple satellites are launched on a single launch all satellites would be placed into the same orbital plane and used to increase the population of the plane, replace failed satellites, or both. When the desired number of operational satellites is exceeded, the excess would be considered as operating spares.
4) The model should allow for the consideration of multiple expendable or reusable launch vehicles with their utilization specified for initial and scheduled replacement flights and an algorithm (to be developed that would consider cost, availability, and other factors) used for selecting launch vehicles in the event of launch failures.
5) Launch vehicles should be described in terms of multiple attributes that include the specification of the time when a vehicle is first available for use, reliability, price, price uncertainty, insurance rate, availability (i.e., the probability that the launch vehicle will be available for use—this reflects delays that can result because of failures that take place when used for other missions), stand-down time statistics, number of satellites that can be placed into orbit per launch, and other factors. It is recommended that all missions in the mission model be considered

together with the utilized launch vehicles. The launches for these missions could then be simulated to establish success on failure and, in the case of failure, the delays and availability for use on the mission under consideration established. This would allow the effects of launch failures that occur during launch of other missions to be taken into account for the specific mission under consideration.

6) An algorithm should be developed and utilized for selecting launch vehicles for the replacement missions (i.e., recovery from a launch failure) that takes into account the availability of multiple launch vehicles operating from multiple launchpads, number of satellites that can be placed into orbit per launch vehicle, cost, reliability, availability, number of launches per year, and minimum time between launches.
7) It is necessary to consider payment schedules for launch services. Therefore, launch payment profiles should be specified via input data and should include provision for down payments that can be fixed in time with other costs floating with launch date.
8) The model should allow launch vehicle characterizations that include delays which can result from a launch failure, minimum time between launches, maximum number of vehicles that can be purchased/year from a particular supplier, and time of successive launches from a single site.
9) Various insurance arrangements should be considered such as taking or not taking insurance and deductible arrangements. Insurance rates would of course be required for each of the considered alternatives.
10) Satellite reliability characteristics are an important determinant of replacement considerations. The model should allow for the considerations of both satellite wear out life and random failure characteristics. Satellite reliability should be modeled by considering the satellite as being comprised of multiple payload subsystems (operating in parallel, from a reliability point of view) and multiple bus support subsystems (operating in series, from a reliability point of view) with each subsystem characterized in terms of expected wear out life, standard deviation of wear out life, and mean time before failure.
11) Because a satellite can consist of multiple payloads (i.e., sensors or transponders), it will be necessary to define what is meant by a failure. Therefore, the model should have the ability to define satellite failure in terms of critical payloads that should be considered individually and in combination (two at a time). Failure of specified combinations of critical payloads would then constitute failure and lead to the consideration of replacement launches.
12) When it is possible to launch multiple satellites simultaneously, there is the possibility that their success/failure characteristics can be correlated. This correlation can result from the satellites being part of the same manufacturing lot or because they share the same ejection/insertion mechanism. Therefore, the model should allow for the consideration of

correlation or probability of success across multiple satellites per launch and the probability of successful ejection (and correlation effects).

13) Nonrecurring and unit recurring costs should be considered. These costs should allow for the consideration of uncertainty and cost spreading over time. In addition, recurring cost-learning effects must be considered.
14) Allowance should be made for the consideration of multiple operating expenses with the name and number specified via input data. Allowance should be made for each to have a fixed and variable component with each considering uncertainty. In addition, multiple capital expenditures should be considered together with cost uncertainties and cost spreading.
15) The model should allow for the consideration of multiple time periods with computations performed on a monthly basis with results displayed on a quarterly and annual basis. To make reasonable estimates of sensor/transponder/satellite availability, it is mandatory that the status of sensors/transponders/satellites be checked on a monthly basis and that all actions be scheduled/taken on a monthly basis.

The input system should be contained within an Excel (or comparable) spreadsheet file with all variables clearly defined and delineated (inclusion within a menu-driven user-friendly input system would be delayed for development until Step 3). Because model results will undoubtedly be required for presentation purposes and for the justification of the continued developed of Steps 2 and 3, the format of the model-generated reports should be consistent with these needs.

The LCS model should develop the following reports:

1) The first report should summarize both quarterly and annual costs and contains expected values and standard deviations with the latter available for major cost elements.
2) Quarterly and annual event statistics reports should be provided. The quarterly results need only indicate expected values, whereas the annual statistics should include the probability distribution of annual events and expected values and standard deviations.
3) A satellite availability statistics report that provides the availability statistics annually and across a specified time period is required. In addition, sensor/transponder availability statistics need to be provided and need to indicate the percentage of the time that each sensor/transponder is available annually.
4) A cause of failure statistics report should be provided. This report should indicate the probability that failures will be caused by each subsystem and considers the entire planning horizon.

Step 2 should seek to eliminate the constraint of orbital plane independence. The Basic LCS model would be modified to allow for multiple planes to be populated by satellites launched from a common launch. This would add the ability to place satellites into a parking orbit that precesses and allows, with appropriate delays and reliability considerations, for satellites that are launched from a single launch to be allocated to different orbital planes. The capability developed during Step 2 should be limited to operating within the Step 1 framework of analyzing launch campaign strategies and alternatives.

The following general capabilities would be added to the LCS model during Step 3. These capabilities would be in addition to the capabilities already developed during Steps 1 and 2 and are aimed at adding a capability to consider alternative sparing and maintenance strategies:

1) Consider satellite replacement in anticipation of wear out failures with anticipatory time considered as a variable.
2) Consider satellite replacement on failure with/without on-orbit spares. An algorithm should be developed that would allow replacement to be considered in terms of individual or multisatellite replacement launches.
3) Consider satellite replacement on failure with/without on-orbit dormant spares. Satellites when in the dormant state might have different failure characteristics than when in the active state. These different failure characteristics would be provided as input to the model as well as the time to move from the dormant location to the active location and the probability that the dormant satellite will operate satisfactorily when called into service.
4) Develop a user-friendly menu-driven input system for facilitating the use of the LCS model.

Many of the preceding features and capabilities already exist in the *SATCAV* model. (A detailed discussion of the *SATCAV* model is presented in Ref. 1.) However, the *SATCAV* model does not have the capability of considering multiple orbital planes, selecting among multiple launches, and utilizing different launch vehicles for placement and replacement missions.

IX. Cost Risk and Life-Cycle Cost/Availability Analysis

Placement, operation, and maintenance of constellations of satellites have become a reality. Satellites that have been placed successfully into low-Earth-orbit constellations have already failed and been replaced in order to maintained the desired continuity of service. (Indications are that by the late Spring of 1999 at least 12 satellites of the Iridium constellation had failed.) As additional constellations are launched, competition for provided services will increase, and the importance of cost minimization within availability objectives will increase. The determination of satellite launch, sparing, and replacement strategies that will minimize the present value of mission life-cycle cost while achieving annual and long-term availability constraints will increase in importance. The analysis of these strategies and the determination of efficient (i.e., cost and risk minimization) strategies is the subject of this section. The determination of efficient strategies requires the consideration of cost and schedule uncertainties, launch and satellite (subsystem) reliability, and delays that might result from failures. Launch, sparing, and replacement strategies include launch (both for placement and replacement) according to a planned schedule, replacement on satellite failure, replacement in anticipation of wearout failures, on-orbit operating and dormant spares, and various combination strategies.

To demonstrate a methodology for minimizing the cost of meeting both short- and long-term availability objectives, results of an analysis that was performed to evaluate the number of satellites that would be required by a remote sensing mission are presented in the following paragraphs. A multisatellite (constellation) mission was simulated utilizing the *SATCAV* stochastic life-cycle cost and availability model (described in detail in Sec. I and summarized briefly in the following paragraphs). This model established annual and present value of life-cycle cost, number of satellites launched, and probability distributions of the number of functioning satellites annually and over a planning horizon when considering a number of launch and sparing strategy alternatives. The result was the determination of the launch and sparing strategy that minimized the present value of life-cycle cost (of the space portion of the overall mission costs) and satisfied overall mission availability requirements.

Prior to discussing life-cycle cost analysis, consideration is given to the estimation of nonrecurring and recurring costs with particular emphasis on the development of cost risk. The resulting expected values and standard deviations can then be used in the *SATCAV* model.

A. Cost Risk

Cost estimation of space missions is usually based upon either a top-down or bottom-up approach. The top-down approach relies upon the use of cost-estimating relationships that relate nonrecurring and recurring costs to system attributes. For satellites the CERs normally relate cost to physical parameters such as mass and power with the result that an estimate of satellite mass and power results in an estimate of nonrecurring and recurring cost. The CERs[29] are established by performing regression analyses on historical data. This has the limitation that innovation and new technologies are inadequately accounted for—only historical data are utilized, and it is the goal of innovation and technology development to deviate from historical performance.

The alternative, or bottom-up approach, attempts to establish cost estimates through an involved process of disaggregating a system into its many subsystems, estimating the labor and materials that would be required for each subsystem, and then aggregating the costs. This has the advantage that considerably more detail is necessary to develop a cost estimate but then has the disadvantage that it requires considerably more effort to develop a cost estimate and is usually not possible in the early stages of program development.

Both the bottom-up and top-down cost-estimating approaches have their place. The top-down approach is normally used early in the planning stage to identify reasonable alternatives rapidly and to establish program reasonableness. The bottom-up approach is normally used when detailed product design has been completed and piece/parts are identifiable and vendor quotes are available. Neither are particularly well suited when considering long-term multiphase programs that require significant research and technology development to achieve program objectives.[30]

Because of the increasing concern with cost overruns and the desire to establish incentives based upon the achievement of cost goals, increased emphasis is being placed on the development of cost risk. Cost risk relates to the chance that actual cost will exceed various predicted levels. The following paragraphs describe a typical method developed and used for estimating the cost risk associated with a proposed multisatellite remote-sensing Earth observation system.

The starting point of the bottom-up approach is the development of a detailed work breakdown structure (WBS). This includes both nonrecurring and recurring cost elements. Figure 4.34 presents a portion of a detailed WBS. (The complete WBS for this example actually has over 900 rows). Rows 6–9 are additive to obtain row 5; rows 5, 10, . . . are additive to obtain row 4; etc. All to frequently the costs identified with each row in the WBS are based upon most likely estimates and should not be, but are usually, added. Significant errors can result from adding most likely values. Mathematically, it is incorrect to add most likely values; only expected values are additive (refer to Fig. 2.2).

The consideration of cost risk tends to eliminate this error by forcing the development of expected values that can then be added to obtain an estimate of the expected cost. This is accomplished by describing the basic cost elements of the WBS in terms of minimum, most likely, and maximum costs as indicated in Fig. 4.35. These estimates take into account subjective judgments relating to "what can go right and what can go wrong." These subjective judgments are then used to establish expected or average costs and related standard deviations and estimates of 20, 50, and 80 percentile costs, as illustrated in Fig. 4.36. The column marked "Proposed" represents the expected value, and "Prop. Std. Dev." is the associated standard deviation.

				PROPOSED
1	1		**Space Based Satellite System**	**6537.73**
2	1.1		**Space Segment**	**2831.23**
3	1.1.1		**Payload**	**1726.89**
4	1.1.1.1	NR	**Payload Development**	**180.04**
5	1.1.1.1.1	NR	**Payload Architecture & Preliminary Design**	**1.45**
6	1.1.1.1.1.1	NR	Payload Requirements & Concept Validation	0.38
7	1.1.1.1.1.2	NR	Payload SRR	0.20
8	1.1.1.1.1.3	NR	Payload Preliminary Design	0.56
9	1.1.1.1.1.4	NR	Payload PDR	0.30
10	1.1.1.1.2	NR	**Payload Detailed Design & Development**	**117.20**
11	1.1.1.1.2.1	NR	SAR Sensor Design & Development	108.20
12	1.1.1.1.2.1.1	NR	SAR Sensor System Engineering (Development)	N/A
13	1.1.1.1.2.1.2	NR	SAR Antenna Design & Development	80.10
14	1.1.1.1.2.1.2.1	NR	SAR Antenna Structure and Mechanisms	31.89
15	1.1.1.1.2.1.2.2	NR	SAR Antenna Initial System Engineering	27.66
16	1.1.1.1.2.1.2.3	NR	SAR Antenna Microcircuit Design & Development	2.65
17	1.1.1.1.2.1.2.4	NR	SAR Antenna Array Design & Development	0.08

Fig. 4.34 Portion of a typical WBS.

				PROPOSED	Minimum	Most Likely	Maximum	PROP Std Dev	20 Percentile	50 Percentile
1	1		Space Based Satellite System	6537.73				194.57	6702	6538
2	1.1		Space Segment	2831.23				129.38	2940	2831
3	1.1.1		Payload	1726.89				124.12	1832	1727
4	1.1.1.1	NR	Payload Development	180.04				5.74	185	180
5	1.1.1.1.1	NR	Payload Architecture & Preliminary Design	1.45				0.07		
6	1.1.1.1.1.1	NR	Payload Requirements & Concept Validation	0.38	0.33	0.33	0.49	0.04		
7	1.1.1.1.1.2	NR	Payload SRR	0.20	0.17	0.17	0.25	0.02		
8	1.1.1.1.1.3	NR	Payload Preliminary Design	0.56	0.48	0.49	0.72	0.05		
9	1.1.1.1.1.4	NR	Payload PDR	0.30	0.26	0.26	0.38	0.03		
10	1.1.1.1.2	NR	Payload Detailed Design & Development	117.20				5.51		
11	1.1.1.1.2.1	NR	SAR Sensor Design & Development	108.20				5.50		
12	1.1.1.1.2.1.1	NR	SAR Sensor System Engineering (Development)	N/A	N/A	N/A	N/A	N/A		
13	1.1.1.1.2.1.2	NR	SAR Antenna Design & Development	80.10				5.37		
14	1.1.1.1.2.1.2.1	NR	SAR Antenna Structure and Mechanisms	31.89	22.32	31.89	41.46	3.91		
15	1.1.1.1.2.1.2.2	NR	SAR Antenna Initial System Engineering	27.66	19.30	27.69	35.99	3.41		
16	1.1.1.1.2.1.2.3	NR	SAR Antenna Microcircuit Design & Development	2.65	2.12	2.65	3.18	0.22		
17	1.1.1.1.2.1.2.4	NR	SAR Antenna Array Design & Development	0.08	0.07	0.08	0.10	0.01		
18	1.1.1.1.2.1.2.5	NR	SAR Antenna RF Module & Time Delay Units	0.08	0.07	0.08	0.10	0.01		
19	1.1.1.1.2.1.2.6	NR	SAR Antenna Sub-Panel Design & Development	13.81	10.16	14.54	16.72	1.36		
20	1.1.1.1.2.1.2.7	NR	SAR Antenna Controller	3.43	2.74	3.43	4.11	0.28		

Fig. 4.35 Consideration of uncertainty for the rows in a WBS.

	PROPOSED	Minimum	Most Likely	Maximum	PROP Std Dev	20 Percentile	50 Percentile	80 Percentile
Space Based Satellite System	6537.73				194.57	6702	6538	6374
Space Segment	2831.23				129.38	2940	2831	2722
Payload	1726.89				124.12	1832	1727	1622
Payload Development	180.04				5.74	185	180	175
Payload Architecture & Preliminary Design	1.45				0.07			
Payload Requirements & Concept Validation	0.38	0.33	0.33	0.49	0.04			
Payload SRR	0.20	0.17	0.17	0.25	0.02			

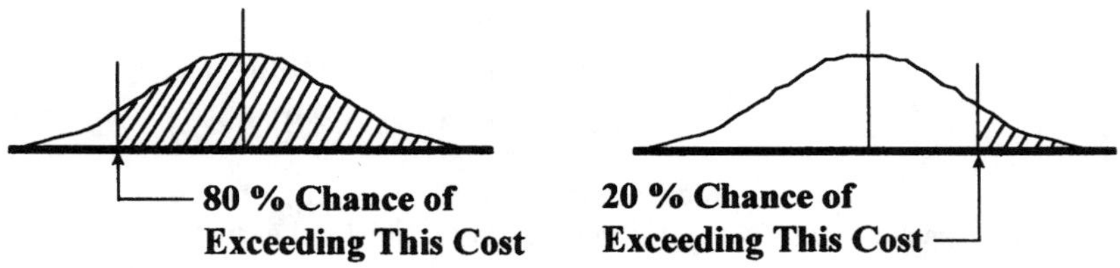

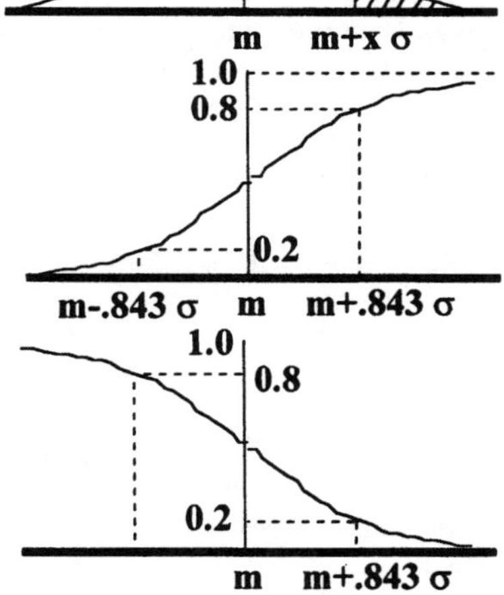

x	Prob. of Exceeding	
-2.0	0.977	
-1.5	0.933	
-1.0	0.841	← 80 percentile
-0.5	0.691	
0.0	0.500	
+0.5	0.309	
+1.0	0.159	← 20 percentile
+1.5	0.067	
+2.0	0.023	

Fig. 4.36 Determination of levels of cost risk.

The basic assumption underlying the employed cost–risk methodology is that as the costs in the WBS are rolled up to higher and higher levels the cost–risk probability distribution, which at the lowest level of detail is represented by a triangular distribution, has a greater tendency toward normality as per the Central Limit Theorem.[31]

All costs are described by estimates that represent minimum, maximum, and most likely values. These values are used to characterize a triangular cost probability distribution for each of the lowest elements used in the WBS. From the triangular distribution characterization expected cost and standard deviation of cost are determined according to the following:

$$\text{Expected Cost}_x = [\text{Minimum Cost}_x + \text{Most Likely Cost}_x + \text{Maximum Cost}_x]/3$$

$$\text{Standard Deviation of Cost}_x = [(\text{Minimum Cost}_x)^2 + (\text{Most Likely Cost}_x)^2 + (\text{Maximum Cost}_x)^2 - (\text{Minimum Cost}_x) * (\text{Most Likely Cost}_x) - (\text{Minimum Cost}_x) * (\text{Maximum Cost}_x) - (\text{Most Likely Cost}_x) * (\text{Maximum Cost}_x)]^{0.5}/[3 * 2^{0.5}]$$

Because it has been assumed that the basic cost elements are described by probability distributions, the addition of costs must take into account these distributions. Because only expected values are additive, the computed expected values are added to establish the expected cost of the next higher level in the WBS. In general, it has been assumed that the elements of the WBS are uncorrelated with the result that the standard deviation of any WBS element is obtained by taking the square root of the sum of the lower-level variances (i.e., σ^2). When WBS elements are correlated (i.e., multiple similar units are utilized but are recorded as separate rows in the WBS), a correlation coefficient of unity can be assumed for the computation of the rolled-up standard deviation.

Based upon the assumption of normality, 20, 50, and 80 percentile costs are obtained as follows and indicated graphically in Fig. 4.37:

$$\text{20 Percentile Cost} = \text{Expected Cost} + 0.843 * \text{Standard Deviation}$$

$$\text{50 Percentile Cost} = \text{Expected Cost}$$

$$\text{80 Percentile Cost} = \text{Expected Cost} - 0.843 * \text{Standard Deviation}$$

The 20 percentile represents the value of cost that has a 20% chance of being exceeded, the 50 percentile represents the value of cost that has a 50% chance of being exceeded, and the 80 percentile represents the value of cost that has an 80% chance of being exceeded.

The described type of cost–risk analysis is becoming widely used when responding to government requirements. When alternatives are compared and decisions made on the basis of costs (i.e., comparisons are for an equal capability), the development of cost risk allows alternatives to be compared taking into account both the expected cost and standard deviation of cost, and appropriate trades can be made between the two. This is indicated in Fig. 4.37, where a number of alternatives are indicated in terms of their expected costs and associated standard deviations. It is clear that alternative 3 is preferable to alternative 1 (same expected value but lower risk), and alternative 4 is preferable to alternative 2 (same expected value but lower risk). Similarly, it is clear that alternative 2 is preferable to alternative 1 (same level of risk but lower expected cost), and alternative 4 is preferable to alternative 3 (same level of risk but lower expected cost). Thus, there is a tendency to find dominant alternatives, and attention can be focused on those

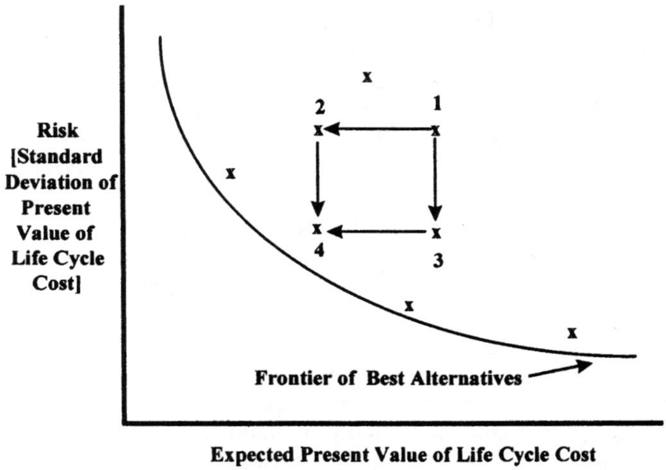

Fig. 4.37 Comparison of alternatives when considering risk.

that lie on or close to a frontier of 'best' alternatives, and a final judgment made taking into account risk-avoidance preferences.

There are still problems with the development and consideration of cost risk. The main problem concerns the lack of understanding of the fact that, in general, it is incorrect to add most likely values. In a recent application [NASA's Space Solar Power (SSP) initiative] minimum, most likely, and maximum value data were carefully developed relating to both cost and technology variables. These data were to be used in an overall system simulation analysis. When it was not possible to perform the simulations, the most likely values were used in the ensuing analyses with the result that meaningless information was developed and used in taking decisions. The good news is that the estimation of minimum, most likely, and maximum values provides additional information for critical review and guides the conscientious analyst so as to provide more thoughtful information.

B. Life-Cycle Cost and Availability Analysis Methodology[8,9,32,33]

As a consequence of the need to consider cost uncertainties and launch and satellite subsystem reliability, life-cycle cost and availability analyses make use of stochastic launch and on-orbit simulation models, such as *SATCAV*. *SATCAV* (described in detail in Sec. I and summarized here) is a dynamic stochastic life-cycle cost and availability model that simulates the launch and on-orbit operations associated with the initiation and continuing operation of a generalized space mission comprising multiple satellites with multiple "sensors." A user-friendly menu-driven input/output system is used to create a data file that is accessed by a compiled FORTRAN Monte Carlo program that performs the life-cycle and availability computations and develops cost, event, availability, and cause of failure statistics reports.

SATCAV simulates the satellite launch operations of a multistage vehicle by considering the reliability associated with the performance of major operations (including payload checkout and testing) and the consequences of a priori specified types of failures in terms of cost-incurring events and time delays. Delays that can result from different types of failures are taken into account in the simulations. Both expendable and recoverable launch vehicles and upper stages can be considered. *SATCAV* simulates the random and wear out characteristics of a multisensor satellite determining when specific failures occur and when maintenance actions are required to respond to critical failures. ("Sensor" is used broadly to either represent actual sensors or communication subsystems or antennas.)

SATCAV encompasses alternative maintenance scenarios that include both ground and on-orbit active or dormant spares. In addition, both launch on failure and launch in anticipation of wear out failure alternatives are available. The model can also be used to analyze scheduled replacement options.

The consideration of subjective uncertainties, transportation system reliability, and satellite subsystem random and wear out failures results in the establishment of the probability distribution of annual events and the probability distributions of annual costs and present value of life-cycle costs. The annual events report includes the probability distribution of launch attempts and transfer vehicles utilized. The cost report summarizes the annual recurring and nonrecurring costs and the present value of costs (at a number of discount rates) in terms of expected values and associated standard deviations.

SATCAV develops sensor availability statistics, which are presented as the probability density function of the number of sensors available (for each sensor type) annually as well as over the mission duration. The statistics take into account sensor and subsystem random and wear out failure characteristics, sparing strategy, transportation scenarios, delays associated with different types of failures, and duration of on-orbit testing. Associated with this report is a cause of failure statistical report that indicates the probability that different subsystem failures were the cause of the sensor failures.

Important attributes of the described version of *SATCAV* include the following: 1) time horizon up to 25 years; 2) consideration of multiple operational satellites; 3) sparing strategies that include a) multiple ground and on-orbit active or dormant spares, b) launch on failure of critical sensors, and c) launch on random failure but in anticipation of wear out failure of critical sensors (Additional strategies, such as scheduled replacement, can be considered through appropriate structuring of input data.); 4) identification of critical sensors (individually and in combination); 5) specification of desired launch dates for initial operational satellites and on-orbit spares (actual launch dates take into account failures and associated delays); 6) specification of completion date of ground spares; 7) specification of the probability that a dormant satellite will operate successfully when activated and the time delay to move the dormant satellite into replacement position; 8) consideration of satellite cost-learning effects; 9) transportation scenario described in terms of the probability of successfully performing each major operation in the launch/recovery sequence taking into account the consequences of failures in terms of cost-incurring events and time delays; 10) a priori defined uncertainty variables (i.e., subjectively specified as ranges of uncertainty and the form of uncertainty) include a) delays (function of failure type), b) transportation cost from Earth to LEO, c) transportation cost from LEO to Earth, d) transportation cost from LEO to P/L orbit, e) satellite unit recurring cost and nonrecurring cost (specified in terms of expected value and standard deviation), f) transfer vehicle repair cost, g) capital expenditures (up to 10), h) satellite control operations cost, and i) engineering expenses; 11) satellite configuration consisting of multiple sensors and multiple satellite bus subsystems (Each sensor and subsystem is described in terms of its random and wear out failure characteristics.); 12) cost spreading for nonrecurring cost of capital items and satellite and launch recurring cost; and 13) development of cost, event, availability, and cause of failure statistics and satellite expected life and associated standard deviation.

The *SATCAV* stochastic simulation analysis allows the effects of 1) utilizing different spacecraft bus and sensor technologies, 2) launch and procurement schedules, 3) achieving of different subsystem reliabilities (random and wear out), 4) alternative sparing strategies, 5) utilizing alternative launch vehicles, 6) utilizing expendable or reusable launch vehicles, and 7) launch vehicle reliability and cost to be evaluated in terms of a) annual cost (expected value and standard deviation), b) present value of annual cost and risk, and c) short- and long-term sensor availability.

A number of reports are developed that include important metrics and associated information including an **Annual Cost Projection**, which summarizes recurring and nonrecurring cost statistics, present value of life-cycle cost statistics, and quantity of satellites produced statistics; **Probability of Indicated Events**, which presents the probability distributions of annual placement and repair (replacement) launches, payloads procured and serviced; **Percentage of Time That Sensors Are Available over Planning Horizon**, which indicates the percentage of time that a specific number of sensors of each of five types are available over the planning horizon; **Percentage of Time That Sensors Are Available Each Year**, which indicates the percentage of time that a specific number of each of five types are available each year; and **Probability That Indicated Subsystem Will Cause Sensor Failure**, which indicates the probability that each subsystem will cause a sensor failure and also indicates expected value and standard deviation of satellite life.

C. Sparing Strategies

To illustrate the methodology, a number of different sparing scenarios are considered in seeking to minimize the present value of life-cycle costs while meeting annual and long-term availability objectives. The *SATCAV* model was used to simulate a number of remote-sensing mission constellation placement and maintenance strategies. The specifics of the considered placement and maintenance strategies are summarized in Table 4.4 with additional clarifications, assumptions, and specifics provided here:

1) All results presented are based upon 1000 Monte Carlo simulations for each of the scenarios (also referred to as cases) and over a 23-year time horizon with satellite launches commencing in the fourth year (i.e., 8/2002). Thus, 20 years of launch and in-orbit operations are considered.
2) The following specific cases are considered:

Table 4.4 Summary of sparing strategies

Case	Description
1	Launch on failure (10-satellite constellation)[a]
2	Launch on failure (12-satellite constellation)[b]
3	Launch on failure (11-satellite constellation)[c]
4	Launch on failure (13-satellite constellation)[d]
5	Launch in anticipation (0.5 yr) of a wear-out failure (10-satellite constellation)[a]
6	Launch in anticipation (0.5 yr) of a wear-out failure (12-satellite constellation)[b]
7	Launch in anticipation (0.5 yr) of a wear-out failure (11-satellite constellation)[c]
8	Launch in anticipation (0.5 yr) of a wear-out failure (13-satellite constellation)[d]
9	Launch in anticipation (1.0 yr) of a wear-out failure (10-satellite constellation)[a]
10	Launch in anticipation (1.0 yr) of a wear-out failure (12-satellite constellation)[b]
11	Launch in anticipation (1.0 yr) of a wear-out failure (11-satellite constellation)[c]
12	Launch in anticipation (1.0 yr) of a wear-out failure (13-satellite constellation)[d]
13	Launch on schedule assuming a 7-yr life satellite (10-satellite constellation)[a]
14	Launch on schedule assuming a 7-yr life satellite (12-satellite constellation)[b]
15	Launch on schedule assuming a 7-yr life satellite (11-satellite constellation)[c]
16	Launch on schedule assuming a 7-yr life satellite (13-satellite constellation)[d]

[a]The basic 10-satellite constellation is launched according to the following schedule: 8/02, 8/02, 4/03, 4/03, 10/03, 10/03, 4/04, 4/04, 8/04, 8/04.
[b]Additional two satellites are added to the constellation according to the following schedule: 0/05, 0/05.
[c]An additional satellite is launched 6/08 and replaced every seven years.
[d]Consists of the basic 10-satellite constellation augmented with the additional two satellites and the single scheduled satellite with all satellites being replaced on a scheduled basis.

a) *Launch on failure* is the first case. A constellation of satellites is launched according to a desired schedule (with launch failures and associated delays altering the desired schedule) with satellite replacements taking place upon satellite failure (both random and wear out failures are considered) and allowing for scheduling delays.

b) *Launch in anticipation of a wear out failure* is the second case. A constellation of satellites is launched according to a desired schedule (with launch failures and associated delays altering the desired schedule) with satellite replacements taking place in anticipation of a satellite wear out failure but upon the random failure of a satellite should such a failure precede the anticipated wear out failures. When random failures are encountered, scheduling delays are considered. Anticipatory (i.e., prior to anticipated wear out failure) times of 0.5 and 1.0 years are also considered.

c) *Launch on a scheduled basis* is the third case. A constellation of satellites is launched according to a desired schedule (with launch failures and associated delays altering the desired schedule) with satellite replacements taking place on a scheduled basis. The specific considered replacement schedule is based upon an assumed seven-year satellite lifetime.

3) Satellite cost-learning rate of 95% is assumed.
4) All present values of life-cycle cost results indicated are based upon a 10% discount rate.
5) One satellite ground spare is assumed and included in the life-cycle cost analysis.
6) The life-cycle cost analysis considers only recurring space system costs (nonrecurring costs cancel out when performing comparisons of the indicated alternatives).
7) Launch vehicle reliability is assumed as 0.98 with a probability of 0.98 that the satellite will function properly when placed into final orbit.
8) Because of a model limitation, single satellite launches are assumed with all launch costs appropriately adjusted. This causes a slight error in the results in that correlated failures are not taken into account. Because of the high assumed launch reliability, it is anticipated that this effect is of little consequence.
9) Launch costs are specified as a range of uncertainty (i.e., 40.9 to 52.7 M$).
10) Launch rescheduling delays (given the occurrence of a launch failure) are assumed to be in the range of 0.1 to 0.5 years with the former being more likely than the latter.
11) Launch rescheduling delays (given the occurrence of a random satellite failure) are assumed to be in the range of 0.5 to 1.0 years with the former being more likely than the latter.
12) On-orbit testing of 1.5 months is assumed. This is considered as a delay with respect to when a satellite becomes operational.
13) The assumed satellite reliability characteristics are as follows:
    a) Payload mean-time-to-failure = 500 years.
    b) Payload expected wear out life = 8.5 years.
    c) Standard deviation of payload wear out life = 0.75 years.
    d) Satellite bus mean-time-to-failure = 55.6 years.
    e) Satellite bus expected wear out = 8.5 years.
    f) Standard deviation of satellite bus wear out life = 0.75 years.

    The reliability model assumes that the payload and bus are in "series" and that the reliability function is described by an exponential (the random failure component) function that is multiplied by one minus the complement of a normal distribution function (the wear out component).
14) Satellite and launch cost spreading is over two years.
15) Expected satellite cost is 72 M$ with a standard deviation of 4 M$.

D. Typical Results of Life-Cycle Cost and Availability Analysis

The results of simulation analyses utilizing the *SATCAV* model are presented first in terms of the long-term availability (i.e., the probability that at least *N* satellites are available over a time span measured from the start of the seventh year through the end of the 23rd year; this corresponds to measuring availability from the start of 2005 through the end of 2021), then as a comparison of long-term availability present value of life-cycle cost and number of satellites (including the assumed ground spare), and finally as the annual availability of each considered case. Both quantitative and graphic results are presented. All indicated cases refer to those shown in Table 4.4. Figure 4.38 illustrates the type of long-term availability results obtained. These results indicate that case 1, which considers a 10-satellite constellation, has only a 0.641 probability that 10 or more satellites will be available over the long term, a 0.883 probability that nine or more satellites will be available over the long term, etc. Similarly, case 2, which considers a 12-satellite constellation, has a 0.953 probability that 10 or more satellites will be available over the long term, a 0.989 probability that nine or more satellites will be available over the long term, etc.

Figure 4.39 presents a summary of the 16 considered cases in terms of the long-term availability [defined as the probability that more than nine (i.e., at least 10) satellites are available during the years 7–23], present value of life-cycle cost, and the expected number of satellites procured for the remote-sensing mission. Based upon only the long-term availability constraint, strategy (case) #7 appears to minimize the expected present value of life-cycle costs. Figures 4.40–4.42 present typical annual availability results.

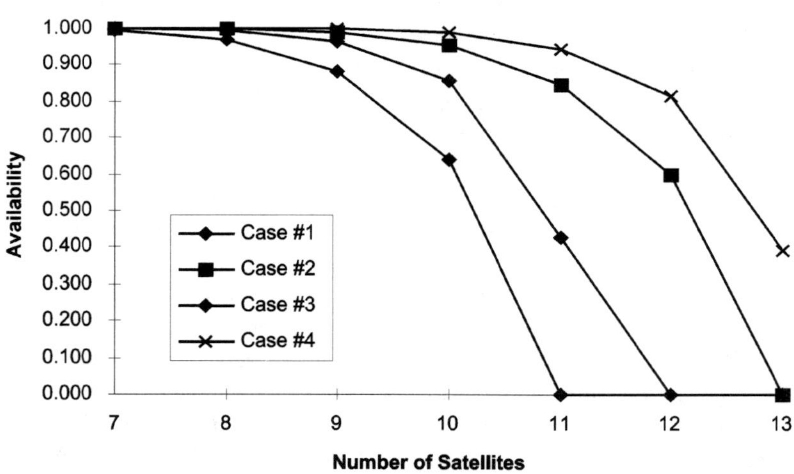

Fig. 4.38 Availability of at least the indicated number of satellites (replacement on failure).

Case	Availability [>9]	Cost [B$]	# of Sats.
1	0.641	1.264	31.9
2	0.953	1.490	37.7
3	0.855	1.356	33.9
4	0.987	1.582	39.7
5	0.711	1.315	33.0
6	0.981	1.567	39.2
7	0.908	1.407	35.0
8	0.996	1.659	41.2
9	0.795	1.335	33.6
10	0.995	1.594	40.0
11	0.949	1.430	35.8
12	0.999	1.689	42.2
13	0.462	1.293	32.3
14	0.928	1.546	38.3
15	0.724	1.380	34.3
16	0.975	1.633	40.3

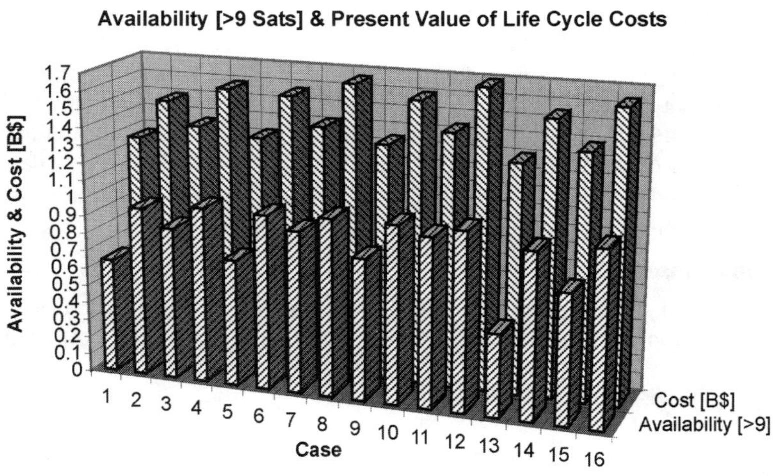

Fig. 4.39 Summary of considered cases in terms of availability, present value of life-cycle cost, and number of satellites procured.

As indicated in Table 4.5, only cases 4, 6, 8, 10, 11, 12, and 16 offer the possibility of achieving both long-term (i.e., over the year 7 through year 23 time period) and short-term (i.e., annual) availability (defined in terms of the probability of more than nine operational satellites) in excess of 0.9. Of these, case 11, launch in anticipation (1.0 years) of a wear-out failure in an augmented 10-satellite constellation, appears to offer the least cost alternative of meeting both a long-term and a short-term 90% availability requirement. Case 11 is based upon

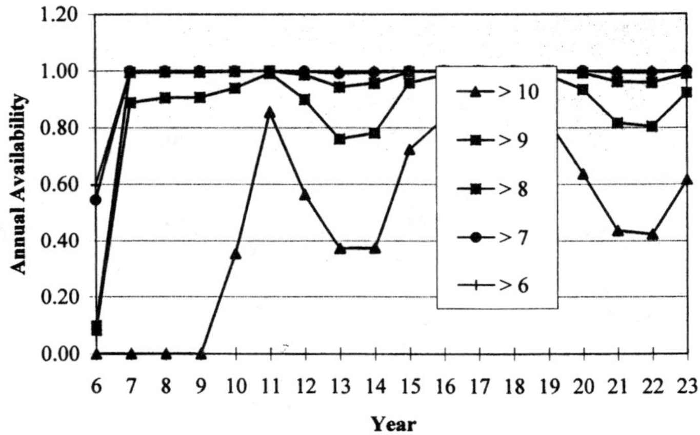

Fig. 4.40 Availability as a function of time (case #7).

a basic constellation of 10 satellites that is augmented by an 11th satellite at the start of the 10th year and replaced at seven-year intervals. This scenario also minimizes the present value of the life-cycle cost and expected number of satellites for achieving a 90% availability requirement. Scenario 11 is an augmentation of scenario 9 that, as seen in Fig. 4.41, achieves the short-term availability in a number of years but falls short in others. Having noted this, another satellite was added to the constellation (becoming scenario 11) to "fill in" the availability shortfall.

E. Some Observations

The preceding analysis demonstrates the use of a stochastic life-cycle cost and availability approach for evaluating the anticipated satellite and launch

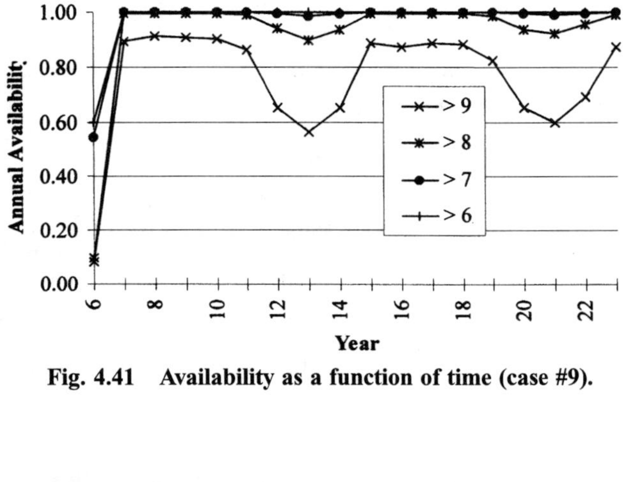

Fig. 4.41 Availability as a function of time (case #9).

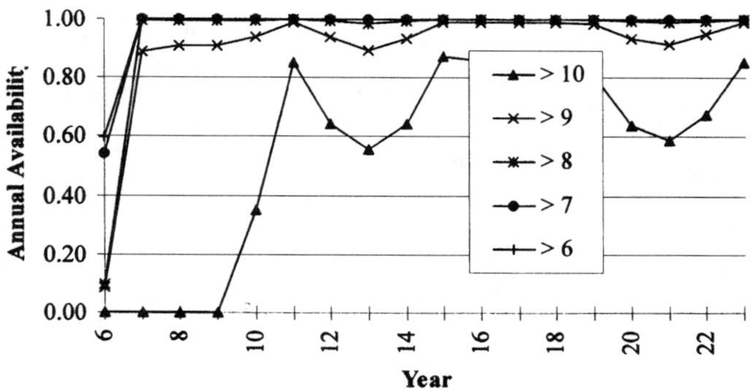

Fig. 4.42 Availability as a function of time (case #11).

requirements for achieving availability constraints imposed by desired constellation performance. The analysis demonstrates the complications associated with determining the minimum life-cycle cost approach to meeting both annual and long-term availability constraints. For illustrative purposes a specific remote-sensing mission was considered and the number of satellites determined that

Table 4.5 Comparison of alternatives

	Availability (>9)			
Case	Long term	Annual, min[a]	Cost, B$	Expected no. of satellites
1	0.641	0.219	1.264	31.9
2	0.953	0.793	1.490	37.7
3	0.855	0.647	1.356	33.9
4	0.987	0.933	1.582	39.7
5	0.711	0.380	1.315	33.0
6	0.981	0.921	1.567	39.2
7	0.908	0.760	1.407	35.0
8	0.996	0.985	1.659	41.2
9	0.795	0.564	1.335	33.6
10	0.995	0.980	1.594	40.0
11	**0.949**	**0.895**	**1.430**	**35.8**
12	0.999	0.998	1.689	42.2
13	0.462	0.276	1.293	32.3
14	0.928	0.822	1.546	38.3
15	0.724	0.486	1.380	34.3
16	0.975	0.945	1.633	40.3

[a]"Annual, min" represents the minimum of the calculated annual availability.

would be required to meet various annual and long-term availability objectives. The multisatellite (constellation) mission was simulated utilizing the *SATCAV* stochastic life-cycle cost and availability model, which established annual and present value of life-cycle cost, number of satellites launched, and probability distribution of number of functioning satellites annually and over a planning horizon when considering a number of sparing alternatives. The result was the determination of the sparing strategy that minimizes the present value of life-cycle cost (of the space portion of the overall mission costs) and satisfies overall mission availability requirements. Although the *SATCAV* model allows the implications of orbital debris, and resulting random failures, to be taken into account, this was not included in the analysis.[34,35] The implications of orbital debris and the form of analysis required to understand these are described in Sec. VII and Sec. III of Chapter 5.

The results summarized in Table 4.5 indicate that scenario 11 minimizes the present value of life-cycle cost while just achieving the availability objectives. This is very significant because nonsimulation analyses performed prior to use of the *SATCAV* model resulted in identifying another scenario as the preferred alternative. Thus the result of performing the mission simulation analysis (utilizing *SATCAV*) was the identification of a superior alternative having the potential of reducing, by several, the expected number of required satellites. This also resulted in a reduction of several hundred million dollars in the expected mission life-cycle cost.

X. Resiliency, Operability, and Launch on Demand

With the increase in the number of satellite constellations that are, or are likely to become operational in the near future for the provision of communication, remote sensing, and position determination services, there has been significant interest in the possibility of reducing or eliminating the need for on-orbit spare satellites, thereby reducing infrastructure costs. Because the need for on-orbit spare satellites is related to transportation architecture robustness (including the availability and scheduling of both launch vehicles and satellites), the following paragraphs describe methods and procedures for analyzing the relationship between transportation architecture robustness and on-orbit sparing. In particular, the concept of launch on demand is discussed with preliminary results presented that indicate the level of capability required of a transportation system in order to eliminate the need for on-orbit spares with the required capability being a function of required space system availability.

A. Resiliency and Operability

Decisions concerning the development and use of space transportation systems have, in the past, placed emphasis primarily upon cost. The lack of other than cost considerations, in particular the lack of consideration of unreliability and its consequences, was a major factor in the inability to launch payloads for a significant period after the loss of the Space Shuttle *Challenger*. The demise of the

U.S. space transportation capability in the late 1980s strongly suggests that other than cost considerations are important. It has been suggested that "resiliency" might be an important consideration in defining robust future architectures.[36,37] Resiliency is defined as the ability to recover readily from misfortune and diversity—in other words, resiliency is concerned with the ability to recover from transportation system failures as manifested in the ability to deliver payloads to orbit.

Resiliency can be thought of simply in terms of the ability to recover from failures. Resiliency can also be thought of in much broader terms so that it is synonymous with "operability." It can be thought of as the overall ability to place payloads into orbit as and when required considering all of the realities including less-than-perfect reliability and consequent failures, stand down time resulting from failures, and constraints such as procurement cycles, number of vehicles maintained in inventory, and many other factors. It is this broader perspective of resiliency that is maintained in the following pages. To emphasize this broader perspective, resiliency is henceforth replaced by operability.

The referenced resiliency studies performed at the Aerospace Corporation and L Systems, Inc., utilized analytic techniques to establish availability as a function of reliability, annual launch rate, surge capability, and stand-down time. Because of the complex nature of the problem, many simplifying assumptions were required in order to achieve an analytic solution. These assumptions relate to the time of failure of the launch vehicle in the overall launch sequence, as well as considering only (the equivalent of) single-stage vehicles with deterministic stand-down times. These limiting assumptions that are required to achieve analytic tractability are not realistic and are not required if computer simulation techniques are employed. In fact, the application of computer simulation to the operability problem allows many operability performance measures to be developed.

It is felt that the primary measure of transportation system operability should be based upon the objective of the transportation system; the placement of payloads into orbit. Thus the primary measure of operability should be concerned with payload queue or backlog as a function of time. Long queues or queues that increase over time indicate that the transportation system is not responding to payload needs.

It is now understood that future transportation architectures should be planned and evaluated explicitly considering the ability to place payloads into orbit as and when required. This requires the consideration of and planning for the recovery from failures resulting from less-than-perfect reliability. A simulation model TRANSOP was therefore developed to perform analysis of transportation system operability.[38] Additional details of the model are presented in the Appendix to this section. TRANSOP is a menu-driven user-friendly Monte Carlo simulation model that performs the following tasks:

1) Simulate the launch sequence of multistage reusable or expendable vehicles.
2) Establish the occurrence of failures and the resulting stand down time with stand down time (an uncertainty variable described as a range of

uncertainty and its associated probability distribution) a function of the specific launch operation failure.

3) Establish the number of payloads likely to transfer to other transportation systems and establishes the payload backlog that results from the stand down time.
4) Adjust the number of vehicles available for use to account for the failures and for the increase in the number of vehicles (resulting from procurements), taking into account procurement schedules.
5) Adjust the flight rate to take into account the number of vehicles available, the failure recovery status, possible surge capabilities, and launch thruput constraints.
6) Adjust payload backlog to account for successful launches at the established flight rate.
7) Simulate (as just indicated) the launch sequence each time a launch takes place to establish success or failure; if a success, decrements the backlog, and if a failure reestablished stand-down time, payload backlog, etc.

The primary measure of operability is concerned with payload queue (or backlog) statistics as a function of time. Long queues or queues that increase over time indicate that the transportation system is not responding to payload needs. Therefore the TRANSOP model develops satellite inventory statistics: the probability distribution of payload inventory (i.e., queue length) as a function of time.

Inventory buildup over time is indicative of a mismatch between payload arrival rates, launch vehicle stage inventories and system thruput, and/or unreliability as manifested in failures and stand down time. Thus, the following additional information is developed, which indicates the likely cause(s) of payload inventory buildup:

1) Launch vehicle inventory statistics are the probability distribution of launch vehicle (by stage) inventory as a function of time.
2) Surge statistics are the probability distribution of the percentage of the time (annually and across the planning time horizon) that the transportation system is operating in a surge mode.
3) Launch statistics are the expected value and standard deviation of the number of launch successes, launch failures, total launches, and desired number of launches (as determined by payload arrival rates and average number of payloads per flight) annually and the expected value across the planning tie horizon.

TRANSOP basically considers one transportation system (distinguished from an architecture that consists of a number of transportation systems) at a time and not an overall architecture, although it allows for payloads to be switched to "other" transportation systems. This specification is in the form of the probability that a payload will switch to another transportation system and is a function of their duration in a payload queue.

An operability analysis was performed for the space transportation system (i.e., the space shuttle). The results demonstrated that the resiliency/operability statistics obtained through the use of analytic models were totally inappropriate because these results indicated only expected values. The simulation results were generally of the form indicated in Fig. 4.43, which indicated a relatively high chance (a function of time but typically in the range of 50 to 60%) of no payload queue but with a disturbingly high chance of a large payload queue (the equivalent of one to three years of payloads). (Because these results were obtained for an analysis of the space shuttle, it is likely that this was the reason that this work was never extended to the broader consideration of transportation architectures.) In such a situation the characterization of operability in terms of expected payload queue has little or no significance. What has considerably more significance are the expected value and standard deviation of the queue given a nonzero queue. In other words, operability can be measured in terms of the probability of having a payload queue and the statistics of the queue should such a queue exist.

B. Launch on Demand: Concerns

Launch on demand (LOD) refers to the ability to place payloads (satellites) into orbit as and when required. Achievement of a desired level of LOD capability must consider transportation system thruput, alternative transportation systems that comprise the transportation architecture, the demand for the transportation alternatives, the reliability and failure recovery characteristics of the alternatives, schedule guarantees, launch delays, payload integration schedules, procurement policies, and other factors. Measures of launch-on-demand capability should relate to the objective of the transportation architecture: the placement of payloads into orbit. LOD capability must be defined in probabilistic terms such as the probability of not incurring a delay in excess of T when it is determined that it is necessary to place a payload into orbit. In fact, there are at least three specific aspects of launch on demand that need to be considered:

1) the ability to recover from adversity (i.e., a launch failure) and to keep up with the steady-state demand for placing satellites into orbit (referred to as operability and resiliency),[38]

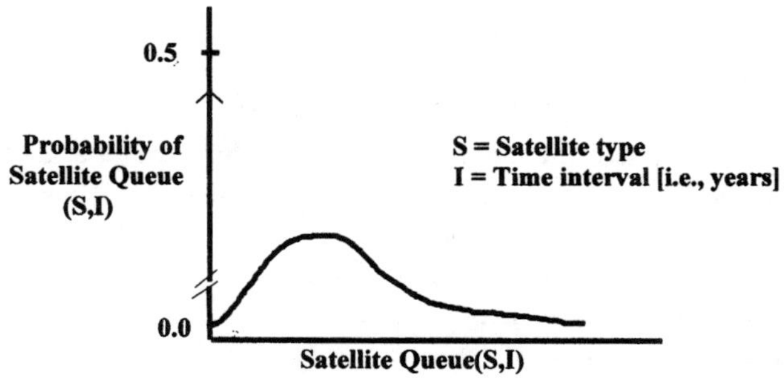

Fig. 4.43 Typical form of annual payload (satellite) queue statistics.

2) the ability to respond to the requirement to launch a satellite when the need arises unexpectedly either because of an unexpected (random) on-orbit satellite failure that requires replacement or because of the sudden recognition of an unanticipated requirement, and
3) the ability to recover from adversity (i.e., a launch failure) during the placement of a constellation into orbit.

The achievement of a LOD capability will affect each of these situations differently. Therefore, specific measures need to be developed so that the performance of an architecture can be measured in terms of each of the preceding aspects.

The following paragraphs outline the general problem of characterizing launch on demand, the required performance measures that can be used to measure the degree of LOD capability, and the characteristics of a LOD simulation model that would develop measures that could be used to analyze and compare the performance of alternative transportation architectures in satisfying LOD requirements. The LOD model would also provide information relating to transportation architecture cost. The model is described in the following paragraphs in terms of inputs and outputs and the general computational flow.

The objective of this discussion is to develop a set of metrics and methods that can be used to evaluate the implications of architecture and policy changes in terms of launch-on-demand requirements. The methods and metrics are aimed at providing answers to such questions as the following:

1) How well does a specified space transportation architecture (comprised of multiple launch vehicles and associated infrastructure) respond to satellite launch demand and changes thereto?
2) How well does a normally functioning and apparently adequate (in satisfying steady-state demand) architecture respond to unanticipated needs?
3) What is the effect of a modification to the architecture (for example, increased thruput, improved reliability, etc.) on its ability to respond to satellite launch demand, including responding to unanticipated needs?
4) What is the cost of the architecture (including facilities, operations, inventory, and satellites)?
5) What is the sensitivity of overall architecture effectiveness and cost to various transportation system delays?
6) What is the effect of adding (or eliminating) a launch vehicle or family of vehicles to (from) the architecture on its effectiveness and cost?
7) What is the value of improving launch vehicle and satellite compatibility and what are the effects on probability of delay statistics and cost of designing for multilaunch vehicle compatibility?
8) What is the effect of increasing launch vehicle inventory and/or launch rate on probability of delay statistics and cost?
9) What is the sensitivity of delay statistics and costs to a range of architecture changes?

10) What is the effectiveness of the architecture in meeting the needs of constellation deployment?

These questions all relate to understanding the cost and effectiveness of alternative architectures and policies on achieving different levels of LOD capability.

The launch infrastructure and response analysis described in the following paragraphs is not a substitute for analyzing the mission cost savings that might be achieved as a result of a LOD capability (i.e., reduced sparing and overbooking costs). It sets the frame of reference for such analyses.

C. Launch on Demand: Performance Measures

Before describing specific performance measures, consider the following example. Assume that a transportation architecture is in place that is capable of providing, and sustaining, a thruput of one launch per week utilizing two different launch vehicles. Assume that the demand for launches is relatively low, say 10 per year. With this level of demand, the assumed transportation architecture has a high level of capability (i.e., probability of a satellite queue is near zero), is likely to be able to respond quickly to unexpected needs (with delays being related more to satellite scheduling and satellite-launch vehicle integration), and because of the dual launch capability, is likely to recover successfully from adversity during the placement of a constellation.

Consider now an increase of demand to say 48 launches per year. With this increased level of demand, the assumed transportation architecture is likely to have a significantly reduced level of operability (i.e., the probability of a resulting payload queue is likely to be high), is likely to have difficulty responding quickly to unexpected needs (with delays being related to satellite-launch vehicle integration and availability of a launch slot), and, even with the dual launch capability, will have difficulty recovering from adversity during the placement of a constellation.

Assume, however, that launch contracts are written so that satellite placements are ensured within a three-week window. This is not likely to affect the overall ability to recover from adversity (i.e., operability) but is likely to have a major beneficial impact on the ability to respond to an unexpected need (with the probability of satisfying multiple needs being a rapidly decreasing function), and the ability to recover from adversity during the placement of a constellation being less positively affected.

Launch-on-demand performance measures need to reflect the effects of many factors including the following:

1) mix of launch vehicles in the transportation architecture;
2) maximum thruput per launch vehicle (and surge capabilities);
3) launch vehicle reliability;
4) fleet inventories;
5) scheduling delay statistics (resulting from satellite arrival delays and launch failures);

6) satellite launch demand (including quantity, mass, launch vehicle compatibility, preferred launch vehicle, and time to switch from one launch vehicle to another in case of a delay);
7) launch contract-scheduling flexibility; and
8) delays from failures and rescheduling of satellites from failed launch to the same or other launch vehicles as well as possible rescheduling of other payloads to other launch vehicles because of delays.

The consideration of these (and other) factors in combination with the three specific aspects of launch on demand requires the development of a number of different performance measures. These are discussed in the following paragraphs.

1. Ability to Recover from Adversity and Keep Up with Steady-State Demand for Placing Satellites into Orbit

Prior resiliency and operability simulation analyses[38] have addressed the problem of developing measures that provide insights into the ability of a transportation architecture to meet the demand for placing satellites into orbit. It has been shown that the operability of a transportation architecture cannot be characterized by a simple measure relating to expected performance.

The prior simulation analyses have shown that the ability of an architecture to meet launch requirements must be described in the form of a probability distribution of satellite queue as illustrated in Fig. 4.43. The specific shape of this distribution is a function of many variables including the mix of available launch vehicles, their thruput capability and probability distribution of stand-down time when failures occur, satellite demand, etc. When a mixed fleet is considered, it is important to understand vehicle capacity and utilization as shown in Fig. 4.44.

Several different performance measures can be developed that describe the responsiveness of the transportation architecture to meeting satellite delivery (to orbit) needs. Because the degree of nonresponsiveness of an architecture is represented by the shape and magnitude of the probability distribution of satellite queue given that a queue exists, a reasonable performance measure can be the expected value of the satellite queue given that a queue exists (normalized and

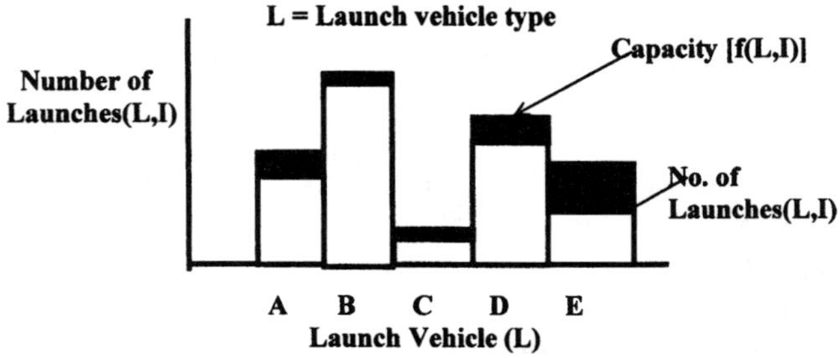

Fig. 4.44 Launch vehicle utilization and launch capacity.

expressed as a percentage of total satellite demand). This effectiveness measure is of the form

$$\text{Effectiveness (\%)} = 100 - \frac{100}{\sum_{i=1}^{T} \text{Desired Sats}(i)} * \sum_{i=1}^{T} \left\{ \frac{1}{p[Q(i,0)]} * \sum_{j=1}^{N} p[Q(i,j)] * Q(i,j) \right\}$$

where

Desired Sats(i)	= demand for satellites as a function of time
$p[Q(i,j)]$	= probability of satellite queue length $Q(i,j)$
$Q(i,j)$	= satellite queue length j at time i
T	= number of years in time horizon

This effectiveness measure represents the degree to which the expected value of satellite queues (given that the queue is greater than zero) exceeds zero, that is, the expected queue when the architecture cannot meet satellite delivery demands. Long- and/or high-probability satellite queues represent a nonresponsive architecture. A highly responsive architecture will have an effectiveness near 100% with less responsive architectures having lower values. The effectiveness measure partially takes into account the risk dimension by considering both the magnitude of the queue and the associated probability distribution. (Additional emphasis could easily be placed upon the risk dimension by multiplying by the square of the satellite queue rather than by the satellite queue.)

2. Ability to Respond to Requirement to Launch a Satellite When Need Arises Unexpectedly

Once a reasonably responsive architecture has been identified, the ability of the architecture to respond to a requirement to launch a satellite when the need arises unexpectedly can be investigated. This investigation entails randomly interjecting one or more unanticipated events (i.e., need arises unexpectedly) and determining the time delay encountered from when the need arises to when a launch can take place. This corresponds to the determination of the transient response of the system as compared with the already described steady-state response (i.e., the operability analysis). The probability distribution of delay, as illustrated in Fig. 4.45, can serve as the performance metric for describing the ability of an architecture to respond

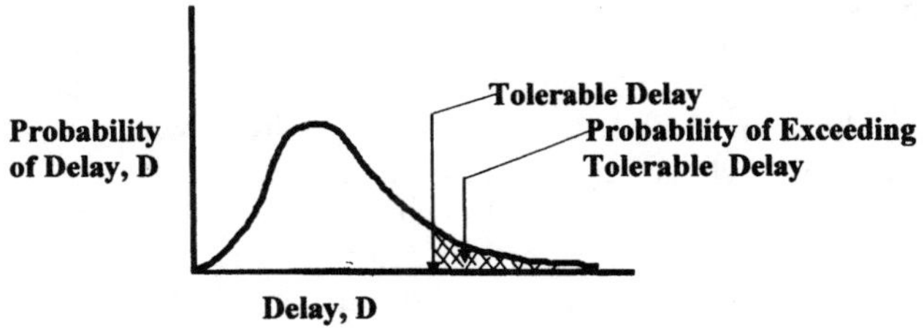

Fig. 4.45 Ability to respond to an unexpected need.

to the requirement to launch one or more satellites when the need arises unexpectedly. This probability distribution is conditional upon the architecture characterization and the steady-state demand assumptions.

Once the ability of an architecture to respond to unanticipated needs has been established, cost-effectiveness issues can be addressed (i.e., on-orbit spares vs ground spares, large on-orbit satellites vs multiple small specifically designed satellites maintained in ground-based inventory, etc.) and cost savings attributable to launch on demand established.

3. *Ability to Recover from Adversity During Placement of a Constellation into Orbit*

The analysis of the ability of a space transportation architecture to recover from a launch failure during the placement of a constellation into orbit must be judged in terms of the probability of not being able to place a constellation into orbit within a specified or desired period of time. Thus a performance measure that describes the ability of a transportation architecture to place a constellation into orbit must develop the probability distribution of the time to place the constellation into orbit and the probability of exceeding specified durations. The development of this performance measure must consider the compatibility of the satellites with multiple launch vehicles.

The constellation analysis can be considered as an extension of the analysis of the ability to recover from adversity (i.e., a launch failure) and to keep up with the steady-state demand for placing satellites into orbit. The steady-state analysis would be performed but with the specified constellation demand subtracted from the overall satellite demand. An architecture with reasonable performance against this demand would be selected, and the additional constellation demand would then be added. The performance of the architecture against this total demand would provide information relating to the change in satellite queue statistics during the time of the placement of the specified constellation.

Once the ability of an architecture to respond to the added constellation has been established, cost-effectiveness issues can be addressed (i.e., on-orbit spares vs ground spares, necessary compatibility of satellites with multiple launch vehicles, satellite procurement and integration policies, etc.) and cost savings attributable to launch on demand established.

4. *Ability to Recover from Adversity and Keep Up with Steady-State Demand for Placing Satellites into Orbit*

To assess the ability to respond to the requirement to launch a satellite when the need arises unexpectedly, it is first necessary to define a space transportation architecture that is responsive to the steady-state demand for placing satellites into orbit. This architecture will serve as a baseline from which change will be measured. Once this architecture has been established, its capability to respond to the unexpected can be evaluated, and changes to the architecture can then be measured in terms of architecture and mission costs and delays.

Both the transportation architecture and the demand for transportation services require characterization. The architecture must be described in terms of the mix of launch vehicles comprising the architecture (for example, Delta, LLV, EELV, RLV, etc.) and a range of variables that describe these launch vehicles. The demand must be described in terms of a mix of satellites and a range of variables that describe the satellites. The basic approach for analyzing the architecture responsiveness in terms of its structure and satellite launch demand is illustrated in Fig. 4.46. Many details have been omitted (for example, the cost dimension has not been shown and requires development) because the intent of the following is to illustrate the type of simulation model that will be required in order to assess the launch-on-demand capability of an architecture and to assess changes that might be necessary in order to achieve a desired level of responsiveness.

The LOD Simulation model must simulate the arrival rate of satellites (disaggregated by mass class) and associated operations of a mixed fleet of expendable and reusable launch vehicles. The LOD model would simulate the success/failure/recovery characteristics of the mixed fleet in responding to a random satellite arrival rate as a function of time. The model would do the following:

1) Select a satellite for launch based upon a specified mission model. The mission model would be structured by satellite type with each type possibly having a different random arrival rate.
2) Place the selected satellite into a queue of satellites awaiting launch.
3) Select the next satellite to be launched from the fly-off queue.
4) Select an appropriate launch vehicle taking into account launch vehicle preferences, launch vehicle operational status and availability (based upon utilization, thruput, etc.), compatibility, and schedule flexibility.
5) Establish the satellite and launch vehicle integration delay (by random sampling of specified delay probability distributions).
6) Establish the time to the next possible launch.
7) After all launch alternatives have been considered, select a launch vehicle that will be used to launch the selected satellite (taking into account the relative importance of schedule and cost, that is, the likelihood of utilizing another launch vehicle in terms of magnitude of the satellite launch queue).
8) Simulate the launch sequence of the satellite utilizing the selected vehicle that can be either a multistage reusable or expendable launch vehicle.
9) Establish the occurrence of failures and the resulting stand-down time with stand-down time (an uncertainty variable described as a range of uncertainty and its associated probability distribution) a function of the specific launch vehicle and launch operation failure.
10) If the launch was a success, increment launch success/failure and time-to-launch (delay) tables, adjust backlog of satellites to account for successful launches, and select another satellite for launch and repeat the preceding process until all satellites in all time periods have been launched.
11) If the launch was not successful, determine the delay to the next possible launch with the selected launch vehicle; increment success/failure and

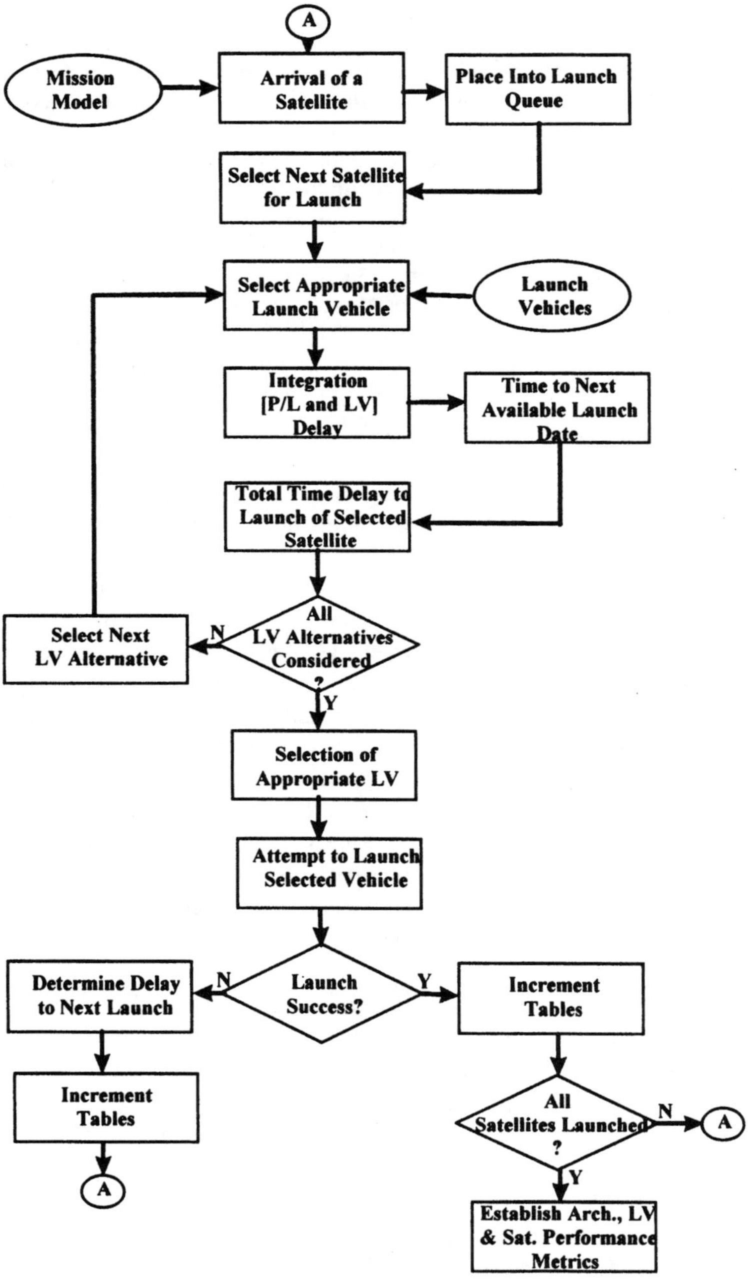

Fig. 4.46 Overview of transportation architecture analysis. (Ability to recover from adversity and keep up with steady-state launch demand.)

time-to-launch (delay) tables; adjust the number of launch vehicles available for use to account for the failure and make necessary adjustments to increase the number of available vehicles to offset the loss (i.e., add to inventory) taking into account procurement schedules and policies; adjust the flight rate to take into account the number of vehicles available, the failure recovery status, possible surge capabilities, and launch thruput constraints; select another satellite for launch and repeat the preceding process until all satellites in all time periods have been launched.

12) When all satellites and time periods have been considered, establish architecture, launch vehicle, and satellite performance metrics.

Although not indicated in Fig. 4.46, the LOD model would, for each launch, establish the satellite, integration, and launch costs. At the completion of all launches within each year, the model would also establish annual operations costs and appropriate infrastructure costs.

The primary measures of operability would be based upon the objective of the transportation architecture: the placement of satellites into orbit in a timely manner. Thus the primary measures are concerned with the magnitude of the satellite queue (backlog as a function of time) and the duration of time in the queue (delay statistics). Long queues or queues that increase over time indicate that the transportation system is not responding to payload needs. Therefore the LOD model would develop 1) *Satellite inventory statistics*, which are the probability distribution of satellite inventory (i.e., queue length) as a function of time, and 2) *Satellite delay statistics*, which are the probability distribution of the time from satellite arrival to the time of launch.

Inventory buildup over time is indicative of a mismatch between payload arrival rates, launch vehicle stage inventories and system thruputs, and/or unreliability as manifested in failure and stand-down time. Thus the following additional information would be developed, which indicates the likely cause(s) of satellite inventory buildup:

1) *Launch vehicle inventory statistics* are the probability distribution of launch vehicle (by launch vehicle type) inventory as a function of time.
2) *Surge statistics* are the probability distribution of the percentage of the time (annually and across the planning time horizon) that each of the launch vehicle types comprising the transportation architecture is operating in a surge mode.
3) *Launch statistics* are the expected value and standard deviation of the number of launch successes, launch failures, total launches, and desired number of launches (as determined by satellite arrival rates and average number of satellites per flight) annually and the expected value across the planning time horizon.

Finally, because achieving different levels of launch-on-demand capability will likely involve different infrastructure costs, the LOD model would develop the probability distribution of annual and the present value of life-cycle costs. This will enable different architectures to be compared taking into account both delay statistics and the infrastructure costs.

This establishment of the various measures of operability and cost would be used to establish the LOD capability of alternative transportation architectures and would take into account many variables including the following: 1) maximum flight rate per launch vehicle type (i.e., the reciprocal of turnaround time); 2) number of vehicle stages per launch vehicle type; 3) quantity of each vehicle stage maintained in inventory; 4) procurement philosophy (i.e., when a launch failure occurs, delay replacements until stand-down interval is completed or continue procurements through standdown interval); 5) vehicle stage procurement rate; 6) vehicle stage procurement time; 7) planned launch vehicle flight rate; 8) launch system reliability for each of the major operations in the launch sequence (including failure/recovery paths); 9) average number of satellites per flight; 10) probability that a satellite will switch to an alternative launch system as a function of the incurred delays; 11) satellite arrival rate (by satellite class); 12) delays incurred because of launch failures (as a function of the type of failure and launch vehicle); 13) satellite compatibility with the different considered launch vehicles; 14) ordered preferences for satellite use of launch vehicles; 15) satellite/launch vehicle integration delays (histogram for each combination of satellite class and launch vehicle); 16) surge capability and duration; and 17) launch operations and infrastructure costs.

A number of the preceding variables would be specified as a function of time and also would be provided as uncertainty variables, that is, they would be specified in terms of their range of uncertainty and the form of the uncertainty within the specified range.

In general, results from the LOD model would 1) lead to a better understanding of the relationship of architecture parameters and operability and delay statistics while providing information relating to architecture life-cycle costs, 2) provide operability- and delay-related information that would allow alternative transportation architectures to be evaluated and compared considering important attributes other than just cost, and 3) allow transportation architecture cost increases to be determined that would result from architecture improvements necessary to achieve LOD delay requirements.

5. *Ability to Respond to Requirement to Launch Satellite When Need Arises Unexpectedly*

Given the capability described in the preceding paragraphs, it would be possible, with a minor change, to achieve the ability to assess the responsiveness of an architecture to respond to an unexpected need. The unexpected need would be stated as the need to launch one or more satellites at specified periods of time or interjected randomly into the architecture configured to satisfy steady-state needs. This would be analogous to specifying another satellite type not already considered in the steady-state demand analysis. The LOD model would then be used to develop the just-described statistics plus the probability distribution of delay for satisfying the unanticipated need (i.e., described by the new satellite type). In other words, results as indicated in Fig. 4.43 and 4.44 would be developed based upon the steady-state plus the unanticipated need demand, with additional results developed in the form of Fig. 4.45 specifically indicating the architecture's ability to respond to the unexpected need.

Additional input data would be required to indicate the number of unanticipated needs, their timing (either deterministic or probabilistic), and satellite/launch vehicle related information as described in preceding paragraphs. An additional consideration needs to be addressed: when an unanticipated need arises, will the satellite take its normal place in an existing satellite queue, or will it be moved to the head of the queue? Both need to be addressed so as to understand the implications on the architecture responsiveness and changes (with related cost implications) that might be required in order to achieve specific responsiveness goals.

In general, results from the LOD model would 1) lead to a better understanding of the relationship of architecture parameters and the ability to respond to unanticipated needs while providing information relating to architecture life-cycle costs, 2) provide operability- and delay-related information that would allow alternative transportation architectures to be evaluated and compared specifically considering the ability to respond to unanticipated needs, and 3) allow transportation architecture cost increases to be determined that would result from architecture improvements necessary to achieve LOD delay requirements.

6. Ability to Recover from Adversity During Placement of a Constellation into Orbit

The implication of the transportation architecture on the ability to recover from adversity during the placement of a constellation into orbit requires specific consideration. In essence, the problem is concerned with isolating a specific multisatellite multilaunch mission and determining the probability distribution of the time to launch all of the requisite satellites. This would require that the steadystate need be comprised of two parts: the general "background" demand and the specific demand created by the constellation mission. The arrival of satellites for launching would then be based upon the sum of the two demands with launch failure induced delays (i.e., stand-down time and reduced thruput) affecting both parts of the demand. The results of the steady-state analysis would include a probability distribution of the time to launch the satellites comprising the constellation as well as related launch vehicle utilization.

To understand the implications of launch failures during the placement of a constellation, it would be necessary to either cause a failure of a launch vehicle or otherwise remove it from service for a specified period of time and then determine the probability distribution of the time to launch the constellation given the sudden change in launch vehicle availability. Modifications could then be made to the launch infrastructure that would negate the prior architecture limitations and the cost implications established.

To accomplish this would require additional input data that would describe the constellation under consideration, including the anticipated launch schedule, satellite mass, launch vehicle compatibility, and launch vehicle preferences.

In general, results from the LOD model would 1) lead to a better understanding of the relationship of architecture parameters and the ability to recover from failures during the placement of a constellation into orbit while providing information relating to architecture life-cycle costs, 2) provide operability- and delay-related information that would allow alternative transportation architectures to be evaluated and compared specifically considering the ability to respond to failures that occur during the placement of constellations into orbit, and 3) allow transportation architecture cost increases to be determined that would result from architecture improvements necessary to achieve tolerable delays for the placement of constellations into orbit.

D. Launch on Demand: Development of Requirements Through Analysis

The preceding paragraphs addressed LOD from a supply-side perspective. The objective was to describe metrics for measuring the ability of a transportation architecture to achieve an LOD capability. It was concluded that LOD capability should be considered from the points of view of 1) the ability to recover from adversity (i.e., a launch failure) and to keep up with the steady-state demand for placing satellites into orbit (referred to as operability and resiliency), 2) the ability to respond to the requirement to launch a satellite when the need arises unexpectedly either because of an unexpected (random) on-orbit satellite failure that requires replacement or because of the sudden recognition of an unanticipated requirement, and 3) the ability to recover from adversity (i.e., a launch failure) during the placement of a constellation into orbit.

The achievement of an LOD capability will affect each of these situations differently. Therefore, specific measures need to be developed so that the performance of an architecture can be measured in terms of each of the preceding abilities. An overall methodology was outlined for considering each of these situations starting with the ability to meet the steady-state demand and then progressing to the ability to respond to unanticipated needs and the ability to place constellations into orbit. Both probabilities of delay and infrastructure life-cycle cost metrics are required to evaluate and compare architectures.

Two specific analyses were performed to assess the potential cost savings that might result from the achievement of an LOD capability. Two missions were considered with one resembling the Global Positioning System (GPS) and the other the Iridium LEO communication system. (It was not possible, because of certain limitations of the *SATCAV* model, which was used to perform the life-cycle cost simulation analyses, to replicate exactly the GPS mission. However, necessary approximations and obtained results provided reasonable estimates of the potential cost savings that can result to a GPS-type mission from achieving an LOD capability.) Both consist of multiple satellites that are placed into orbit and maintained in a constellation over an extended period of time. The *SATCAV* life-cycle cost and availability model was used to simulate the placement and maintenance of such constellations. The analysis considered launch on demand parametrically (i.e., launch delays were varied from 0 to 0.2 tenths of a year) and developed the present value of mission life-cycle costs and satellite availability as a function of the launch time delay. As will become evident from the presented results, impacts of LOD must be considered in terms of both cost and availability; in fact, as will be shown, it is possible to establish both the effects of launch on demand on cost and availability as well as to determine the required responsiveness (i.e., launch on demand capability) that must be obtained in order to reduce or eliminate the need for on-orbit spare satellites without altering satellite availability.

First, the GPS mission is discussed. GPS is a medium-altitude global positioning satellite mission consisting of 21 satellites (plus 3 spares) that are placed into 6 orbital planes. The satellites have a recurring cost of about $41 million per satellite. Limitations imposed by the *SATCAV* model result in the need to consider each orbital plane independently. However, for the current analysis it was possible to consider two planes together. Each group of two planes can have one spare satellite. The impact of launch on demand would be to eliminate the need for this spare.

Specific assumptions related to the analysis (mainly imposed by the structure of the *SATCAV* Model) and data used in the analysis are as follows:

1) Two orbital planes were considered in the simulations with the other planes considered as being independent.
2) Satellites were considered as being uncorrelated and launched one at a time.
3) Availability is defined as the fraction of the time that at least seven satellites are operational within the two simultaneously considered orbital planes.
4) All results are based upon 1000 Monte Carlo simulation runs (i.e., each time a variable such as number of spares, delay, etc., is changed, 1000 simulation runs were performed in order to develop the cost and availability results).
5) The mission was considered over a 25-year time frame with both availability and life-cycle costs considered over this time frame. Availability was measured starting in the fifth year.
6) A 95% cost-learning curve was assumed for satellites.
7) Satellites were launched at the start of the third year followed by others at intervals of 0.2 years.
8) Four ground spares were procured, but it was assumed that a spare was always available as and when required (i.e., there were spares always residing in inventory, and their cost was accounted for as well as the costs for all satellite replacements).
9) Launch vehicle reliability was 0.98.
10) Probability that a satellite operates successfully when placed into orbit was 0.96.
11) Launch cost was $45 million.
12) It was assumed that delays would result in the range of 0.1 to 0.5 years as a result of a launch vehicle failure. (The shorter delays were assumed to be more likely.)
13) The satellite was assumed to consist of a single payload that had an expected wear out life of 15 years with an associated standard deviation of 0.5 years and an overall mean time between failures (mtbf) of

approximately 60 years (with the overall result of achieving an expected satellite life of slightly more than 12 years).

14) Payload control operations costs and engineering expenses were included (in satellite cost).
15) Satellite nonrecurring costs were not considered.
16) Cost spreading was assumed (over a two-year period) for satellite recurring and launch costs.
17) Nominal on-orbit test and acceptance time was 0.25 years.

The analysis of the GPS-like mission was concerned with establishing the present value of life-cycle cost and availability statistics as a function of the number of on-orbit spares, launch delays (the on-orbit test and acceptance time was varied to simulate changes in launch delays), and maintenance strategy. Two specific strategies were considered: replacement on failure (ROF) and launch in anticipation (LIA) of a payload failure. The obtained results are illustrated in Fig. 4.47 that indicates the relationship between availability and the present value of life-cycle cost (at a 10% discount rate) for each of the situations considered.

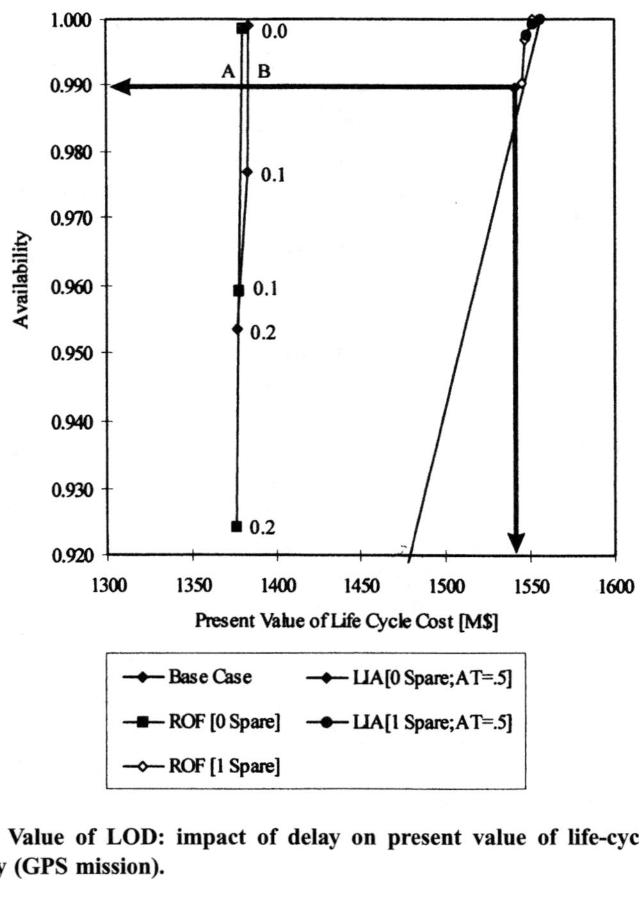

Fig. 4.47 Value of LOD: impact of delay on present value of life-cycle cost and availability (GPS mission).

The base case (as just described) is characterized by an availability of 0.99 and a present value of life-cycle cost (PVLCC) of \$1540 million. (Nonrecurring costs have not been considered, nor have ground costs.) On average, 71 satellites (including ground and on-orbit spares) will be acquired over a 25-year period. The curve indicated as *ROF[0 Spare]* indicates the availability and life-cycle cost relationship when there are no spares and delays range from 0.0 to 0.2 years. A cost savings (present value of life-cycle cost) in the amount of approximately \$160 million would likely result if an instantaneous launch capability could be achieved. Increasing delays tend to reduce the present value of costs (i.e., costs are delayed and the determination of present value takes into account their timing) but with significant impacts on availability. By interpolating the *ROF[0 Spare]* curve (indicated as A), delays of less than 0.025 years must be achieved (with savings on the order of \$150 million) if availability is not to be compromised.

It is possible to launch in anticipation of satellite wear out failures. Results are shown that consider launching in anticipation with anticipatory times (i.e., AT) of 0.5 years. The results indicate that launching in anticipation will generally increase availability but will reduce potential savings. If satellites are launched 0.5 years in anticipation of wear out, delays of up to 0.04 years (point B) are tolerable without compromising availability and with likely savings on the order of \$145 million per two orbital planes. Increasing anticipatory time has little effect on the present value of life-cycle cost because anticipatory time is small compared to satellite life. Further increases in anticipatory time will affect allowed delays.

The GPS mission consists of six orbital planes with four satellites in each of three planes and three satellites plus one spare in each of the other orbital planes (a total of 24 in-orbit satellites). For analysis purposes it was assumed that the system could function normally with seven operating satellites per two orbital planes. Thus the analysis considers availability as the probability that seven or more satellites were operational in a group of two orbital planes.

Similar results were obtained for the Iridium system. Iridium is (was?) a LEO communication satellite mission consisting of 66 satellites (plus 6 spares) that were placed into 6 orbital planes. The satellites have a recurring cost of about \$10 million per satellite. Limitations imposed by the *SATCAV* model result in a maximum of 10 satellites being considered per orbital plane. Thus, in the developed example it is assumed that there are nine operational satellites per plane plus (when spares are considered) one spare per plane. The impact of launch on demand would be to eliminate the need for this spare.

Specific assumptions related to the analysis (mainly imposed by the structure of the *SATCAV* model) and data used in the analysis are as follows:

1) Orbital planes were considered as being independent.
2) Although Iridium launched multiple satellites per launch, which introduces correlation effects, satellites were considered as being uncorrelated and launched one at a time.
3) Availability is defined as the fraction of the time that at least nine satellites are operational within an orbital plane.
4) All results are based upon 1000 Monte Carlo simulation runs (i.e., each time a variable such as number of spares, delay, etc., is changed, 1000

simulation runs were performed in order to develop the cost and availability results).

5) The mission was considered over a 20-year time frame with both availability and life-cycle costs considered over this time frame. Availability was measured starting in the fourth year.
6) A 95% cost-learning curve was assumed for satellites.
7) Five satellites were launched at the start of the third year with another four (or five) launched six months later.
8) One ground spare was procured, but it was assumed that a spare was always available as and when required (i.e., there were spares always residing in inventory, and their cost was accounted for as well as the costs for all satellite replacements).
9) Launch vehicle reliability was 0.96.
10) Probability that a satellite operates successfully when placed into orbit was 0.96.
11) Launch cost was $15 million.
12) It was assumed that delays would result in the range of 0.1 to 0.5 years as a result of a launch vehicle failure. (The shorter delays were assumed to be more likely.)
13) The satellite was assumed to consist of a single payload that had an expected wear-out life of 6.5 years with an associated standard deviation of 0.5 years and an overall mtbf of approximately 20 years.
14) Payload control operations costs and engineering expenses were included.
15) Satellite nonrecurring costs in the amount of $100 million were assumed.
16) Cost spreading was assumed (over a 3-year period) for nonrecurring, satellite recurring and launch costs.
17) Nominal on-orbit test and acceptance time was 0.25 years.

The analysis of the Iridium-like mission was concerned with establishing the present value of life-cycle cost and availability statistics as a function of the number of on-orbit spares, launch delays (the on-orbit test and acceptance time was varied to simulate changes in launch delays), and maintenance strategy. Two specific strategies were considered: ROF and LIA of a payload failure. The obtained results are illustrated in Fig. 4.48, which that indicates the relationship between availability and the present value of life-cycle cost (at a 10% discount rate) for each of the situations considered.

The base case (as just described) is characterized by an availability of 0.953 and a PVLCC of $1610 million. The curve indicated as *ROF[0 Spare]* indicates the availability and life-cycle cost relationship when there are no spares and delays range from 0.0 to 0.2 years. A cost savings (present value of life-cycle cost) in the amount of approximately $105 million would likely result if an instantaneous launch capability could be achieved. Increasing delays tend to reduce the present value of costs (i.e., costs are delayed and the determination of present value takes into account their timing) but with significant impacts on availability. By

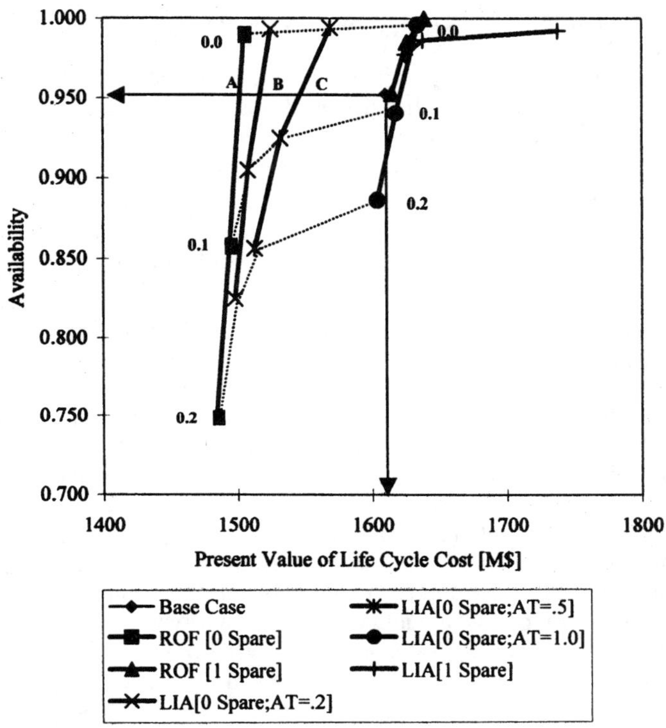

Fig. 4.48 Value of LOD: impact of delay on present value of life-cycle cost and availability (Iridium mission).

interpolating the *ROF[0 Spare]* curve (indicated as A), delays of less than 0.03 years must be achieved (with savings on the order of $110 million) if availability is not to be compromised.

It is possible to launch in anticipation of satellite wear out failures. Results are shown that consider launching in anticipation with anticipatory times (i.e., AT) of 0.2, 0.5, and 1.0 years. The results indicate that launching in anticipation will generally increase availability but will reduce potential savings. If satellites are launched 0.5 years in anticipation of wear out, delays of up to 0.06 years (point C) are tolerable without compromising availability and with likely savings on the order of $65 million per orbital plane.

E. Launch on Demand: Conclusions

Reducing launch delays can have significant impacts on mission life-cycle cost and satellite availability by enabling on-orbit spare satellite requirements to be reduced or eliminated. For the GPS- and Iridium-like missions, delays of less than 0.05–0.06 years (i.e., several weeks) must be achieved in order to eliminate on-orbit spares without compromising availability. (Delays incurred for on-orbit checkout have been taken into account in the simulation analysis.) With the assumption that orbital planes are independent (i.e., placement, sparing, and replacement of failed satellites in one orbital plane will not affect actions in another orbital plane), the LOD savings for the considered GPS-like mission can be on the order of $450 million when delays of less than one month (actually in the range of 0.6 to 1 month) are achieved. Similarly, for the Iridium-like mission, when delays of less than one month are achieved, savings can be on the order of $350 to $400 million.

F. Appendix: Description of the TRANSOP Resiliency/Operability Model

TRANSOP is a Monte Carlo simulation model that simulates launch operations and, by taking into account launch vehicle and payload arrival rates, launch vehicle reliability and failures, launch constraints, stand-down time resulting from failures, and other factors, establishes measures such as the probability distribution of payload backlog as a function of time, which are indicative of transportation system operability. An overview of the general flow of the TRANSOP model is presented in Figs. 4.49 and 4.50.

Referring to Figure 4.49, input data (provided via a user-friendly menu-driven system) are read, a number of variables (primarily those associated with establishing expected values and standard deviations that require results from all Monte Carlo simulation runs) are initialized, and the Monte Carlo loop is entered. Another set of variables (for example, inventory levels) is initialized inside the Monte Carlo loop for each Monte Carlo run. From input data the initial payload (P/L) and launch vehicle stage (LSV1 and LSV2 refer to stages 1 and 2, respectively) inventories are established. Also, based upon input data (i.e., procurement rates), P/L, LSV1, and LSV2 arrival times are established. The model time frame is based upon years (I) and tenths of years (II)—the result is that all events are established to the nearest tenth of a year, and an II loop is maintained within the I loop, which in turn is within the Monte Carlo run (R) loop.

The number of payloads that switch to other transportation systems are established by generating a random number for each P/L in inventory and comparing this value with the probability of P/L switching as provided by input data. Because this is accomplished for each II time interval, the probability of a P/L switching is related to the length of time that it remains in inventory and is basically of the form:

$$\text{Prob. of Switching} = 1 - [1 - PS]^N$$

where PS is the probability of a P/L switching if delay is one-tenth of a year and N is the number of time periods that the P/L is likely to remain in inventory.

After the determination is made with respect to P/L switching, P/L, LVS1, and LVS2 inventories are adjusted. A determination is made with respect to whether or not a launch takes place within each specific I and II time interval. If the time of next launch (initialized to 1.0, the first time period, to start the process) is greater than I + 0.01 * II, then no launch will take place, and the I and/or II counters are incremented. When the time of next launch is equal to or less than I + 0.1 * II, a launch is required, and the flight rate is established and is a function of thruput-related data (i.e., facility thruput, LVS1 and LVS2 turnaround time, launch vehicle

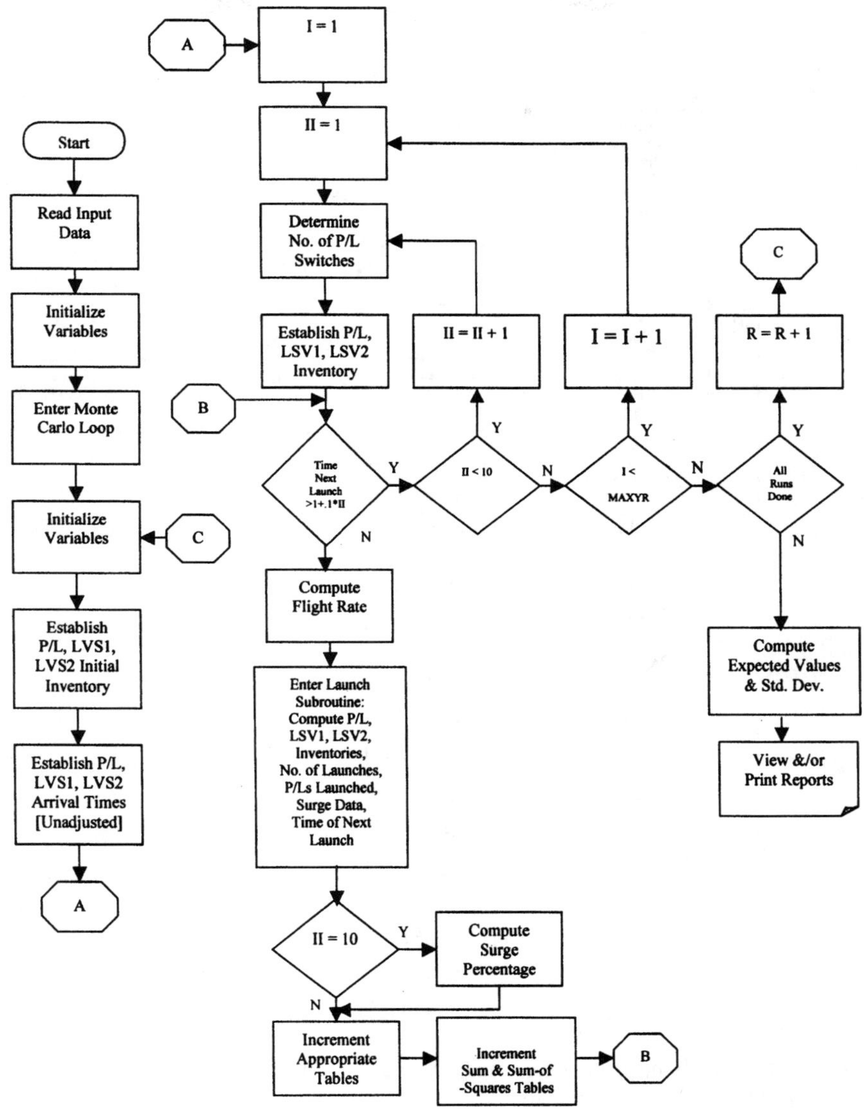

Fig. 4.49 Overview of TRANSOP model.

integration and checkout time, LVS1 and LVS2 procurement time, and surge capability and duration). Once flight rate is established, the launch subroutine is entered and P/L, LVS1, and LVS2 inventories computed, as well as number of launches, P/L launched, surge data, and the time of the next launch. Upon return from the subroutine, all tables are incremented and the process then repeated by comparing the time of the next launch with $I + 0.1 * II$. When the maximum

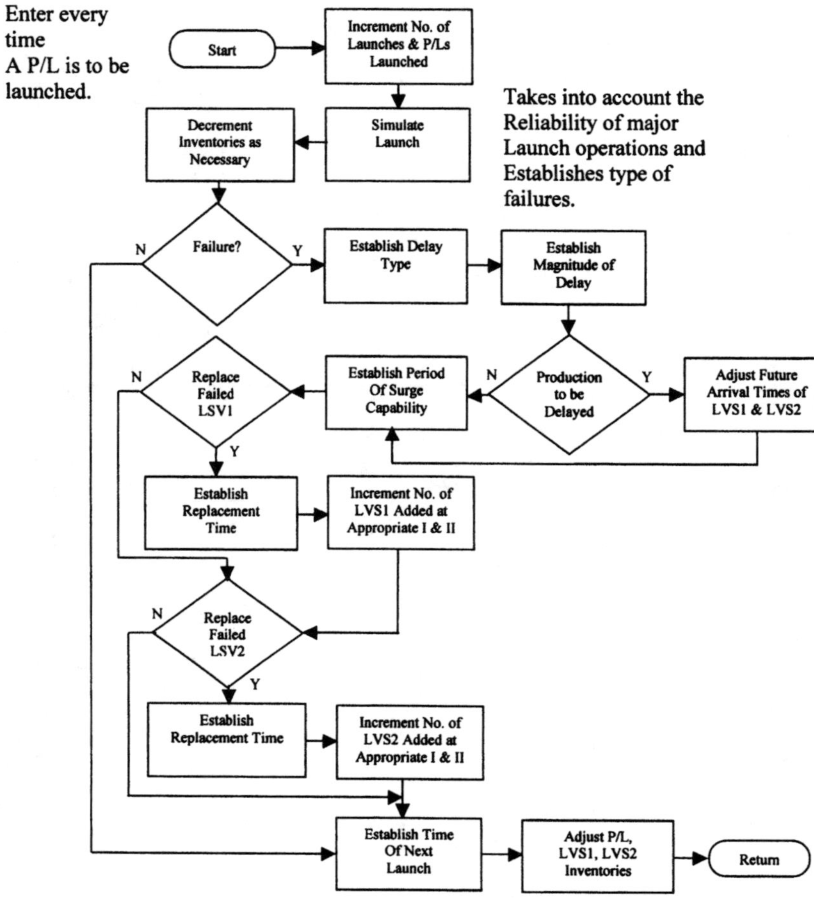

Fig. 4.50 Overview of launch subroutine.

values of I and II have been reached, the next Monte Carlo run is started. The preceding process is repeated for all Monte Carlo runs, and, when the process is completed, expected values and standard deviations computed and reports generated and made available for printing.

Every time a P/L is to be launched (i.e., time of next launch is equal to or less than $I + 0.1 * II$), the launch subroutine is entered (Fig. 4.50). Upon entry, the number of launches and P/L launched are incremented, and the launch is simulated. The launch simulation takes into account the reliability of many launch operations and establishes the type of failure (if any). Inventories are decremented as appropriate. If a failure occurs, the type of delay associated with the failure is established, and its magnitude determined. Because the delays are considered as uncertainty variables, random sampling is used to establish the specific delay value to be utilized. If production is to be delayed because of a failure (as specified via input data), the arrival times of the next LVS1 and LVS2 are adjusted. When a failure has occurred, the period of surge capability is established (this period starts at the end of the delay period) and a decision made with respect to whether or not to replace the failed or lost stages. (This decision is specified via input data—a reusable vehicle would likely be replaced.) If LVS1 or LVS2 is to be replaced, the number added is incremented at the appropriate value of I and II.

The time of next launch is now established as

$$\text{Time of Next Launch} = \text{Time of Current Launch} + \text{Delay} + 1.0/\text{Flight Rate}$$

Finally, inventories are adjusted, and the launch subroutine is exited.

The input data required to operate the TRANSOP Model are as follows:

1) Global data include a) the number of years to be considered; b) the number of Monte Carlo simulation runs to be performed in the analysis; c) the probability that a satellite in the payload queue will switch to another transportation system (The probability is per one-tenth of a year time interval in the queue and results in an increasing chance of a payload switching as a function of the length of time in the queue, for example, a probability of switching equal to 0.1 results in a 41% chance of switching after remaining in the queue for 0.5 years.); and d) the average number of satellites placed into orbit each flight of the transportation system.
2) Initial inventories include a) the starting of initial satellite inventory, which represents the number of payloads awaiting launch at the start of the analysis time frame; b) the starting or initial stage 1 inventory (a two, stage launch vehicle is considered), which represents the number of first stages available for use at the start of the analysis time frame; and c) the starting or initial stage 2 inventory (a two-stage launch vehicle is considered), which represents the number of second stages available for use at the start of the analysis time frame.
3) Thruput-related data include a) the average or normal thruput capability (when not operating under surge conditions) of the launch complex (flights per year); b) the time required getting stage 1 ready for its next flight (years), which does not include integration and checkout time (LVS1 turnaround time is assumed overlapped with LVS2 turnaround time); c) the time required getting stage 2 ready for its next flight (years), which does not include integration and checkout time (LVS2 turnaround time is assumed overlapped with LVS1 turnaround time.); d) the time required to integrate the launch vehicle stages and the payloads (years); e) indicator of whether or not a replacement should be acquired if and when a stage 1 fails and is not recovered; f) the time to procure stage 1 of the launch vehicle (years); g) indicator of whether or not a replacement should be acquired if and when a stage 2 fails and is not recovered; h) the time to procure stage 2 of the launch vehicle (years); i) the maximum launch rate of the launch complex that can be maintained for relatively short intervals of time

(launches/year); and j) the maximum reasonable duration of maintaining a surge capability (years).

4) Acquisition rate data include a) the maximum and minimum anticipated stage 1 annual acquisition rate (number of vehicles per year) and the identity of the associated uncertainty profile; b) the maximum and minimum anticipated stage 2 annual acquisition rate (number of vehicles per year) and the identity of the associated uncertainty profile; c) the maximum and minimum anticipated satellite annual arrival or acquisition rate (number of payloads per year) and the identity of the associated uncertainty profile; and d) indicator of whether or not the procurement of launch vehicles will be interrupted in the event of a failure and delayed by the stand-down time (i.e., a hiatus in the delivery schedule equal to the stand-down time).
5) Launch vehicle reliability data include a) the probability of success of stage 1 for each year of the analysis; b) the probability of recovering LVS2 given the loss of LVS1, which is set equal to zero when considering an expendable launch vehicle; c) the probability of recovering LVS1 given a successful flight of LVS1, which is set equal to zero when considering an expendable launch vehicle; d) the probability of success of stage 2 given that stage 1 was successful; e) the probability of recovering LVS2 given an LVS2 abort, which is set equal to zero when considering an expendable launch vehicle; and f) the probability of recovering LVS2 given an otherwise successful mission, which is set equal to zero when considering an expendable launch vehicle.
6) Stand-down time-related data include a) the maximum and minimum anticipated stand-down time that is likely to result from an LVS1 failure with LVS2 being recovered (years) and the identity of the associated uncertainty profile; b) the maximum and minimum anticipated stand-down time that is likely to result from an LVS1 failure with LVS2 not being recovered (years) and the identity of the associated uncertainty profile; c) the maximum and minimum anticipated stand down time that is likely to result from an LVS2 failure with recovery of LVS2 (years) and the identity of the associated uncertainty profile; d) the maximum and minimum anticipated stand down time that is likely to result from an LVS2 failure where the LVS2 is not recovered (years) and the identity of the associated uncertainty profile; e) the maximum and minimum anticipated stand down time that is likely to result from a failure to recover LVS2 given an otherwise successful mission (years) and the identity of the associated uncertainty profile; and f) the maximum and minimum anticipated stand down time that is likely to result from a failure to recover LVS1 given an otherwise successful LVS1 flight (years) and the identity of the associated uncertainty profile.

XI. On-Orbit Service/Repair (Government Missions)

Over the past several decades, many studies have addressed the technical and economic feasibility of on-orbit service/repair. (This discussion is based upon a study performed by the Naval Research Laboratory and reported in Ref. 39.)[40–46] These studies considered a range of missions including the use of the space shuttle for performing repair of satellites orbiting in close proximity to the shuttle (i.e., the rendezvous, docking, and repair of the Hubble Space Telescope), the placement of a satellite and return of a failed satellite to a transportation node (i.e., on Earth or in LEO) for repair and eventual reuse, servicing of a satellite in its operational orbit, etc. The impetus for these studies was the desire to reduce space system costs by repairing and reusing satellites rather than simply replacing failed satellites. This has therefore been the focus of most economic analyses. However, the potential benefits of satellite servicing extend beyond the reduction of life-cycle costs (a result of extended satellite life through service/repair). An important aspect normally overlooked when analyzing on-orbit service/repair economics is the effect that this will have on overall mission performance (for example, increased sensor or transponder availability). The economic analysis described in the following paragraphs demonstrates the importance of considering both expected life-cycle costs and sensor availability when evaluating the economic implications of on-orbit satellite service and repair. Because government missions differ from commercial missions in that they normally are not concerned with profit and rate of return metrics, life-cycle cost and availability metrics serve as the basis for the comparison of alternatives.

A basic truism that drives on-orbit service/repair economic considerations is the following: the greater the cost of the satellite that would have to be replaced (without the availability of an on-orbit service/repair capability), the lower the cost of space transportation, and the greater the required level of availability, the more attractive on-orbit service/repair is likely to be. With this in mind, a study[39] was undertaken by the Naval Research Laboratory to examine and understand the implications of on-orbit satellite servicing. The study was referred to as Spacecraft Modular Architecture Design (SMARD) and had the objective of evaluating the potential costs and benefits of on-orbit satellite servicing. This study is used to illustrate the economic implications of achieving an on-orbit satellite service/repair capability.

The SMARD study was primarily concerned with evaluating the potential costs and benefits of on-orbit satellite servicing for an existing flight architecture. For this purpose an existing classified satellite, referred to generically as LeoSat, was used as the baseline for the SMARD study. The basic concept was to modularize the design of the existing satellite so that important (in the reliability sense) components could be replaced on orbit through an attached replacement. The large expensive satellites would be placed into orbit with the necessary large and expensive launch vehicles. The required replacement attachments, when determined to be necessary, would be carried into orbit utilizing a smaller and less expensive launch vehicle. Rendezvous and docking would take place, allowing the replacement attachments to be made and thereby maintaining the functioning of the satellite.

The LeoSat vehicle's electrical and mechanical systems were evaluated to determine the potential impacts of servicing, and the costs associated with modifications to the design of the satellites. It was concluded that approximately one-third of all LeoSat components could be practically replaced, and many more could be replaced by adopting a more modular bus and payload design. On-orbit replacement of components was considered as being performed functionally, not physically, by a highly autonomous rendezvous/docking (R/D) servicer. All of the replacement components are packaged in a single module, which the R/D servicer attaches to a docking interface on LeoSat. This functional replacement strategy minimizes the cost and complexity of the servicing mission.

A conceptual design for the R/D servicer was developed as part of the SMARD study. The R/D servicer design effort was extensive and included a detailed technical assessment of enabling technologies required for the mission (i.e., laser metrology guidance/docking system, high-reliability docking mechanism, and autonomous GPS-based navigation). The conceptual design was developed in such detail that a credible bottoms-up cost estimate could be made. The nonrecurring and recurring costs of the R/D servicer development were included in the system-level life-cycle cost evaluations.

Several costing models were developed for the SMARD study. A purely deterministic analysis was used to evaluate the cost and availability impacts of scheduled satellite servicing. An advanced Monte Carlo simulation (the *SATCAV* model, described in Sec. I) was used to gauge the impacts of scheduled servicing as well as unscheduled servicing (i.e., launch on demand).

The baseline LeoSat mission consisted of a time-phased constellation of relatively large (massive) multisensor satellites placed into orbit with an Atlas II class launch vehicle. These satellites would normally be replaced on a scheduled basis with a specific number of satellites procured. With the modular design concept under consideration, the baseline satellite would be redesigned to allow on-orbit maintenance so that a large class of failures would be correctable by launching an R/D servicer, when appropriate (depending upon the maintenance scenario), rendezvousing and docking with the satellite upon which maintenance is to be performed, and attaching a permanently docked payload module. This module would functionally replace a large class of subsystems that have high failure rates. The R/D servicer would be placed into orbit with a Pegasus class launch vehicle. The intent is therefore to extend the expensive satellite life through a series of low-cost maintenance flights carrying low-cost service modules, thereby foregoing the direct replacement of the expensive satellites and their associated expensive launches.

The comparison of alternative spacecraft modular architecture designs considered both costs and benefits. The costs considered for comparison purposes were the life-cycle costs computed as the present value of the annual costs (discounted at 10%) considered over the mission-planning horizon. The development of annual costs considered launch costs, launch vehicle reliability, payload and repair kit nonrecurring and unit recurring costs (taking into account cost spreading over time), and satellite subsystem reliability characteristics (both random and wear out failures). All costs were developed in terms of constant dollars.

When comparing alternatives having different reliability, scheduling, and delay characteristics, system availability must be considered in addition to the cost considerations. Availability is defined as the probability that at least n sensors (out of N possible) will be available or operational during a specified period of time. Because the architectures under consideration were found to have significant impacts on availability, both the present value of life-cycle cost and availability measures were developed for each considered architecture and associated variants. Benefits of one architecture relative to another were measured in terms of the change in net present value of expected life-cycle cost and the change in sensor availability. The selection of a desired architecture must consider both the cost and availability dimensions and tradeoffs between these two dimensions. [Changes in risk (i.e., the variability of the present value of expected life-cycle costs) should also be considered. In actuality, it was found that for the cases considered, risk levels were comparable, and their explicit consideration would tend to cloud the main thrust of this discussion—that multiple attributes (at least cost and availability) must be considered when comparing alternatives. In general, however, the comparisons need to consider both the expected and risk dimensions of cost and both annual and long-term availability.]

The objective of the performed analyses was to develop and then compare life-cycle cost and availability statistics for a number of alternative approaches for performing a mission comprising eight satellites in four orbital planes. In all cases satellites were launched according to a specified schedule with the mission lasting for an extended period of time. The following three basic scenarios or architectures were considered (with a number of alternatives considered within each as described in following paragraphs):

Scheduled satellite replacement without refurbishment/repair is the first scenario. A number of large expensive satellites are placed in orbit on a scheduled basis and replaced on a scheduled basis (every eight years). This is considered as the base case.

Refurbishment/repair on an as-required basis is the second scenario. A number of large expensive satellites are placed in orbit on a scheduled basis (same as just stated), and as critical sensors fail refurbishment/repair missions are undertaken to extend the life of the large satellites. The nonrecurring cost of the large satellites is increased because of the need to incorporate the refurbishment/repair concept into its design. The large satellite recurring cost is reduced (along with mean time to failure) because of the reduction in redundancy. The reductions in overall reliability are offset by the use of a refurbishment/repair module with its associated nonrecurring, recurring and transportation costs.

Refurbishment/repair on a scheduled basis is the last scenario. This is similar to the preceding case (refurbishment/repair on an as required basis) with the exception that refurbishment/repair is performed on a scheduled basis (i.e., every two, three and four years).

The analysis of these architectures considered both a deterministic and a stochastic approach. The deterministic approach utilized both bottoms-up and CERs to evaluate all elements of nonrecurring and recurring costs. Satellites were considered to fail at the end of their expected lives with service/repair flights undertaken at specified times. This deterministic analysis was used as the basis for developing the data requirements for the stochastic approach. Uncertainty estimates were associated with the various cost estimates, and reliability estimates were made for the various subsystems associated with both those that were serviceable and those that were not. Reliability estimates were also made for each of the important launch and on-orbit operations.

The stochastic approach used the already developed *SATCAV* Life-Cycle Cost and Sensor Availability Model.[10,33] As described in Sec. I, *SATCAV* is a dynamic stochastic life-cycle cost and availability model that simulates the launch and on-orbit operations associated with the initiation and continuing operation of a generalized space mission comprising multiple satellites with multiple sensors.

SATCAV was used to simulate satellite launch operations using expendable launch vehicles and upper stages and took into account the consequences of a set of defined failures in terms of cost-incurring events and associated time delays. *SATCAV* simulated the random and wear out of the multisensor LeoSat satellites determining when specific failures occurred and when maintenance actions were required to respond to critical failures. Because multiple sensors were considered, it was necessary to identify critical failures that initiated cost-incurring events (such as replacement or repair). These were defined in terms of both individual sensor failures and combinations (one or two) of sensor failures. For example, if either sensor A or sensor B or sensors C and D failed, a satellite replacement or repair flight was initiated.

SATCAV's capability to simulate alternative maintenance scenarios was used to analyze both launch on failure and launch in anticipation of wear out failure scenarios. Different transportation scenarios were, as necessary, selected (for placement and maintenance flights) from a number of general scenarios that included both ground- and space-based assets.

The consideration of subjective uncertainties, transportation system reliability, and satellite subsystem and sensor random and wear out failures resulted in the establishment of the probability distribution of annual events (i.e., launch attempts, payloads serviced and repaired and transfer vehicles utilized) and the probability distributions of annual costs (both nonrecurring and recurring) and present value of life-cycle costs. *SATCAV* also developed sensor availability statistics, which are presented as the probability density function of the number of sensors available (by type) annually as well as over the mission duration. The statistics took into account sensor and subsystem random and wear out failure characteristics, sparing strategy, transportation scenarios, delays associated with different types of failures, and duration of on-orbit testing.

Important attributes of *SATCAV* that made possible the analysis of the LeoSat service/repair mission and the development of the present value and sensor availability statistics include the following

1) consideration of time horizons of up to 25 years;
2) consideration of multiple operational satellites;
3) sparing strategies that include multiple ground spares, launch on failure of critical sensors, launch on random failure but in anticipation of wear-out failure of critical sensors;

4) identification of critical sensors (individually and in combination);
5) specification of desired launch dates for initial operational satellites and on-orbit operational satellites and on-orbit spares (actual launch dates take into account failures and associated delays);
6) specification of completion date of ground spares;
7) consideration of satellite cost-learning effects;
8) consideration of satellite test and checkout time on orbit;
9) choice of transportation scenario for initial placement and maintenance flights (for example, direct placement, placement and return, on-orbit repair, replacement/return/repair);
10) transportation scenarios described in terms of the probability of successfully performing each major operation in the launch/recovery sequence taking into account the consequences of failure in terms of cost-incurring events and time delays;
11) a priori defined uncertainty variables (i.e., subjectively specified as ranges of uncertainty and the form of uncertainty), which included delays (function of failure type), transportation cost from Earth to LEO, transportation cost from LEO to P/L orbit, satellite unit recurring cost, satellite repair kit cost, transfer vehicle repair cost, capital expenditures, satellite control operations cost, and engineering expenses;
12) satellite configuration consisting of multiple sensors each with multiple support systems and multiple satellite bus subsystems, with each sensor and subsystem described in terms of its random (mean-time-to-failure) and wear-out (expected wear out life and the standard deviation of wear-out life) failure characteristics;
13) cost spreading for nonrecurring capital items and satellite, satellite repair kit, and launch recurring costs; and
14) development of cost, event, and availability statistics, and satellite expected life and associated standard deviation.

The preceding features were used directly in the analysis of the spacecraft modular architecture design concepts. Some additional features were created by "fooling" the model through the use of appropriately structured input data. For example, scheduled replacement scenarios were developed by creating a fictitious critical sensor having a wear out life equal to the interval between scheduled replacements and treating the other actual sensors as noncritical. This mechanism allowed sensor availability statistics to be developed accurately because the actual sensors were still allowed to fail according to their random and wear-out failure characteristics.

In general the *SATCAV* model provided a reasonable, but not exact, fit to reality. A major limitation of the model was that a constellation of eight satellites exceeded its capability. To get around this restriction, two separate analyses were performed for constellations of four satellites each having appropriate launch dates. Nonrecurring costs were assigned to the first constellation, and total present value of cost was obtained as the sum of the present values of the two missions. Sensor availability statistics were developed for each constellation (i.e., the probability of 0, 1, 2, 3, or 4 sensors of each type) and the statistics combined outside of the *SATCAV* model to indicate the probability of 0, 1, . . . , 7, 8 sensors being available and at least N (for example, four or more) being available.

The PVLCC at a discount rate of 10% is indicated in Table 4.6 for each of the considered scenarios and a number of variants of the main scenarios. The three basic scenarios were 1) scheduled satellite replacement *without* service/repair (referred to as the base case), 2) refurbishment/repair on an as-required basis, and 3) refurbishment/repair on a scheduled basis.

Also indicated in Table 4.6 is the availability of sensors #2 and #3 in terms of the percentage of time that at least 4 (i.e., ≥ 4), 6, or 8 of each of these sensors will be available. Sensor #1 was used to help define the mission scenarios as already described and to understand the sensitivity of the obtained results to definition of critical failures. Its statistics are therefore of no import.) All results shown are based upon 500 simulation runs (i.e., the mission was "flown" 500 times each time, launching the satellites, determining launch success or failure, and in the case of a failure establishing delays and relaunch requirements, establishing time of sensor and satellite bus subsystem failures and initiating appropriate maintenance strategy, determining costs, etc.), allowing for the statistical nature of failures as described by the set of input data and delays that might occur as a result of the failures.

Table 4.6 Summary of present value of life-cycle cost and sensor availability results

		Availability ($\geq N$), %					
		≥ 4		≥ 6		≥ 8	
Scenario[a]	PVLCC (10%), M$	#2	#3	#2	#3	#2	#3
Scheduled satellite replacement without service/repair							
Base case	2688	70.6	88.7	20.8	44.8	0.8	4.0
Refurbishment/repair on as-required basis							
Base case	2134	99.8	99.0	92.0	80.1	33.3	18.2
Critical sensors (1 or 2 or 3)	2214	99.0	99.9	93.2	96.6	34.8	45.9
Critical sensors (2 or 3)	2215	99.9	99.9	97.2	98.1	46.1	50.3
Critical sensors (1 or 3)	2020	86.6	99.9	38.6	96.6	2.5	45.3
Critical sensors (1 and 2)	1952	82.8	98.3	33.1	75.7	2.1	14.5
Critical sensors (1 and 3)	1952	82.2	98.3	32.6	75.1	2.0	13.9
Refurbishment/repair on a scheduled basis							
Base case	2752	99.3	99.9	84.0	95.0	21.2	38.2
Servicing every 3 yr	2439	98.7	99.9	77.2	94.1	14.9	36.3
Servicing every 4 yr	2263	97.6	99.8	68.7	92.1	10.4	32.5

[a]All results are based upon a satellite having three sensors, where the first sensor is assumed to have a very high reliability and/or to be of little or no concern from the point of view of availability. In actual fact, the data provided indicate that this sensor has a relatively low reliability.

In addition to the three basic scenarios, a number of variants were considered in order to understand the implications of certain assumptions. In the case of refurbishment/repair on an as-required basis, the effect of definition of critical sensors was considered. The base case considered the failure of either sensor #1 or sensor #2 as being critical. For this scenario sensor #1 was for all practical purposes defined as a sensor that would not fail. The minimum cost situations resulted when both sensors 1 and 2, in one case, and sensors 1 and 3, in another case, must fail in order to initiate a cost-incurring event. Because sensor #1 would not fail of its own accord, a bus failure must occur, which then turned off both sensors and the subsequent initiation of a cost-incurring event. In the case of refurbishment/repair on a scheduled basis, the effect of service interval was investigated. The base case represents a two-year service interval.

The implications of these various scenarios and their variants are summarized in Table 4.6. It was found that the business as usual scenario (i.e., scheduled satellite replacement) results in a present value of life-cycle cost (replacement of satellites after approximately eight years of life) of approximately $2700 million with relatively low levels of availability (e.g., there is little chance of achieving a capability of eight operational sensors; there is a 20% chance of achieving a capability of six or more operational sensors of type #2, etc.). Of course, there are many ways within this scenario of increasing levels of availability but all are likely to further increase the present value of life-cycle cost.

It was also found that, in general, refurbishment/repair consideration are likely to reduce the mission life-cycle cost while having the potential to significantly increase sensor availability. Performing refurbishment/repair on a scheduled basis requires careful consideration of scheduling in order to achieve both cost and availability benefits. Finally, through the consideration of alternative critical sensor definitions it is shown that the definition of criticality can have a reasonable impact on life-cycle cost but a very significant impact on system availability.

The base case is a representation of an actual planned (or ongoing) mission. It appears that either the mission planners are willing to accept a very low level of availability, have not performed an adequate analysis of availability and are being misled by inadequately performed analyses, and/or budgets are likely to be too low to adequately handle the likely increase in costs as the realization of inappropriate levels of availability are observed.

XII. Economic Implications of Orbital Debris

Just as pollution is reaching a critical point here on Earth, near-Earth space continues to grow crowded with human-generated orbital debris. The problem has sparked increasing concern, and international policy makers are beginning to give serious consideration to various solutions involving both satellites and launch vehicles. However the solutions have economic impacts: the consideration of these impacts is the subject of the following paragraphs.

Orbital debris is of concern for a number of reasons. Increases in the debris population will tend to increase, through debris-satellite impacts, satellite failure rates, thereby adding to space operations costs and will affect mission strategies aimed at maintaining system availability. For example, the Iridium system, a LEO constellation of 66 communication satellites, planned to deorbit its satellites near their end of useful life, so as to ensure that the environment within which Iridium was to operate remained relatively clean for long-term system operations. Because of concern with the potential growth of the orbital debris population, there is increasing concern with measures that can be taken to control debris population growth. Both voluntary and mandatory mitigation measures are being considered (see Sec. III of Chapter 5). The fear is that inadequate actions in the near term might lead to prohibitive costs in the long term. The economic implication of orbital debris are also of interest in the setting of design standards for satellites and launch vehicles aimed at reducing the debris population.

In general, orbital debris will increase the costs of space missions through costs incurred to conform to debris mitigation policies and/or losses that result from debris impacts. It is likely that orbital debris mitigation policies will, in the long term, reduce mission costs (below those that would result from pursuing a policy of no orbital debris mitigation) but are likely to increase mission costs in the short term. This is the typical investment dilemma: Is it desirable to make an investment in the near term in order to achieve long-term returns?

The space community, in general, and policy makers, in particular, have not been willing to accept the fact that there will be costs associated with debris mitigation. For example, a recently published International Academy of Astronautics position paper[47] failed to discuss, or even mention, the cost and economic implications of mitigation.

The objective of the following paragraphs is to establish a method that can be used for placing the economic impacts of orbital debris into a quantitative (rather than a qualitative) regime so that informed decisions can be made with respect to mitigation alternatives and related decisions. Because industry and government concerns differ, this section is concerned with the economic impacts on both costs and profitability. A later section (Sec. III of Chapter 5) discusses government concerns relating to orbital debris from the perspectives of the possible need for, and the form of, a regulatory regime and concerns with overall economic benefits. In the following paragraphs of this section, emphasis is placed upon the economic impacts of orbital debris on the life-cycle costs of performing typical LEO missions. Economic impacts are developed in terms of forecasted orbital debris environments that are specified in terms of the probability of debris impact per unit area per year as a function of time and the time of mission initiation within this time-changing environment. Geosynchronous communication satellite missions are also briefly considered. For these missions, where each satellite can be considered as a profit center, economic impacts are developed in terms of profit and rate of return financial performance measures. (To a first-order approximation, geosynchronous communication satellites can be considered as individual profit centers, whereas low-Earth-orbiting satellites are frequently part of a functioning constellation. It is this distinction that leads to the consideration of profit impacts for GEO satellites and life-cycle cost impacts for LEO satellites.) The developed economic impacts can be used to provide insights into the affordability of debris mitigation measures as well as the need for mitigation.

A. Introduction

Space activity is placing debris in orbit faster than the natural effects of drag remove it. As a result, the probability of satellites impacting and being damaged or destroyed by orbital debris is increasing. In the absence of debris mitigation policies, it is likely that continuing or increasing space activity will lead to an increasing probability of collisions between orbital debris and operating satellites. This has led government agencies and industry to seriously consider the orbital debris problem and debris mitigation policies.[48–50] Although serious consideration has been given to orbital debris and mitigation policies, little or no analysis (with the reported work by this author being an exception) has been done regarding the cost implications of orbital debris and remediation policies.

Although government agency policies concerning orbital debris are only just forming, orbital debris considerations have already caused changes in the plans and activities of some agencies.[51–53] Some policies and activities are motivated by the need to protect a spacecraft; others are designed to prevent debris proliferation. Efforts have come in three distinct areas. First, research is underway to model and measure the debris environment more precisely. Second, operational procedures are being considered to limit growth in the debris population. Third, the design process for future missions and spacecraft are beginning to address debris considerations.

Not taking action to reduce orbital debris will result in an increase in the likelihood of satellite damage with a consequent reduction in satellite life and additional satellite and launch costs incurred to maintain desired levels of capability (for example, sensor or transponder availability). Taking actions such as hardening of satellites can maintain satellite life but lead to increased satellite cost and mass, increased transportation cost, or reduced satellite capability and loss of benefits. Requiring upper stages to reenter the Earth's atmosphere and to burn up can affect payload delivery capability, which will lead to increased transportation costs. Not taking actions to reduce growth of the orbital debris environment will result in reduced satellite useful life and higher costs or other financial and/or availability consequences.

Actions taken in the near term can have significant impacts on the long-term orbital debris environment. However the consequences of actions or inactions will not significantly affect satellite failure rates resulting from debris impacts for perhaps two or more decades. Thus it is necessary to justify costs incurred for instituting debris mitigation measures that are being considered today. Random failures caused by debris impacts will result in reduced satellite life. Mitigation may reduce these random failures. The effects of actions/inactions are summarized in Table 4.7.

The objectives of the following paragraphs are to 1) describe a method for evaluating the life-cycle cost impacts on space missions that might result from different orbital debris mitigation policies, including the potential impacts on costs if a "no mitigation" policy were adopted, 2) present results of an analysis of the likely impacts of orbital debris on the costs of performing LEO missions in order to develop insights into the affordability of near-term mitigation measures, and 3) to indicate the financial impacts that mitigation policies might have on the performance of GEO communication satellite missions.

B. GEO Satellite Missions

Orbital debris mitigation has been an ongoing activity for GEO communication satellites. As per international agreements, GEO communication satellites are

Table 4.7 Consequences of space debris and mitigation policy

No mitigation	Mitigation
Short-term	*Short-term*
+ No significant impact on cost	+ Increased mission cost
	+ Increased nonrecurring cost
	+ Increased unit recurring cost
+ No significant impact on level of service	+ Little/no effect on level of service
Long-term	*Long-term*
+ Increased mission cost [satellite replacements] and/or	+ Reduced mission cost [reduced number of satellite replacements] and
+ Reduced level of service [i.e., availability]	+ Increased level of service [i.e., availability]

normally moved to higher altitudes near end of useful life in order to reduce the probability of collisions and other forms of interference. At least two mitigation measures are being discussed that would affect the financial performance of GEO communication satellite business ventures. One would require GEO satellites to be moved to higher altitudes (higher than the several hundred miles above GEO altitude to which they are currently raised) upon reaching the end of their useful life; the other would place constraints on transfer orbits by requiring, for example, that transfer stages reenter the Earth's atmosphere and burn up rather than remain in orbit for an appreciable length of time. However, both of these solutions would have a negative impact on the financial performance of communication satellite business ventures.

The direct effect of either alternative would be a reduction in satellite performance, increased transportation costs, or possibly both. Moving satellites to higher altitudes requires mass in the form of propellant and/or propellant and thrusters. In either case it would amount to shortening the satellite's on-orbit stationkeeping life (the amount of life reduction being heavily dependent upon the utilized technology) or utilizing a launch vehicle with greater payload delivery capability and higher cost, both of which have financial impacts.

Constraining transfer trajectories would most likely reduce the mass delivery capability of existing vehicles. This would lead to diminished satellite capability and possibly less stationkeeping propellant, which would reduce on-orbit life; there would be less subsystem redundancy, which would affect failure rates and so forth.

The financial impact of these alternatives has been evaluated through the use of a communication satellite business financial planning model. This stochastic simulation model[22] (described in Sec. I) simulates a generalized communication satellite business and is specifically configured so that a large number of different business situations can be identified and analyzed. Examples might include the use of alternative launch vehicles, various satellite technologies, space-based facilities, and alternative sparing and service/repair/replacement concepts. The effects of each choice on financial performance can then be established.

The model also allows for multiple communication services, demand and cost uncertainties, subsystem unreliability (both random and wear out) and sparing concepts, and expendable and/or reusable transportation systems and associated reliabilities. An important feature of the model is that it also allows for time delays that are a function of failure type and for insurance costs related to specific operations or functions.

The model thus enables a business venture to be planned and financial documentation developed. It allows satellite, launch vehicle, market, and general business attributes to be varied. Their consequences can then be evaluated in terms of changes in financial performance measures such as present value of cash flow, profit, ROI and payback period. It is possible, for example, to change satellite life characteristics in ways that might result from the orbital debris reduction approaches just described. The effects of those changes on a business's financial performance can then be estimated.

A typical GEO communication satellite fixed satellite service business scenario and associated database were developed and used as the basis for financial analyses of communication satellite ventures in terms of satellite life characteristics. Typical results of analyses are indicated in Fig. 4.51 in terms of the change in expected NPV of cash flow (at a 10% discount rate) and changes in expected ROI as a function of satellite life reduction. The results of these analyses indicate that as satellite life decreases both NPV and ROI fall very quickly.

The results are shown in Fig. 4.51 in terms of satellite life reduction. The satellite life reduction is assumed to be directly related to the propellant mass required to move the satellite to a different altitude, that is, the mass required to

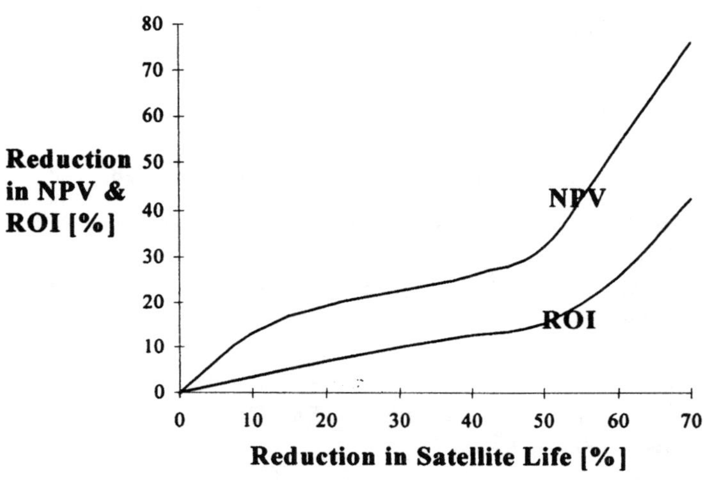

Fig. 4.51 Effect of satellite life reduction on financial performance of GEO communication satellite business ventures.

move the satellite from its initial altitude to a new altitude is no longer available for orbit maintenance and leads to a reduction in useful satellite life. The amount of mass reduction depends upon the technology utilized and the desired altitude change. For relatively small altitude changes Δr the change in mass Δm required for achieving the small altitude change can be computed by first establishing the change in velocity Δv required to achieve the desired altitude change and then establishing the change in mass to achieve the required velocity change.[54]

The change in velocity is accomplished as follows:

$$\Delta v = 0.5 * [\Delta r/r] * v_i$$

$$v_i = [\mu/r]^{0.5}$$

where

$$r = \text{initial satellite altitude, km}$$
$$\Delta r = \text{change in satellite altitude, km}$$
$$v_i = \text{initial satellite velocity, km/sec}$$
$$\mu = 3.986 * 10^5 \text{ km}^3/\text{s}^2$$

and the change in mass is then obtained from

$$\Delta m = \{1 - \exp[-\Delta v/g * I_{\text{SP}}\} * m_i$$

where

$$m_i = \text{initial satellite mass, kg}$$
$$I_{\text{SP}} = \text{specific impulse, s}$$

The change in satellite useful life ΔL, expressed as a percentage of the satellite life if their were no altitude change, is obtained from

$$\Delta L = [\Delta m/m_{\text{ip}}] * 100$$

where

$$m_{\text{ip}} = \text{initial orbit maintenance propellant, kg}$$

Once specific orbital debris remediation policies are identified, for example, moving communication satellites to a specified altitude beyond GEO, their particular effects on satellite useful life and business performance can be estimated with more precision.

C. Conceptualization of Cost Impacts (Non-GEO Earth-Orbiting Satellite Missions)[55]

Orbital debris will increase the costs of space missions through costs incurred to conform to debris mitigation policies and/or losses that result from debris impacts. Cost-effective orbital debris mitigation policies will, in the long term, reduce mission costs (below those that would result from pursuing a policy of no orbital debris mitigation) but can increase mission costs in the short term. This is the typical investment dilemma: Is it desirable to make an investment in the near term in order to achieve long-term returns? or What can be afforded in the near term to achieve desired long-term objectives? Normally, to answer this question the net present value of the resulting cash flow (in this case the cash outflows are the costs incurred to conform to a mitigation policy, and the cash inflows are the cost savings that are the result of conforming to the mitigation policy) as a function of time is developed at an appropriate cost of capital or discount rate. If the resulting NPV is positive (at an acceptable level of risk), then the investment should be made; otherwise, the investment should not be made.

Decisions regarding orbital debris mitigation policies are quite similar to most investment decisions (i.e., spend now for future rewards) except that the time frame is considerably greater (measured in terms of perhaps 50 to 100 years or more) than that encountered in most investment decisions. Therefore simple discounting models (which discount future cash flows CF_i according to $CF_i/[1 + DR]^i$, where DR is the discount rate and i is the year of the cash flow. When i becomes large, the contribution of CF_i to the NPV approaches zero.) are not likely to represent effectively the economic impacts of alternative orbital debris mitigation policies because the cost savings occur well into the future. To avoid the long-term discounting problem, an alternative approach is to use discounting over the relatively short mission horizon and then to consider different mission start dates. This allows the present value of mission costs to be developed based upon different debris mitigation policies (including the no debris mitigation alternative) and relative costs established for missions starting at different points in time and under the influence of different debris mitigation practices initiated at different points in time.

The basic measure of merit, indicated as K, is the present value of mission life-cycle cost given an orbital debris mitigation policy, relative to the present value of mission life-cycle costs in the absence of orbital debris. This relative measure allows all debris mitigation alternatives (including the no-debris mitigation alternative), as well as the timing of their initiation, to be compared on a common basis.

For the analysis of the economic impact of orbital debris mitigation alternatives, the orbital debris environment can be described in terms of the probability of impact with debris per unit spacecraft surface projected area per unit time as illustrated in Fig. 4.52. This figure shows three projected environment alternatives: 1) environment growth with no mitigation, 2) environment growth with mitigation policy X, and 3) environment growth with initial mitigation policy X and additional mitigation policy Y beginning in year T_y. The probability of impact is a function of debris size, models of debris sources, projected space traffic and debris mitigation practices, and satellite orbit and design characteristics. It is a function of both satellite design and configuration as well as launch vehicle configuration and trajectories. As will be discussed in following paragraphs, the probability of impact, assumed to be a direct measure of satellite subsystem failure rates, is used to establish satellite failure and replacement rates and to establish mission life-cycle cost impacts.

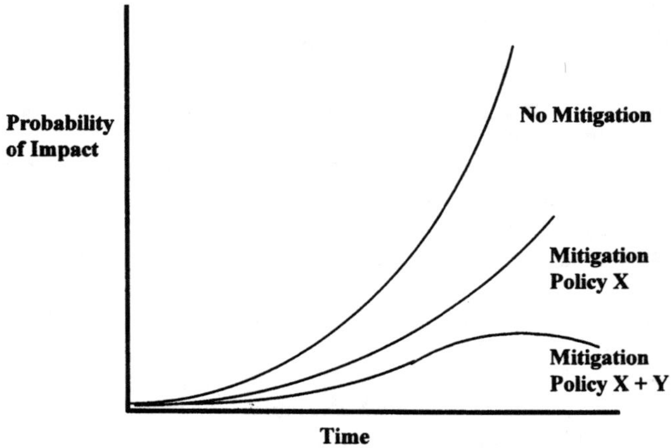

Fig. 4.52 Description of the orbital debris environment: probability of impact scenario.

The basic measure K, representing the relative net present value of mission life-cycle cost (relative to the no debris case), is illustrated conceptually in Fig. 4.53 for the "no mitigation" scenario. The two points on the curve represent the relative present value of life-cycle cost K for the same mission but initiated at different points in time (for example, a 15-year remote-sensing mission initiated in 2005 and the same mission initiated in 2025). The different values of K depend on projected differences in the debris environment (i.e., the probability of impact as a function of time). The solid curve indicates the locus of the values of K for the same mission but initiated at different points in time. The general shape of

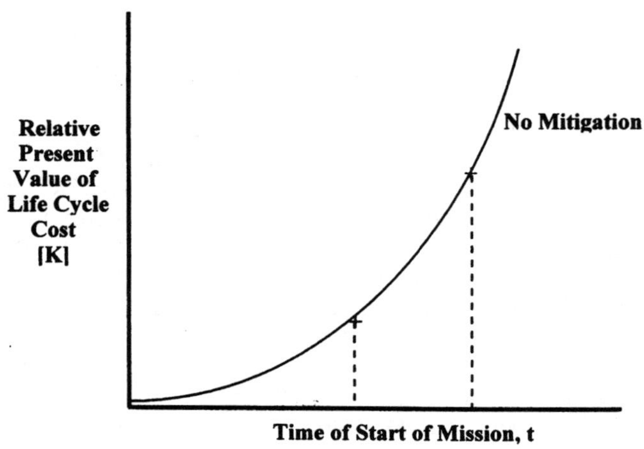

Fig. 4.53 Relative present value of mission life-cycle cost as a function of time of mission initiation: no debris mitigation.

the curve indicates that in the absence of debris mitigation, the economic impacts are likely to increase dramatically; however, the specifics of the timescale are uncertain.

Figure 4.54 illustrates conceptually the relative present value of mission life-cycle cost (relative to the no debris case) for a specified debris mitigation policy and the no mitigation policy alternatives. (The form of the "specification" is discussed in the following paragraphs.) The curve indicated as "no mitigation" indicates the relative cost (relative to the mission cost if there were no orbital debris) for a specified mission initiated at different points in time. For example, a mission initiated at time T_3 has a present value of cost $K(T_3)$ times that of the same mission if there were no orbital debris and the same mission initiated at time T_4 will have a present value of cost $K(T_4)$ times that which would result if there were no orbital debris. The curves in this figure depend upon both the projected debris environment and on mission specifics.

The dashed curve indicates the locus of relative present values of mission costs if the specified debris mitigation policy were initiated at various points in time (i.e., T_0, T_1, T_2, etc.) and for missions that commence at the same point in time. From this dashed curve emanates a series of curves marked $t = T_0$, $t = T_1$, $t = T_2$, etc. These curves indicate the value of K, the relative present value of life-cycle cost, that will result for missions which originate at any time t after the initiation of the debris mitigation policy at $t = T_x$. The intersection of the dashed curve with the ordinate $K(T_0)$ represents the relative increase in the present value of mission life-

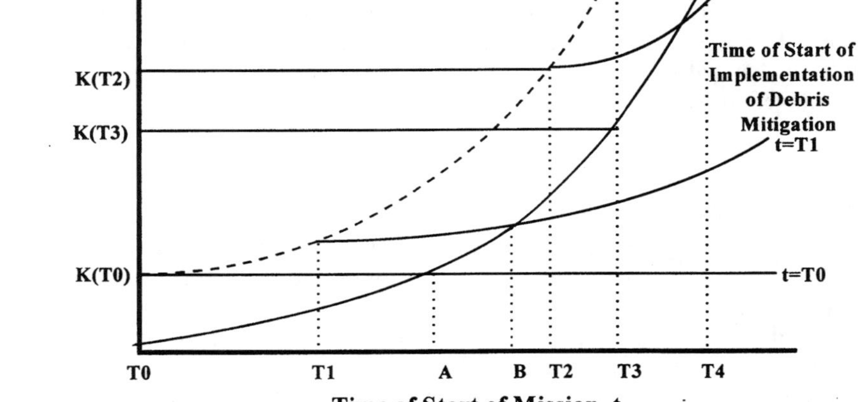

Fig. 4.54 Relative present value of mission life-cycle cost as a function time of implementation of debris mitigation policy and mission initiation.

cycle cost that results from the mission conforming to the stated orbital debris mitigation policy. When the debris mitigation policy is initiated immediately, i.e., $t = T_0$, all missions initiated in the future will have relative costs K that lie on the curve indicated by $t = T_0$. This is so because it is assumed in the example that if mitigation were to commence immediately there would be no growth in the debris population and the failure rate would remain constant over time. Missions initiated prior to time $t = A$ will have larger values of K than the same missions under a no mitigation policy. After time $t = A$ the situation changes and the no mitigation policy will result in higher mission costs. When the debris mitigation policy is initiated at $t = T_1$, missions initiated prior to this time will have values of K that lie upon the no mitigation curve and after this time will have values of K that lie upon the $t = T_1$ curve. Prior to $t = B$ the value of K will be greater for missions with the debris mitigation policy in place, and after $t = B$ the value of K will be less for missions with the debris mitigation policy in place. [The specific shape of the $t = T_x$ curves is not known. Depending upon the specifics of the debris mitigation policy and its impacts on orbital debris as a function of time, it is possible that these curves could reach a maximum (value of K) and then decrease asymptotically as they approach the $t = T_0$ curve.]

Again, it must be emphasized that each orbital debris mitigation policy will have its associated set of curves. In addition, each mission will have its associated curves. The intent of the following paragraphs is to present preliminary results that reflect typical economic impacts that can be expected for LEO missions if mitigation is or is not pursued. When government policy regarding orbital debris mitigation is being considered, the orbital debris environment as expressed in terms of the probability of impact as a function of time must reflect the debris mitigation policy. When the owner of a constellation considers the implications of orbital debris on business decisions, it may be reasonable to consider the debris environment as being independent of the business decision. (This of course will depend on the general population characteristics of the mission altitude and inclination angle regime of concern.)

Several important assumptions should be noted. First, future missions and operations are assumed to be basically the same as those being planned for the near future. It is also tacitly assumed that no new technology will be developed and utilized that will significantly alter the orbital debris projections in ways not considered in the current assessment of debris mitigation measures.

D. Evaluation Methodology

The evaluation methodology is aimed at establishing the present value of life-cycle cost of a typical or representative mission initiated at several different points in time and given a number of orbital debris forecasts (scenarios based upon specified orbital debris mitigation policies) characterized in terms of the probability of debris impact as a function of time and satellite orbit. The probability of impact affects satellite replacement timing in order to meet sensor or transponder availability requirements. (The methodology described in the following paragraphs also allows for the consideration of satellite propulsion and/or mass changes that might be required to satisfy different mitigation policies and the effect on both nonrecurring and recurring costs. Presented results consider just the probability of impact dimension.)

The evaluation methodology has been specifically developed so that both the impacts of debris scenarios as well as the combination of debris scenarios and mitigation policies (i.e., requirements imposed upon satellite configurations and/or launch vehicles) can be evaluated utilizing a common set of equations or model. Thus, the basic approach is to specify a set of sensors (or transponders) that then impose requirements upon satellite bus subsystems, which are then configured and costed. Because mitigation policies can be specified in terms of changes to bus subsystem requirements, the same set of equations can then be used to reconfigure and recost the subsystems. This approach minimizes errors by basing calculations and costs on a common set of equations and related data and assumptions. Relative cost and availability are of primary concern with absolute values being of secondary importance.

The analysis described in the following paragraphs is but a first step to establish the economic impacts of orbital debris and debris mitigation policies. Only the impacts of orbital debris, as specified by probability of impact scenarios, have been considered to date. The SMALLSAT model described briefly in the following paragraphs served as the computational tool for evaluating the economic impacts of orbital debris. (SMALLSAT is a computer-aided system, developed by Princeton Synergetics, Inc., and User Systems, Inc., for Phase A spacecraft design and technology evaluation. The development of SMALLSAT was supported by NASA's Lewis and Langley Research Centers.) The model structure incorporates the impacts of orbital debris (as specified by the probability of impact as a function of time) so that remediation requirements could be evaluated in a totally consistent manner. The SMALLSAT satellite design and life-cycle cost model can be used for the assessment of both the impacts of orbital debris and orbital debris mitigation policies.

SMALLSAT is a computer-aided Phase-A design and technology evaluation tool for small satellites. (A similar but earlier version was developed for midsize satellites.) SMALLSAT enables satellite designers, mission planners, and technology program managers to observe the likely consequences of their decisions in terms of satellite configuration, nonrecurring and recurring cost, launch vehicle selection, and mission life-cycle cost and availability statistics.

In SMALLSAT a satellite is configured in terms of sensor attributes, such as dynamic mass, dimensions, data and power profiles, and pointing and stabilization requirements, and bus subsystem technology. SMALLSAT, illustrated in Fig. 4.55, models the satellite in terms of nine interrelated subsystems: attitude reference and control, power, thermal, orbital maintenance, propulsion, data handling, on-board computer, TT&C, and structure. The design of these subsystems is iterated so as to take into account their mass and power interactions. The results of the satellite design can then be passed to an integrated stochastic life-cycle cost and availability model, which places the configured satellite into a mission setting and establishes the mission life-cycle cost and sensor availability statistics.

SMALLSAT contains sensor, launch vehicle, and bus subsystem technology databases. The user can select input from these databases during satellite configuration. The user can also selectively override this input to model new technology developments. This enables analysis of technology impacts on satellite

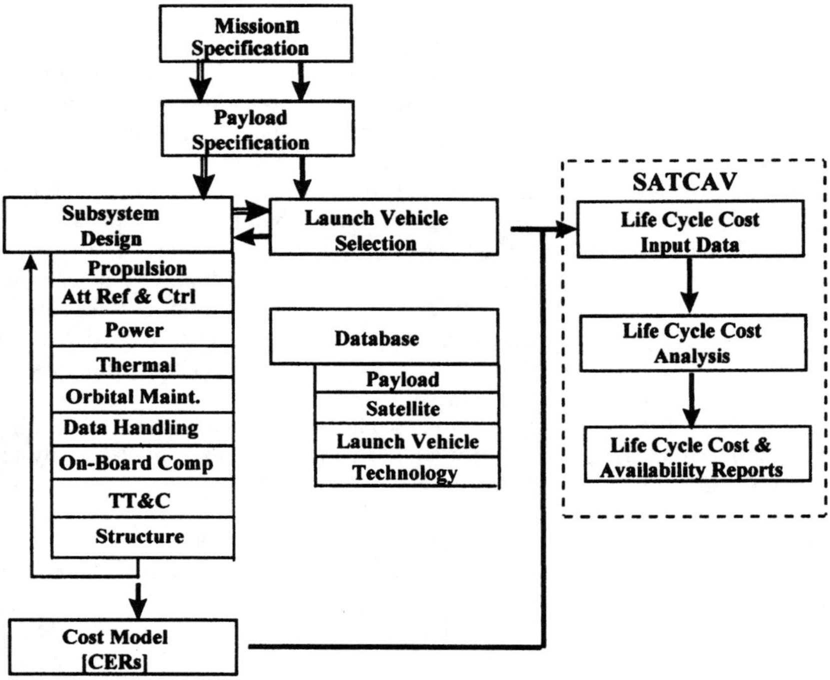

Fig. 4.55 Overview of the SMALLSAT model.

configuration and costing. For example, launch vehicles can either be selected from the database, in which case they place constraints on satellite configuration, or they can be selected based upon the resulting satellite configuration. (Thus there is a mechanism for assessing the impact of launch vehicle debris mitigation requirements both in terms of transportation cost as well as on satellite design and cost.) Modifying launch vehicle attributes allows the effect of new launch capability to be ascertained in terms of satellite configuration and cost, and mission life-cycle cost. In addition, bus subsystem technology levels, such as solar-cell efficiency, can be selectively altered to reflect likely technology advances or the technology advance that is sought through a specific technology program. This allows the likely effects of technology programs to be observed in terms of satellite configuration, cost, and the ability to switch launch vehicles.

The impact of the resulting satellite configuration on mission life-cycle cost and sensor availability statistics can also be determined using SMALLSAT. A set of CERs (specifically developed for small satellites by the Aerospace Corporation) contained within SMALLSAT is utilized to provide estimates of both nonrecurring and unit recurring costs at the subsystem level. The estimated costs can be selectively overridden by specifying expected values and standard deviations. This allows the uncertainty of costing associated with the introduction of new technology to be explicitly and quantitatively considered. An integrated Monte Carlo simulation model determines the resulting life-cycle cost and availability statistics based on user-specified mission parameters such as number of satellites required, their desired launch dates, satellite sparing and replacement strategies, and cost spreading functions.

SMALLSAT incorporates the Monte Carlo simulation model *SATCAV* (described in more detail in Sec. I) to determine life-cycle cost and availability statistics. The *SATCAV* model utilizes satellite design and technology data, mission parameters such as number of satellites required, their desired launch dates, sensor and subsystem random and wear out failure characteristics, launch vehicle reliability and cost, satellite sparing and replacement strategies and cost-spreading functions, and other information to establish the life-cycle cost and availability statistics.

To utilize SMALLSAT for the evaluation of alternative orbital debris mitigation policies, it is necessary to specify typical missions in terms of sensor or transponder complement and orbital parameters. SMALLSAT configures and costs the resulting satellite. Mitigation policies that require satellite design changes can be specified in terms of subsystem parameters such as Δv requirements (as determined by policies to alter orbits near end of satellite design life) and structural mass requirements to provide shielding. These parameters alter satellite configuration and resulting nonrecurring and unit recurring cost.

The probability of a random impact is considered as a critical subsystem (the failure of which causes the loss of the satellite) with a failure rate equal to the probability of impact that varies with time depending upon the specified debris mitigation policy and assumptions regarding debris flux as a function of time. [Because of the small probability of impact per unit time, Poisson statistics apply, and the reliability can be characterized by $\exp(-\lambda t)$ with λ being the debris impact probability.] Different mission initiation times and debris mitigation policy initiation times can be accounted for through the appropriate specification of failure rates as a function of time. In other words, failure rate as a function of time is a reflection of the assumed mitigation policy and launch and payload forecasts.

The first step is thus the specification of a mission. This then results in the determination of satellite configuration, nonrecurring and unit recurring cost, and resulting present value of life-cycle cost of the specified mission in the absence of orbital debris. This satellite configuration, and associated nonrecurring and unit recurring costs, is then used for the analysis of the mission life-cycle cost given different orbital debris scenarios (i.e., probability of impact as a function of time). The relative costs of the mission when there is orbital debris are then established.

The next step would normally be the specification of the debris mitigation policy in terms of probability of impact as a function of time and satellite Δv and/or mass changes. SMALLSAT would be again used to establish nonrecurring and unit recurring costs and present value of mission life-cycle cost. The relative costs of the mission with different debris mitigation policies initiated at different points in time would then be established.

E. Analysis

To demonstrate typical results and to place the orbital debris problem into perspective, the results of a preliminary analysis are described in the following paragraphs. This analysis deviates slightly from the methodology just described in that for a first-order approximation a mission has been defined as placing a single satellite into orbit and maintaining an operational capability in orbit (i.e., replacing failures). Thus all presented results are for a single satellite. The values of K presented in following figures are assumed to be independent of the number of satellites in a LEO constellation. Again, K is the ratio between the present value of mission life-cycle cost in the presence of orbital debris and in the absence of orbital debris. K remains constant, but the present value of mission life-cycle cost for a constellation of N satellites is assumed to be N times that of the single satellite mission.

The base case simulates a generic mission (described in the following paragraphs) in the absence of orbital debris; in other words, satellite failures are caused entirely by satellite subsystem random or wear out failures. The same mission was then simulated within specified orbital debris environments so that failures were caused by the combination of satellite subsystem failures and those caused by orbital debris impacts.

It was assumed that debris impacts, independent of the debris size considered, would cause a satellite failure and would necessitate a satellite replacement. An approximate scaling algorithm is presented to allow the obtained results to be modified to take into account estimates of probability of failure as a function of debris size. The difference between the present value of mission life-cycle cost with and without orbital debris represents the cost impact of the considered orbital debris environments for the same mission initiated at various points in time. The same missions were considered starting in 2000, 2025, 2050, and 2070. All presented results are at a discount rate of 10%. (Discounting is only over the mission 20-year life.)

In all cases considered, the planning horizon for the missions was 20 years, 1000 Monte Carlo simulation runs were performed (i.e., all analytic results are based upon simulating each mission 1000 times), and all are based upon placing and maintaining a single satellite in orbit over the planning horizon.

The following is a general summary of the assumed base-case mission: 1) planning horizon is 20 years; 2) learning rate (recurring cost of replacement satellites) is 90%; 3) probability of launch success (including payload operation) is 0.92; 4) launch cost is $10 million; and 5) satellite is considered as a "black box" with a) mean time to failure of 50 years, b) expected wear-out life of 6 years, c) nonrecurring cost is taken as zero because it is assumed to cancel when comparing alternatives, d) unit recurring cost of $10 million, and e) effective satellite cross-sectional area of 10 m^2. In the absence of orbital debris, this single satellite mission has a net present value (at 10% discount rate) of $73 million. Scaling to other situations is discussed in following paragraphs.

The failure characteristics of the satellite are described in terms of mean time to failure, expected value, and standard deviation of wear out life (i.e., a combined exponential distribution for random failures and a normal distribution for wear-out failures). The debris "subsystem" is characterized by a discontinuous exponential distribution as illustrated in Fig. 4.56 with the λ representing the probability of impact (probability of impact per unit area multiplied by the satellite cross-sectional area perpendicular to its direction of flight) per year associated with the debris scenario. It is assumed that a debris impact will cause total satellite failure independent of the size of the debris particle. The time of satellite failure is thus

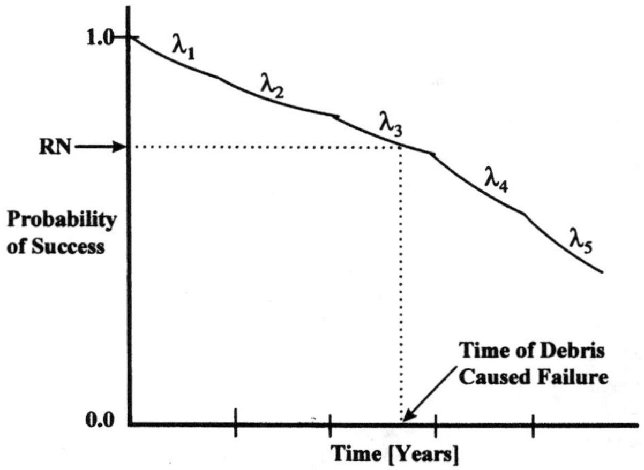

Fig. 4.56 Orbital debris subsystem reliability considerations.

established by generating a random number for each of the satellite subsystems and for the debris subsystem, interrogating the associated reliability curves and establishing the earliest time of failure. This earliest time of failure then becomes the satellite failure time initiating a replacement launch with its associated delays, probability of success and satellite and launch costs.

F. Characterization of Orbital Debris Environment

The Lockheed Engineering and Space Company has under contract to NASA developed a mathematical model for forecasting the orbital debris environment (i.e., probability of impact) as a function of time, altitude, inclination angle, and traffic and mitigation assumptions. [EVOLVE was developed by Dr. Robert Reynolds at Lockheed–Martin Engineering and Sciences Company for NASA Johnson Space Center as an evolution model that defines the environment as a time-varying ensemble of intact space systems and debris fragments each in an orbit and having mass, cross-sectional area, and other associated data. EVOLVE uses as input data the historical record of launch traffic (satellites, upper stages, and operational debris) and mission model data for future traffic, and uses break-up models to determine the distribution of debris fragments resulting from collisions or explosions occurring in orbit. The orbit propagator accounts for atmospheric drag and perturbations associated with the oblateness of the Earth and, for highly elliptical orbits, lunar and solar perturbations.] Results from this model were provided and used in the reported analysis. The results from this model were utilized because this model provided debris forecasts for the very long time frames (50 to 100 years) that are necessary to judge the long-term implications of orbital debris and mitigation policies. The specific forecasts that were utilized are summarized in Figs. 4.57–4.59 (for the business-as-usual debris scenario) and Figs. 4.60–4.62 for an assumed mitigation scenario. These figures present the results of the EVOLVE model in terms of probability of impact per unit area per

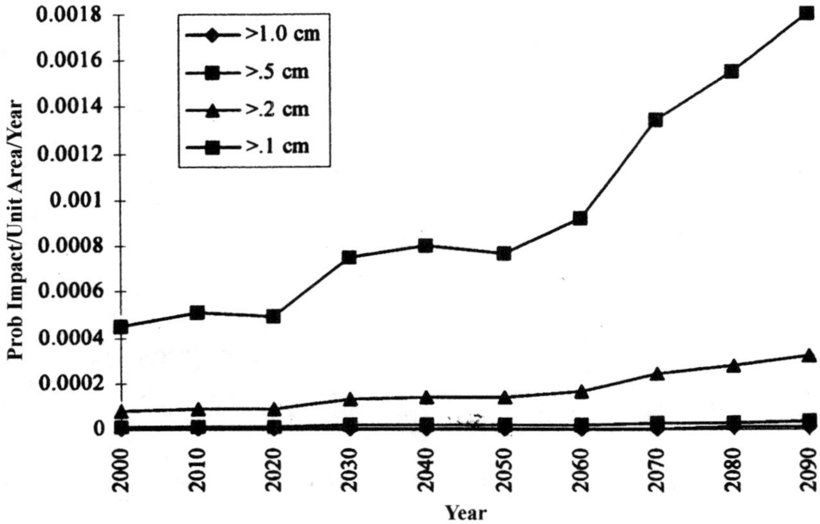

Fig. 4.57 Orbital debris projections (from EVOLVE model) as a function of debris size: business as usual (400 km).

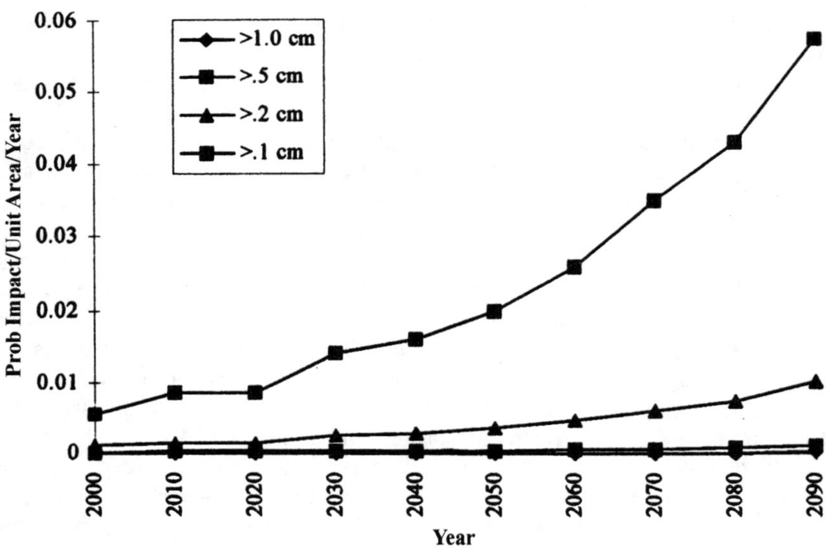

Fig. 4.58 Orbital debris projections (from EVOLVE model) as a function of debris size: business as usual (700 km).

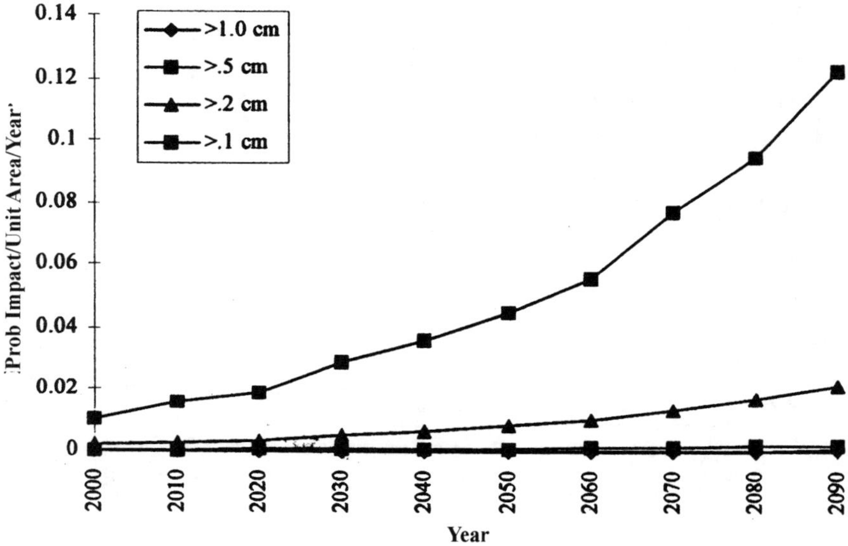

Fig. 4.59 Orbital debris projections (from EVOLVE model) as a function of debris size: business as usual (1000 km).

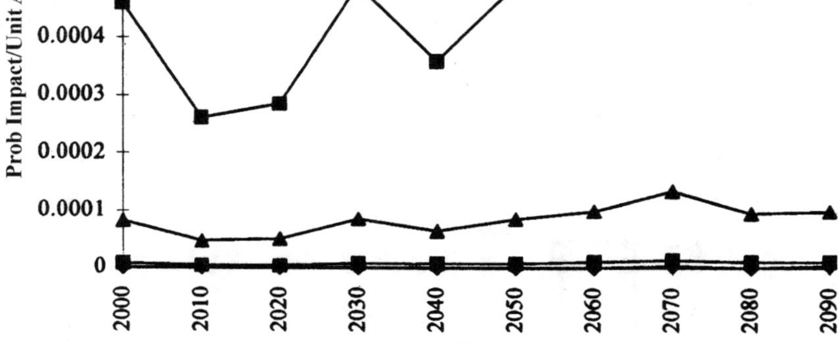

Fig. 4.60 Orbital debris projections (from EVOLVE model) as a function of debris size: a mitigation scenario (400 km).

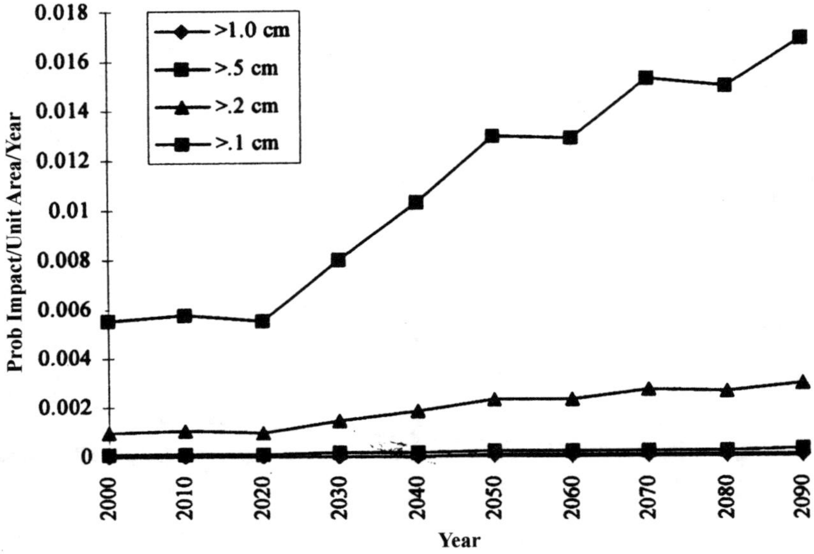

Fig. 4.61 Orbital debris projections (from EVOLVE model) as a function of debris size: a mitigation scenario (700 km).

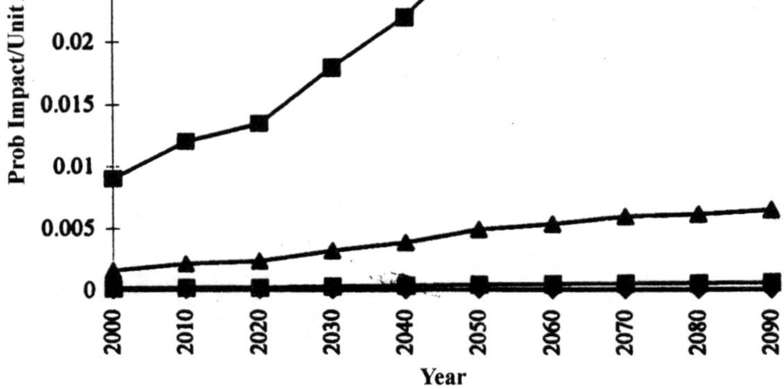

Fig. 4.62 Orbital debris projections (from EVOLVE model) as a function of debris size: a mitigation scenario (1000 km).

year as a function of time and as a function of orbit altitude (400, 700, and 1000 km) for near-polar orbits. Data were provided for each year through 2090. This data therefore provided the specific values of λ as indicated in Fig. 4.56. Each figure indicates the probability of impact for debris greater than 1.0, 0.5, 0.2, and 0.1 cm. Debris data for less than 1.0 cm were scaled according to $d^{-2.5}$. This scaling rule is relative to the 1-cm debris environment and is a reasonable extrapolation of the debris environment to smaller debris sizes (Reynolds, R., private communication Lockheed–Martin Engineering and Sciences Co., Sept. 1996).

The business-as-usual scenario assumes a constant launch rate based upon historical data. It also assumes that explosions will continue to occur at the historical rate and that no mitigation will take place. The mitigation scenario assumes that after 2000 upper stages will have all stored energy removed (i.e., no explosions). This eliminates small particles that result from breakups with collisions being the source of small particles. The scenario also assumes deorbiting of upper stages and payloads starting in the year 2000 and 2030, respectively.

G. Obtained Results

Computed results are presented in Figs. 4.63–4.65 for the business-as-usual debris scenario with satellites at 400, 700, and 1000 km. Computed results for the mitigation scenario are presented in Figs. 4.66–4.68 with satellites at 400, 700, and 1000 km.

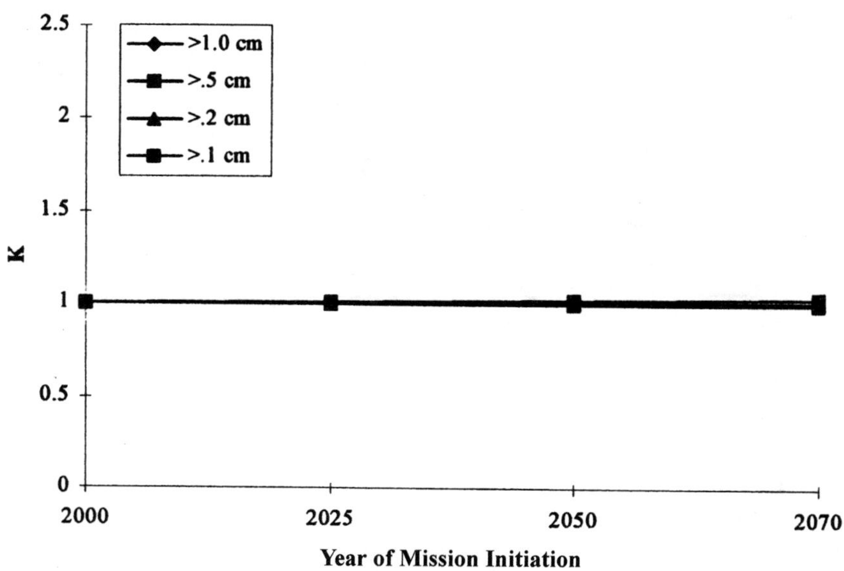

Fig. 4.63 Ratio of mission life-cycle cost *K* with and without debris as a function of debris size and time of mission initiation: business as usual (400 km).

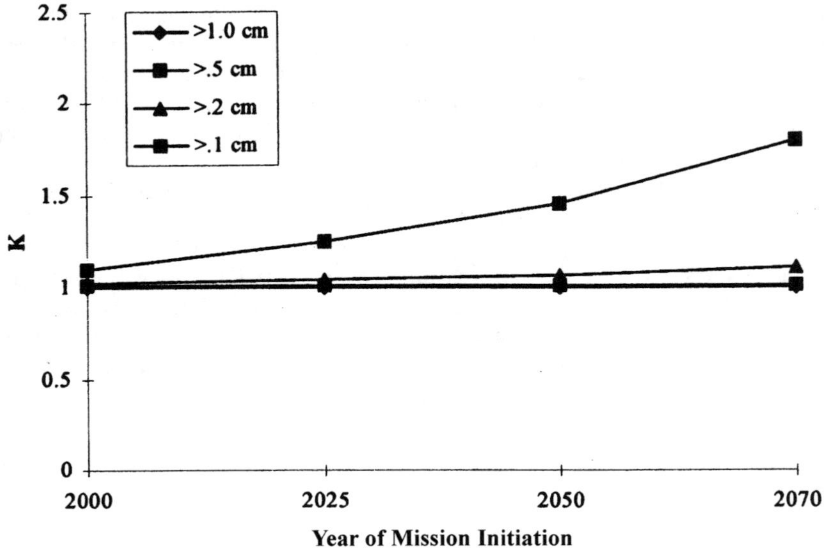

Fig. 4.64 Ratio of mission life-cycle cost *K* with and without debris as a function of debris size and time of mission initiation: business as usual (700 km).

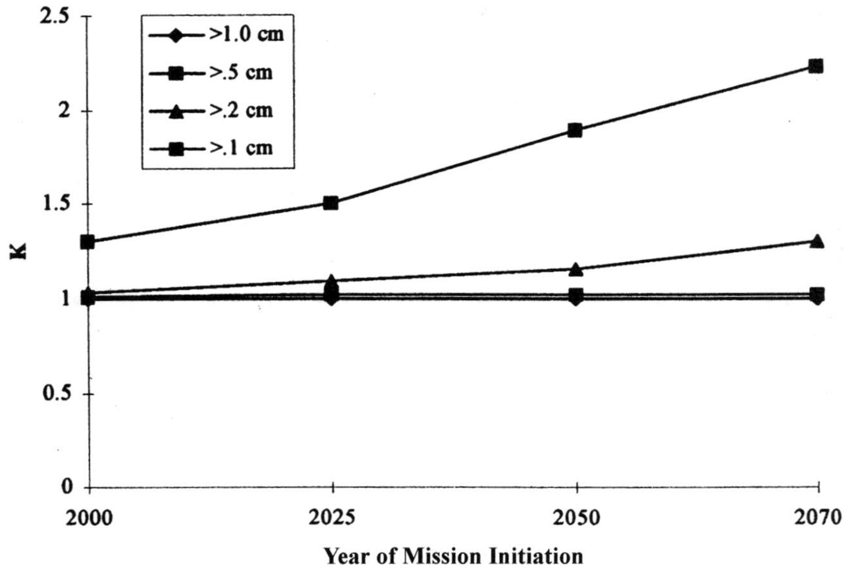

Fig. 4.65 Ratio of mission life-cycle cost *K* with and without debris as a function of debris size and time of mission initiation: business as usual (1000 km).

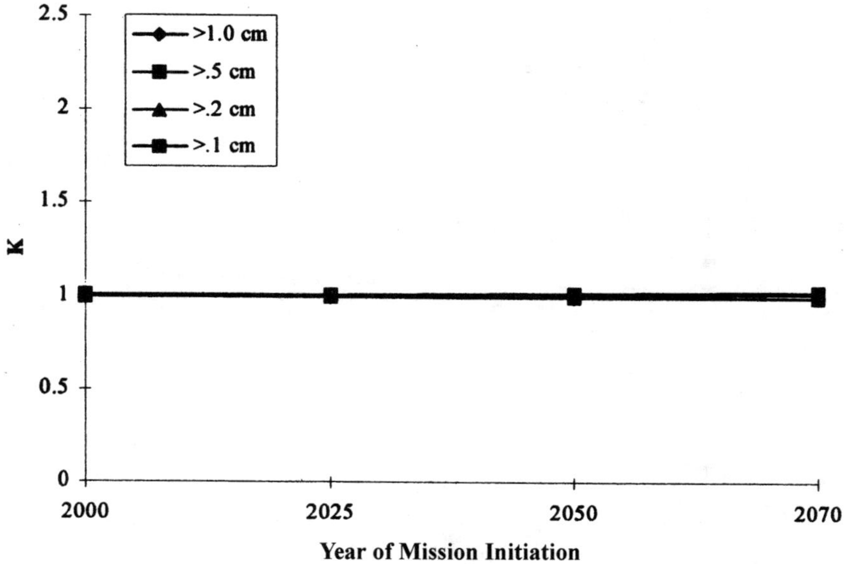

Fig. 4.66 Ratio of mission life-cycle cost *K* with and without debris as a function of debris size and time of mission initiation: a mitigation scenario (400 km).

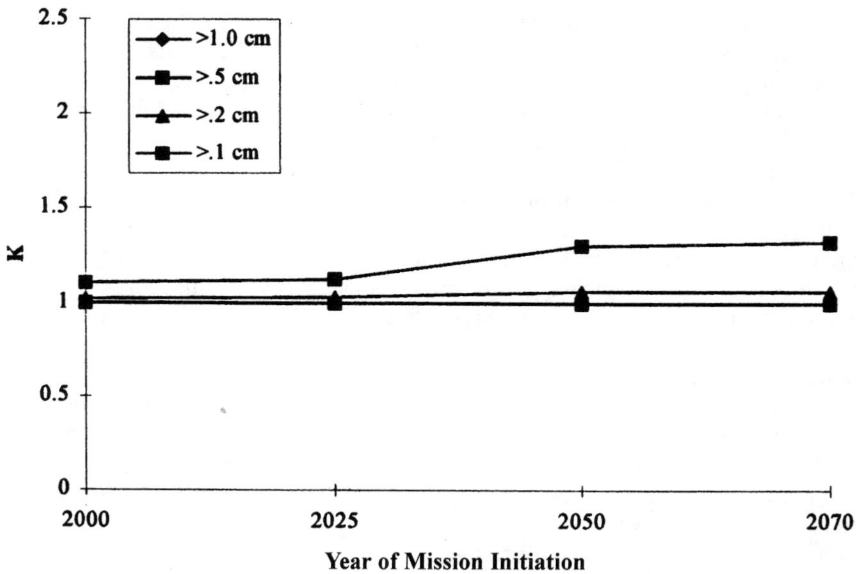

Fig. 4.67 Ratio of mission life-cycle cost *K* with and without debris as a function of debris size and time of mission initiation: a mitigation scenario (700 km).

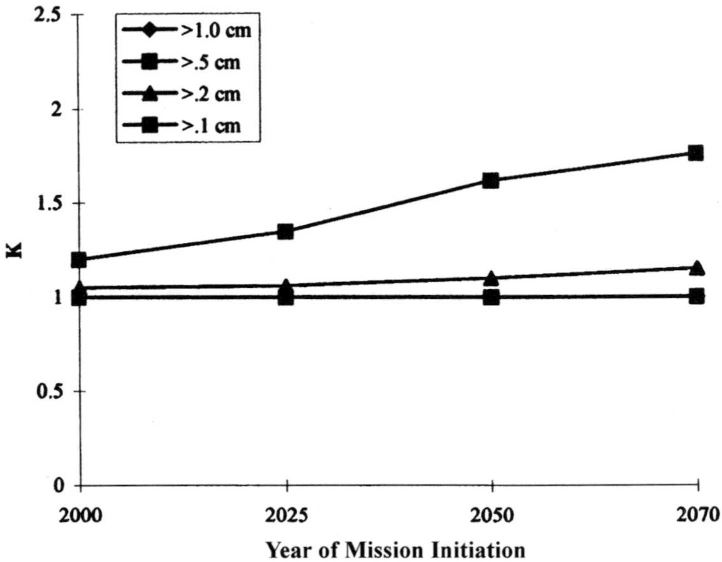

Fig. 4.68 Ratio of mission life-cycle cost *K* with and without debris as a function of debris size and time of mission initiation: a mitigation scenario (1000 km).

All computed results are in terms of K, the ratio of the present value of mission life-cycle costs with orbital debris relative to the present value of mission life-cycle cost if there where no orbital debris. The value of K is a function of when the mission is initiated. The reported results are based upon missions initiated in year 2000, 2025, 2050, and 2070. Values of K for other mission initiation dates can be obtained by interpolation.

The obtained results indicate little or no economic impacts at low (400 km) altitudes, i.e., values of K remain close to 1.0. As altitude increases, values of K increase. (A value of $k = 1.1$ indicates a 10% increase in the present value of mission life-cycle cost.) For the business-as-usual case, values of K increase exponentially with time as driven by the debris forecast that is increasing exponentially with time. The exponential growth is curtailed in the assumed mitigation scenario.

All of the values of K presented in the figures are based upon the assumption that a debris impact, immaterial of size, causes a satellite failure with a subsequent cost-incurring replacement. This restriction can be eliminated, and the probability of destruction given an impact can be considered to be a function of debris size as follows, that is, a new value if K, K^1, can be computed that reflects the probability of satellite failure in terms of debris size.

$$K^1 = 1 + \left\{ \sum_{i=1}^{4} (p_i - p_{i-1}) * (P_i) * (K_i - 1) \right\} / p_4$$

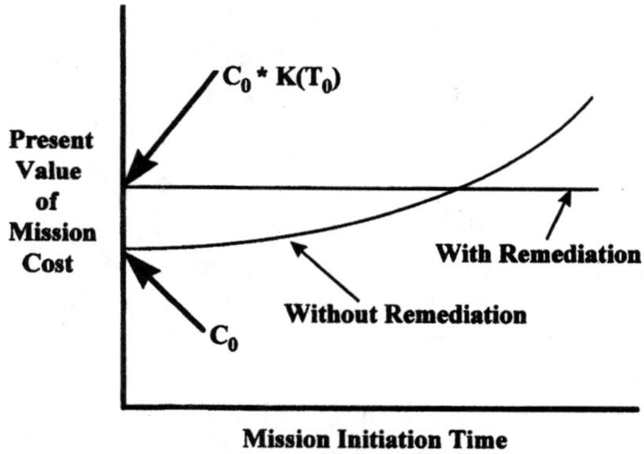

Fig. 4.69 Establishing remediation affordability.

where

p_i = probability of impact of debris (average over time frame of interest) greater than size range i (from the data presented in Figs. 4.57–4.62 with $P_0 = 0$

K_i = value of K in debris size range i (from the results presented in Figs. 4.63–4.68)

P_i = probability of destruction given an impact as a function of size range i

The size ranges i are defined as shown in Table 4.8.

The recurring cost of a satellite (including launch) is assumed to be \$20 million resulting in a net present value of mission life-cycle cost (for the single satellite mission) of \$73 million. This present value scales linearly with satellite recurring cost. In the context of this preliminary analysis, it is assumed that this value will also scale linearly with constellation size.

All of the present results are based upon a satellite cross-section area of 10 m^2. The results can be scaled to smaller satellites in the range of 1 to 10 m^2 as follows. With A representing cross-section area (m^2),

$$K^1 = 1 + 0.1 * A * [K - 1.0]$$

where K represents the value based upon a cross-section area A of 10 m^2 and is obtained from the appropriate figure (Figs. 4.63–4.68).

Table 4.8 Classification of debris size ranges

Size range, i	Debris diameter, cm
1	$1.0 \leq d < \infty$
2	$0.5 \leq d < 1.0$
3	$0.2 \leq d < 0.5$
4	$0.1 \leq d < 0.2$

Note that all scaling rules are consistent with the preliminary nature of the presented results and are provided so that the reader may obtain an understanding of the methodology for assessing the economic impacts of orbital debris while at the same time obtaining a "feel for the numbers."

H. Remediation Affordability (?)

With but small impacts well into the future, how can expenditures be justified today that will reduce the economic consequences of debris that might not become consequential for 50 to 100 years or more? This is basically the same problem that confronts other environmental issues. Delays in making investments in the near term might have dire, and possibly irreversible, consequences in the long term. (This is obviously a function of altitude and inclination angle.)[56] Preliminary results relating to remediation affordability can be developed by establishing the maximum economically justifiable investment that can be made in the near term to solve the long-term problem. (In the near term, mission costs will be increased so as to have reasonable mission costs in the long term.)

The problem, as depicted in Fig. 4.69, is to establish the maximum mission incremental cost increase for debris remediation that will result in the present value of mission costs with and without remediation being equal over the long term. To accomplish this, let C_i represent the present value of mission costs when a mission is initiated in year I, and assume the following:

1) A single new mission (for example, the Landsat mission) is initiated each year.
2) All missions have similar costs.
3) Because of debris (if not remediated), mission costs will increase with time of initiation at a constant or average rate R, such that the $C_i = [1 + R] * C_{i-1}$. (The rate is a function of altitude and inclination angle because the debris environment and satellite traffic are a function of altitude and inclination angle. If the rate of increase is not constant over time, then the equations obviously must be modified to reflect this. In general, if the rate of increase increases with time, then remediation affordability level will increase.)

The present value of mission costs (with debris), PVC, is therefore

$$PVC = C_0 * \sum_{i=1}^{\infty} [(1 + R)/(1 + R)]^i$$

where C_0 is the present value of the life-cycle cost of a mission initiated at the present time.

When $R \geq r$ (that is, the rate of increase in cost is equal to or greater that the discount rate r)

Then

$$PVC = \infty$$

When $R < r$

Then

$$PVC = C_0 * [(1 + r)/(r - R)]$$

Referring to Fig. 4.69, it is evident that the problem is to find the maximum value of $K(T_0)$, such that $[C_0 * K(T_0)$ remaining constant over time is assumed for the sake of simplicity. The actual shape of the C_i curve will depend upon the scope and timing of the remediation policy under consideration and the effect of the remediation policy on the future debris environment and resulting probability of impact as a function of time.]

$$C_0 * K(T_0)/r \leq C_0 * [(1 + r)/(r - R)]$$
$$\text{Max } K(T_0) \leq r * [(1 + r)/(r - R)]$$

If it is assumed that the primary effect of debris remediation will be on unit recurring cost (not nonrecurring cost), then Max $K(T_0)$ represents the allowable increase in unit recurring cost for debris remediation. When r, the real cost of capital, is on the order of 3%, then

$$\text{Max } K(T_0) = 3.09/(3 - R)$$

R is a function of mission, debris forecast, and debris size of concern. (R was taken as the rate of change that fits both the current cost C_0 and the cost in year 2070 C_{70}.) For Landsat-type missions in the altitude and inclination angle regime of Landsat and for small debris,

R, %	Max $K(T_0)$
0.01	1.033
0.02	1.037
0.03	1.040

and it may therefore be concluded that it might be appropriate to increase satellite costs (with all of the preceding caveats) by up to 3 to 4% starting now in order to eliminate the future growth of orbital debris—in other words, remediation actions are economically attractive (for the Landsat regime) if they result in less than 3 to 4% increase in satellite cost.

The major determinant of affordability is the rate of increase of the costs caused by debris. When R is small, little can be afforded for debris remediation, whereas when R is relatively large then vast sums can be affordable for debris remediation. This is similar to saying that when R is small debris has not yet become a problem, and when R is large debris will become a problem. It might therefore be desirable to focus more attention on those missions (altitude and inclination angle) that are likely to have large R values than those with small R values.

I. Summary

The objectives of the preceding paragraphs were to describe a methodology that has been developed for assessing the potential economic impacts of orbital debris and to present preliminary results of an analysis of the potential economic impacts of orbital debris on the life-cycle cost of performing typical LEO missions. Economic impacts were developed in terms of forecasted orbital debris environments that are specified in terms of the probability of debris impact per unit area per year as a function of time and the time of mission initiation.

LEO near-polar missions at 400, 700, and 1000 km were considered. Two debris scenarios and forecasts were considered at each of these altitudes and were obtained from the EVOLVE debris forecasting model. These forecasts were used to establish the ratio of the present value of mission life-cycle cost when the mission encounters the forecasted debris environment relative to the present value of mission life-cycle cost in the absence of orbital debris. This ratio is referred to as K.

Rules were developed for scaling the obtained values of K to different mission initiation dates and different satellite cross-section areas. A method for considering the probability of satellite failure caused by impacting debris was also developed in terms of orbital debris size. All obtained results were for a mission consisting of a single satellite that is placed into orbit and then replaced as random, wear out, and debris-caused failures occur. The obtained values of K, when multiplied by the present value of this mission life-cycle cost and the number of satellites in a constellation, results in a preliminary estimate of the total debris economic impact on a constellation of LEO satellites.

Finally, a preliminary methodology was developed for estimating the affordability of satellite remediation measures. Preliminary results indicate that unit recurring costs can be increased, for Landsat regime missions, by 3 to 4% if there is a resultant large decrease in the orbital debris environment.

References

[1]Greenberg, J.S., and Hertzfeld, H. (eds.), *Space Economics*, Progress in Astronautics and Aeronautics, Vol. 144, AIAA, Washington, DC, 1992, Chap. 2.

[2]Greenberg, J.S., "Reliability, Uncertainty and Risk Assessment of Space Systems—A Methodology for Decision Making," Princeton Univ., AMS Rept. 1085, Princeton, NJ, Dec. 1972.

[3]Greenberg, J.S., and Hazelrigg, G.A., Jr., "Methodology for Reliability-Cost-Risk Analysis of Satellite Networks," *Journal of Spacecraft and Rockets*, Vol. 11, No. 9, 1974, pp. 650–657.

[4]Greenberg, J.S., "A Simulation Analysis of Space Operations," International Academy of Astronautics, Paper 87-621, Oct. 1987.

[5]Dole, S.H., et. al., "Establishment of a Long-Range Planning Capability," Rand Corp., RM-1651-NASA, Santa Monica, CA, Sept. 1969.

[6]Quade, E.S., and Boucher, W.I., (eds.), *System Analysis and Policy Planning*, Elsevier, New York, 1968.

[7]Greenberg, J.S., *Investment Decisions: The Influence of Risk and Other Factors*, American Management Association, New York, 1982.

[8]Greenberg, J.S., "Satellite Configuration, Transportation, Maintenance and Sparing Decisions: The *SATCAV* Model," Third Space Logistics Symposium, May 1990.

[9]Greenberg, J.S., "Satellite Reliability Requirements: Effects of Transportation System, Sparing and Maintenance Strategies," *Acta Astronautica*, Vol. 21, No. 6/7, 1990.
[10]Greenberg, J.S., "*SATCAV*: A Space System Life Cycle Cost and Availability Model," International Academy of Astronautics, Paper 89-694, Oct. 1989.
[11]Greenberg, J.S., McCandless, W., "Design Tool for Small Satellites: SMALLSAT," Space Technology and Applications International Forum, Jan. 1996.
[12]Princeton Synergetics, Inc., "SMALLSAT: A Computer-Aided Phase A Spacecraft Design System," NASA, Final Rept., Princeton, NJ, July 1995.
[13]Greenberg, J.S., and Nichols, R., "Economic Impact of New Technology on Domestic Satellite Communications," Princeton Univ., AMS Rept. 1285, Princeton, NJ, March 1976.
[14]Greenberg, J.S., "Evaluating the Economic Impact of Design Alternatives on Domestic Communication Satellite Ventures," Paper 78-A-38, Oct. 1978.
[15]Greenberg, J.S., "The *DOMSAT* III Communications Satellite Business Venture Financial Planning Model," Ford Aerospace, Final Rept., Princeton, NJ, Jan. 1988.
[16]Greenberg, J.S., "Communication Satellite Business Ventures (Measuring the Impact of Technology Programs and Related Policies)," *Space Policy*, Feb. 1986.
[17]Greenberg, J.S., and Gaelick, C., "Evaluating Satcom Investment Decisions," *Satellite Communications*, Jan. 1988.
[18]Ford Aerospace, "Communication Satellite Systems Operations with the Space Station," NASA CR-179527, 1987.
[19]Price, K.M., Dixson, J.E., and Weyandt, C.J., "Economic Benefits of the Space Station to Commercial Communication Satellite Operators," International Academy of Astronautics, Paper 87-622, Oct. 1987.
[20]Davis, Bob, "Untitled," U.S. Dept. of Defense, The Newsletter of the Space Transportation Association, Space Transportation Association, Vol. 7, No. 2, March/April 1997.
[21]Greenberg, J.S., "Selecting a Launch Vehicle for GEO Missions," AIAA Paper 96-1112, Feb. 1996.
[22]Greenberg, J.S., "Financial/Investment Analysis," *Space Economics*, edited by J.S. Greenberg and H. Hertzfeld, Progress in Astronautics and Aeronautics, Vol. 144, AIAA, Washington, DC, 1992, pp. 3–33.
[23]Greenberg, J.S., and Gaelick, C., "When a Launch Goes to the Competition," *Aerospace America*, Aug. 1995, pp. 36–39.
[24]"Memorandum of Agreement Between the Government of the United States of America and the Government of the People's Republic of China Regarding International Trade in Commercial Launch Services," U.S. Dept. of Commerce, Washington, DC (undated).
[25]Hazelrigg, G.A., and Greenberg, J.S., "Cost Estimating for Technology Programs," International Academy of Astronautics, Paper 91-638, Oct. 1991.
[26]Greenberg, J.S., "Space Transportation Architecture Comparisons Explicitly Considering Uncertainty and Risk," International Academy of Astronautics, Paper 96-IAA.1.1.05, Oct. 1997.
[27]Mankins, J.C, "Maglifter: An Advanced Concept for Reducing the Cost of Earth-to-Orbit Transportation," NASA White Paper, 5 Dec. 1993.
[28]Mankins, J.C., "The Maglifter: An Advanced Concept Using Electro-Magnetic Propulsion in Reducing the Cost of Space Launch," AIAA Paper 94-2726, June 1994.
[29]Mandell, H.C., Jr., "Cost-Estimating Relationships for Space Programs," *Space Economics*, edited by J.S. Greenberg and H.R. Hertzfeld, Progress in Astronautics and Aeronautics, Vol. 144, AIAA, Washington, DC, 1992, pp. 57–87.

[30]Hazelrigg, G.A., "Cost Estimating for Technology Programs," *Space Economics* edited by J.S. Greenberg and H.R. Hertzfeld, Progress in Astronautics and Aeronautics, Vol. 144, AIAA, Washington, DC, 1992, pp. 97–115.
[31]Cramer, H., *The Elements of Probability Theory and Some of Its Applications*, Wiley, New York, 1955.
[32]Davinic, N., Arkus, A., Chappie, S., and Greenberg, J.S., "Cost-Benefit Analysis of On-Orbit Satellite Servicing," *Journal of Reducing Space Mission Cost*, Kluwer Academic Publishers, Netherlands, 1998, 1:27–52.
[33]Greenberg, J.S., "Space System Life-Cycle Cost and Availability Analysis," *Space Economics*, edited by J.S. Greenberg and H.R. Hertzfeld, Progress in Astronautics and Aeronautics, Vol. 144, AIAA, Washington, DC, 1992, pp. 117–147.
[34]Greenberg, J.S., "Economic Implications of Orbital Debris Mitigation (LEO Missions)," International Academy of Astronautics, Paper 97-IAA-6.5.08, Oct. 1997.
[35]Greenberg, J.S. "Economic Analysis Requirements in Support of Orbital Debris Regulatory Policy," International Symposium on Optical Science, Engineering and Instrumentation, Society of Photo-Optical Instrumentation Engineers and The International Society for Optical Engineering, Aug. 1996.
[36]Bernstein, H., and Abbott, A.D., "Space Transportation Architecture Resiliency," Aerospace Corp., Los Angeles, Aug. 1987.
[37]L Systems, Inc., "Space Transportation Analysis," U.S. Air Force, Space Div., Contract F04701-85-C-0055, Los Angeles, 2 March 1987.
[38]Greenberg, J.S., "Space Transportation: An Operability Analysis," International Astronautical Federation, 89-241, Oct. 1989.
[39]Davinic, N., Chappie, S., Arkus, A., and Greenberg, J.S., "Spacecraft Modular Architecture Design Study: Cost Benefit Analysis of On-Orbit Satellite Servicing," International Academy of Astronautics, 97-IAA.1.4.07, Oct. 1997; also *Journal of Reducing Space Mission Cost*, Vol. 1, No. 1, 1998, pp. 27–52.
[40]Greenblat, E., (project manager) "Economic Analysis of Advanced Propulsion Systems," NASA, April 1973.
[41]Greenberg, J.S., and Price, K., "The Economics of Satellite Retrieval," AIAA Paper 88-0843, March 1988.
[42]*AIAA Guide for the Serviceable Spacecraft Grasping/Berthing/Docking Interface*, AIAA, Washington, DC, 1992.
[43]Graves, C.D., and Rosen, A., "Orbital Servicing from an Expendable Launch Vehicle," AIAA Paper A90-13398, 1990.
[44]Hendrick, R., and Teoh, B., "On-Orbit Servicing: Technology and Logistics Integration," AIAA Paper 91-4081, 1991.
[45]Levin, F.M., and Erwin, H.O., Jr., "An Overview of the Office of Space Flight Satellite Servicing Program Plan (NASA)," International Astronautical Federation, Paper 87-35, 1987.
[46]Yasaka, T., and Yoshitsugu, Y., "Geostationary Service Vehicle (GSV) for Economical On-Orbit Servicing," International Astronautical Federation, 89-696, Oct. 1989.
[47]Space Debris Subcommittee of the IAA, "Position Paper on Orbital Debris, Edition 2001," International Academy of Astronautics, Paper No. 24.11.2000, Sept. 2001.
[48]*Orbital Debris Mitigation Techniques: Technical, Legal, and Economic Aspects*, AIAA, Rept. No. SP-016-1992, Washington, DC, 1992.
[49]Committee on Transportation Research and Development, "Interagency Report on Orbital Debris," National Science and Technology Council, Washington, DC, Nov. 1995.

[50]Ad Hoc Expert Group of the IAA Committee on Safety, Rescue, and Quality, "Position Paper on Orbital Debris, Edition 1995," International Academy of Astronautics, 1995.
[51]Reynolds, R.C., and Loftus, J.P., "A Handbook to Support the NASA Policy to Limit Orbital Debris Generation," International Academy of Astronautics, Paper IAA.3.3-93-700, Oct. 1993.
[52]Interagency Group (Space), "Report on Orbital Debris," National Security Council, Washington, DC, Feb. 1989.
[53]"Space Station Program Natural Environment Definition for Design," SSP 30425, Rev. A, NASA, Washington, DC, June 1991.
[54]Boden, D.G., "Introduction to Astrodynamics," *Space Mission Analysis and Design*, edited by J.R. Wertz and W.J. Larson, Kluwer Academic, Norwell, MA, 1991.
[55]Greenberg, J.S., "Orbital Debris and Remediation Policies: Economic Considerations," International Academy Astronautics, IAA.3.3-93-702, Oct. 1993.
[56]Greenberg, J.S., and Reynolds, R.C., "Orbital Debris: Economic Impacts on Remote Sensing Missions," International Astronautical Federation, Paper No. IAA-94-IAA. 1.3.602, Oct. 1994.

Chapter 5

Licensing and Regulatory Issues

A NUMBER of licensing and regulatory issues and processes relating to commercial space launches and commercially operated launch sites are discussed in the following paragraphs. Emphasis is placed upon the need for and form and scope of economic impact analyses that are required in support of regulatory actions. Although a number of different agencies regulate space-related activities, the discussion presented in the following pages considers primarily the Federal Aviation Administration's (FAA) role in regulating launch and related activities. [The Federal Communications Commission (FCC) is responsible for regulating, among other things, communication satellites, and the National Oceanic and Atmospheric Administration (NOAA) regulates remote-sensing satellites. This is discussed further in Sec. III with regard to orbital debris.] Section III broadens the scope somewhat via a discussion of orbital debris and the scope and methodology of related regulatory economic analysis.

To demonstrate regulatory economic impact analysis, the FAA's rule making relating to financial responsibility is described. This rule making was in response to a statutory requirement for establishing insurance requirements for commercial space launches. The insurance requirements consider both insurance to cover potential damage to government facilities and potential third-party losses (i.e., persons and property that are not participants in the space launch and, in general, includes the general public). The statutory indemnification requirements and interparty waiver of claims are discussed, as are the establishment of current practice, the use of maximum probable loss in the setting of insurance requirements, the form of and role of risk assessment used to establish the probability of and associated levels of damage, the setting of threshold probabilities (i.e., the acceptable levels of risk for third parties and for government facilities), the identification of alternatives and the development of economic impacts (i.e., costs and benefits).

As the number of Earth-orbiting objects increases, so does the potential for generating orbital debris with the consequent increase in the likelihood of impacting and damaging operating satellites. Various debris remediation approaches are being considered, which encompass both in-orbit and return-to-Earth schema and have varying degrees of operations, cost, international competitiveness, and safety implications. Because of the diversity of issues, concerns, and long-term impacts, there is a clear need for the setting of government policies that will lead to an orderly abatement of the potential orbital debris hazards. These policies might require the establishment of a supportive regulatory regime. This section describes the various agencies that might be involved in orbital debris regulatory actions, the myriad of issues concerning the need for establishing government policies relating to orbital debris remediation, proposed technical solutions, and their economic and safety implications. Particular emphasis is placed upon addressing cost-effectiveness and economic analysis methods and techniques as they relate to economic impact analysis in support of regulatory impact analysis.

Finally, the government's role in providing third-party liability indemnification is considered. The space insurance industry has, for many years, provided an array of insurance products for the space industry. These products have included facility insurance, third-party liability insurance, launch and satellite insurance, on-orbit insurance, as well as other products. Of particular concern is the role of the government in augmenting the provision of third-party liability insurance by providing indemnification for damage to third parties from space launch and re-entry operations. Recent legislation has extended indemnification but has raised questions relating to its need and other possible alternatives. (H.R. 2607, the *Commercial Space Transportation Competitiveness Act*, extended launch indemnification to the U.S. commercial launch industry for four more years, through the end of 2004. This and its predecessor legislation has placed the government in a risk-sharing role with the space launch industry.) Special emphasis is placed upon third-party liability indemnification issues, the need to develop a long-term course of government action within a competitive environment, and possible alternatives to the current indemnification regime. This is likely to lead to the need for additional regulatory economic impact analyses.

I. Regulatory Economic Impact Analysis[1,2]

A. Requirements

A regulation proposed by a government agency will usually require changes in the way individuals, firms, and others act or conduct business. Therefore, the government agency must carefully determine who will be affected by the regulations and what will be the costs and benefits that are the result of the regulation. FAA regulatory evaluations and assessments that are prepared as part of the rulemaking process are used to answer these questions and must comply with Executive Order 12866, "Regulatory Planning and Review" (1980), U.S. Department of Transportation Order 2100.5, and the *Regulatory Flexibility Act*.

Executive Order (E.O.) 12866 establishes a regulatory philosophy and the definition of a significant regulatory change. A "significant" regulatory action is deemed to have an annual effect on the economy of $100 million or more or adversely affect in a material way the economy, a sector of the economy, productivity, competition, jobs, the environment, public health or safety, or state, local or tribal governments or communities. If a regulatory change is significant, then E.O. 12866 requires that an assessment be prepared that includes the following:

1) Identifying the problem that the regulation addresses and the significance of the problem,
2) Examining whether existing regulations have created or contributed to the problem the regulation is intended to correct and whether those regulations should be modified,
3) Identifying and assess available alternatives, including economic incentives and performance-based approaches,
4) Designing regulations in the most cost-effective manner,
5) Assessing the costs and benefits of the intended regulation.

If a regulatory action is judged to be significant, then an assessment must be made of the costs and benefits of the intended regulatory action and potentially effective and reasonably feasible alternatives identified. If a regulation is not judged to be significant, U.S. Department of Transportation (DoT) Order 2100.5 requires that a regulatory evaluation (as opposed to a regulatory assessment) be prepared. The evaluation consists of a cost and benefit analysis of the intended regulatory action and of possible international trade implications, but does not require an assessment of alternatives.

The *Regulatory Flexibility Act of 1980* requires that a regulatory flexibility analysis be prepared when there is a significant impact on a substantial number of small entities. A small entity can be a small business, small organization, or small governmental jurisdiction. The intent of the Regulatory Flexibility Act is to ensure that government regulations do not unduly restrict the ability of small firms or other entities to compete. If a proposed rule is not expected to have a significant economic impact on a substantial number of small entities, either adversely or beneficially, the agency is not required to perform an initial regulatory flexibility analysis.

A typical benefit/cost analysis (assessment) performed in support of a proposed rulemaking includes a discussion of alternatives, analysis of costs, benefits, comparison of costs and benefits, economic impact assessment, regulatory flexibility analysis, international trade impact assessment, and unfunded mandates assessment. The analysis of costs and benefits includes the identification of the base case or baseline to which the postregulation situation is compared, the types of costs and benefits expected, all of the assumptions and methodologies used in the calculation of the benefits and costs, and presents the quantitative results of the calculation of the costs and benefits.

The discussion of alternatives defines the rational and resulting alternatives that have been selected for analysis of costs and benefits and cites reasons for rejecting other possible alternatives. The regulatory flexibility analysis measures the impacts on small entities, the international trade analysis includes an assessment of the impact of proposed and final rules on trade opportunities for the U.S. abroad, as well as for foreign firms in the U.S., and the unfunded mandates assessment is concerned with the effects of the proposed rulemaking on expenditures by state, local, and tribal governments.

B. Establishing a Base Case

To estimate the costs and benefits that can result from a regulatory action, it is necessary to "draw a line in the sand" from which the costs and benefits will be measured. The line in the sand is referred to as the base case and represents the scenario or situation that would likely result in the absence of the establishment of the proposed regulatory regime. The base case is usually grounded in current practice but can reflect conditions and decisions that are likely to develop in the absence of the proposed action. Two examples, drawn from recent FAA regulatory actions, are presented in the following paragraphs to illustrate the diverse considerations that have been employed in the establishment of base cases utilized in recent economic impact analyses. A third example illustrates base-case considerations should it be necessary to establish an orbital debris regulatory regime.

1. Financial Responsibility Requirements for Licensed Launch Activities[3]

Because of safety concerns and international agreements to which the U.S. government is a party, legislative action was taken [Commercial Space Launch Act (SLA) of 1984], which made it a requirement for commercial entities to obtain a license from the DoT (in 1995 this authority was delegated to the FAA) in order to undertake launch and related activities. The resulting licensing process addresses these safety concerns and establishes procedures for demonstrating compliance with those requirements and for implementing risk allocation provisions. The establishment of a financial responsibility regulatory regime is, to a large measure, dictated by the provisions of legislation; in particular, 49 U.S.C. Subtitle IX, Chapter 701, *Commercial Space Launch Activities*, formerly the *Commercial Space Launch Act of 1984* as amended. The financial responsibility requirements levied on licensees require that an FAA-determined amount of third-party liability insurance (or equivalent) be obtained to cover potential damage to third parties with the amount not to exceed $500 million or the maximum amount of liability insurance available on the world market at a reasonable cost. In addition, insurance is required in an amount determined by the FAA, but not to exceed $100 million, to cover potential damage to government facilities. The risk-based method for determining the required levels of insurance is specified in the legislation. (This is described in Sec. II.) The legislation also directs that cross-waivers be established between the licensee, its contractors, subcontractors, and customers under which each party to each waiver agrees to be responsible for any property damage or loss it sustains or for any personal injury to, death of, or property damage or loss sustained by its own employees resulting from activities carried out under the license.

The DoT issued general license orders that included financial responsibility requirements for insurance to protect the U.S. government from claims by third parties and for damage or loss to government property. These orders were issued in 1988, shortly before the enactment of the *Amendments to the Space Act*, for the three major commercial launch vehicles at the time (Atlas, Delta, and Titan). For establishing a base case, it was assumed that these orders represented DoT's intent with regard to financial responsibility requirements. In other words, the DoT developed a risk-based approach for determining financial responsibility requirements. Also, the DoT elected to exercise its authority to issue financial responsibility requirements to protect the government, as well as fulfilling the Act's directive that it set requirements to cover claims by third parties. Thus, the base case is considered as current practice that was established by the DoT prior to the 1988 amendments and would thus, in all likelihood, be continued even if there were no formal requirement for establishing a formal regulatory regime. The base case also included a launch forecast (disaggregated by type of launch vehicle) that was provided by the DoT.

2. Financial Responsibility Requirements for Licensed Reentry Activities

The *Commercial Space Act of 1998* (CSA), Public Law 105-303, granted the DoT with the authority to license and regulate reentry vehicle operators and the operation of reentry sites by a commercial or non-Federal entity and thereby extended the DoT's authority beyond their previous authority associated with expendable launch vehicles. The statutory objective in licensing reentry activities is to ensure that public health and safety and the safety of property are not jeopardized, as well as to ensure consistency with U.S. national security and foreign policy interests. As a result of the statutory requirement, the FAA proposed implementing rulemaking that specifies the financial responsibility requirements which are applicable to commercial reentry activities authorized under a license issued by the FAA.

Because the financial responsibility rule making associated with space launch activities associated with expendable launch vehicles is a *fait accompli*, there exists a considerable history of licensed commercial space launches, and reentry licensing will draw significantly upon the existing current practice for licensing of space launches. *The economic impact analysis is concerned with assessing the potential economic impacts of the current rule making relative to current practice, and the economic impacts are the result of a comparison of licensed reentry with the proposed regulatory action vs licensed reentry without regulatory action.*

Because there is little history associated with the licensing of reentry activities [with the possible exception of the FAA's efforts to license the Commercial Experiment Transporter (COMET) and the related reentry vehicle system (METEOR); the COMET program was terminated, and the METEOR reentry vehicle was destroyed as a result of a launch failure], there is no definitive "current practice" that can be pointed to as the basis for establishing a base case. However, it is reasonable to assume that the FAA's financial responsibility methods and procedures developed for the licensing of commercial space launches could be used to identify the likely FAA practice that would be employed for establishing financial requirements for the licensing of reentry activities and that the same levels of safety and related practices would be considered.

The following factors were considered in establishing a base case (i.e., FAA practice in the absence of a formal rulemaking) for licensing of reentry activities:

1) An FAA study, completed in 1995, evaluated the sufficiency and applicability of CSLA financial responsibility requirements to licensed reentry operations. The study evaluated the appropriateness of using the maximum probable loss (MPL) risk-based methodology (described in following paragraphs) and appropriateness of statutory ceilings on launch requirements (i.e., $500 million for liability and $100 million for government property) for establishing financial responsibility requirements for reentry activities. The study concluded that the use of the MPL methodology was valid and appropriate and that statutory

ceilings established for launch operations were adequate for reentry operations. [The study also concluded that if the $500 million liability ceiling were not sufficient to adequately address reentry liability then perhaps it would suggest that the reentry proposal under review (i.e., a license application) was too hazardous and risky to be authorized by the Agency.]

2) The FAA has in the past and currently utilizes successfully an MPL risk-based approach for establishing the level of financial responsibility for launch operations.
3) Because the financial responsibility requirements for reusable launch vehicle (RLV) launch activities will be considered by the FAA in the same manner as expendable launch vehicle (ELV) launch activities, and because both launch and reentry actives must be licensed, and because launch and reentry activity can be included in a single license, it would seem reasonable that in the absence of a rule making the FAA would use a common approach (MPL risk-based methodology) for establishing financial responsibility for the launch operations and the reentry operations.
4) Because both RLVs and ELVs can launch reentry payloads, the base case should be amenable to considering both RLVs and ELVs.
5) There is a statutory mandate to utilize an MPL-based methodology for establishing financial responsibility for licensed commercial reentry operations.
6) Even though reentry operations have not been licensed previously and financial responsibility determinations have not been made in support of license applications, safety is still of concern, and risk assessments would be required by the FAA in support of license applications. Even in the absence of a regulatory requirement, the FAA would most likely base their financial responsibility analyses upon an MPL-based methodology as directed by the statute. Thus, with or without regulation, safety concerns and statutory responsibility would necessitate the establishment of a formal method for establishing financial responsibility requirements.

It was therefore concluded that the proposed rule making, with but minor exception, codifies existing practice. That is, in the absence of the proposed rule making current practice of the FAA has evolved to the point that the following applies:

1) The same MPL risk-based methodology as used for establishing financial responsibility requirements for space launch activities would be utilized for establishing financial responsibility requirements for reentry-related activities.
2) The same threshold probabilities (for third-party and government property damage) would be used for determining insurance requirements for reentry and launch-related activities.
3) The same statutory requirements concerning insurance availability at reasonable prices and levels of government property and third-party liability indemnification would apply for reentry and launch-related activities.

4) Cross-waivers would be required for all parties involved in space launch and reentry activities. (Because much of the reentry traffic would likely involve RLVs and because launch operations and reentry operations success can be correlated, the interparty waivers would include both those involved in launch and reentry activities.)

The duration of required insurance is also of concern. Because current practice for licensing launches is to require insurance to remain in force for a period of 30 days, current practice, in the absence of the proposed rulemaking, would likely have evolved to include a requirement for insurance coverage for reentry activities to remain in force for a period of 30 days from the initiation of reentry operations. This differs from the requirement contained in the proposed rulemaking, which states that the MPL methodology will be extended to determine the required duration of insurance for reentry operations. That is, the proposed rulemaking states that the MPL risk-based methodology will be utilized to determine both the required amount of insurance and the duration for which the insurance will be required. The duration of coverage will be established, utilizing risk analysis techniques, such that the threshold probability will not be exceeded. This determination would be made at the same time that the required level of insurance would be determined and would thus be mission, vehicle, etc., dependent. Because of the difference between the rulemaking requirement and current practice, resulting costs and benefits need to be established.

3. *Orbital Debris Remediation Requirements*

The establishment of a base case for orbital debris remediation related rulemaking is complicated by the number of agencies that appear to have some degree of regulatory authority with regard to orbital debris. The FAA regulates launch and reentry and related activities. NOAA regulates remote-sensing satellites, and the FCC regulates communication satellites. Because launch vehicle design, launch and reentry operations, as well as satellite design and operations all are potential contributors to the orbital debris environment, the individual and combined regulatory regimes of these three agencies will ultimately affect the debris environment, and their proposed rulemakings will require supportive economic impact analyses with their required establishment of base cases. Both the FAA and NOAA have taken initial steps toward establishing regulatory regimes.

The FAA has issued a final rulemaking, which states:

> To obtain safety approval, an applicant must demonstrate for any proposed launch that for all launch vehicle stages or components that reach Earth orbit [a] There will be no unplanned physical contact between the vehicle or its components and the payload after payload separation; [b] Debris generation will not result from the convergence of energy sources into energy that fragments the vehicle or its components. Energy sources include chemical, pressure, and kinetic energy; and [c] Stored energy will be removed by depleting residual fuel and leaving all fuel line valves open, venting any pressurized system, leaving all batteries in a permanent discharge state, and

> removing any remaining source of stored energy. Other equivalent procedures may be approved in the course of the licensing process.[4]

According to the FAA, these requirements can be adequately implemented utilizing current practice. Thus the base case consists of debris remediation measures that are standard industry practice and include maneuvering launch vehicle orbital stages after payload separation to minimize the likelihood that the orbital stage will recontact the payload, rendering liquid fueled orbital stages as inert as possible by expelling all propellants and pressurants and protecting batteries from spontaneous explosions, etc.

The approach taken by the FAA was to develop regulatory requirements that implement current practice and not to establish requirements that achieve a stated orbital debris remediation goal such as achieving a specified probability of impact per unit time per unit cross-sectional area as a function of time. (This is discussed further in Sec. III.) In other words, the approach taken states that you cannot do worse than you are currently doing but does not state how much better you have to do to solve the problem. In any event, because of the regulatory approach taken the base case is simply current practice, and, as a result, there are no economic impacts.

The Department of Commerce (NOAA) issued a Notice of Proposed Rulemaking (NPRM) in 1997.[5] This NPRM, aimed at establishing a regulatory regime for the licensing of private land remote-sensing space systems, stated:

> A Licensee shall dispose of any space platforms owned or operated by the licensee upon termination of operations under the license in a manner satisfactory to the President. The licensee shall obtain approval from the Administrator of all plans and procedures for the disposition of such platforms, e.g., uncontrolled reentry, burn on reentry or controlled de-orbit, providing sufficient notification to allow determination that the proposed procedures will minimize orbital debris and not jeopardize safety.

NOAA apparently did not conduct an economic impact assessment prior to issuance of its NPRM, which establishes its orbital debris remediation requirements. This, when pointed out to NOAA, resulted in a delay in the issuance of formal requirements.

But what should the base case be for remote-sensing satellites? Because current practice for remote-sensing satellites, as well as with but rare exceptions all other satellites, is not to dispose of the satellites at end of life, a strong case can be made for assuming that such a situation would likely be continued in the absence of the proposed regulatory regime. However, to measure the costs and benefits of the proposed regulatory regime it is necessary to go further. It is necessary to forecast the probability of debris impact per unit time per unit cross-sectional area as a function of time without the proposed regulatory regime and with the proposed regulatory regime. It can immediately be seen that there is a further complication. Because the current environment is such that there are three agencies that are involved in establishing orbital debris regulatory regimes, it is necessary to include in both the base-case forecast the impacts that the other agencies will have on the probability of impact as a function of time. The importance of the forecast of the probability of impact without and with the proposed regulatory regime will be clarified in Sec. III.

II. Financial Responsibility

A number of U.S. states and foreign countries have developed, are developing, or thinking about developing spaceports (for example, Florida, California, Alaska, Idaho, Australia, Canada) from which space launches will take place. Because spaceports provide services that can place government property and personnel and third parties at risk, there is a need to understand and control risk and ensure that desired levels of public safety are achieved and to understand the current regulatory regime that is imposed by the FAA. In addition, if damage to persons and/or property results from or because of launch operations/services it is necessary to ensure that adequate resources are available to provide necessary compensation.

The objective of this section is to describe methods and considerations for establishing financial responsibility requirements for launch and related operations. Specifically discussed are methods and considerations for establishing the magnitude of resources that might be required to provide compensation for damages that can be the result of launch operations and the use of insurance for ensuring the availability of the compensation resources. Because of the significant experience of the U.S. Department of Transportation in licensing commercial space launches and in setting financial responsibility requirements, a summary is presented of U.S. policy and DoT/FAA's methods/procedures for establishing financial responsibility requirements. Finally, a spaceport's insurance concerns are addressed in terms of financial responsibility requirements, government policy impacts on commercial endeavors, and implications of insurance policies on launch service providers and their utilization of the spaceport.

A. International Treaties Relating to Liability

Government participation in international treaties, concern for public safety, and the desire of the government to be reimbursed for damage caused by commercial launches from government facilities have resulted in the setting of policies and procedures relating to licensing and financial responsibility. The 1967 Outer Space Treaty and Liability Convention have been influential in the setting of financial responsibility requirements. Both the 1967 Outer Space Treaty and the Liability Convention require that a launching country be liable for any damages caused by an object that is launched from that country. These treaties provided some of the impetus for NASA's requirement that commercial users of the space shuttle and ELV's procure liability insurance:

1. Treaty on Principles Governing the Activities of States in the Exploration and Use of Outer Space, Including the Moon and Other Celestial Bodies

The so-called "Moon Treaty" states that each State party to the Treaty that launches or procures the launch of an object into outer space, or from whose territory or facility an object is launched, is internationally liable for damage caused by the launched object (or its component parts) on the Earth, in air space or in outer space, to another State party to the Treaty or to its citizens.

2. Convention on International Liability for Damage Caused by Space Objects, United Nations, 9 October, 1973

The Liability Convention allocates liability for damage caused by space objects. It states that a launching State is absolutely liable for damages caused by its space objects (or space objects launched by its nationals) on the surface of the Earth or to craft in flight. In the event of damage being caused elsewhere, the launching State is liable if fault can be shown. The Convention also provides for the establishment of a claims commission to regulate and decide these claims. This treaty formed the basis for consequent NASA action regarding third-party liability insurance requirements.

Thus, international treaties and conventions imply that governments will ultimately be liable for damage caused by launched objects. It is thus in the government's interest to ensure that commercial endeavors associated with space launches have adequate financial resources should an accident occur (that could cause a government liability) in keeping with the level of risk that the government is willing to accept.

B. Definitions

Prior to embarking upon a discussion of the financial responsibility regulatory process it is helpful to define a number of commonly utilized terms. These definitions are presented here.

Third party is defined in the *Commercial Space Launch Act Amendments* as "any person or entity other than the United States, its agencies, or its contractors or subcontractors involved in launch services; the licensee or transferee; the licensee's or transferee's contractors, subcontractors, or customers involved in launch service; or, any such customer's contractors or subcontractors involved in launch services." This definition, by not listing the employees of the U.S. government with those excluded from the list of third parties, defines government employees involved in providing launch services as third parties.

For the purposes of calculating maximum probable loss (to be discussed in following paragraphs), third party includes any person other than on-range employees of an entity involved in the licensed commercial support and launch activities under consideration. The term also includes on-range government and government contractor employees; although these persons might be involved in the licensed activities, they are by statutory definition considered third parties. Other on-range third parties include the personnel of commercial launch firms not involved in the activities under consideration and any other persons or entities that are not private party participants (where private party participants include contractors and subcontractors) in the licensed activities. (Range policies generally prohibit the exposure of such third parties to significant support and launch risks.) All persons off of the launch range are considered to be third parties.

Third-party liability claims refer to claims by a third party for death, bodily injury, or loss of or damage to property resulting from activities carried out under the license in connection with any particular launch.

MPL is the maximum magnitude of loss such that there is less than a specified probability (i.e., the threshold probability) of exceeding this level, as illustrated in Figs. 5.1 and 5.2. This is also referred to as the threshold accident.

Threshold probability represents the probability that loss or damage will exceed a specified level. The threshold probability is a quantitative measure selected by the FAA as representing the probability of occurrence associated with "unlikely" events of levels of damage caused by launch and launch-related activities. The values currently in use by the FAA are 10^{-5} for government property and 10^{-7} for third-party damage. These values were selected so as to be less than other risks faced by the general public.[6] The threshold probability is illustrated in Figs. 5.1 and 5.2 as the cross-hatched areas under the loss probability density function.

Licensed activities as used in the following paragraphs refers to the activities performed on a launch range by a commercial firm that are or are expected to be covered by an FAA license. Such activities can include launch activities, support activities in preparation for a specific launch, or the commercial activities of a company with a permanent facility on a launch range. Licensed activities can also include activities that are specifically required prior to launch for the preparation of a reentry vehicle. For example, at federal launch ranges it is the FAA's intent to cover all activities conducted by commercial firms engaged in commercial launch and related activities.

Finally, licensed activities are not necessarily limited to a launch range, for example, airborne launches can include as licensed activities the preparation and takeoff of associated aircraft from an air field not generally used for space launches. With respect to reentry activities, licensed activities are assumed to include operations from preparation of the reentry system for reentry through impact and removal of hazardous materials (if any).

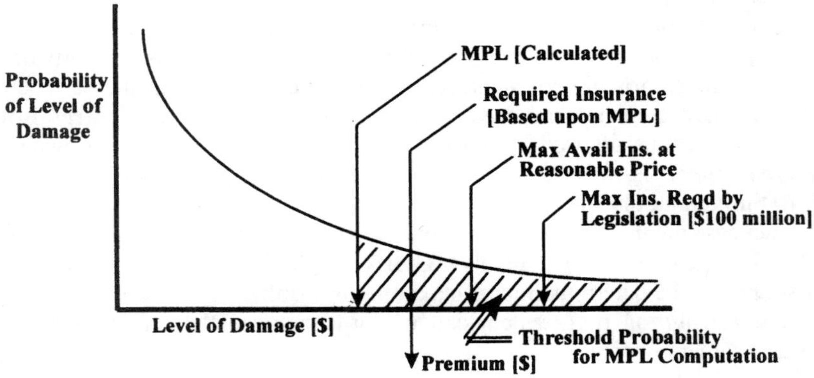

Fig. 5.1 Hypothetical probability distribution of government facility damage (function of launch vehicle and range).

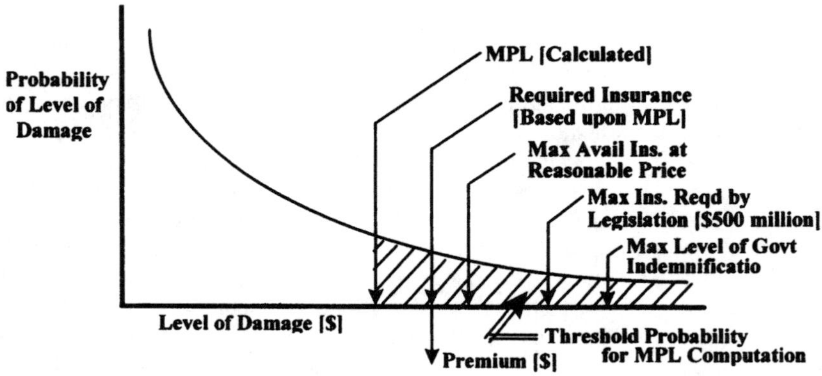

Fig. 5.2 Hypothetical probability distribution of third-party damage (function of launch vehicle and range).

Launch activities are those activities that occur during the launch period for a specific launch. The launch period generally begins at ignition and continues through flight and orbit. For orbital missions, launch activities are considered to be completed 30 days after the launch in most cases. For suborbital launches launch activities are considered to be completed when the launch vehicle returns to Earth, or when payload and motor impact occur when payload recovery does not take place and the completion of payload recovery operations when recovery does take place, whichever occurs later.

Support activities begin with the arrival of a firm (its vehicle components, payload, or personnel) on the launch (or reentry) range in preparation for commercial launch or reentry operations. Support activities end when the licensed entity leaves the launch range or reentry operations have been completed.

Government property includes fixed property, such as buildings and launchpads, and nonfixed property such as equipment and launch vehicle and spacecraft components, which are owned by the government or government contractors on the launch or reentry range and are considered to be government property for the purposes of determining maximum probable loss. Property of the launch range is normally not included, except, for example, special cases such as airborne space launches and specifically identified government property. In the former case the aircraft used to launch the space vehicle (if the aircraft is owned by the government) is to be considered government property (for the purposes of MPL) throughout the launch, even when the aircraft is not on the launch range. In the latter case the intent is to consider only those government properties that are of high value and exposed to relatively high risk. In addition, transient property is not considered in the determination of maximum probable loss because the exposure of the transient property (e.g., payload and/or launch vehicle on an adjacent pad) is a government choice.

Casualty refers to a person suffering death or serious injury as the result of an accident associated with licensed activities. The cost of a casualty is estimated as $3 million for the purposes of determining MPL.

Property damage (for government property) refers to damage to fixed and nonfixed property owned by the government or its contractors. For property owned by third parties, property damage is defined as any damage to any property. The cost of property damage is calculated when possible based on the replacement value of the property. As in the preceding definition of government property, transient property is not considered.

C. Establishing Financial Responsibility Requirements

The methodology for setting financial responsibility requirements for commercial launch activities was developed to protect third parties and government property exposed to potential damage or loss during commercial launch activities. The goals were to protect the general public and government against losses that might occur from launch and reentry activities, establish conditions that make it very unlikely that the government would be called upon to augment the financial requirements imposed by the government upon industry, and to provide financial responsibility requirements upon industry that are not onerous and within which industry can comfortably operate. (This discussion is based upon methods and experience in dealing with commercial launches from government facilities. When dealing with commercial spaceports, issues relating to indemnification emerge. This is discussed in subsequent pages.) In the following paragraphs emphasis is placed upon launch operations. Similar methods and procedures are appropriate for establishing the financial responsibility requirements and economic impacts for reentry operations.[7]

These goals were achieved for the commercial launch activities through legislation (the *Commercial Space Launch Act of 1984*, as amended in 1988). The *1988 Amendments to the Act* set the basic requirements for insurance for commercial space transportation activities at federal and other facilities. These requirements encompassed financial responsibility to cover both damage to government facilities and third parties (persons and property) resulting from the licensed activities. The insurance requirements are based upon the computation of maximum probable loss, MPL (the maximum magnitude of loss such that there is less than a specified probability, referred to as the threshold probability, of exceeding this level), consideration of the maximum available insurance at reasonable prices, and the maximum insurance required as stated in the enabling legislation. ("Threshold probability" represents the probability that loss or damage will exceed a specified level. The threshold probability is a quantitative measure selected by the FAA as representing the probability of occurrence associated with "unlikely" events or levels of damage caused by launch activities.) A typical relationship between MPL, required insurance (based upon MPL), maximum available insurance at reasonable prices, and maximum insurance requirements and maximum level of government indemnification are illustrated in Figs. 5.1 and 5.2. The relative position of these are determined by many factors including the characteristics of the launch vehicle, mission, launch facility and range, and the host of market factors that affect insurance availability and price.

Figure 5.1 illustrates the probability (in the form of a probability density function) of government facility damage as a function of the level of damage. This probability distribution indicates the chance that different levels of damage will result to the government launch facilities from a licensed activity. This probability distribution is conceptual (i.e., insufficient data are available to establish the specific shape for each launch vehicle) and is presented only to serve as the basis for illustrating general concepts. The damage probability distribution is independent of MPL, threshold probability, required level of insurance, etc. and is a function of many factors relating to the specific activity (e.g., characteristics of the launch vehicle, mission, and launch facility and range). The calculation of the maximum probable loss establishes the level of damage that, on average, will be exceeded with a stated probability. This level of probability, or risk, is referred to as the threshold probability. The threshold probability was conservatively selected by DoT[8] as 10^{-5} for government property damage and 10^{-7} for third-party damage and is illustrated as the cross-hatched area to the right of the MPL level of damage. The threshold probability represents the level of risk, that is, the chance that damage will occur that is in excess of the computed MPL (discussed in following paragraphs)[9,10] that has been chosen as being acceptable. A threshold probability of 10^{-7} implies that damage will exceed the MPL value, on average, once in every 10,000,000 launch attempts.

The selection of appropriate values of threshold probability (i.e., the acceptable level of risk) was based upon a review of levels of risk associated with a wide range of commonly faced activities. This is summarized in Fig. 5.3. DoT/FAA set the threshold probabilities such that the level of risk that was deemed acceptable for launch activities would be an order of magnitude less than existing levels of risk from non-space-related activities.

Referring to Fig. 5.1, a required level of insurance is established that is based upon the MPL but less than the maximum available insurance at a reasonable price as established by the Secretary of Transportation. At no time can the amount of insurance exceed the maximum amount as stipulated in the enabling legislation (i.e., $100,000,000). Setting the required level of insurance in excess of the computed MPL is tantamount to changing the acceptable threshold probability. (It is possible that MPL and required insurance levels might differ and are illustrated in this manner in Figs. 5.1 and 5.2. Under normal circumstances it is anticipated that the MPL and required insurance levels will be the same.)

In a similar manner, the probability distribution of third-party damage is illustrated in Fig. 5.2. The relationship between MPL, required level of insurance (based upon MPL), maximum available insurance at a reasonable price, and the limit placed upon maximum insurance required by the enabling legislation ($500,000,000) is illustrated. In addition, the maximum level of government indemnification ($1,500,000,000 above the insurance requirement) is also indicated. The level of indemnification as stipulated in the Act (as amended) has a sunset provision. This is discussed in detail in Sec. IV, which discusses the need (?) for continuing indemnification. Figure 5.4 illustrates the general framework of the payouts and level of risk associated with insurers, providers of launch services, and the government for government property and third-party damage, respectively.

Figure 5.4 disaggregates the probability of level of third-party damage into probability distributions of cost from the points of view of the licensee, the insurance industry, and the government. Figure 5.4a illustrates the probability distribution of insurance company payout. This probability distribution is

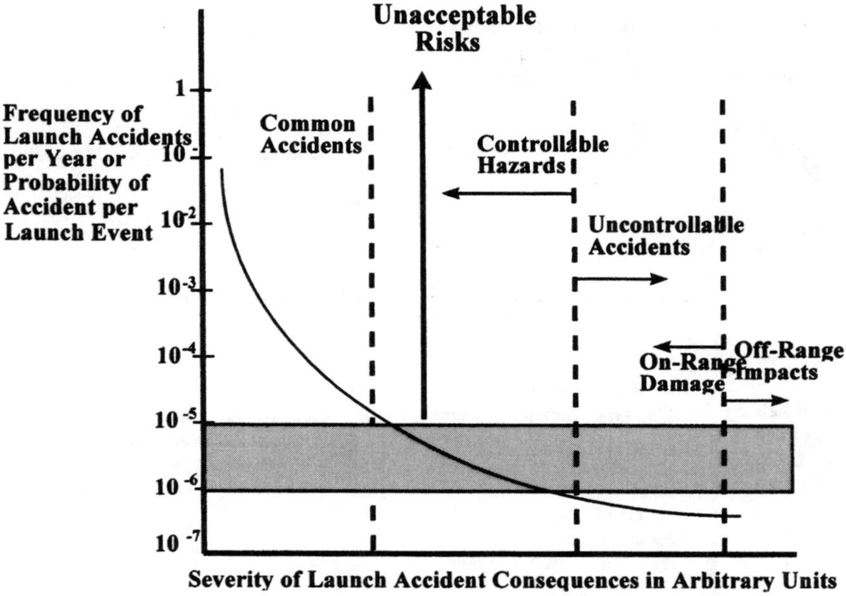

	Number of Deaths In Representative Year	Individual Risk/Year
Coal Mining		
Accident	**180**	**1.3×10^3 or 1/770**
Black Lung Disease	**1,135**	**8×10^3 or 1/125**
Fire Fighting		**8×10^4 or 1/1,250**
Motor Vehicle	**46,000**	**2.2×10^4 or 1/4,500**
Truck Driving	**400**	**10^4 or 1/10,000**
Falls	**16,339**	**7.7×10^5 or 1/13,000**
Football [ave. over participants]		**4×10^5 or 1/25,000**
Home Accidents	**25,000**	**1.2×10^5 or 1/83,000**
Bicycling	**1,000**	**10^5 or 1/100,000**
Air Travel [1 trip/year]		**2×10^6 or 1/500,000**

Threshold Probability:
10^{-5} for Government Property
10^{-7} for Third-Parties

Fig. 5.3 Determination of threshold probability (Source: Refs. 6 and 11).

truncated at the level of required insurance with this occurring with a probability equal to the threshold probability (i.e., if a loss occurs in excess of the required level of insurance, the insurance company payout will be equal to the level of required insurance).

Figure 5.4b illustrates the probability distribution of the licensee's cost that can result from third-party damage. Because the licensee is covered by insurance up to the required level of insurance (and often beyond) and the government then indemnifies (in actuality the licensee will be held harmless and government will

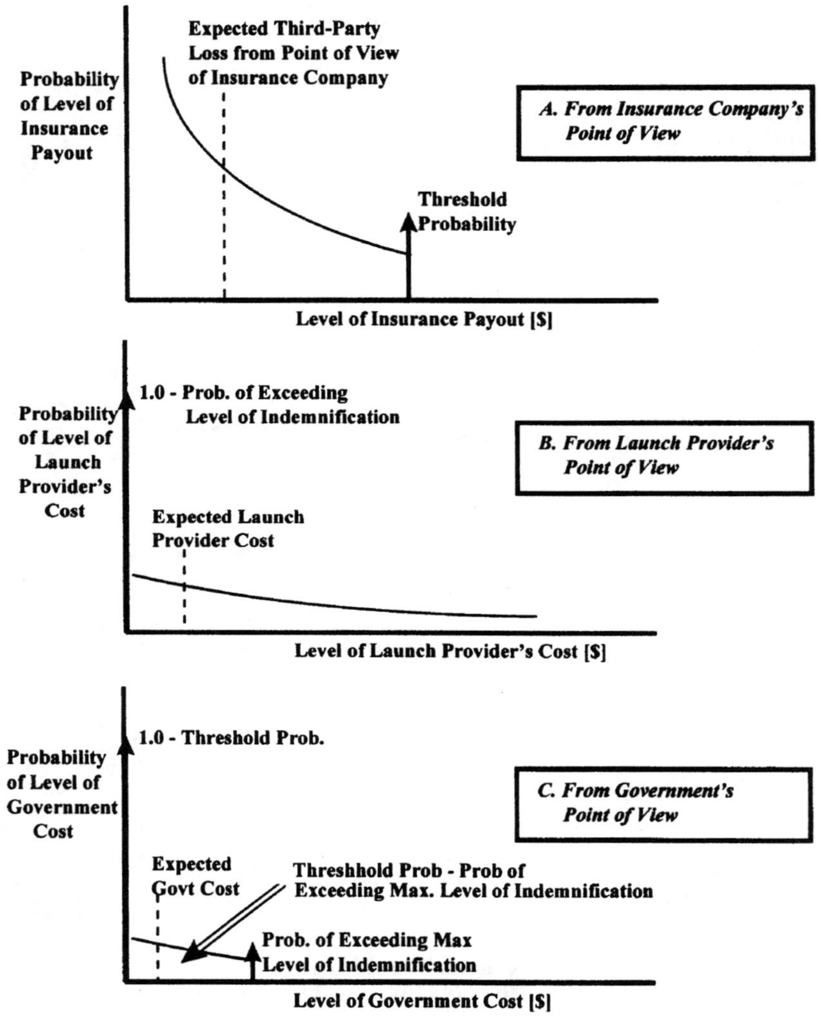

Fig. 5.4 Insurance payout and government cost as a function of required level of insurance: third-party damage.

make payment if appropriations are available) to a maximum level of $1.5 billion above this, only claims in excess of the government's level will be passed on to the licensee. The probability that the licensee will incur a cost is extremely small and unknown and likely to be significantly less than the threshold probability. Because there is a chance, albeit very small (i.e., the threshold probability), that damage will exceed the level of required insurance, the government is also at risk and can incur costs above the insured level but not exceeding the indemnification level. The probability distribution of government cost is illustrated in Fig. 5.4, where there is a high probability that the government will incur no cost (i.e., 1 − threshold probability) and a very small chance of incurring cost above the indemnification level of $1.5 billion. Of course, if damage to foreign entities exceeds the level of indemnification and the licensee does not have the ability to satisfy these claims, then the U.S. government, because of the international agreements, has the final responsibility for satisfying these claims. The impacts on insurance companies, launch providers, and the U.S. government can be evaluated in a similar fashion when considering government facility damage.

An approach used for the determination of MPL for launch operations is described in the following paragraphs (Details of the methodology can be found in Refs. 9, 10, and 12.):

MPL is the maximum magnitude of loss such that there is less than a specified probability (i.e., the threshold probability) of exceeding this level. Exposure of government property during launch support and launch activities is routinely accepted as part of doing business. In determining the maximum probable loss for government property, focus of the evaluation is on those highest cost facilities exposed at or just below the threshold level. The replacement costs of each facility are listed (illustrated in Fig. 5.5) in descending order and the individual probabilities of loss accumulated until the probability threshold level (i.e., 10^{-5}) is reached. The facility loss at this level is determined to be the maximum probable loss. Losses of lesser amounts, no matter what their likelihood of occurrence, will be covered, and the likelihood of losses exceeding this level is the threshold probability. Given a commercial launch rate of 10 per year, the losses are expected to exceed the maximum probable loss on the average somewhere between once every 1000 and 10,000 years. This is a very conservative interpretation of the concept of maximum probable loss.

		Probability of Damage	
Asset at Risk	Restoration Cost [% of Facility]	Asset	Cumulative
$Asset_1$	\$ (Restore $Asset_1$) [% Total Value $Asset_1$]	P(Damage to $Asset_1$)	P(Damage to Asset $_1$)
$Asset_2$	\$ (Restore $Asset_2$) [% Total Value $Asset_2$]	P(Damage to $Asset_2$)	P(Damage to $Asset_2$)+ P(Damage to $Asset_1$)
$Asset_3$	\$ (Restore $Asset_3$) [% Total Value $Asset_3$]	P(Damage to $Asset_3$)	P(Damage to $Asset_3$)+ P(Damage to $Asset_2$)+ P(Damage to $Asset_1$)
		. . .	
$Asset_N$	\$ (Restore $Asset_N$) [% Total Value $Asset_N$]	P(Damage to $Asset_N$)	P(Damage to $Asset_N$)+ . . . P(Damage to $Asset_3$)+ P(Damageto $Asset_2$)+ P(Damage to $Asset_1$)

Note: **\$ (Restore $Asset_1$) > \$ (Restore $Asset_2$) > \$ (Restore $Asset_3$) > ...> \$ (Restore $Asset_N$)**

Fig. 5.5 Method for determining MPL (example: government property).

The following example is presented to illustrate the methodology described in Fig. 5.5:

If

$$P(\text{Damage to Asset}_2) + P(\text{Damage to Asset}_1) < 10^{-5}$$

And

$$P(\text{Damage to Asset}_3) + P(\text{Damage to Asset}_2) + P(\text{Damage to Asset}_1) > 10^{-5}$$

Then

$$\$(\text{Restore Asset}_2) = \text{MPL}$$

In looking at the fatalities that might occur, various forms of the following equation are often used for expected casualties E_c:

$$E_c = P_i * LA * P_d$$

where

P_i = probability of impact in a specific area
LA = lethal area of the impacting debris
P_d = population density

Given the correct inputs (for example, areas containing different population density and different probability of impact must be considered separately with results being additive), the approach provides a fairly reasonable estimation of the "expected number" of casualties. To obtain conservative results, the highest population density region can be used for the entire exposed region. (This approximation requires some care because in theory it can lead to unrealistically high insurance requirements with accompanying higher insurance costs with little or no added value.) When the necessary data are available, complex simulation models are also used, which reflect the flight dynamics of the vehicle in question, failure probabilities, and modes over the flight profiles and the vehicle's debris characteristics.

Expected (average) loss, however, does not directly provide the information necessary to determine the probability of a particular loss level being exceeded. Knowing the expected losses and using the definition of expected value,

$$E_c = \sum_{n=1}^{n=\infty} n * P_n$$

where P_n is the probability of n casualties occurring; conservative estimates of the upper bounds of probabilities of an event or range of events can be made by assuming the probability of all other events are zero.

In developing the MPL values for launch support activities, the information about the hazardous activities that are performed, the policies and procedures followed, and the number of personnel and facilities (including replacement costs) that will be exposed are considered. For example, on government ranges the siting of many facilities is based on a quantity/distance relationship to ensure that a prelaunch accident in one facility will not affect another. There are various policies in place that correct past deficiencies and limitations on the number of people that can be present during the performance of certain hazardous activities.

Estimates of the likelihood of major accidents occurring during various phases of prelaunch activities have been made by range safety experts. These estimates are based on their years of experience and the refinement over the years of the policies and procedures in response to accidents or near accidents that have occurred.

For the flight of a launch vehicle, a more complex analytic approach is used to derive maximum probable losses for third parties and government property. An approach is to use a failure event tree like that indicated in Figs. 5.6 and 5.7. Figure 5.6 describes and identifies the process with Fig. 5.7 being a restatement of Fig. 5.6 but with typical probabilities indicated. Over the years safety systems and procedures have been developed which assume that every vehicle launched is going to fail. For these reasons systems like the flight termination system (FTS), the purpose of which is to stop a vehicle before it or its debris could reach a populated area, are designed and acceptance tested to have very high reliabilities (failure rates on the order of 10^{-3} or less). In performing the analyses for maximum probable loss, the reliability figure used for the FTS is the lesser of the historically demonstrated reliability and the design requirement.

The failure event tree is used to estimate the probability of each of the possible outcomes, and the consequences of each of the outcomes is then estimated.

In summary, several different approaches might be appropriate for the determination of MPL, with the selected approach dependent upon the circumstances involved. In general, as uncertainty increases and/or the stakes get higher, the more rigorous techniques (as just described) apply that employ formal methods, procedures, and models. The approach used can depend on the information available, the sensitivity of the situation (e.g., one or more very high cost facilities that might be exposed or effect MPL value, highly concentrated population centers vs low-density widely distributed population, etc.), and can employ a range of analytic techniques that include complex models and data analysis or direct approaches that rely more upon judgment, experience, and estimation (e.g., if one or more facilities are exposed and it is obvious that they each are exposed to risks that exceed the threshold probability, then it is appropriate to just select the largest loss as being the value of MPL if the loss of each is independent of the others). The overall objective is to identify the largest loss that could occur within a reasonably remote likelihood (the threshold probability) and to accomplish this in a conservative manner (i.e., when assumptions or estimates are made, conservatism is the rule) by using the tools and techniques that are appropriate for analyzing the specific task at hand.

Typical government property and third-party liability insurance requirements (as based upon MPL determinations) for different launch vehicles at different ranges are illustrated in Table 5.1. A number of these vehicles are no longer under consideration but have been included to illustrate the magnitude of the MPL determinations for a considerable range of launch vehicles.

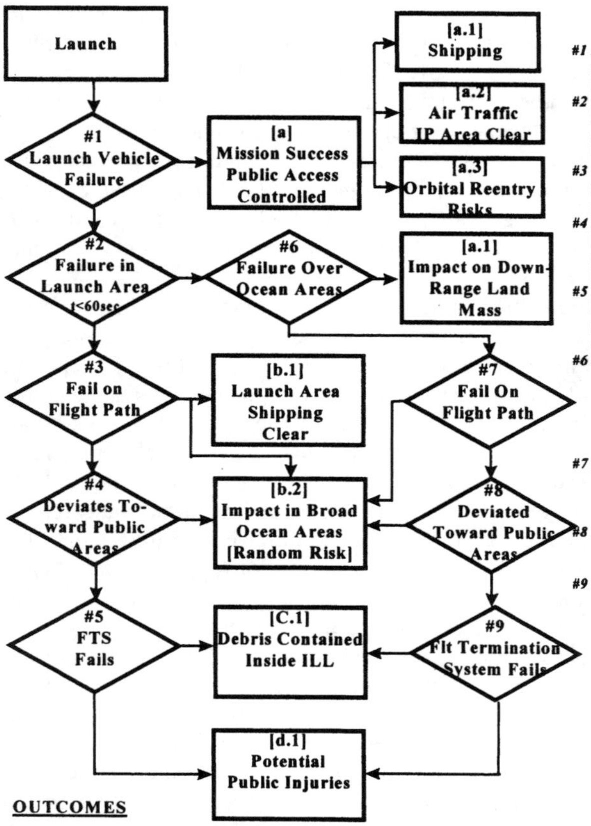

EVENTS

#1 Launch Vehcle Failure: **Probability that launch vehicle will fail [I.e., launch will not be successful].**
#2 Failure in Launch Area [t<60 sec]: **Probability of failure in the launch area during the early launch phase [I.e., 0 to 60 seconds after launch].**
#3 Fail on flight Path: **Probability of failure occurring on original flight path.**
#4 Deviates Toward Public Areas: **Probability of deviation toward populated areas protected by impact limit lines.**
#5 Flight Termination System Fails: **Probability that flight termination system will fail [I.e., 1-reliability of FTS system].**
#6 Failure Over Ocean Areas: **Probability that launch failure will occur over ocean areas [I.e., 1-probability that failure will occur during over-flight of populated downrange land masses].**
#7 Fail on Flight Path: **Probability that vehicle will fail on original flight path [not deviated toward public areas].**
#8 Deviates Toward Public Areas: **Probability that vehicle will deviate toward public areas.**
#9 Flight Termination System Fails: **Probability that flight termination system [FTS] fails.**

OUTCOMES

[a] ***Mission Success, Public Risks Controlled*** **May directly follow event node #1.**
[a.1] ***Shipping Risks*** **to shipping from boster stages and other discarded debris impacting within planned areas.**
[a.2] ***Air Traffic, IP Area Clear*** **Planned air traffic exposures.**
[a.3] ***Orbital Reentry Risks*** **Risks to third-parties from eventual decay from orbit of all orbiting hardware.**
[b] May directly follow event nodes #3, #4, #7, and #9.
[b.1] ***Launch Area Shipping Clear*** **Risks to shipping in the ocean near the launch site.**
[b.2] ***Impact in Broad Ocean Areas - Random Risk Shipping and Air Traffic*** **Random risk to shipping and air traffic in broad ocean areas. [Note: also reflects random risk to range facilities due to impact within range boundaries.]**
[c] May directly follow event nodes #5 and #9.
[c.1] ***Debris Contained Inside ILL*** **Risks associated with debris which is contained within Impact Limit Lines. [Note: May also reflect risks associated with debris contained within range boundaries.]**
[d] May directly follow event nodes #5 and #9.
[d.1] ***Potential Public Injuries*** **Risks to public in areas outside the Impact Limit Lines.**

Fig. 5.6 Public launch hazard tree (Source: Ref. 6).

Risk analysis is concerned with the identification of possible hazardous events, the probability of occurrence, and the level of the likely consequences. The setting of maximum probable loss utilizes this information together with the threshold probability to establish the appropriate level of financial responsibility so that government property and third-party losses are adequately covered by insurance

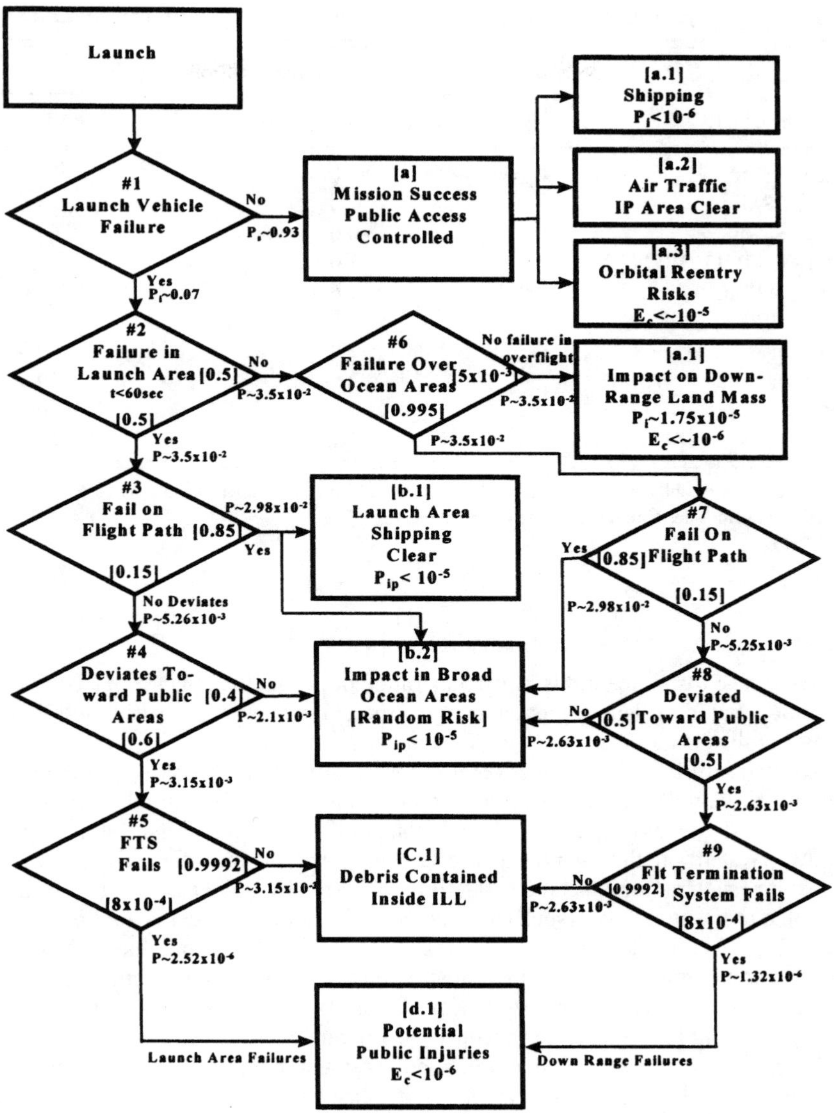

Fig. 5.7 Typical launch failure event tree (Source: Ref. 6).

should a hazardous event actually occur. The magnitude of the coverage or compensation must be based upon the likely loss to be sustained and not the expected value of the loss. This is specifically the role of the MPL methodology, which establishes the maximum probable loss *(not the maximum possible loss)* that requires insurance coverage.

Table 5.1 Typical insurance requirements as determined by MPL considerations

	Insurance, millions of dollars	
Launch vehicle	Government property	Third party
Titan, ESMC	80.0	215.0
Atlas, ESMC	75.0	164.0
Delta, ESMC	80.0	164.0
AMROC SMLV, WSMC[a]	1.0	12.0
Scout, WSMC[a]	1.6	10.0
Scout, WFF[a]	21.6	10.0
SSI Consort 1, WSMR	1.0	10.0
SSI Consort 2–5, WSMR	2.5	10.0
Conatec	1.0	10.0
Joust	1.0	12.5

[a]MPL has been established at the higher of the site operations and launch operations costs. In some cases these values are tentative and not fully evaluated.

D. Overview of the Space Insurance Industry

The space insurance industry offers a range of products to insure space-related activities including launch insurance, government property insurance, and third-party liability insurance.

Launch insurance includes cover for launch failures (encompassing failures of the launch vehicle and transfer stages and usually failures of initial satellite on-orbit operations) and is purchased by the launching firm and/or its customer. The policy pays off in the event that the launch fails. The payoff can be in cash, hardware (such as a replacement satellite, if the policy holder is not the satellite manufacturer), or services (such as a relaunch, if the policy holder is not the launch firm). There is no federal requirement that launch insurance be purchased. Policies are also sometimes written to cover lost revenue from premature communication transponder failures or other system degradations (such as loss or unplanned use of station-keeping propellant) that result in lost revenue.

A *government property insurance* policy must be purchased, or financial responsibility demonstrated, by the entity licensed by the FAA to conduct commercial space transportation activities. The policy must be at least for the amount specified by the FAA and pays in the event that government property on a commercial launch range is damaged by licensed commercial space transportation activities. Government property insurance for commercial space launches is required by law.

A *third-party liability insurance* policy must be purchased, or financial responsibility demonstrated, by the entity licensed by the U.S. Department of Transportation to conduct commercial space transportation activities. The policy must be at least for the amount specified by the DoT, and pays in the event that third-party claims for property damage, injuries, or casualties are made in association with licensed commercial space transportation activities. Third-party liability insurance (or demonstration of adequate financial responsibility) for commercial space launches is required by law.

Figure 5.8 indicates the estimated current and historic prices of third-party liability insurance. As will be discussed in more detail in Sec. IV, the cost of liability insurance is small relative to the cost of other forms of launch and on-orbit related insurance. To place third-party liability insurance premiums in perspective, consider an Atlas launch of a communication satellite with the launch plus satellite cost in the range of $200 million. The associated launch insurance (i.e., launch vehicle plus payload) premium is on the order of $35 million, and the third-party liability insurance premium (for $100 million of cover) is on the order of $0.2 million. On-orbit insurance is on the order of $2 million per year.

E. Costs Incurred by Launch Providers, Insurance Providers, and Government

The expected cost per launch for third-party damage is comprised of an expected payout from the insurance industry C_L (or the equivalent from the licensee if it seeks to self-insure), an expected cost that will be borne by the government L, and an expected cost that will be borne by the licensee X. The sum, $C_L + L + X$, is a constant and represents the expected third-party damage per launch. Using different values of threshold probability or MPL will shift the burden imposed on the different parties to the transaction but will not alter the expected third-party damage per launch.

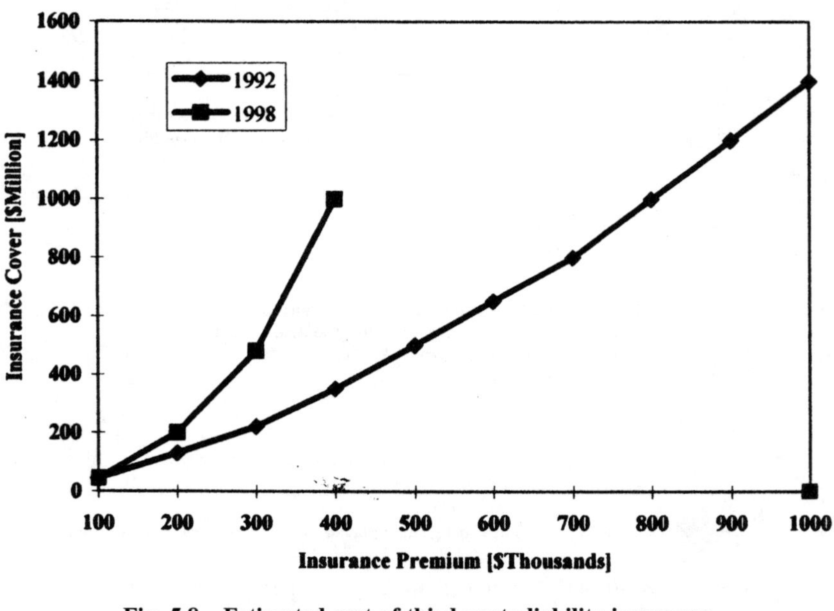

Fig. 5.8 Estimated cost of third-party liability insurance.

Figures 5.9 and 5.10 illustrate the cost pass-through among launch participants for government property damage and third-party damage, respectively. Two scenarios are considered in each: scenario A represents a situation where the licensee passes insurance cost on to the payload manufacturer who, in turn, passes insurance cost on to the satellite owner or service provider (ultimately passed on to the user of the services provided by the payload); scenario B represents a situation where the licensee absorbs the insurance cost.

Considering first the situation of government property damage, Fig. 5.9 indicates the revenue and expense associated with damage and insurance for the various involved parties: satellite owner, payload manufacturer, launch company (assumed to be the licensee), insurance company, and the government. The insurance company incurs a "paperwork" cost C_{PW} associated with putting

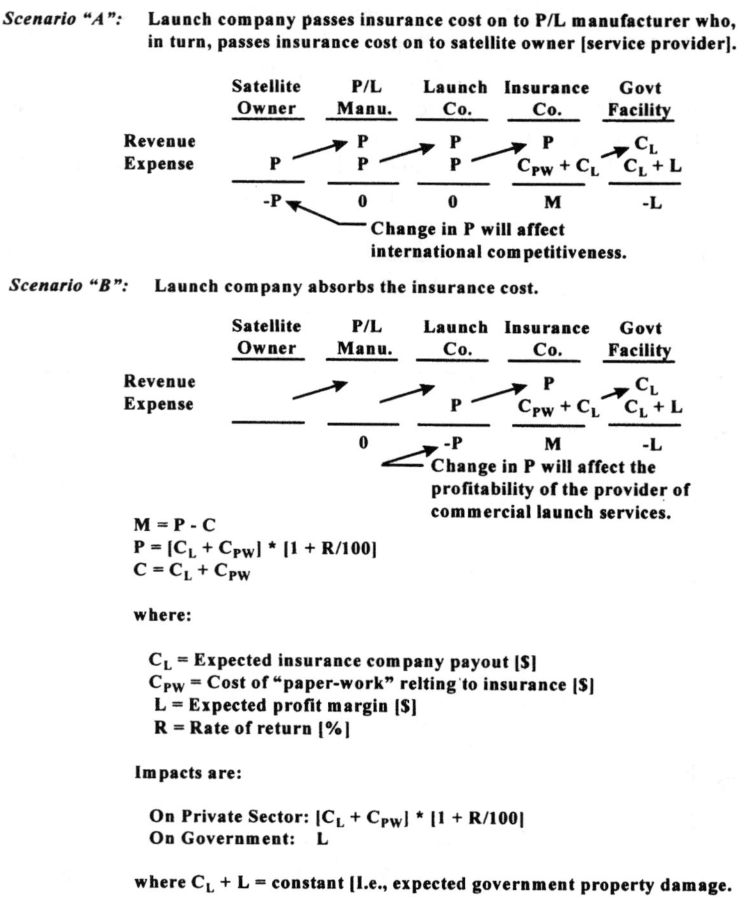

Fig. 5.9 Cost pass-through among launch participants (government property damage).

Scenario "A": **Launch company passes insurance cost on to P/L manufacturer who, in turn, passes insurance cost on to satellite owner [service provider].**

	Satellite Owner	P/L Manu.	Launch Co.	Insurance Co.	Third Parties	Govt Facility
Revenue		X+P	X+P	P	C_L+X+L	
Expense	X+P	X+P	X+P	C_{PW}+C_L	C_L+X+L	L
	-X -P	0	0	M	0	-L

Change in X and P will affect international competitiveness.

Scenario "B": **Launch company absorbs the insurance cost.**

	Satellite Owner	P/L Manu.	Launch Co.	Insurance Co.	Third Parties	Govt Facility
Revenue				P	C_L+X+L	
Expense			X+P	C_{PW}+C_L	C_L+X+L	L
			-X -P	M	0	-L

Change in X and P will affect the profitability of the provider of commercial launch services.

M = P - C
P = [C_L + C_{PW}] * [1 + R/100]
C = C_L + C_{PW}

where:

C_L = Expected insurance company payout [$]
C_{PW} = Cost of "paper-work" relting to insurance [$]
M = Expected profit margin [$]
L = Expected govt payment [in excess of required insurance] [$]
X = Expected industry payment [in excess of gove indemnification] [$]
R = Rate of return [%]

Impacts are:

On Private Sector: X + [C_L + C_{PW}] * [1 + R/100]
On Government: L

where C_L + X + L = constant [I.e., expected third-party loss.

Fig. 5.10 Cost pass-through among launch participants (third-party damage).

together the insurance package and is a function of the magnitude of the required insurance. The insurance company is also faced with an expected cost or payout C_L, for government property damage per launch. Because the insurance company is in business to make a profit, an insurance premium P will be set that covers the insurance company's expected costs plus an adequate profit margin to compensate for its risk. The premium will be passed on as indicated: in scenario A the premium expense is passed on to the satellite owner with a change in P resulting in, among others, an impact on international competitiveness; in scenario B the premium expense is absorbed by the licensee with a change in P affecting the profitability of the provider of commercial launch services. In both cases the insurance company's expected payout to the government to cover damage is in the amount C_L. Because the expected level of damage to government property is $C_L + L$, the government expected damage per launch is L.

The direct cost impacts relating to the setting of insurance requirements for government property damage are as follows:

On the private sector:

$$(C_L + C_{PW}) * (1 + R/100)$$

On government:

$$L$$

where R is the desired rate of return by the insurance industry and $C_L + L$ is a constant and is equal to the expected damage (per launch) to government property.

The consideration of the costs and their pass-throughs for third-party damage is similar (to government property damage) and is illustrated in Fig. 5.10. Third parties are added to the list of involved participants. As in the case of government property damage, the insurance company incurs costs associated with paperwork and expected payout. After these costs are increased to include profit, they are passed on to the licensee. The licensee incurs this cost plus the expected industry payment X, which results from the chance of losses exceeding the level of government indemnification. These costs are passed on in scenario A and absorbed in scenario B. In the former case the sum of X and P will have an effect on international competitiveness, and in the latter case will affect the profitability of the provider of commercial launch services.

Government will also incur an expected cost per launch (in the amount of L). This is as a result of the small but finite chance (i.e., the threshold probability) that third-party losses will exceed the maximum probable loss. Expected payments will be made to the damaged third parties from the insurance company (in the amount of C_L), the licensee (in the amount of X), and the government (in the amount L).

The direct cost impacts relating to the setting of insurance requirements for third-party damage are as follows:

On the private sector:

$$X + (C_L + C_{PW}) * (1 + R/100)$$

On government:

$$L$$

where $C_L + X + L$ is equal to the expected damage (per launch) to third parties.

The computation of expected cost is made difficult by the lack of knowledge (as a result of extremely limited data) of the shape of the damage probability distributions. The expected value of a loss resulting from damage is

$$\text{Expected loss} = \int_o^D x\,p(x)\,\mathrm{d}x$$

where x is the magnitude of damage and $p(x)$ is the probability associated with x and D is the maximum level of damage. For example, $p(x)$ can be the probability distribution shown in Fig. 5.4c and D is the maximum level of damage in excess of the required insurance. This represents the situation of third-party damage in excess of the level of insurance set to protect government and third parties. There is a small chance (the threshold probability) of this occurring.

Because $p(x)$ is not known, in order to proceed it is necessary to make assumptions about $p(x)$. A very conservative assumption can be made: namely that $p(x)$ is approximated by a two-point probability distribution where there will be no damage (in excess of the level of required insurance) with a probability equal to $1 -$ threshold probability and there will be maximum possible damage with a probability equal to the threshold probability. This results in an expected loss or cost equal to the threshold probability multiplied by the maximum level of damage. Any other choice of $p(x)$ would lead to less conservative results. For example, if $p(x)$ were uniform in the range of zero to the maximum possible level of damage the expected loss would be reduced by a factor of two, and if it were a triangular distribution (with maximum probability of a loss at zero) the expected loss would be reduced by a factor of 3.

F. Government Spaceport Considerations

To a certain extent, the U.S. and foreign governments have similar problems with respect to the establishment of spaceports; their solutions can differ significantly. It is assumed that foreign governments have signed, as has the U.S. government, the international treaties and conventions that deal with liability in the case of damage or injury to third parties. Thus, by having a spaceport located in a foreign state, that state (for example, Canada, Australia, etc.) is exposed to potential liability stemming from space launches emanating from that soil. Government assumption of this potential liability can be viewed as an inducement (i.e., a subsidy) for space launch from, for example, a "Spaceport Canada." or a "Spaceport Australia."

On the other hand, if the foreign state chooses to limit its potential liability it can establish a financial responsibility regime similar to that established in the U.S. and described in the preceding pages. The U.S. has adopted an approach that limits the exposure of the government at a tolerable level of risk while providing encouragement to the U.S. space transportation industry.

U.S. launch vehicles that can be launched from a foreign state would most likely have to be licensed by the U.S. government. Obtaining a launch license requires the determination of maximum probable loss and the setting of financial responsibility requirements (i.e., levels of insurance). The U.S. currently indemnifies the launch operator against catastrophic losses as described in the preceding pages. Because both the state from which the launch takes place and the state that provides the launch vehicle are potentially liable for damage that can result from a space launch, it appears that the foreign and U.S. governments should establish agreements with respect to their respective shares of liability should an accident occur.

A launch operator licensed by a foreign government might be liable for damage to both the launching and vehicle initiation states. Differences between financial responsibility requirements and indemnification policies between these states can influence the desirability of utilizing one spaceport vs another spaceport. For example, if a foreign spaceport does not provide indemnification similar to that provided by the U.S., will a U.S. launch operator possibly jeopardize its indemnification status in the U.S.? (Congress provided the indemnification at the behest of the launch operators as a necessary step to maintain a commercial space launch industry.)

It must be pointed out that the U.S. has not as yet developed policies with respect to licensing of foreign launches of U.S. vehicles. This is of concern because it is not yet clear as to what the U.S. policy will be with respect to indemnification. The elimination of government indemnification for foreign launches of U.S. vehicles could impact the utilization of foreign launch sites by U.S. providers of space transportation services.

G. Foreign Spaceport Insurance Concerns

A spaceport located in a foreign state must consider the risk of damage to spaceport property and of damage to third parties when considering financial responsibility and insurance requirements. The problem is basically the same as that faced by U.S. launch ranges, but the severity of the problem is likely to be different. A foreign spaceport will require sufficient resources to repair/replace launch facility equipment in the case of an accident with insurance being a possible way of providing the required resources. If the insurance route is selected as the approach for providing the necessary resources, then it is recommended that the methods developed by the FAA (using the threshold probability and the determination of maximum probable loss) be used in the setting of insurance requirements by the spaceport.

Because a foreign state (government) is likely to be responsible for third-party claims made by foreign governments and both the foreign state and the spaceport will be liable for other potential third-party damage, it is presumed that the commercial spaceport and the launching organization will be required, by the foreign government, to demonstrate financial responsibility such that the foreign government will not be exposed to potential liability. Again, it is recommended that the MPL methodology developed by the FAA be used to establish the third-party liability financial responsibility and insurance requirements for the foreign spaceport. Calculated values of MPL can be relatively low with careful site selection and facility placement. It is likely that the foreign government will impose financial responsibility requirement on the commercial spaceport.

III. Orbital Debris and Regulatory Analysis

As discussed in Sec. VII, we cannot escape the fact that orbital debris will increase the cost of space missions either through costs incurred to conform to debris remediation policies or losses that result from debris impacts. It is therefore important to choose cost-effective orbital debris remediation policies that will, in the long-term, reduce mission costs below those that would result from pursuing a policy of no orbital debris remediation. [The International Academy of Astronautics (IAA) recently published a position paper entitled "*Position Paper on*

Orbital Debris" (September 2001),[13] which discussed orbital debris and remediation policies. This paper, unfortunately like many other papers on the subject, failed to discuss, or even mention, the cost and economic implications associated with the proposed debris remediation measures. When this author raised the issue of the cost of and benefits from remediation alternatives with the IAA, it was pointed out that estimated costs and benefits were extremely difficult to measure in quantitative terms and that the IAA-recommended measures were low in cost and therefore were cost effective. The IAA apparently was unaware of methods that were developed and published (by the IAA) that described methods and procedures for evaluating costs and benefits of remediation alternatives and presented results of analyses (funded by NASA) of the economic impacts of orbital debris on both medium- and low-altitude satellites. With regard to cost effectiveness, the authors of the IAA paper were apparently unaware that low cost does not imply cost effectiveness, that is, cost is not likely to be directly related to efficiency.] Undertaking debris remediation is not likely to be free but can, in fact, lead to increased mission costs in the short term. Thus we have a typical investment dilemma: is it desirable to make an investment and incur costs in the near term in order to achieve benefits in the long term? or, stated another way, how much can we afford to spend to achieve anticipated savings? Decisions regarding orbital debris remediation policies are quite similar to most investment decisions (i.e., spend now for future rewards) except that the time frame is considerably longer (measured in terms of perhaps 50 to 100 years or more) than that encountered in most investment decisions. In fact, it is likely that the generation that incurs the costs will not be the generation that obtains the benefits. This long time frame opens up the distinct possibility that normal and prudent business decisions might not produce results that are in the public interest and might result in the need for the development of a regulatory environment. The promulgation of a regulatory regime requires a regulatory impact analysis. Economic analysis in support of an orbital debris regulatory impact analysis is discussed in the following paragraphs.

The need for regulatory consideration is the result of likely market failure caused by externalities. An externality is said to occur when one party's actions impose uncompensated benefits or costs on another outside of the marketplace. In the case of orbital debris, market failure will likely be the result of the long time delays between cause and effect and the lack of economic incentive to make near-term investments by those who will not be around to be affected by the long-term debris environment. Environmental problems are a classic case of externality. The clutter of space with orbital debris is an environmental problem. The consequences of orbital debris might be of minor import today, but if appropriate measures are not taken in the near term to restrict or reverse the growth of the orbital debris population long-term effects of orbital debris might be irreversible and could be disastrous.

There are, of course, certain types of space missions that already perceive economic incentives that encourage taking of cost-incurring actions in the near term so as to eliminate or substantially reduce the economic consequences of orbital debris, for example, geosynchronous orbit (GEO) communication satellites are moved to higher altitudes near the end of their useful lives and low Earth orbit (LEO) communication satellites are likely to be moved to other orbits (including reentry) near the end of their useful lives. Currently, these are business decisions that reflect the importance of moving satellites out of harm's way so that business operations can proceed unencumbered by ones own failed satellites.

It is because of the lack of uniformity of economic incentives coupled with the long time between cause and effect that externalities appear to exist of sufficient concern to warrant the consideration of the need for regulation.

As the number of Earth-orbiting objects increases, so does the potential for generating orbital debris with the consequent increase in the likelihood of impacting and damaging operating satellites. Various debris remediation approaches are being considered that encompass both in-orbit and return-to-Earth schema and have varying degrees of operations, cost, international competitiveness, and safety implications. Because of the diversity of issues, concerns, and long-term impacts, there is a clear need for the setting of government policies that will lead to an orderly abatement of the potential orbital debris hazards. These policies might require the establishment of a supportive regulatory regime.

Various studies[14–17] have concluded that the orbital debris environment will pose a threat to future space operations if remedial measures are not taken to control the rate of growth of the debris population. Currently, orbital debris poses a safety problem for large space structures operating in LEO. Accordingly, the space station program has taken steps to maximize protection from debris penetration by implementing state-of-the-art shielding; using existing ground radar to track and subsequently maneuver to avoid larger debris; and actively developing operational and design options that will minimize the risk to the crew and the station.

Since the publication of the National Security Council report in 1989, the U.S. and a number of national and international space-faring organizations have begun to address orbital debris concerns. The U.S. and other space-faring nations have taken voluntary design measures (i.e., tethering of orbital debris such as lens caps and the use of debris free devices for separation and release) as well as operational procedures to reduce the generation of debris. GEO communication satellites are currently moved to higher altitudes near satellite end of useful life, LEO communication satellites (such as those of the Iridium system) are designed to be deorbited near end of useful life, and various LEO communication missions have set up cooperative procedures for minimizing collisions of operational satellites in low Earth orbits.

It has generally been concluded that closer international cooperation is necessary for dealing effectively with orbital debris. It is in the interest of the U.S. to continue to maintain a leadership role in international considerations relating to orbital debris. The U.S. considers the development of technical cooperation and consensus to be a prerequisite for any potential international agreements, regulatory regimes, or other measures relating to orbital debris. The unilateral application of debris mitigation measures could put U.S. satellite and launch vehicle industries at a competitive disadvantage.

Part of the orbital debris problem is that even though there is reasonable knowledge of the larger LEO debris population insufficient data exist to accurately characterize and forecast the debris population as a function of time. This is caused by a combination of data availability and uncertainty associated with debris forecast modeling techniques, particularly taking into account collisions and resulting fragmentation. This uncertainty leads to uncertainty with respect to the form and cost of mitigation measures that are justifiable and otherwise required.

NASA has made substantial progress in documenting and defining safety requirements and specific design measures that can be taken into account during the development of satellites and launch vehicles in order to minimize or eliminate debris generation. Design guidelines and agency requirements are being established.[18] Because NASA and DoD are major players with respect to satellite and launch vehicle procurement, it is likely that requirements imposed upon satellites and launch vehicles procured for government agencies will carry over to commercial practices. This could have implications with respect to international competitiveness if other nations do not set similar requirements.

To reduce the potential long-term impacts of orbital debris, it is necessary to undertake cost-incurring actions in the short term. Actions of commercial firms normally are based upon financial motivations. Because costs might have to be incurred in the short term without clear-cut short- or even mid-term benefits to the present decision makers, it is not clear whether or not long-term profit motives (vs the incurrence of short-term costs) will be sufficient to obtain the form and timing of remediation measures deemed necessary. In the absence of clear-cut and continuing industry participation in remediation actions, it is likely that a regulatory regime will be required.

A. Orbital Debris Environment

The estimated mass of man-made orbiting objects (orbital debris) within 2000 km of the Earth's surface is about 2,000,000 kg. (The discussion of the orbital debris environment is based upon Ref. 17.) These objects are in mostly high inclination orbits and pass one another at an average relative velocity of 10 km/s. Most of the mass is contained in about 3000 spent rocket stages, inactive satellites, and comparatively few active satellites. A smaller amount of mass, about 40,000 kg, is in the remaining 4000 objects currently being tracked by space surveillance sensors.

Information about the current debris environment is limited by the inability to track and catalog small objects utilizing radar and optical techniques. [Relatively low-cost and technically feasible bistatic forward-scatter techniques can be utilized for detecting small objects (less than 10 cm) near GEO satellites utilizing the undisturbed transmissions of communications satellites and Earth-based receivers employing wideband autocorrelation techniques.[20]] Although the mission of the Space Surveillance Network (SSN) is to track all man-made orbiting objects, technology, and natural constraints serve to limit the effective tracking of objects smaller than 10 cm. With only limited knowledge of the current space debris environment,[21] which can serve as input to orbital debris forecasting models coupled with future traffic uncertainties and fragmentation and other modeling uncertainties, it is difficult to develop good forecasts of the long-term orbital debris environment. The need for, and the value of, orbital debris remediation policies must be considered within this framework.

Debris of concern include the following:

1) Objects larger than 10 cm in diameter, commonly referred to as large objects, are routinely detected, tracked, and cataloged.
2) Objects between 1 and 10 cm in diameter, commonly referred to as risk objects, cannot normally be tracked and cataloged. Depending on their relative impact velocities, these risk objects can cause catastrophic damage.
3) Objects smaller than 1 cm in diameter are most commonly referred to as small debris. These objects cannot be tracked but their existence is known via observed damage to returned space objects.

The population of debris objects smaller than 10 cm is derived from statistical measurements made either in situ or from ground-based sensors. The interaction among these three classes of objects, combined with the long residual times in orbit of the larger fragments, leads to further concern that there might be collisions producing additional fragments and causing the total debris population to grow. The existing and likely future orbital debris populations vary significantly with altitude and inclination angle.

The effects of orbital debris impacts depend on velocity, angle of impact, and mass of the debris. For debris of sizes less than about 0.01 cm, surface pitting and erosion are the primary effects. Over a long period of time, the cumulative effect of individual particles colliding with a satellite might become significant because the number of particles in this range is very large in LEO. Debris of sizes 0.01 to 1 cm produce significant impact damage, which can be serious, depending upon system vulnerability and defensive design provisions. Objects larger than 1 cm can produce catastrophic damage.

For debris larger than about 0.1 cm, structural damage to the satellite becomes an important consideration. The kinetic energy in an aluminum sphere with a diameter of 1.3 mm at 10 km/s is the same as that in a 22-caliber-long rifle bullet. It is currently practical to shield against debris particles up to 1 cm in diameter, a mass of 1.46 g. For larger sizes of debris, current shielding concepts become impractical. Because debris damage is a function of relative velocity and the velocities at geosynchronous altitudes are relatively low, one-tenth those in LEO, the consequences are less dramatic but could still be significant.

Operational debris is composed of inactive payloads and objects released during satellite delivery or satellite operations. This class of debris is diminishing as designs are adopted that no longer release such objects. Of the cataloged objects in Earth orbit, 95% can be considered orbital debris as opposed to operational spacecraft.

Of particular concern is the sustained rate of fragmentation events despite the active efforts of space-faring nations to reduce the probability of such events by making all of their systems passive at mission end by expelling residual propellants and other forms of stored energy. Despite the introduction of procedures to eliminate stored energy, there has not yet been a change in the rate of breakups.

Very small debris particles are created by the gradual disintegration of spacecraft surfaces as a result of exposure to the space environment. This deterioration includes paint flaking and plastic and metal erosion. Solid rocket motors typically are used to transfer objects from LEO to GEO, and they eject thousands of kilograms of aluminum oxide dust into the orbital environment. This ejected dust is very small, with characteristic sizes believed to be less than 0.01 cm. Nonetheless, long-term exposure of payloads to such particles is likely to cause erosion of exterior surfaces, chemical contamination, and can degrade operations of vulnerable components such as optical windows and solar panels.

To project the future debris environment, assumptions have to be made concerning debris sources and sinks. With regard to debris sources, assumptions have to be made concerning launch and fragmentation rates. Uncertainties arise from traffic model predictability, observational limitations, unmodeled sources, limitations of break-up models, debris propagation and lifetime models, and variability in solar activity. In addition, the advent of large LEO satellite constellations could present a significant new issue for the orbital debris environment, modeling, and projections. For example, although the Iridium satellites were designed for reentry near end of useful life an inordinately large number of its satellites have failed in orbit, and it is not clear whether or not these satellites can be deboosted as planned. If they cannot, then there is an increased likelihood of collisions within the Iridium constellations because the failed satellites are likely to be in close proximity to the replacement satellites.

Models have been developed for forecasting the orbital debris environment over long periods of time. Because the economic impacts of debris might not be observable in the near term, models must be used that are capable of forecasting for at least 50 to 75 years. A particular model, EVOLVE, developed by NASA, is capable of the required long-term forecasting and was reportedly used to assess the effect of deploying several LEO satellite constellations. EVOLVE was developed by Dr. Robert Reynolds while at Lockheed–Martin Engineering and Sciences Company for NASA Johnson Space Center. It is an evolution model that defines the environment as a time-varying ensemble of intact space systems and debris fragments each in an orbit and having mass, cross-sectional area, and other associated attributes. EVOLVE uses as input data the historical record of launch traffic (spacecraft, upper stages, and operational debris) and mission model data for future traffic, and uses break-up models to determine the distribution of debris fragments resulting from collisions or explosions occurring in orbit. The orbit propagator accounts for atmospheric drag and perturbations associated with the oblateness of the Earth and, for highly eliptical orbits, lunar and solar perturbations. The analysis assumed that five launches a year would deploy multiple spacecraft and examined the effect of such an increase in LEO activity and the influence of a spectrum of mitigation strategies in the long-term future environment. Mitigation options ranged from actions to eliminate future explosions to removing upper stages and spacecraft from orbit at end of mission lifetime. Figure 5.11 illustrates typical resulting projections of the cumulative cross-sectional flux (i.e., probability of debris impact per unit cross sectional area). Similar results have been used in the economic analyses reported in Sec. VII of Chapter 4. Because of the long time frames involved and the many uncertainties in the measurement and modeling processes, the indicated results should be considered as indicating trends with less reliance placed upon specific numbers. As the curves in Fig. 5.11 indicate, failure to take any action will likely lead to

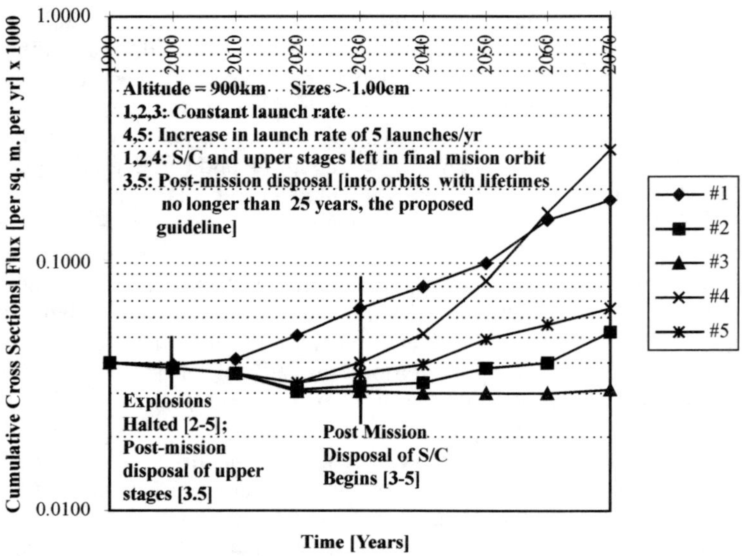

Fig. 5.11 EVOLVE projection of the future environment with increased launch and spacecraft operation in LEO (Source: Ref. 18).

significant increase in orbital debris during the next century, but relatively modest active measures (as identified in cases 3 and 5) are likely to keep the environment essentially as it is today. Both the Teledesic and Iridium LEO communication satellite constellations planned to deorbit their upper stages and spacecraft at their end of life (at least for the current system configurations), but in-orbit satellite failures could negate this intended action.

The debris scenarios are summarized in general terms in Fig. 5.11. In addition, it should be noted that in cases (scenarios) 2–5, the decline of the debris environment through the year 2020 reflects a cleaning out of explosion fragments from pre-2000 explosions, and, also in cases 2–5, all debris environment growth after year 2000 is caused by random collisions. Halting explosions is effective in the near term, but is insufficient by itself for long-term control of the debris environment. In addition, it is assumed that postmission disposal of upper stages begins (for cases 3–5) and explosions are halted (cases 2–5) in year 2000 and that postmission disposal of spacecraft begins in year 2030 (cases 3 and 5).

B. Debris Remediation: Economic Implications

As discussed in Sec. VII of Chapter 4, orbital debris will increase the cost of space missions through costs incurred to conform to debris remediation policies and/or losses that result from debris impacts. Cost-effective orbital debris remediation policies will, in the long term, reduce mission costs (below those that would result from pursuing a policy of no orbital debris remediation) but might increase mission costs in the short term. The methodology for assessing the economic impacts of orbital debris and remediation alternatives was also developed in Sec. VII of Chapter 4.

For the analysis of the economic impact of orbital debris remediation alternatives, the orbital debris environment is described in terms of the probability of impact with debris per unit spacecraft surface projected area per unit time. The probability of impact is a function of debris size, models of debris sources, projected space traffic and debris remediation practices, and satellite orbit and design characteristics. The probability of impact affects satellite failure and replacement rates and, as a result, mission life-cycle costs.

The following is a summary of the economic impacts of orbital debris on Landsat and Topex types of missions.[22] A number of orbital debris scenarios were considered for the Landsat (705 km at an inclination angle of 98.2 deg) and Topex (1336 km at an inclination angle of 65.1 deg) orbits. A typical scenario[22] assumes an increased use of space with easily achieved debris mitigation measures employed. It is assumed that the number of launches increases by approximately five per year, explosions are eliminated in launches after year 2000, and all spacecraft, upper stages, and launch debris are abandoned in their final mission orbits. The specifics of this typical scenario are depicted in Fig. 5.12 in terms of the probability of debris impact per unit area per year for debris of diameters greater than 0.316, 0.631, and 1.0 cm. It is evident that the debris environment is likely to be significantly different for these orbits, and as a general rule it should be assumed that the debris environment will be a function of altitude and inclination angle.

The indicated probability of debris impacts was utilized in mission simulations, which considered the details of the satellite configuration, satellite subsystem failure rates, launch vehicle reliability, satellite recurring and nonrecurring costs, launch costs, etc. The simulations considered launch and on-orbit operations of the satellites over extended periods of time replacing satellites that failed (as a result of launch and/or on-orbit failures). The consideration of orbital debris caused an increase in the number of satellites requiring replacement with an accompanying increase in costs. The results of the economic analyses are summarized in Fig. 5.13 for the Landsat mission. The indicated cost impacts represent the increase in life-cycle cost that results from the assumed orbital debris environment and are relative to the no debris case. The cost curves indicate the cost increases given that the mission is initiated in the indicated year. Topex impacts are similar but of lesser magnitude than those obtained for Landsat because of the combined effect of smaller debris flux and smaller satellite cross-sectional area (6.9 m^2 for Topex vs 12 m^2 for Landsat). The obtained results indicate that economic differences will result from the consideration of different satellite altitudes and inclination angles and therefore might have a significant bearing on the need for and timing for the initiation of regulatory actions.

C. Remediation Schema

The following is a summary of the commonly practiced techniques and those favored for future implementation (as per Ref. 16). Design and operational

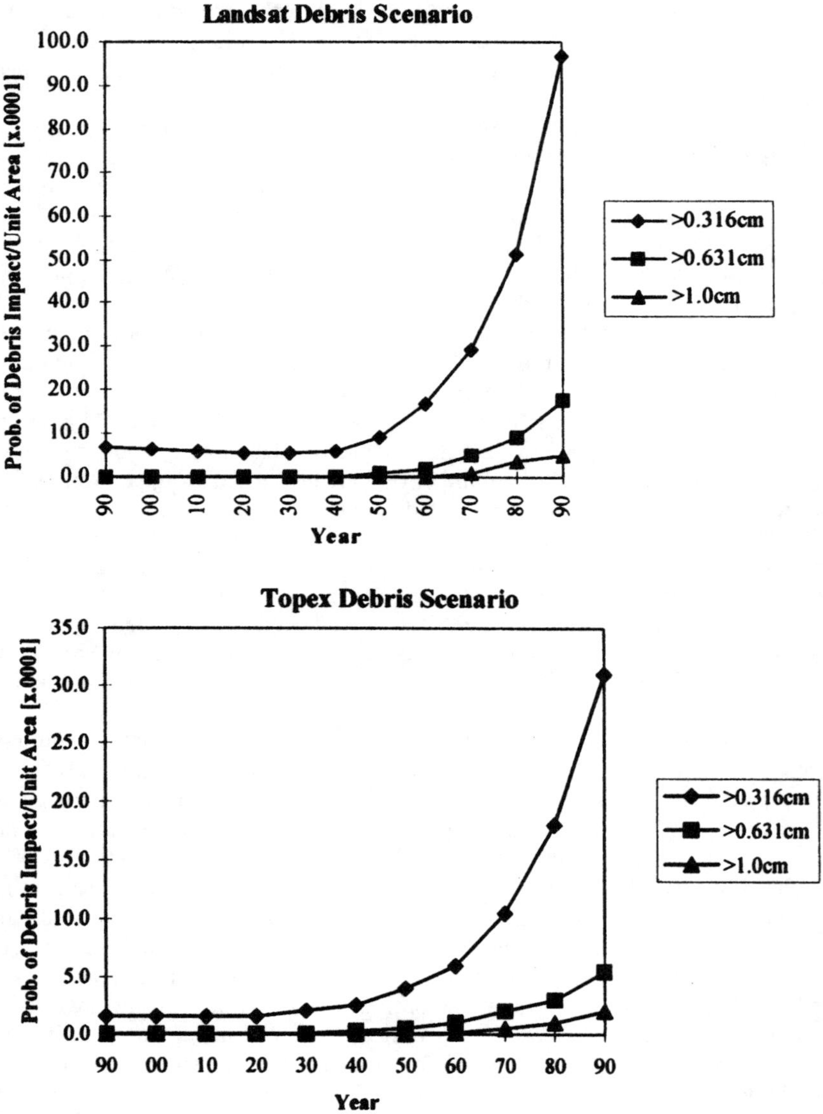

Fig. 5.12 Probability of orbital debris impact: typical debris scenarios for Landsat and Topex orbits.

techniques already used for reducing the creation of orbital debris include the following: 1) discarded rocket bodies, with a) expulsion of excess propellant, b) expulsion of excess pressurants, c) minimization of independent launch vehicle parts allowed to reach orbit, and d) securing parts to the upper stages; 2) spacecraft that have terminated their missions, with a) deorbit and controlled reentry—LEO spacecraft, and b) orbit maneuvering to shift spacecraft or components into

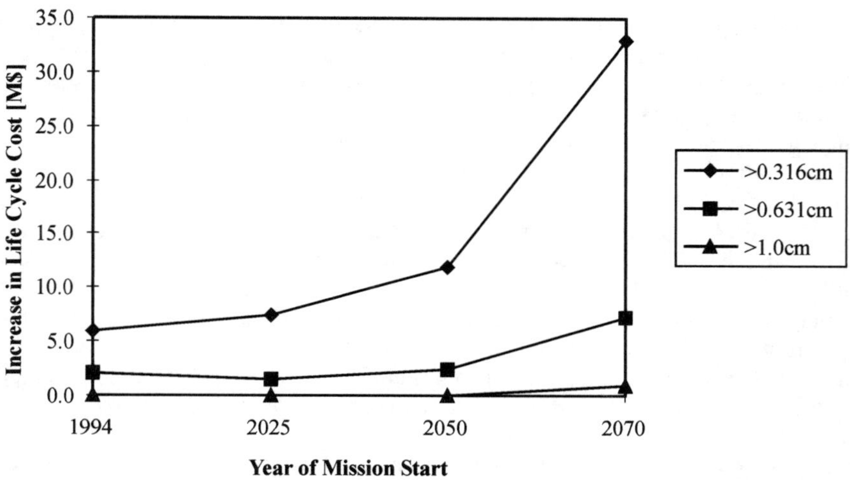

Fig. 5.13 Typical Landsat orbital debris induced life cycle cost impacts.

disposal (graveyard) orbits (not a long-term solution); 3) operational debris released from spacecraft during their missions, with a) lanyards attached to all potentially releasable items such as camera lens and instrument covers, equipment used by astronauts in extravehicular activities, and other material, and b) structural attachment of otherwise detachable elements; and 4) fragments originating from explosions or collisions, with increased shielding.

In addition to the preceding techniques, other measures appear to have widely acknowledged potential. Among these are the following: 1) discarded rocket bodies, including a) use of separation devices, b) use of the COMBO program (a program developed by the U.S. Air Force Space Command that performs the calculation of miss between orbits), c) enhancement of the accuracy of the COLA collision avoidance program, and d) selection of launch times and dates to exploit natural forces for more rapid reentry of debris into the atmosphere; 2) spacecraft that have terminated their mission, including a) retrieval and/or reuse of spacecraft, and b) use of active beacons for spacecraft detection and avoidance; 3) operational debris released from spacecraft during their missions, including a) storing of trash and human waste and return with logistics flights, and b) deorbiting trash and human waste for incineration in the atmosphere; and 4) fragments originating from explosions or collisions, including a) protecting and preventing hardware elements from exploding, b) designing for graceful degradation of components and systems, c) incorporating adequate shielding, and d) use of low fragmentation materials.

D. Regulatory Issues[23]

To understand how government regulation will play a role in the commercial space sector's debris-reduction effort, it is necessary to understand the federal regulatory approach to the commercial sector, as well as the different types of regulation. Most federal regulation falls within one of the following categories: 1) direct control of commerce and trade under a program of economic regulation, 2) protection of public health and safety and the environment, and 3) proper management and control of federal funds and federal property.

The authority of both the FCC and NOAA is related to the first category. The DoT/FAA and FCC are charged by statute with carrying out the second category: the DoT/FAA regulates the commercial launch sector to protect public health and safety, and the FCC regulates communications for the purpose of promoting safety of life and property. The FCC's authority also falls into the third category in that it manages and controls the private sector's use of the national radio frequency spectrum, a public good. The FCC's role in regulating commercial space activities involves licensing providers of telecommunications services (which can include satellites), assignment of orbital positions consistent with international treaties, and establishment of standards governing transmitter design and operation to ensure appropriate frequency usage (e.g., spacecraft control pointing accuracy and position tolerance). The FCC authorizes the construction, launch, and operation of U.S. commercial communication satellites in geostationary and nongeostationary satellite orbits, while recognizing the DoT/FAA's responsibility for safety issues associated with payload launch operations and launch mission.

NOAA is responsible for licensing private remote-sensing space systems. Its licenses include a condition under which the licensee must "upon termination of operations under the license, make disposition of any satellites in space in a manner satisfactory to the President." This clearly provides adequate authority to require that a spent spacecraft not be left in a position that contributes to the space debris problem. NOAA's authority does not extend to activities that are part of the launch. NOAA currently implements its statutory requirement in individual licenses to remote sensing satellite operators.

In accordance with the statutory requirement for satellite disposal, on 3 November, 1997 NOAA released an NPRM proposing the following rule:

A licensee shall dispose of any space platforms owned or operated by the licensee upon termination of operation under the license in a manner satisfactory to the President. The licensee shall obtain approval from (NOAA) of all plans and procedures for the disposition of such platforms, e.g., uncontrolled reentry, burn on reentry, or controlled deorbit, providing sufficient notification to allow determination that the proposed procedures will minimize orbital debris and not jeopardize safety.

NOAA received approximately 17 comments on the NPRM; a final rule has not been issued. Although required by government policy to perform a regulatory impact analysis, a part of which is an economic impact analysis, no such economic impact analysis was performed. Considerable uneasiness was evident when this oversight was pointed out to NOAA and might have led to the significant delay in the rulemaking.

The principal purpose of the authority granted to the Secretary of Transportation under the *Commercial Space Launch Act of 1984*, as recodified as 49 United States Code Subtitle IX, Chapter 701 (the Act), is to oversee and coordinate the conduct of commercial space launch operations in a manner that protects the important national interests associated with such activities: public health and safety, safety of property, U.S. national security, and foreign policy interests. The Secretary is empowered to issue licenses authorizing the conduct of commercial launch activities and to establish the regulatory regime for ensuring that they are conducted safely and responsibly.

The DoT's charter as a safety regulatory agency encompasses all nongovernment launches conducted by U.S. citizens or from U.S. territory, payloads involved in launches subject to the DoT licensing requirements, and non-U.S. government launch sites (e.g., privately operated or state-run spaceports). With specific regard to nongovernment payloads on nongovernmental launch vehicles, proposals that are not subject to licensing by another U.S. Government agency must be regulated by the DoT from the standpoint of the national interests that the DoT is charged with protecting. The DoT/FAA has no licensing or regulatory jurisdiction over FCC and DoC-licensed satellites (except for the DoT's authority for licensing launch and related operations).

Regulatory oversight of the commercial space launch sector for the purpose of preventing and controlling orbital debris falls into the "safety" category of regulatory functions. As just noted, the DoT is expressly authorized to regulate commercial launch activities, including reentry operations, in terms of public safety and other public interests, and the FCC is expressly authorized to regulate the use of radio to make available an efficient nationwide, worldwide, radio communication service. It is the DoT's plan to evaluate space debris issues consistent with its Congressional authorization to license and regulate commercial launch activities in a manner that ensures protection of public health and safety, safety of property, and other U.S. interests.

Pursuant to this authority, the FAA (DoT) has proposed to restrict the generation of orbital debris from commercial launch operations. In March 1997 the FAA adopted a NPRM proposing new rules that essentially codify existing debris mitigation practices among mature launch vehicle operators, including fuel depletion and venting of upper stages. The new FAA rules would require commercial launch vehicle operators to design and operate their vehicles in a manner which ensures that there is "no unplanned physical contact between the vehicle and its components and the payload after separation." Furthermore launch licensees must ensure that "debris generation will not result from conversion of energy sources into energy that fragments the vehicle or its components." In other words, launch operators must take care to prevent explosion and breakup of upper stages.

The National Space Policy requires that orbital debris mitigation measures be "consistent with mission requirements and cost effectiveness." Debris mitigation design solutions will result in some added cost or payload penalty. By implementing these solutions during the system design process, these penalties can be kept to a minimum. A requirement to deorbit upper stages, for instance, entails weight and performance changes that increase launch costs. In determining what steps the U.S. government should take to address the orbital debris problem, it is necessary to consider the economic impact of these commercial regulations on the domestic launch and satellite industries. Unlike government, the private sector functions in a highly competitive environment. The costs associated with orbital debris remediation measures are passed on to the customer. If the same launch requirements are not imposed on foreign competitors in the launch industry, U.S.

launch firms might have to operate at a competitive disadvantage. Similarly, added costs can have a direct bearing on the competitiveness of space-based technologies (e.g., satellite communications), as compared to terrestrial alternatives (e.g., fiber-optics communications).

E. Regulatory Economic Impact Analysis

E.O. 12866 of 30 September 1993 states that "federal agencies should promulgate only such regulations as are required by law, are necessary to interpret the law, or are made necessary by compelling public need, such as material failures of private markets to protect or improve the health and safety of the public, the environment, or the well-being of the American people." In deciding whether and how to regulate, agencies need to assess all costs and benefits of available regulatory alternatives, including the alternative of not regulating. In the case of orbital debris regulatory action, regulatory alternatives are likely to include the imposition of different levels of requirements, different timing for the setting of regulation, and different requirements and timing of regulations for different satellite altitudes and inclination angles. Costs and benefits need to include both quantifiable and qualitative measures. In choosing among alternative regulatory approaches, it is desirable to select those approaches that maximize net benefits (including potential economic, environmental, public health and safety, and other advantages; distribution of impacts; and equity). For orbital debris, net benefits can be measured in terms of cost savings for achieving the same level of operational performance (for example, on-orbit level of availability).

When considering the imposition of a regulatory regime, it is necessary to perform a regulatory impact analysis (RIA) to demonstrate that 1) there is adequate information concerning the need for and consequences of the proposed action; 2) the potential benefits to society outweigh the potential costs; and 3) of all the alternative approaches to the given regulatory objective, the proposed action will maximize net benefits to society.

Therefore, an RIA normally contains a statement of the potential need for the proposed regulation, an examination of alternative approaches, an analysis of benefits and costs, the rationale for choosing the proposed regulatory action, and a statement of statutory authority. It is necessary to identify and assess available alternatives to regulation including providing economic incentives to encourage the desired behavior, such as user fees or marketable permits, or providing information upon which choices can be made by the public. Each agency must tailor its regulation so as to impose the least burden on society, including individuals, businesses of differing sizes, and other entities (including small communities and government entities) consistent with obtaining the regulatory objectives, taking into account, among other things, and to the extent practicable, the costs of cumulative regulations.

To establish the potential need for the proposal, the analysis needs to demonstrate that market failure exists that is not adequately resolved by measures other than federal regulations. In other words, left to its own devices, a normally functioning marketplace will not cause the desired (and necessary) changes to take place. Three major types of market failure need to be considered: externality, natural monopoly, and inadequate information.

The orbital debris regulatory problem appears to fall in the category of market failure resulting from externalities. An externality is said to occur when one party's actions impose uncompensated benefits or costs on another outside the marketplace. Environmental problems are a classic case of externality. Another example is the case of common property resources that can become congested or overused, such as fisheries or the broadcast spectrum. The clutter of space with orbital debris is an environmental problem. The consequences of orbital debris might be of minor import today, but if appropriate measures are not taken in the near term to restrict or reverse the growth of the orbital debris population long-term effects of orbital debris can be irreversible and disastrous. Market failure is the result of the possibly long time delays (perhaps 50 to 100 years) between cause and effect and the lack of economic incentive to make near-term investments by those who will not be around to be affected by the long-term debris environment.

There are, of course, certain types of space missions that already perceive economic incentives that encourage taking of cost-incurring actions in the near term so as to eliminate or substantially reduce the economic consequences of orbital debris—for example, GEO communication satellites are moved to higher altitudes near end of useful life and LEO communication satellites are likely to be moved to other orbits (including reentry) near end of useful life.

It is because of this lack of uniformity of economic incentives coupled with the long time between cause and effect that externalities appear to exist of sufficient concern to warrant the consideration of the need for regulation.

F. Economic Impact Analysis in Support of Establishing a Regulatory Regime

In support of establishing a regulatory regime for orbital debris, economic analysis must first demonstrate the need for regulatory action. The case must be made that the normally functioning marketplace will not properly safeguard the environment and will ultimately adversely effect the well-being of the American people (i.e., market failure as the justification for regulatory action). This must be accomplished in light of GEO and LEO communication satellite actions and other voluntary debris remediation actions.

The economic analysis must identify cost-effective approaches to satisfy requirements while indicating the importance, or lack thereof, of the regulatory requirement on the approaches. In other words, it is necessary to identify the least costly approach to satisfying the requirement and the sensitivity of the result to level of the requirement.

For the cost-effective alternatives it is necessary to develop benefits and costs that will result from regulatory action. In accomplishing this, the economic analysis must consider different missions (altitudes and inclination angles) initiated at different points in time and must consider different regulatory alternatives that can be imposed at different points in time and for different orbit altitude and inclination angle regimes. Benefits are, in general, the result of longer satellite life (i.e., fewer satellite replacements) as a result of fewer debris impacts.

The analysis must identify who bears the costs and who obtains the benefits and when the costs and benefits occur so that the equity of impacts can be observed. Costs must take into account all effects of changes in launch vehicle capability (resulting from meeting requirements) as they affect user "value." For example, requiring transfer stages to reenter the Earth's atmosphere within a specified period of time can result in nonoptimal trajectories with resulting reduction in lift capability. A reduction in lift capability can be translated into shorter satellite life time, which will affect mission economic value. This reduction in economic value must be contrasted with the effects of orbital debris on increased mission costs if the remediation measure were not taken. Similarly, if the regulatory action will affect satellite capability, then the related costs (for example, increased satellite mass resulting in less station-keeping propellant and reduced satellite life) must be contrasted with those that would be incurred in the absence of the regulatory action.

The bottom line is that the economic analysis should develop the case that debris, if the regulatory action is not taken (i.e., no remediation), will lead to costs that will exceed remediation costs and that, because of market failure, regulatory action is required.

The economic analysis must consider the implications of the regulatory action on international competitiveness. The economic analysis must consider the impacts of the proposed regulatory actions on market share for launch services, satellites, and associated products and services. This must take into account both likely foreign actions and inaction. Inaction must be considered because this will allow maximum competitiveness impacts to be developed.

A first step in performing the economic analyses just identified is the establishment of a set of requirements (i.e., the specific goal or objective of the regulatory action as manifested in terms of specific operational requirements). NASA has already taken a step in this direction. NASA has developed a set of guidelines and methods to comply with NASA policy to limit debris generation.[18] The guidelines serve to help ensure that launch vehicles, upper stages, and payloads meet acceptable standards for limiting debris generation. These guidelines are imposed upon NASA missions and have been recommended for industry adoption with regard to the generation of debris. The guidelines are specific with regard to orbital debris size, mass, and duration in orbit. The economic impacts of these agency self-imposed guidelines have not been analyzed. It is this set of guidelines that can serve as a reasonable starting point for regulatory economic impact analysis. The economic analysis is concerned with the impacts of a base-case set of requirements as well as alternatives (which can be the provision of incentives as well as variations to the base-case set of requirements).

Another important input for the regulatory economic impact analysis is a forecasted relationship of the orbital debris environment (i.e., probability of impact per unit cross-sectional area as a function of time) without and with regulatory action as a function of time and as a function of the time of initiation of the regulatory action. The debris scenarios are the result of the regulatory actions imposed upon satellites, launch vehicles, and related operations as forecasted to affect the orbital debris environment.

It is also necessary to postulate and consider regulatory alternatives, for example, timing of initiation and different regulatory requirements for different altitudes and inclination angles. For example, require satellites to move to a different location (or reentry) near end of life; this might reduce useful life but will lead to a less severe environment. Tradeoffs must be developed and evaluated with respect to those impacted in the short term vs those impacted in the long term. Benefits (for example, cost savings) should exceed costs but are not likely to for missions initiated in the short or midterm, whereas benefits are likely to exceed costs for missions initiated in the longterm. The economic analysis is concerned with how these should be traded off and whether or not all parties involved are impacted equitably.

In the event that damage is caused to operational satellites by orbital debris, it might not be possible, because of the long time periods involved, the size of the orbital debris that can cause the damage, etc. to pinpoint the responsible party, if indeed the party still exists. Insurance might be available to cover such situations. Another alternative might be the development of a fund created by taxing all launch providers and satellite owners in proportion to payload delivery capability, payload mass, etc. The intent would be to create a sufficient fund in the long term to provide the resources for compensating for damage caused by orbital debris of unknown ownership.

G. Summary

The use of the space environment has become extremely important to the functioning of the world's economy. The rate of growth of the orbital debris population, if allowed to continue unabated, might in the long term substantially reduce the ability to utilize the space environment. Debris remediation measures might have to be introduced sooner rather than later to protect this scarce resource. Much has been accomplished through voluntary remediation measures such as moving GEO communication satellites to higher altitudes near their end of useful life, the planning by several LEO communication satellite organizations to cause their satellites to deorbit near end of useful life, venting of upper stages, etc. It is not yet clear whether voluntary action will be consistent across all space-faring nations and whether these actions, if undertaken, will be sufficient, and whether these voluntary actions will be continued in the future.

Consideration is currently being given to the imposition of a regulatory regime in order to obtain the necessary remediation actions. The DoT's regulatory responsibilities relating to the protection of public health and safety, safety of property, and U.S. national security and foreign policy interests provide the authority for developing a regulatory regime relating to orbital debris that can result from commercial launches and from reentry operations. NOAA has the authority to regulate remote-sensing missions, and the FCC has the authority to regulate communication satellite missions. Because orbital debris remediation policies and resultant remediation actions are not likely to be separable, it might not be wise to separate the orbital debris regulatory responsibilities.

IV. Space Launch Indemnification

Commercial Space Launch Activities—49 U.S.C. Subtitle IX, Chapter 701, formerly the *Commercial Space Launch Act of 1984* as amended by Public Law 100-657 (15 November, 1988, also known as the "Commercial Space Launch Act

Amendments of 1988") or in the following discussion simply referred to as "the Act," among other things, established specific insurance and financial responsibility requirements that had to be met by a launch provider in order to qualify for a FAA license that would permit the launching of an object into space. The Act specifically identified the U.S. Government as a potential sharer in the financial risks that might result from damage to third parties caused by a licensed commercial space launch and related operations. This risk sharing takes the form of holding the licensee harmless for damages to third parties in excess of FAA-determined insurance requirements but less than $1.5 billion above this amount. This is often referred to as the *indemnification* provision of the Act. (This is actually a misnomer in that government payments are not guaranteed because they cannot be made without the specific authorization of Congress.) The risk sharing also takes the form of holding the licensee harmless for damages to government facilities in excess of FAA determined insurance requirements but not to exceed $100 million.

The indemnification provision of the Act played a significant role in the transitioning from government ownership and operation of space launch services to the expanding commercial launch industry of today, which provides commercial space launch services from federal, state, and privately owned launch facilities. The Act states that the indemnification provision "shall apply to each license issued or transferred under the Act for which a complete and valid application has been received by the Secretary (of DoT) to the date that is 5 years following the date of enactment of the Commercial Space Launch Act Amendments of 1988." This *sunset* provision of the Act was revisited by Congress and extended for another five years (through December 1999). It has recently again been extended for another five years. The following paragraphs provide background information and identify issues relating to third-party indemnification that can prove useful in establishing future government policies relating to the indemnification and sunset provisions of the Act. [Based upon the Act, the term "third party" has been defined (Federal Register Part IV, Department of Transportation, 14 CFR Part 440, Financial Responsibility Requirements for Licensed Launch Activities; Final Rule, 26 August, 1998) "as any person or entity other than the United States, its agencies, or its contractors or subcontractors involved in launch services; the licensee or transferee, the licensee's or transferee's contractors, subcontractors, or customers involved in launch services; or, any such customer's contractors or subcontractors involved in launch services." This definition, by not listing the employees of the U.S. government, defines government employees involved in providing launch services as third parties.]

A. Legislation in Support of the Space Industry

To avoid confusion, insurance, liability, and indemnification are defined as follows. **Insurance** "is an arrangement for transferring and distributing risk. It is an arrangement under which one [called an insurer] contracts to do something that is of value to another [usually called an insured] upon the occurrence of a specified harmful contingency."[24] **Liability insurance** "is that form of insurance which indemnifies against liability on account of injuries to the person or property of another. It may be issued to cover the liability of, for example, carriers, contractors, employers, landlords, manufacturers, owners, and railroads."[25] **Indemnity**

> "is a collateral contract or assurance by which one person engages to secure another against an anticipated loss or to prevent him from being damnified [that is, suffering loss or damage] by the legal consequences of an act or forbearance on the part of one of the parties or of some third person. The term is also used to denote a compensation given to make the person whole from a loss already sustained; as where the government gives indemnity for private property taken by it for public use.... It means, also, restitution or reimbursement."[25]

The *Commercial Space Launch Act Amendments of 1988* established specific insurance and financial responsibility requirements that had to be met by a launch provider in order to qualify for a FAA license that would permit the launching of an object into space. The Act specifically identified insurance and/or financial responsibility requirements to compensate for damage to third parties as well as to government facilities.

The Act required each licensee or transferee to obtain liability insurance or to demonstrate financial responsibility in an amount sufficient to compensate the maximum probable loss (as determined by the DoT) from claims by a thirdparty for death, bodily injury, or loss of or damage to property resulting from activities carried out under the license in connection with any particular launch. The Act further stated that in no event shall a licensee or transferee be required to obtain insurance or demonstrate financial responsibility in an amount that exceeds $500 million or the maximum liability insurance available on the world market at a reasonable cost.

The Act also required each licensee or transferee to obtain liability insurance or to demonstrate financial responsibility in an amount sufficient to compensate the maximum probable loss (as determined by the DoT) from claims against any person by the United States for loss of or damage to property of the United States resulting from activities carried out under the license in connection with any particular launch. The Act further stated that in no event shall a licensee or transferee be required to obtain insurance or demonstrate financial responsibility for any particular launch in an amount that exceeds $100 million or the maximum liability insurance available on the world market at a reasonable cost.

Specifically with respect to third-party liability, the Act further stated that should claims exceed the level of required liability insurance or financial responsibility then the U.S. government would make payment for such excess claims up to an amount not to exceed $1.5 billion with the payment of these claims requiring the enactment of additional legislative authority. This is often referred to as the indemnification provision of the Act. This portion of the Act has a five year sunset provision, which states that it "shall apply only to each license issued or transferred under the Act for which a complete and valid application has been received by the Secretary [of the DoT] prior to the date that is 5 years following the date of enactment of the *Commercial Space Launch Act Amendments of 1988*." This sunset provision was extended for a subsequent five-year period that expired in December 1999 and was then extended again for another five years.

The Report of the Senate Committee on Commerce, Science, and Transportation on H.R. 4399 provides an interesting discussion of the rational behind the indemnification provisions of the 1988 Amendments at the time of their passage. This discussion is summarized next and serves as a general background for the consideration of current issues concerning the continuing need for indemnification or other forms of assistance for the space launch industry.

According to the Senate report, two of the primary purposes of the legislation were 1) to establish a risk-sharing regime to facilitate the transition of the Nation's launch industry from a government activity to a commercial activity and 2) to maintain the competitive position of domestic launch services in the world markets. This transition began in earnest in August 1986, when NASA was instructed by the President to terminate most commercial space launch activities. However, despite this policy proclamation the prevailing conditions of the marketplace were not geared to commercial launch operations. For example, there was no commercial launch insurance industry to protect against losses of government property. Such insurance had never been required, either in the U.S. or elsewhere in the world. In addition, the market for launch liability insurance had undergone severe strain in the late 1980s, and there were significant losses (because of launch and/or satellite failures but not because of third-party damages) within that industry. Insurance costs, therefore, were extremely high and were expected to remain high until the insurance market recouped its losses. Furthermore, space launch liability insurance had previously never been required by the U.S. without some form of limitation and indemnification being provided by the Government. [When commercial payloads were launched aboard the space shuttle, they were required to obtain $500 million of third-party liability insurance (or the maximum amount available at reasonable rates). The U.S. government indemnified losses above the amount of required and obtained insurance.]

The Senate Committee therefore believed that there would be a period of adjustment necessary to develop and mature the world insurance markets to accommodate the domestic commercial launch industry. It was felt that insurance was a fact of commercial life and no operations could be conducted without adequate protection of corporate assets. An effective and competitive domestic launch industry was deemed to be in the public interest, and the Committee was convinced that some risk sharing for a limited period of time was necessary to sustain the growth of the industry.

A 10-year period of time was proposed for a sunset provision (which was changed to five-years in the ensuing legislation and a five-year extension was granted in 1993) and was considered as adequate to permit the insurance industry to develop actuarial data on which to base policies and risk. The Committee expected that the experience in the commercial industry would be such that at the conclusion of this period the world liability insurance markets would be sufficiently developed to accommodate the needs of the domestic industry. In addition, it was expected that rules of the road concerning pricing and business practices associated with the commercial expendable launch vehicle industry would be negotiated with other launching states during this same period to permit competition on a more equitable basis.

It was felt that the termination provisions incorporated a full marketing cycle for the commercial launch industry and would permit them ample time to prepare for the sunset of the government risk-sharing mechanism.

Recent (1998) legislation concerning NASA and indemnification relating to the X-33 and X-34 RLV research and development programs is also of interest. Public Law 85-804 authority allows NASA, among others, to indemnify contractors when such an action facilitates the national defense. But when the purpose of a contract is to demonstrate technologies needed to lower purely commercial risk and lead to a commercial vehicle designed to meet the needs of the commercial marketplace, a national defense link cannot be credibly claimed. It has thus been thought that the authority of P.L. 85-804 is not available to protect contractors in commercially focused programs like NASA's reusable launch program. But because the RLV flight-test program is being conducted for the public good as part of an experimental technology demonstration effort, the Administration and many outside experts believed that this liability risk would be an unfair expense for the company to bear. As a result, Congress has given NASA new authority regarding indemnification such that a claims mechanism has been established for R&D vehicles that is essentially identical to that provided by DoT for commercial launches. That is, the legislation provides the promise of expedited consideration by Congress of third-party claims in excess of insurance available at a reasonable price. In other words, the government provides indemnification above the level of obtained insurance. The authority to enter into such agreements is discretionary with NASA.

But what about indemnification for launch vehicle research and development programs that is truly commercial (i.e., no NASA involvement)?

B. Precedents

The following discussion is aimed at identifying a range of historical precedents that might have bearing on the indemnification issues and possible flexibility with respect to alternatives to the current indemnification regime.

1. Nuclear Power Industry

Because the *Price-Anderson Act* in support of the nuclear power industry is often cited as a precedent for launch industry indemnification, the following paragraphs provide a brief summary of the Act and its evolution and implications over time. Following sections discuss the nuclear industry insurance premiums and payout history.

The nuclear power industry continues to be characterized as an industry with the possibility of catastrophic accidents but with low probability of occurrence. Originally, the fear of potentially large losses led to industry reluctance to participate in nuclear power generation. To overcome this reluctance, as well as to protect the public if such an incident were to occur by ensuring the availability of funds to satisfy public liability claims and avoid delay in such compensation to the public, the *Price-Anderson Act of 1957* (amending the *Atomic Energy Act of 1954*) was enacted.

Price-Anderson required that a nuclear power reactor operator submit proof of financial protection, either through private insurance, private contractual indemnities, self-insurance, or other financial responsibility, covering liability claims for bodily injury or property damage losses caused by nuclear materials. The required amount of protection was equivalent to all of the nuclear liability insurance that was available from private insurers. This initially amounted to a maximum of $60 million (1957 dollars) per facility in 1957 when the Act was first passed. The government agreed to be liable for $500 million beyond the amount privately available. The Act limited each nuclear facility's maximum liability to $560 million, the amount then available from the private market plus the $500 million the government would indemnify in case of loss exceeding private market capacity. For the extra protection from the government, reactor operators had to pay an annual fee to the government. This was the first time any such protection had been offered by the federal government for any particular industrial hazard, and the amount of protection it allowed was unprecedented in the history of liability insurance covering a commercial activity.[26]

A 1975 amendment to the Act required the creation of a secondary level of financial protection that would be provided through payments of *retrospective premiums* ($5 million per large operating reactor) when and if a nuclear catastrophe occurred at one of the plants that exhausted the primary level of financial protection.[27] If a nuclear accident were to occur causing damages in excess of the primary layer, which is the amount available from private sources and which has grown from $60 million to more than $160 million maximum (in the early 1980s) per nuclear reactor, each nuclear plant operator would be billed up to $5 million for each incident for each reactor. There was a limit of $10 million for each reactor in a calendar year. The intention was to phase out government indemnification, and this has actually occurred as the financial protection available from private sources reached (in the mid-1980s) $700 million ($160 million primary and $540 million secondary—108 power plants providing $5 million each of retrospective premium).[28] Under the Amendments, the maximum liability limit floats upward to meet the amount of protection both primary and secondary available from private sources.[29] Currently, the retrospective premium is on the order of $70 million, making the amount of coverage in excess of $7 billion.[30] Should losses ever exceed this maximum limit on liability, Congress has obligated itself to review the situation and take appropriate action to protect the public.

2. Military and Other Government Agencies

Man-made catastrophic accidents resulting from government programs can occur. Government contractors might find themselves liable for some or all of the claims that could exceed any insurance policy that the contractor would have in force in the normal course of business. Not only could a contractor's business be in jeopardy, but also the people injured would likely not be able to recover in full.

Since 1952, the Department of Defense (DoD) has been able to indemnify its research and development contractors against claims arising out of direct performance of their contracts that result from risks defined in the contracts as "unusually hazardous." Because of definitional ambiguities in the drafting of the legislation and because it applies only to research and development activities in the DoD, this type of indemnification authority has not been extensively used.

In 1958, Public Law 85-804 was enacted (50, U.S.C., Section 1431-35). It permits indemnification by the DoD for procurement contracts where the "magnitude of the risks involved...in these areas have rendered commercial insurance either unavailable or unlimited in coverage" and where indemnification would further the national defense. The Act specifically mentions both nuclear risks and those from the use of highly volatile fuels in the missile program. And, the Act draws specific analogy to the indemnification authority of the Atomic Energy Commission in the *Price-Anderson Act*.

There are no limits to the indemnification that would be provided under this Act. Furthermore, this Act was not limited to the DoD. As amended, the Act applies to other selected agencies, including NASA and the DoT.

NASA has applied the law in only a limited number of cases, whereas the FAA has not used the provisions of the law. One possible reason for their sparing application of this law might be that they did not want to have the work they were doing labeled as unusually hazardous activity. The Act provides no provisions for any special relief for the injured public, such as the waiver of defenses that has become part of the *Price-Anderson Act*. There is also no definition of how much insurance the contractors must carry, and each agency is able to set its own limits.

3. *NASA/Space Shuttle*

Since the late 1970s, NASA has required commercial users of the space shuttle to obtain insurance to protect the federal government from potential tort liability resulting from injury to third parties. The required level of insurance was set at the amount that was available from private sources at reasonable prices but not to exceed $500 million. NASA provided indemnification above this amount. When there was more than one commercial user aboard a shuttle flight, each was required to obtain up to the $500 million amount. However, as a result of market conditions this was not always possible; typically, liability coverage for two payloads amounted to $750 million.

By the authority provided to NASA via Section 308 (1979) of the *National Aeronautics and Space Act of 1958*, the NASA Administrator may "provide liability insurance for any user of a space vehicle to compensate all or a portion of claims by third-parties for damage resulting from activities carried on in connection with the launch, operations or recovery of a space vehicle." The users are to reimburse the Administration to the maximum extent "practicable" for this insurance. Subsection [b] of Section 308 states:

> Taking into account the availability, cost and terms of liability insurance, any agreement between the Administration and a space vehicle user may provide that the United States will indemnify the user against third-party liability claims resulting from activities carried on in connection with the launch, operations or recovery of the space vehicle to the extent that the claims are not compensated by the liability insurance of the user.[31]

Section 308 was implemented under Executive Order No. 10789, a nonmajor, final regulation titled "Insurance and Indemnification of NASA Space Vehicle Users" (14 CFR Part 1214 Subpart 1214.13). The regulation states that all users of a NASA space vehicle, with the exception of those flying small self-contained payloads, and users providing payload specialist services for NASA missions must obtain insurance protecting themselves, the U.S. government, and other parties identified in the launch agreement. However, this insurance requirement can be waived for a particular user or flight if it is determined that such action is in the "public interest." NASA and the user, taking into consideration the insurance available in the world market at a "reasonable premium," must agree to the amount of insurance and the terms and conditions of insurance. If NASA determines that adequate third-party liability insurance is not available on reasonable terms and conditions or at a reasonable premium on the private market, or that the availability of insurance prevents an orderly and equitable allocation of liability risk, NASA will provide adequate liability protection for users and the U.S. government.

4. *FAA/Aviation Industry*

The U.S. aviation industry is another example of Congress creating the potential for unfunded liability generated by the actions of a commercial firm. If insurance is unavailable from the private sector on reasonable terms, the federal government can insure or reinsure any aviation risk. However, this insurance is only available for aircraft engaged in foreign air commerce and if there is a determination that the insurance scheme is needed for carrying out overall U.S. foreign policy.

In actual operation aviation insurance provided by the DoT has been used rarely. There are two types of insurance provided. Nonpremium insurance can be provided for international flights under government contract. For example, insurance might be provided if the DoD chartered aircraft from a carrier and was operating in a risky zone where commercial insurance would be unavailable. Arrangements for this type of insurance would be made with the U.S. government agency involved signing an indemnification agreement with DoT.

Premium aircraft insurance can be provided when the legislation criteria (as just stated) are met. DoT has no clear definition for what would be deemed reasonable terms for a private premium, although they have attempted to define it as "a premium set so that a carrier could make a profit on the flight." DoT sets its limits of liability for a flight equal to the carrier's commercial insurance—typically in the range of $600 to $800 million. Most of the requests received for DoT are for hull insurance, not third-party liability. This premium insurance legislative authority has been utilized infrequently. The most recent use has been for commercial flights to Tel Aviv at the time of the Gulf War. DoT utilized this authority to maintain reasonable insurance rates for the commercial flights.

5. *Other Precedents*

Many other precedents relating to indemnification and the establishment of unfunded government liabilities exist. Many relate to environmental concerns such as marine transportation of oil, which can result in large and destructive oil spills with very high clean-up costs. There are a number of statutes dealing specifically with oil spills, which are aimed mainly at ensuring that funds for spill clean up will be available and that the sole source of those funds is not the U.S. government. Two aspects of the variety of general and specific laws that govern liability for oil spills are relevant as precedents for space-related third-party liability regulations. The first is the *Limitations of Liability Act*, which limits the liability of nonnegligent ship owners in the case of an accident. The second is a group of pollution related statutes passed since 1970.

The government has set in place a number of measures that place the government in risk-sharing situations. These include the FDIC, Export-Import

Bank, and Federal Emergency Management Administration (FEMA). (Other programs that place the government at risk or in a risk-sharing mode include the Federal Crop Corporation, and the National Credit Union Insurance Corporation, Social Security, Medicare/Medicaid, Catastrophic Medical Insurance, and Pension Benefits Guarantee Administration.) The FDIC is a corporation set up by the government. It charges premiums for insurance on individual bank deposits up to $100,000 per account. If a bank cannot cover its liabilities, then the FDIC is obliged to act to protect individual account holders. Many actions are possible, from taking over the bank operations on a temporary basis to selling or forcing a merger of the bank with another institution. As a last resort, it might have to honor the insurance commitments and pay from the FDIC's own account enough money to cover the liabilities up to the insurance limits.

The Export–Import Bank, by contrast, has a limitation on loans, guarantees, and insurance and is organized to help U.S. companies sell their products abroad by providing financial assistance to foreign buyers to purchase U.S. products. The Bank operates under program-specific limitations but normally has a significant total liability (in 1988 this exceeded $10 billion).

The U.S. government provides flood insurance through FEMA. Insurance is sold to individuals in flood-prone areas, where private insurance is not available. Congress has limited the amounts of notes and obligations outstanding at any one time to $500 million without the approval of the President and to $1 billion with the President's approval.

C. Space Insurance Industry

Insurance for commercial space launches is now in part regulated by the federal government. The FAA, as part of the DoT, requires that insurance (or an approved alternative) be obtained to cover damage to government property and property damage or injury to third parties that might occur as a result of commercial space transportation activities. Launch insurance, protecting the launching firm and its customers against launch or payload failure, is not required or regulated. There are four main types of insurance for commercial space launch activities: prelaunch insurance, launch insurance, in-orbit insurance, and liability insurance.

Prelaunch insurance covers satellites and launch vehicles while they are on the ground and prior to the commencement of hazardous launch operations. This insurance covers risks during transportation from the factory to the launch site as well as during assembly, test, and checkout on the launchpad, including risks during inspection and waiting for liftoff. Usually the period of cover begins when the satellite is accepted from the manufacturer and ends with either ignition or liftoff of the launch vehicle.

Launch insurance coverage begins when the prelaunch insurance ends. It includes cover for launch failures (encompassing failures of the launch vehicle and transfer stage and usually of initial satellite on-orbit operations) and is purchased by the launching firm and/or its customer. The policy pays off in the event that the launch fails. The payoff can be in cash, hardware (such as a replacement satellite, if the policy holder is not the satellite manufacturer), or services (such as relaunch, if the policy holder is not the launch firm).

In-orbit insurance coverage begins after a satellite reaches orbit and completes the initial functionality testing. The owner or operator of the satellite is the insured, and the insurance cover is usually based upon one-year renewable policies. The policies basically insure the revenue generation ability of the satellite.

Liability insurance includes both government property and third-party liability insurance. A government property insurance policy covers damage to government facilities that results from commercially licensed launches. This insurance must be purchased, or financial responsibility demonstrated, by the entity licensed by the FAA to conduct commercial space transportation activities. The policy must be at least for the amount specified by the FAA, and pays in the event that government property on a launch range is damaged by licensed commercial space transportation activities. Government property insurance for commercial space launches is required by law. In addition, a third-party liability insurance policy must be purchased, or financial responsibility demonstrated, by the entity licensed by the FAA to conduct commercial space transportation activities. The policy must be at least for the amount specified by the FAA and pays in the event that third-party claims for property damage, injuries, or casualties are made in association with licensed commercial space transportation activities. Third-party liability insurance for commercial space launches is required by law.

Launch and in-orbit premiums and claim history are summarized in Fig. 5.14. The data presented are in current dollars and are based upon published data obtained from COMSTAC (the FAA's Commercial Space Transportation Advisory Committee).[31] It appears that the space insurance industry has been in a profitable position for six of the last 10 years. 1998 was a particularly bad year with payouts (approximately $1.5 billion) being almost double the value of received premiums (approximately $800 million). There is little correlation between launch insurance

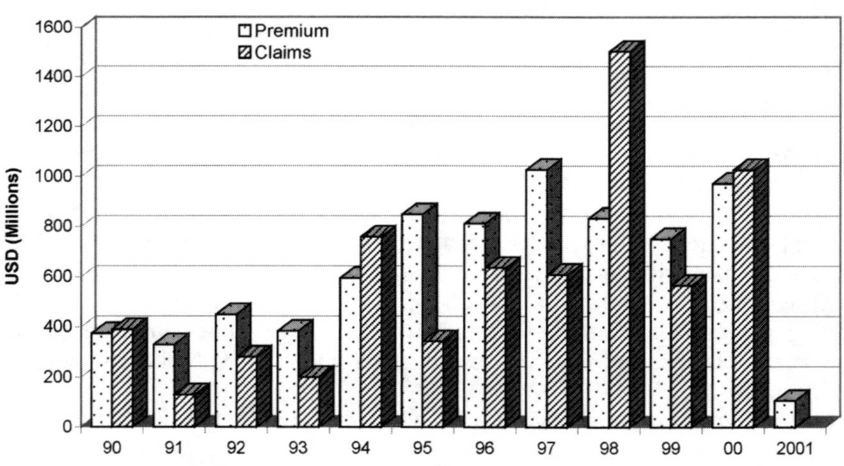

Notes:
1) Premiums are Gross 2) Claims figures are before salvage

Fig. 5.14 Launch and in-orbit premiums and claims (current dollars) (Source: Ref. 32).

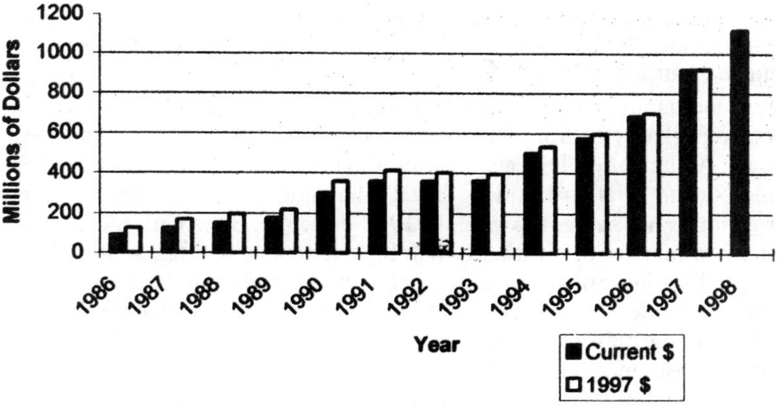

Fig. 5.15 Launch insurance market capacity in current and constant (1997) dollars.

availability and premiums and third-party liability insurance availability and premiums because aviation insurers, a different market segment, provide the latter.

The ability of the space insurance industry, based upon industry-provided data, to provide launch insurance cover is summarized in Fig. 5.15 in both current and constant (1997) dollars. It appears that market capacity has grown significantly in recent years and hovers around the $1 billion dollar amount.

The insurance industry also provided statistics on the maximum and minimum rates for launch insurance (which includes 180 days of in-orbit testing) and failure rates for the years 1986–1998. These are indicated in Fig. 5.16 and represent the range of insurance rates across launch vehicles and missions and average failure rates across launch vehicles and missions. The years 1997 and 1998 indicate a significant drop in the minimum insurance rates and presumably reflect launches

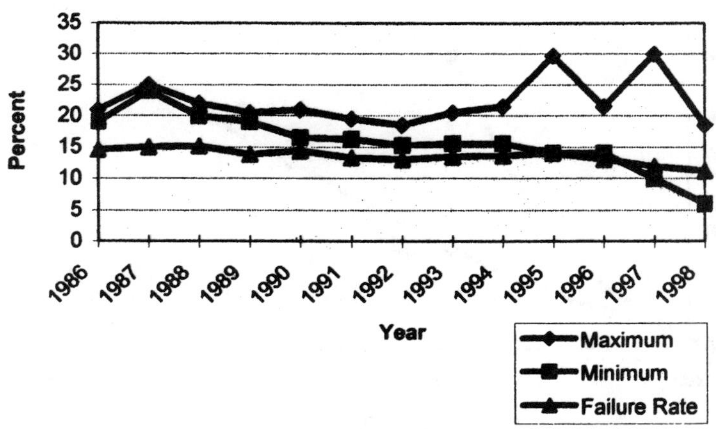

Fig. 5.16 Launch insurance premium rates (launch plus 180 days) and failure rates.

to low Earth orbit. It is also likely that failure rates in 1997 and 1998 include launches to low Earth orbit.

Third-party liability insurance has historically been provided by aviation insurers and liability insurance for space activities included as part of the aviation insurance package for firms that are in both the aviation business and the space business. A number of the recent entries into the space transportation business are not also in the aviation business. Thus, obtaining third-party liability premium information is not straightforward. In 1992 Princeton Synergetics, Inc., compiled information relating to third-party liability premiums as a function of insurance coverage. This information was recently (1999) provided to a limited number of insurance providers who were asked to indicate the change that had taken place from 1992 to the present. The results are indicated in Fig. 5.17. Indications are that the cost of insurance has decreased significantly over the past several years. The curves indicated in Fig. 5.17 as "thin lines" actually have considerable variability associated with them because the actual premium is established on a case-by-case basis. The liability insurance comes from the aviation pool and, in theory, should be affected by the aviation accident history. However, the recent significant aviation accidents have not affected the price or the availability of aviation liability insurance. (The possible impacts of the events of 11 September 2001 on third-party liability insurance price and availability are not yet known.)

It was estimated by representatives of the insurance industry that \$1.0 to \$1.5 billion of liability insurance could be available as and when required. One respondent indicated that the capacity could be \$2 billion, whereas another felt that the current capacity is \$500 million but that could probably reach up to \$1 billion.

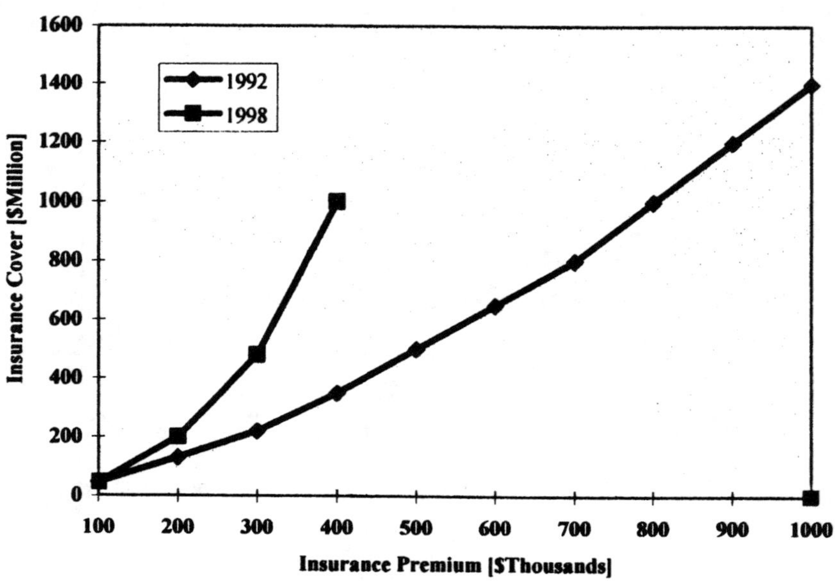

Fig. 5.17 Estimated cost of third-party liability insurance.

D. Summary of Accident History

The following paragraphs summarize the accident history of the U.S. space launch industry, the nuclear industry, and the aviation industry. To place these accident histories into perspective, the world's 18 most costly insured losses (prior to 11 September 2001) are summarized in Table 5.2, which indicates that natural catastrophes dominate the list of largest insured losses, although two man-made events make the list.

1. *U.S. Space Launch Industry*

The accident history (both government and commercial) of the U.S. space launch industry is summarized in Table 5.3 for 1963–1990 for both support prelaunch and expendable launch vehicle launch activities. During this nearly thirty 30-year period, there have been a number of accidents associated with prelaunch and launch activities that have resulted in facility damage and casualties.

Prelaunch activities have resulted in a small number of fatalities, several injuries (some serious), and minor damage to facilities. Perhaps the worst of the accidents involved solid rocket motor (SRM) activities and resulted in three deaths, five injuries, and less than $200,000 damage to facilities. A number of other SRM and fueling accidents resulted in lesser levels of damage. A number of industrial-type accidents also occurred with the worst resulting in one fatality, one injury, and minor damage to facilities. Several toxic leak and related accidents have also occurred without fatalities and in the worst case three casualties and minor damage to facilities.

Table 5.2 World's most costly insured losses (Source: Ref. 33)

Date	Catastrophe	Location	Insured damages
8/92	Hurricane Andrew	U.S.	17.95
1/94	Northridge earthquake	U.S.	13.28
9/91	Typhoon Mireille	Japan	6.42
1/90	Winter storms	Europe	5.53
9/89	Hurricane Hugo	U.S.	5.33
10/87	Autumn windstorm	Europe	4.15
2/90	Winter storm	Europe	3.84
7/88	*Piper Alpha* explosion	U.K.	2.66
1/95	Hanshin earthquake	Japan	2.55
10/95	Hurricane Opal	U.S.	2.17
3/93	Winter storm	U.S.	1.91
9/92	Hurricane Iniki	U.S.	1.80
10/89	*Phillips* explosion	U.S.	1.68
9/79	Hurricane Frederic	U.S.	1.63
9/96	Hurrican Fran	U.S.	1.60
9/74	Hurrican Fifi	Honduras	1.60
9/88	Hurrican Gilbert	Jamaica	1.48
12/83	Winter storms	U.S.	1.39

Table 5.3 Summary of U.S. space launch industry accident history (then-year dollars)

Accident	Result
Support prelaunch activities (1963–1990)	
SRM activities (5)	3 deaths, 5 injuries, < $200,000
Industrial activities (5)	1 death, 1 injury, $small
Toxic leaks, etc. (several)	0 deaths, < 3 injuries, $small
Significant ELV launch accidents (1957–present)	
Atlas/Centaur (1965)	Pad damage, $1M
Titan/Centaur (1975)	Pad and building damage, $10M
Atlas/Centaur (1987)	Centaur stage replacement, $14M
Titan 34D (1986)	Minor injuries, facility, $40M
Delta II (1997)	Facility and car damage, $3M

The most significant expendable launch vehicle launch accidents are also summarized in Table 5.3. None of the accidents resulted in third-party fatalities with one accident causing some minor injuries. Significant facility damage occurred in five accidents and ranged from $1 million caused by an Atlas/Centaur (1965) to approximately $40 million caused by a Titan 34D in 1986. This latter amount is approximately $80 million in 1998 dollars.

Significant accidents are indeed a possibility, but history has shown that with good safety procedures in place they are rather unlikely events. However, this excellent safety record with resulting low levels of accident damage could be compromised if commercial launch sites do not maintain the same levels of safety as the government ranges currently and historically have exhibited.

2. Nuclear Industry

Figure 5.18 summarizes the claims experience in the nuclear liability insurance for 195 alleged nuclear incidents reported from inception in 1957 through 31 December 1997 (Ref. 30). The figure indicates that 137 payments were in the range of $0 to $100,000; 43 payments were in the range of $100,000 to $1 million; 13 payments were in the range of $1 to $10 million; 1 payment was in the range of $10 to $50 million; and 1 payment (Three Mile Island) was in the range of $50 to $100 million. Total insurance payments of $131 million for indemnity and defense costs were made during this period. (Because payments were made over a period of time, it is actually inappropriate to add these amounts without having first made adjustments to convert to constant dollars. This was not done by the reporting agency.) Of this amount $70 million was paid in connection with the Three Mile Island accident that occurred in 1979. (This is equivalent to approximately $200 million in 1998 dollars.)

3. Aviation Industry

Recent airline premiums and claims history of the airline industry are summarized in Figs 5.19 and 5.20 (Ref. 34). Recent premium experience indicates

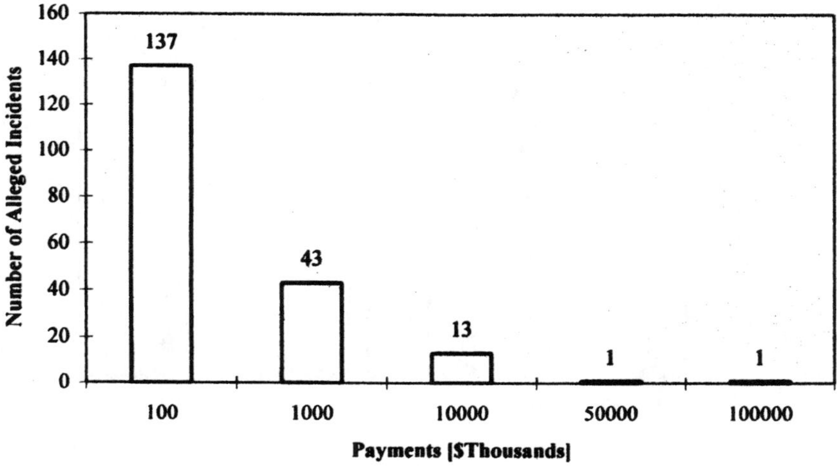

Fig. 5.18 Nuclear industry experience: alleged incidents and payments.

that rates are falling in spite of several major losses (i.e., the Federal Express MD-11 estimated as $115 million; the Garuda B747-300 loss at Guam estimated as $300 million; and PanAm Flight 103 estimated as $800 million). This is the same market that provides third-party liability insurance for space launches.

The airline industry is primarily concerned with providing insurance coverage to cover passenger liability with apparently little concern given to liability coverage for potential nonpassenger liabilities. The magnitude of airline industry insurance coverage has increased significantly with the introduction of wide-bodied aircraft. A new wide-bodied aircraft might be required to be insured for

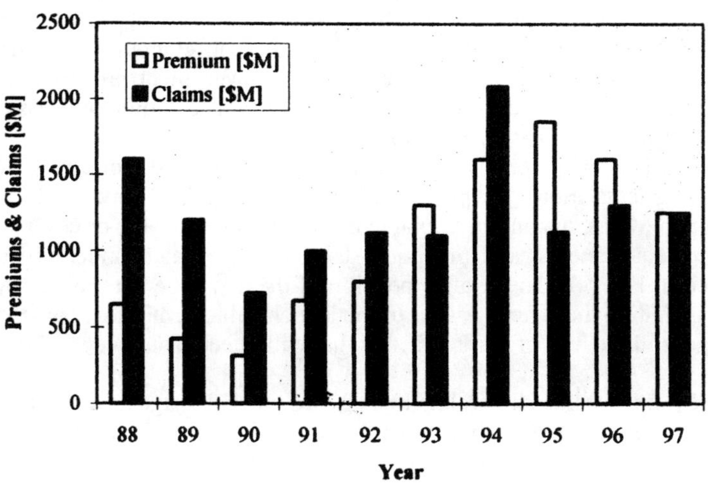

Fig. 5.19 Airline experience: hull/liability premiums/claims.

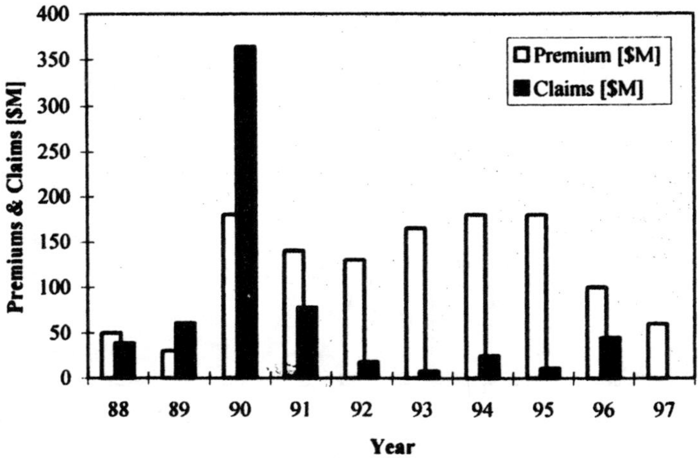

Fig. 5.20 Airline experience: worldwide airline hull war and allied perils business.

$170 million or more for physical damage plus liability coverage of up to $1 billion per occurrence. Some have suggested that given the potential of increased passenger liability it might not be long until the typical airline policy liability limit will be $1.5 to $2.0 billion.

E. Economic Implications of Indemnification

The 1988 Amendments to the *Commercial Space Launch Act of 1984* and its subsequent codification as 49 U.S.C. Subtitle IX, Ch. 701 set the basic requirements for insurance for commercial space transportation activities at federal and other facilities. These requirements encompass insurance to cover both damage to government facilities and third parties (persons and property) resulting from the licensed activities. (Because the legislative sunset provisions do not apply to the indemnification of government property, government property insurance requirements are not discussed further.) The insurance requirements are to be based upon the computation of MPL, consideration of the maximum available insurance at reasonable prices, and the maximum insurance required as stated in the enabling legislation. A typical relationship among MPL, required insurance (based upon MPL), maximum available insurance at reasonable prices, maximum insurance requirements, and maximum level of government indemnification is illustrated in Fig. 5.2. The relative position of these will be determined by many factors, including the characteristics of the launch vehicle, mission, launch facility and range, and the host of market factors that will affect insurance availability and price.

Figure 5.2 illustrates the probability of third-party damage as a function of the level of damage. This probability distribution indicates the chance that different levels of damage will result to third parties from a licensed activity. This probability distribution is conceptual (i.e., insufficient data are available to establish the specific shape for each launch vehicle) and is presented only to serve as the basis for illustrating general concepts. The damage probability distribution is independent of MPL, threshold probability, required level of insurance, etc. and is a function of many factors relating to the specific activity (e.g., characteristics of the launch vehicle, mission, and launch facility and range, population density and distribution, reliability, etc.). The calculation of the maximum probable loss establishes the level of damage that, on average, will be exceeded with a stated probability. This level of probability, or risk, is referred to as the "threshold probability". The threshold probability, 10^{-7} for third-party damage, is illustrated as the cross-hatched area to the right of the MPL level of damage. The threshold probability represents the level of risk, that is, the chance that damage will occur that is in excess of the computed MPL, that has been chosen as being acceptable.

The relationship among MPL, required level of insurance (based upon MPL), maximum available insurance at a reasonable price, and the limit placed upon maximum insurance required by the enabling legislation ($500 million) is illustrated. In addition, the maximum level of government indemnification ($1.5 million above the insurance requirement) is also indicated. The Statute provides for an automatic inflation adjustment to be made to this amount.

To summarize, the expected cost per launch for third-party damage is comprised of an expected payout from the insurance industry exposure (or the equivalent from the licensee if it seeks to self-insure), an expected cost that will be borne by the government, and an expected cost that will be borne by the licensee. The sum of these amounts is a constant and represents the expected third-party damage per launch. Using different values of threshold probability or MPL will shift the burden imposed on the different parties to the transaction but will not alter the expected third-party damage per launch.

All parties associated with a launch share in the financial risk associated with the occurrence of third-party damage. (A more detailed discussion of who bears the risk and the magnitude of potential impacts is presented in Ref. 35.) The insurance industry's liability is finite and is limited to the amount of required insurance. The government's exposure is in the range of $0 to $1.5 billion with the probability of having to make a payout being extremely small (less than 10^{-7} of a payout in this range occurring, on average, once in every 10,000,000 launches). The licensee's exposure starts if the third-party damage exceeds $1.5 billion plus the required level of insurance and is in the range of $0 to some unknown amount. This exposure occurs with a very small probability (less than 10^{-7} or a payout in this range occurring, on average, once in every 10,000,000 plus launches). Typically, the licensees have purchased more insurance than the required level of insurance, thus, in essence, further reducing the likelihood of a payout by the licensee.

F. Pertinent Issues When Reexamining Indemnification

Discussions were held with a large number of organizations that are stakeholders in the debate over the continuation or termination of government indemnification of third-party damage as set forth in the 1988 Amendments. Discussions were held with launch providers (ELV and RLV), satellite manufacturers, insurance brokers, insurance underwriters, operators of non-federally owned launch sites, government agencies, and consultants. In addition,

Congressional testimony and other published literature were reviewed. In the following paragraphs concerns and issues associated with third-party liability indemnification and its possible elimination are described and where deemed appropriate elaborated upon and discussed. The discussion presents both the views of the space transportation industry as well as the author's elaborations and interpretations. The issues and concerns include both those identified by the organizations contacted and those indicated by about 30 representatives from the commercial space industry.[36] The two continuously recurring (both in time and among stakeholders) and interrelated issues have to do with "betting the company" and "international competitiveness."

1. Issue: Betting the Company

The providers of space transportation have indicated, currently and in the past, that the elimination of indemnification would place their businesses at significant unknown and unbounded risk that might be associated with third-party damage from a licensed activity (most recently at the 10 May 2001 and again at the 18 October 2001 COMSTAC meetings held at the FAA Headquarters in Washington, DC). In fact, the risk is of significant concern and has resulted in the conclusion that without the government indemnification the launch companies would be betting the company each time a launch occurred.

This conclusion has been reached even though there has been no third-party damage resulting from more than a decade of licensed commercial launches. There remains, however, no matter what history has shown, the ever-present but very low probability of an event that could result in significant third-party damage.

Figure 5.21 attempts to reconcile the qualitative and quantitative aspects of the implications of having or not having indemnification. With the current indemnification legislation the insurance company bears the risk for damage up to the required level of insurance (based upon a maximum probable loss with a probability of 0.0000001 that this level of damage will be exceeded). If additional insurance is taken, then the insurance company is also at risk for damage that is in excess of the required level of insurance plus the amount of government indemnification up to the insured level of damage. The licensee is responsible for

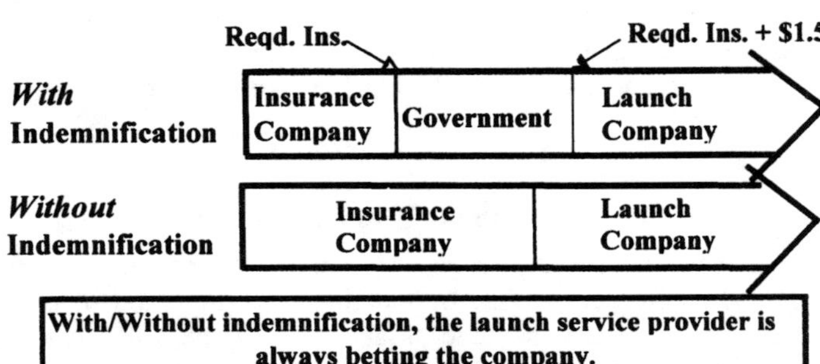

Fig. 5.21 Risk bearing with/without indemnification.

liability claims above this insured and indemnified amount. Even with indemnification the licensee is betting the company.

If the current indemnification legislation were not renewed, the licensee would be slightly worse off in that the level at which the licensee's liability would have to be confronted and would occur at a lower level of third-party damage. This can be either partially or fully offset by obtaining additional insurance should such insurance be available at a reasonable price. A phasing out over several years of the government's indemnification liability might be required for there to be an orderly transition to the ensurance of large amounts of insurance at affordable prices (i.e., such that there is inconsequential impact on international competitiveness considering all of the attributes of launch competitions).

With/without indemnification the licensed launch service provider is always betting the company. It is a matter of degree. Cost and availability of insurance, particularly for small firms and/or new launch service providers, are of continuing concern. Also of concern is the government's liability if a firm's assets are insufficient to compensate third-party claimants and the impacts on third parties should adequate compensation not be provided. (Whenever there is third-party damage, no matter what the cause, there is the possibility that insufficient assets will be available to cover liabilities and third parties might not be adequately compensated.)

2. *Issue: Availability and Cost of Insurance*

It has been stated that with the absence of indemnification with the associated need to increase insurance coverage, insurance costs will increase, and there is no assurance that adequate insurance will be available at affordable prices.

During the past several years, liability insurance costs have fallen significantly (refer to Fig. 5.17) while availability has increased. Indications are that coverage in the range of \$1 to \$1.5 billion could be obtained if and when required. The premium for \$200 million of third-party liability coverage was on the order of \$300,000 in 1992 and currently is comparable, possibly as low as \$200,000. The most significant change has been at the high end with the level of availability increasing with rates decreasing. To place the insurance costs in perspective, consider obtaining insurance cover of \$200 million with indemnification and \$100 million in the absence of indemnification. In the former case, in 1992 a premium of about \$300,000 was expected, whereas in the latter case in 1998 a premium of about \$400,000 might be expected.

3. *Issue: Effect of Increased Launch Cost on International Competitiveness*

It is argued that the absence of indemnification would harm international competitiveness because the providers of space transportation services, in order to remain competitive in internationally competed launches, would have to purchase additional insurance. Because Ariane is provided with government (France) indemnification above 400 million French francs (approximately \$60 million), the current U.S. indemnification helps to "level the playing field," and its elimination would place the U.S. launch industry at a competitive disadvantage. In prior years indemnification was not provided for Proton nor Long March. Both vehicles utilized pricing offsetting indemnification, and the users acquired additional insurance (typically $300 to $500 million). Currently, the Peoples' Republic of China offers insurance protection in the amount of $100 million, with full indemnification for claims exceeding the liability insurance. Russia (Khrunichev) provides up to $200 million of insurance protection against third-party claims and indemnifies the launch customer against third-party claims in excess of the amount of insurance.

Purchasing additional insurance, should indemnification be eliminated, will increase the cost basis of U.S.-provided space transportation services. The purchase of additional insurance, if available, will likely increase the cost basis of U.S. launch services. The increased costs will likely be on the order of several hundred thousand dollars with the largest percentage cost increases being incurred by small low-cost launch vehicles. (This is the result of lower MPLs and resultant lower amounts of required insurance.) Because the procurement of launch services considers multiple attributes such as cost (price), reliability, availability, scheduling, etc., other factors might actually have an offsetting effect. U.S. government nonrecovery of full launch service costs increases U.S. competitiveness as does the reduction in launch costs that can result from investments in technology development and demonstration (i.e., the development of new launch vehicles such as the EELVs and the government partnered X-33 and X-34 programs and resulting RLVs).

4. *Issue: Effect of Accident (Major Loss) on Availability and Cost of Insurance*

Concern has been voiced that a catastrophic event (i.e., an accident resulting in significant third-party casualties and damage and resulting liability), in the absence of indemnification, would adversely affect the amount of insurance available and premium rates for related industries including airlines and general aviation. Third-party liability insurance for space launches is normally provided from a pool that provides insurance coverage for aviation. Recent airline experience has been that catastrophic events like Lockerbie with potential liability close to or possibly in excess of $1 billion have not led to rate increases. This is the result of a soft insurance market that might not last forever. The impact of the 11 Septmber 2001 events is still uncertain. In addition, because the space launch industry has never encountered a catastrophic loss the first such loss might have a transient impact on availability and pricing of insurance.

It would seem highly desirable that if indemnification were eliminated or phased out steps would be taken to ensure the existence of a stable insurance industry capable of providing the required affordable insurance.

5. *Issue: Impact on Small Company Entry*

Concern has been indicated with regard to the potential impact of the elimination of indemnification on small companies that are or are attempting to become providers of space launch services. If it is assumed that "small company" is synonymous with small and low-cost launch vehicle, then the elimination of indemnification is likely to result in the need for additional insurance (beyond that currently required as a result of the MPL determination), which could have a disproportionally large impact on profits and/or competitiveness if insurance premiums rise. But all firms in a competitive market place face cost and market uncertainties. If it is assumed that "small company" is synonymous with low capitalization and assets, then should indemnification be eliminated there is likely to be a need to acquire additional insurance in order to ensure that adequate resources are available to compensate third parties should there be a catastrophic loss. In any case the impact on small companies is likely to be an increase in insurance costs that might be disproportionally large when compared to "large companies."

6. *Issue: Industry/Customer Expectations*

It has been pointed out that the customers or users of space transportation services have, over the last decade, developed expectations with respect to indemnification. Customers seem to feel that the current indemnification scheme is important to their well being. This is a perception that can have an impact on international trade should indemnification be instantaneously eliminated but could likely be negated through a phased (over time) approach to elimination.

7. *Issue: Ability to Finance*

Immediate elimination of indemnification can cause concern and uncertainty in financial markets, making it more difficult and costly to obtain financing for new space-related ventures. This would likely be a short-lived or transient phenomenon and could likely be eliminated through an orderly transition to a no indemnification state.

8. *Issue: Migration to Off-Shore Launching*

It has been stated that without indemnification the nation's space transportation industry will leave the U.S. and launch from foreign states (i.e., Australia, Brazil). Kistler is currently making plans to launch from Australia (which might be more a result of bureaucracy than indemnification), and Australia has established a regulatory regime modeled after the U.S. and provides indemnification for launches under its control.

The decision to utilize a foreign spaceport involves many factors including, of course, indemnification. Other issues include its compatibility with launch vehicles, provided services, safety regime, and its effect on availability and cost of insurance, range charges, effect on payload delivery capability (a function of launch site location and payload inclination angle, additional costs incurred to provide support at a distant location, scheduling and scheduling flexibility, "red tape," etc.) It is not clear how important indemnification is relative to these other factors that will also have an impact on decisions to launch from foreign states.

9. *Issue: Effect on R&D and Incorporation of New Technology*

It has been stated that the U.S. space transportation industry, without indemnification, is less likely to develop new, more efficient hardware. This issue is more likely to be related to indemnification during the development and demonstration phase of new technology than with respect to indemnification of commercial operations. Separate recent legislation has dealt with development and demonstration flights and authorizes NASA to provide indemnification for X-33 and X-34 RLV development and demonstration flight testing that will not be FAA-licensed activities. Several private sector initiatives, funded by private sector investments, are currently developing reusable launch vehicles. These investments have been made even though there is no guarantee that indemnification will be continued.

10. Issue: Reusable Launch Vehicles and Reentry Payloads

The 1988 Amendments deal with expendable launch vehicles. The recently passed *Commercial Space Act of 1998* (CSA) deals with reusable launch vehicles and reentry activities and among other things extends the indemnification provisions or the 1988 Amendments to reusable launch vehicles and to licensed reentry. The sunset provisions of the 1988 Amendments also apply to CSA. Therefore, elimination of indemnification, unless appropriate actions were to be taken, would also most likely mean the elimination of indemnification for the RLVs and reentry operations.

If indemnification were to be eliminated or appropriately phased out, the RLV and reentry licensees would most likely be required to, or would otherwise choose to obtain, third-party liability insurance well in excess of the levels established utilizing the MPL methodology. This may slightly increase the RLV cost per flight but is not likely to affect international competitiveness because the RLVs currently under development are aimed at significantly reducing the cost of space transportation, and they are not likely, at least in the near term, to encounter competition for reusable launch services from foreign competitors.

11. Issue: Implication of "Exclusions"

The 1988 Amendments acknowledge that certain risk situations can arise that are beyond the control of the licensee and for which the licensee might be unable to obtain adequate insurance coverage at reasonable cost (for example, to protect against damage that might have been caused as a result of the Y2K problem). In the event that such situations arise, the government agrees to pay claims from the first dollar of loss; in other words, the government indemnifies the licensee. The wording of the 1988 Amendments is such that the "exclusion indemnification" and the "financial responsibility indemnification" (i.e., the $1.5 billion third-party damage) are both under the sunset provisions. Accordingly, if indemnification is eliminated via the sunset provision both forms of indemnification will be terminated. This would cause an unwanted situation that would significantly increase the risk perceptions associated with licensed operations. This unwanted situation could be remedied by legislative action that would clearly separate the two forms of indemnification.

G. Where to from Here? Several Thought-Provoking Scenarios

The ultimate liability for third-party damage outside of the U.S. rests with the U.S. government. Therefore, if severe damage were to be incurred because of space launch and/or reentry activity, where the level of damage exceeds the amount of insurance in force, the U.S. government would be at risk with the amount at risk depending upon the level of sustained damage, the amount of insurance, the level of indemnification, and the launch service providers ability to pay claims in excess of the level of insurance plus the amount of indemnification. This is depicted in Fig. 5.22. The ultimate safeguard to limit the U.S. government amount at risk is the amount of insurance coverage taken by the launch provider.

A similar situation exists for third-party domestic damage with the exception that third parties bear the ultimate risk in the event of damage that exceeds a launch provider's ability to pay. Without indemnification then, insurance is the salvation, with the exception that taking additional insurance will increase costs and can have an impact on international competitiveness.

Several scenarios are now described with the intent of stimulating debate. They present a number of hypothetical situations that could possibly result from a planned phaseout of the currently provided indemnification. Several of the scenarios are structured so as to minimize the issues and concerns already identified.

1. Scenario A: Phaseout of Indemnification

It is assumed that all space transportation providers establish subsidiaries so that significant corporate assets are not at risk. (It is possible that the asset base could differ significantly between ELV and RLV endeavors if title to the launch vehicles rests with the launch provider.) Lockheed Martin indicated that its proposed RLV business venture, VentureStar, would have been conducted as a separate corporate entity.

It is assumed that the launch providers indemnify their customers with/without an increase in transportation price. Because launch providers normally already have insurance coverage that is in excess of the required insurance (based upon the calculation of maximum probable loss) and because, in general, the cost of insurance coverage decreases per unit of increased cover (refer to Fig. 5.17), the increase in price is likely to be small or possibly absorbed, that is, zero. Therefore the effect on international competitiveness will likely be nothing as a result pricing, but there might be an impact because of perceptions of risk: the difference

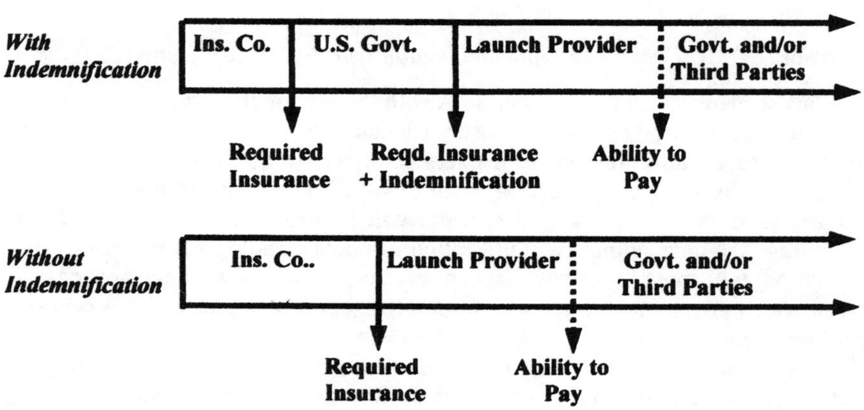

Fig. 5.22 At risk with/without indemnification.

between the maximum probable loss plus the $1.5 billion level of indemnification and the amount of cover (without indemnification) taken by the launch provider. The effect of this perception is not likely to be quantifiable nor of significance particularly for new ELVs that might be the outgrowth of the EELV program and for RLV transportation providers that claim to be able to significantly reduce transportation prices. [It has been reported that at least one payload might have been lost to foreign competition (Ariane) because of differences in indemnification policies.] In any event it is assumed that if international competitiveness becomes of major concern it can be dealt with by providing equivalent offsets for services (i.e., use government facilities, funding of R&D, etc.) that would have an offsetting effect of lowering launch costs.

It is assumed that indemnification is phased out over a period of time so that providers' and users' risk concerns are not jolted, but sufficient time is provided for appropriate adjustments to take place.

It is assumed that MPL computations are performed, and that over a period of time the required insurance amount is increased from the MPL amount to some multiple (over time, the multiple might approach a factor of seven to eight or more making the required insurance for large vehicles closer to the $1.5 billion indemnification amount). The value of the multiple will reflect government concern with the ability of third parties to be compensated for damage. At the same time that the multiple is being increased, it is assumed that the level of government indemnification is reduced so that the indemnification amount plus the required level of insurance is maintained or reduced over a period of time with the level of government indemnification eventually reduced to zero. However, sight must not be lost of the cost of insurance and its effect, coupled with other launch vehicle attributes on international competitiveness.

To ensure the availability of insurance at reasonable prices, it is assumed that a federal insurance program is put in place that will provide the higher tier insurance at reasonable prices if and when it is not available from commercial sources. This would be patterned after the FAA's authority to provide liability insurance for the airlines when flying on international routes and providing it is in support of national policy.

2. *Scenario B: Phaseout of Indemnification with New Entry Option*

This scenario is similar to scenario A with the exception of the realization that the purchase of large amounts of insurance can be burdensome for new entrants into the space transportation launch service arena. To assist new entrants while adequately ensuring the protection of third parties, it is assumed that established launch services and new launch services will be treated differently in order to encourage the development and utilization of new technology. It is assumed that scenario A will apply for existing launch services, but that new launch services will be provided government indemnification for several years after which scenario A will be applied.

It is assumed that reusable launch vehicles and reentry operations would be considered as new launch services, whereas the existing launch vehicles and their close derivatives and upgrades would be considered as established and scenario A would be applied.

3. *Scenario C: Continuation of Indemnification but with Payment for Indemnification*

It is assumed that the existing government indemnification schema is continued but that the licensee has an option to pay the government for its provision of $1.5 billion of indemnification or to purchase additional insurance at market rates and to receive government indemnification in the amount of $1.5 billion less the amount of insurance purchased above the required amount with prorated payments. The rate paid to the government should establish a reasonable rate in the market place and would ensure the availability of insurance at reasonable rates should problems develop in the aviation insurance market.

4. *Scenario D: Continuation of Current Indemnification*

It is assumed that the current form of indemnification is extended and expanded to cover all licensed launch (i.e., ELV and RLV) operations and reentry operations. The amount of required insurance could vary depending upon the specific operations (for example, reentry vs launch) and the associated computed value of maximum probable loss.

5. *Scenario E: Termination of Indemnification*

It is assumed that the indemnification aspects of the 1988 Amendments are terminated at a specified point in time in accordance with a desire to eliminate the government's risk participation. Even though the space transportation industry has been on notice that indemnification is only an option and could have been terminated in December 1999, few if any launch providers have made plans should indemnification be eliminated in the future. If indemnification is to be eliminated, a date should be set at some point in the future when this is to occur so that adequate plans and provisions can be made, taking into account in new launch contracts and pricing policies and existing contracts and contractual arrangements not placed in jeopardy.

H. Some Questions That Need Consideration

1) Are there other ways in which the government can offset cost increases associated with taking additional insurance without sharing in the risk?
2) Should the required level of insurance be changed if indemnification is eliminated?
3) Can/should the government assist in ensuring that adequate insurance is available at a reasonable price?
4) If RLVs are indemnified, at what point will they be considered as airplanes and not require indemnification?
5) At what point will the space transportation industry be considered mature, and how is a "mature industry" defined?
6) Should government remain forever in the risk-sharing mode with the space transportation and related industries?

I. Some Final Thoughts

When considering indemnification, it is interesting to note that the risk perceived by the FAA (based upon safety analyses) is on the order of 10^{-7} (for third parties). [The risk is the chance that third-party loss (per launch) will exceed the level of required insurance.] This translates into an expected cost per launch to the government of $150 assuming that any accident would result in a loss of $1.5 billion. However, there seems to be a disagreement with respect to the level of risk. The insurance companies have stated that $1 billion of insurance (per launch) above the FAA-determined required level of insurance would likely cost about $400 thousand. If insurance price were about two times the expected loss, then the insurance industry is stating, through its pricing, that the risk perceived by the insurance industry is on the order of 4×10^{-4}. Thus there appears to be about three orders of magnitude difference in government and industry risk perceptions. If the government is correct in its assessment, then the launch industry should happily pay the government some nominal amount for this insurance coverage. If the insurance industry is correct in its price-demonstrated assessment of risk, then the government is likely to receive a rude shock. Another possibility is that the insurance industry has significantly overpriced its product.

If the U.S. government is to reduce its space transportation risk-sharing role (i.e., indemnification) in the reasonably near future, it is apparent that a plan needs to be developed for the orderly phaseout of indemnification. The plan should consider the following: 1) expendable launch vehicles (operational), 2) reusable launch vehicles (operational), 3) R&D vehicles (including those developed independently as well as in partnership with the U.S. government), 4) reentry vehicles and payloads, 5) federal and nonfederal ranges, 6) large and small launch vehicles, and 7) large and small firms. The objective is to establish an approach for phasing the government out of the space transportation risk-sharing business without "shocking" the space transportation industry.

J. True Issue: Competitiveness

Competitiveness is the true issue, not whether or not the government should provide indemnification. The issue boils down to the cost of insurance to cover the level of indemnification (currently $1.5 billion), the implication of this added cost on international competitiveness, and whether the government can, and is willing to, offset this added cost through the provision of other services, for example, range fees charged commercial users of government launch ranges. [A methodology was described in preceding sections that allow the effects of changes in one or more of the multiple attributes of a launch service (for example, price, reliability, insurance cost, etc.) to be established in terms of launch service market share. Applying this methodology would allow the effects of elimination of indemnification on market share to be established. It is unfortunate that the recent FAA report entitled "Liability Risk-Sharing Regime for U.S. Commercial Space Transportation: Study and Analysis"[37] concluded that it was not possible to relate indemnification to market share. The FAA is correct in this conclusion, but the wrong problem was addressed. The correct problem was to establish a relationship between *a change in the indemnification scheme and a change in launch service market share*. This can be done by utilizing the methods described in the preceding sections noting that indemnification can be replaced with insurance, the cost of which can be reflected in launch service pricing, which in turn affects market share.]

According to *Space News*,[38] the Air Force may ask Congress to amend the fee structure for charging commercial launch customers for the use of government ranges—the objective is to achieve full cost recovery and not incremental cost recovery. It is likely that changing fees charged for federal launch range services will affect the competitiveness (as measured by U.S. market share) of U.S. commercial space transportation services. Increasing fees will tend to reduce market share, whereas reducing fees will tend to increase market share. As a result, a change in launch fee policy should not be considered just in terms of Air Force budgets but should be considered in terms of its impact on the U.S. economy, for example, in terms of the change in U.S. Treasury revenue and jobs that might result from a change in market share (refer to Sec. IX in Chapter 2).

Not achieving full cost recovery can be considered as a subsidy. However, the U.S. government already subsidizes the U.S. commercial space transportation industry through the provision of third-party liability indemnification and the funding of technology development. Policies relating to these are currently under review, and changes could also affect competitiveness. It seems to be fundamentally wrong to evaluate separately each of the actions that can alter competitiveness. It would seem more reasonable to determine what investment the U.S. government is willing to make to support the U.S. commercial space transportation industry and then to determine the package of mutually supporting programs and policies that should be undertaken to accomplish the government's goal (for example, a reduction in launch fees can be used to offset the cost of additional third-party liability insurance should indemnification be eliminated, an increase in launch fees accompanied by an increase in R&D expenditures, a delay in payment of launch fees so that the change in the present value of payment equates to the additional cost of insurance, etc.).

It is recommended that an overall government policy be developed with respect to government efforts to maintain or increase the competitiveness of the commercial space transportation industry so that supportive fee structure, indemnification, and other policies can be developed. Government programs and policies need to be identified that can affect competitiveness and should be considered individually and in combination. The goal is to identify each of these alternative's impacts on competitiveness (likely changes in market share), establish resulting economic impacts (for example, changes in jobs and U.S. Treasury revenue), and thence to select that mix of programs and policies which maximizes net benefits.

Finally, decisions concerning indemnification should not be made without the consideration of other programs and policies that can affect international competitiveness.

References

[1]Becker, G.S., "Cost/Benefit Analyses in the Department of Transportation: A Perspective from the Federal Aviation Administration," Western Economic Association, Las Vegas, NV, July 1998.

[2]"Regulatory Assessment and Regulatory Evaluation Handbook," Office of Aviation Policy, Plans and Management Analysis, Federal Aviation Administration, Washington, DC, March 1994.
[3]"14 CFR Part 440 Financial Responsibility Requirements for Licensed Launch Activities, Final Rule," *Federal Register*, 26 Aug. 1998.
[4]"14 CFR Parts 401 et al. Commercial Space Transportation Licensing Regulations, Final Rule," *Federal Register*, Vol. 64, No. 76, 21 April 1999.
[5]DoC/NOAA, "15 CFR Part 960, Licensing of Private Land Remote-Sensing Space Systems," *Federal Register*, Vol. 62, No. 212, 3 Nov. 1997.
[6]"Hazard Analysis of Commercial Space Transportation (Vol. I, Operations; Vol. II, Hazards; Vol. III, Risk Analysis)," Dept. of Transportation, Office of Commercial Space Transportation, Washington, DC, May 1988.
[7]"Economic Impact Assessment of Financial Responsibility Requirements for Licensed Reentry Activities (14 CFR Part 450)," Dept. of Transportation/Federal Aviation Administration, Office of Commercial Space Transportation, Washington, DC, 9 March 1999.
[8]Princeton Synergetics, Inc., "Maximum Probable Loss: Assessment of Conservatism," Dept. of Transportation, Office of Commercial Space Transportation, Washington, DC, 29 April 1991.
[9]Gress, R.K., "Derivation of Maximum Probable Losses for Commercial Launch Operations," Dept. of Transportation, Office of Commercial Space Transportation, Washington, DC.
[10]Princeton Synergetics, Inc., "Maximum Probable Loss: Rationale and Supporting Information," Dept. of Transportation, Office of Commercial Space Transportation, Washington, DC, 29 April 1991.
[11]Hutt, P.B., *Food, Drug, Cosmetic Law Journal*, Vol. 33, 1978, pp. 558–589.
[12]Princeton Synergetics, Inc., "Guidelines for Determining Maximum Probable Loss," Dept. of Transportation, Office of Commercial Space Transportation, Washington, DC, 29 April, 1991.
[13]"Position Paper on Orbital Debris, Edition 2001," International Academy of Astronautics, Sept. 2001.
[14]Interagency Group (Space), "Report on Orbital Debris," National Security Council, Washington, DC, Feb. 1989.
[15]"Orbiting Debris, A Space Environmental Problem," Office of Technology Assessment, OTA-BP-ISC-72, Washington, DC, Sept. 1990.
[16]*Orbital Debris Mitigation Techniques: Technical, Legal, and Economic Aspects*, Special Project Report of AIAA, Rept. No. SP-016-1992, AIAA, Washington, DC, 1992.
[17]"Interagency Report on Orbital Debris," Office of Science and Technology Policy, Washington, DC, Nov. 1995.
[18]"NASA Safety Standard: Guidelines and Assessment Procedures for Limiting Orbital Debris," NASA NSS1740.14, Aug. 1995.
[19]Committee on Transportation Research and Development, "Interagency Report on Orbital Debris," National Science and Technology Council, Washington, DC, Nov. 1995.
[20]Greenberg, J.S., and Jackson, C., "On the Possibility of Using Comsats to Detect Small GEO Orbital Debris," International Academy of Astronautics, 97-IAA.6.3.04, Oct. 1997.
[21]Loftus, J.P., and Portree, D.S.F., "Orbital Debris: A Chronology," NASA/TP-1999-208856, Jan. 1999.

[22]Greenberg, J.S., and Reynolds, R.C., "Orbital Debris: Economic Impacts on Remote Sensing Mission," International Astronautical Federation, 94-IAA.1.3.602, 1994.
[23]"MEO/LEO Constellations: U.S. Laws, Policies, and Regulations on Orbital Debris Mitigation," AIAA Rept. SP-016-2-1999, Reston, VA, 1999.
[24]Keeton, Robert E., *Insurance Law, Basic Text*, West Publishing Co., Eagan, MN, 1971.
[25]*Black's Law Dictionary*, West Publishing Co. Eagan, MN.
[26]Rockett, Laurie R., et al., *Financial Protection Against Nuclear Hazards: Thirty Years' Experience Under the Price-Anderson Act*, Trustees of Columbia Univ. of New York, 1984.
[27]"The Price-Anderson Act: The Third Decade," U.S. Nuclear Regulatory Commission, Rept. to Congress, Washington, DC.
[28]"Reactor Information Report," Atomic Industrial Forum, 27 Sept. 1984.
[29]"Nuclear Liability Insurance: Protection for the Public," American Nuclear Insurers, Rept. No. 3, Washington, DC, March 1983.
[30]"The Price-Anderson Act—Crossing the Bridge to the Next Century: A Report to Congress," Nuclear Regulatory Commission, NRC FIN A0400, Aug. 1998.
[31]Deem, C. L., "Liability of Private Space Transportation Companies to Their Customers," *Insurance Council Journal*, July 1984.
[32]Vinter, J., "COMSTAC: An Update on the Space Insurance Market," Federal Aviation Administration, May 2001.
[33]*Business Insurance*, 30 Oct. 1997.
[34]Viccars, P., "A View from London: An Update of Insurance Issues," Airfinance Legal Conf., Nov. 1997.
[35]Princeton Synergetics, Inc., "Economic Impact Assessment of Financial Responsibility Requirements for Licensed Launch Activities," Federal Aviation Administration, 14 CFR Part 440, Washington, DC, 16 March 1998.
[36]"Minutes of Commercial Space Competitiveness Meeting," Washington, DC, 28 July 1998.
[37]"Liability Risk-Sharing Regime for U.S. Commercial Space Transportation: Study and Analysis," Federal Aviation Administration, April 2002.
[38]"Air Force May Ask Congress to Amend Range Fee Rules," *Space News*, Vol. 12, No. 15, 16 April 2001.

Chapter 6

Beyond Space: Energy and Gaming

THE preceding chapters were concerned with the application of economic concepts and associated analytic techniques for improving decisions relating to a broad range of issues facing the space transportation, satellite, and related industries. Other sectors of the economy are faced with similar issues and concerns. The general economic and decision analysis methods already described for the space industry have wide application. The following paragraphs illustrate the diversity of application of economic analyses and associated computer simulations and decision analyses in support of both tactical and strategic decision situations. Emphasis is on structured analyses that 1) clearly indicate the variables considered and their interrelationships, 2) provide the ability to rapidly change the values considered in order to ascertain their relative and absolute importance (i.e., sensitivity analyses), 3) provide the ability to explicitly consider uncertainty and risk, and 4) allow rapid and real-time consideration of alternative decisions and likely consequences. The diversity of economic and related analyses in support of decision making is demonstrated by considering the energy sector and the gaming industry.

In the **energy category** two specific applications are described. The first considers government decisions concerning investing in technology and the setting of policies that affect manufacturer and end-user investment decisions relating to building envelope and equipment. The government's objective is the achievement of energy conservation goals. The emphasis is on a stochastic market penetration model that relates government technology programs and policies to new product attributes that then affect end-user multiattribute and multi-product replacement and purchase decisions. User decisions are considered as probabilistic. The second application considers manufacturer product design, pricing, and investment decisions in the area of manufactured housing (i.e., panel construction). Multiproduct decisions are considered that are aimed at achieving investment objectives and are influenced by related government investments and policies whose goal is to enhance energy conservation. Stochastic supply-side and demand-side models are utilized, with probabilistic product choices considered in the demand-side model. These two models are then utilized to establish supply–demand equilibrium.

The two considered energy applications demonstrate stochastic techniques that have been used in the energy sector for forecasting product demand and resulting energy consumption. The methodology provides a means for relating demand to product attributes and government policies and programs that can affect these attributes or other factors (such as policies aimed at altering taxes by providing specific tax credits related to product attributes or policies aimed at forcing technology change via the regulatory environment). Thus, methods for evaluating the effect of government programs and policies aimed at altering private sector investment decisions are described that have application in other sectors of the economy, including the space sector.

In the **gaming category** applications focus upon the table games of blackjack, craps, and baccarat with the games described and issues, resulting analyses, and decisions discussed briefly. This is then followed by a discussion of high-roller play and the use of computer simulation and decision analysis discussed in terms of awarding complementaries and negotiating with high rollers to obtain their play. The discussion of the table games first considers decisions that rely upon the long-term statistics of the games (for example, setting of game rules and procedures). This is then followed with a discussion of the short-term statistics of the games (for example, the long-term statistics of coin tossing indicates a probability of 0.5 of obtaining a head on the next coin toss; the short-term statistics are concerned with developing the probability distribution of the number of heads obtained as a function of the number of coin tosses where the number of tosses is relatively small). Finally, both are combined by considering the problem of real-time negotiations with high rollers in order to develop an incentive package that will result in the high-roller play but with the casino being able to achieve profitability objectives within specific risk constraints.

The selected gaming applications indicate the use of stochastic techniques to evaluate the impact of altering product attributes (the rules and procedures of the table games) on both the providers (the casinos) and end-users (the players). The gaming application also demonstrates the use of formal methods that consider risk and its acceptable level, when negotiating with customers (the high rollers). Similar techniques can be used by the space industry, for example, when vying, in a competitive environment, for a launch contract.

In general, the same methods and techniques developed and used in the energy and gaming sectors of the economy are applicable to other sectors of the economy, including space.

I. Energy Conservation: Government Investment in Technology and Setting of Policies

A. Introduction

Long-term energy consumption goals (such as no net increase in nonrenewable energy consumption in the buildings sector in the long term) together with forecasted energy consumption play a major role in the setting of overall energy policy and decisions relating to investments in technology programs that can result in new energy efficient products and thus allow the achievement of stated goals. A key element of energy policy and technology investment decisions is the ability to forecast changes in energy consumption that are likely to result from the policy and investment decisions. This implies being able to link technology programs and related policies (such as the setting of efficiency standards and various information dissemination and incentive programs) to market penetration rates.

Historically, penetration rates have been projected by analogy, that is, "s-shaped" penetration curves have been used that historically resulted for similar products. There was no linkage between the shape of the assumed s-shaped penetration curve and technology programs and policies and product attributes. To provide an alternative to the s-shaped curve method, the U.S. Department of Energy (DOE) sponsored the development of the Market Penetration Model (MPM).[1] The MPM was developed to provide a mechanism for forecasting the rate of penetration of building envelope and end-use equipment products into new building/establishment and retrofit residential and commercial markets and to establish the overall impacts on energy consumption. The MPM links product attributes to market penetration rates taking into account impacts of building envelope products on end-use consumption, interactions between the various end uses, interaction between current and future purchase decisions, the setting and timing of the introduction of standards, products that can affect multiple energy types, potential impacts of other technology transfer activities, forced (i.e., product failure) and economic choice decisions, perceptions of future energy prices, and uncertainty perceptions.

The MPM is a stochastic simulation model developed to forecast energy consumption as a function of time in terms of residential and commercial new building and retrofit purchase decisions. The Model is predicated on the assumption that decisions to utilize new energy efficient products are based upon perceived product economics, but decisions also take into account other economic, and noneconomic factors. Thus, purchase decisions are described in probabilistic terms and are a function of multiple attributes.

B. Analysis

The basic approach, as illustrated in Fig. 6.1, is to develop residential and commercial end-user (i.e., decision maker) perceived economics in terms of expected payback period, risk (described as the standard deviation of payback period), and relative magnitude of the investment, and to then use these results to enter data sets that establish the likelihood of investment (i.e., purchase) in terms of these same decision variables. Separate data sets are used to characterize new and retrofit residential and commercial decisions. In practice, input data specify the probability of product purchase at different combinations of expected payback (of a new product relative to a current product), standard deviation of payback period (of a new relative to a current product), and magnitude of investment in the new product relative to that of the current product. The MPM computes values of these decision variables and then uses these values to interpolate the appropriate decision database.[2] The resulting probability of purchase information, when combined with the number of establishments (i.e., potential users or decision makers) that have not already purchased the energy efficient products, and awareness (the effect of technology transfer programs) result in the determination of expected values and standard deviations of annual purchases, annual energy consumption, cumulative purchases and consumption as a function of time (and product and energy type).

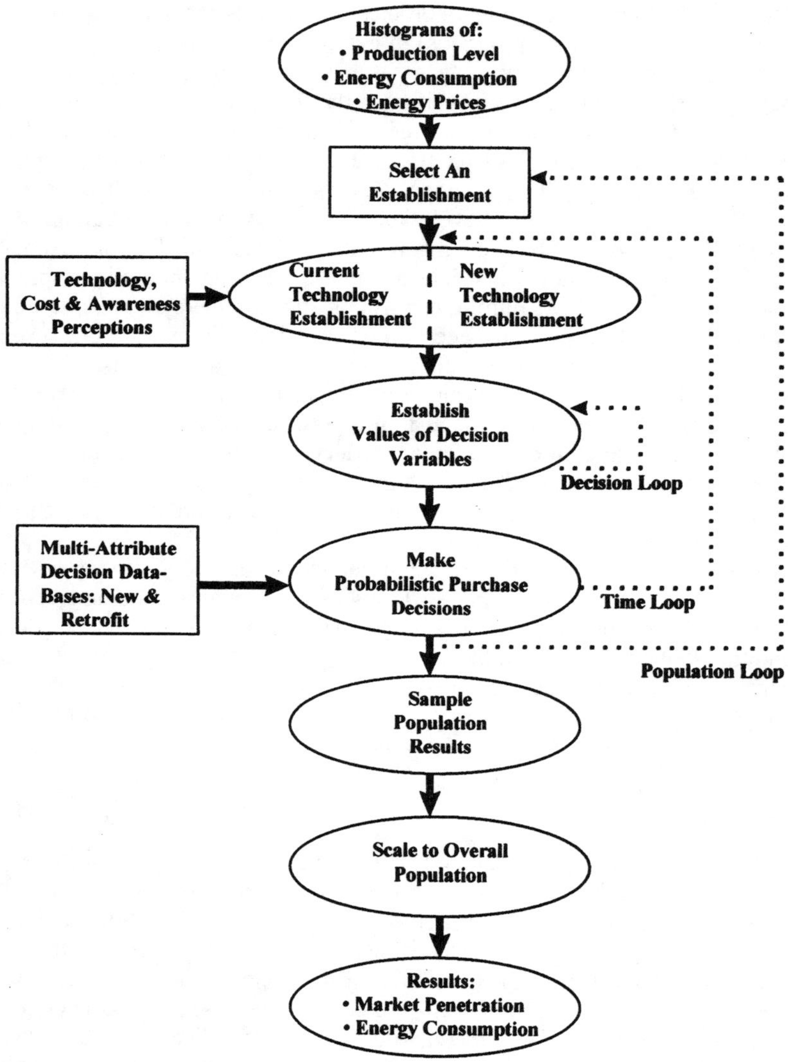

Fig. 6.1 Overview of market penetration model simulation process.

The residential and commercial computations are the same, with minor exceptions, but utilize different decision data sets to characterize the probabilistic purchase decisions. Both consider that multiple product purchase decisions are interdependent such that prior investment decisions will affect current and future investment decisions. For example, a prior decision to increase the R value of a dwelling by adding additional insulation will reduce the energy savings potential of a new high efficiency furnace and will thus reduce the likelihood of such an investment. Both consider the interactions between building envelope and end-use equipment energy consumption as well as the interactions between end-use equipment (for example, situations such as the purchase of energy efficient windows that reduce heating/cooling consumption but increase lighting consumption, with lighting consumption in turn affecting heating/cooling consumption can be taken into account).

The MPM, utilizing Monte Carlo simulation techniques, allows a number of histograms to be sampled, which results in the selection of a building. This building is then tracked over time with all purchase decisions and resulting energy consumption noted. The results are saved and another building selected. This is accomplished for a large number of sampled buildings and the results of the selected sample scaled to the overall building population.

The Monte Carlo MPM incorporates a number of variables (for example, end-use energy consumption, energy price, product efficiency) that can be described by histograms or probability distributions (referred to as uncertainty profiles). The MPM uses uncertainty profiles that are described in terms of minimum and maximum values of the range of uncertainty and the form of the uncertainty within this range. The uncertainty profiles are described as five-point discontinuous distributions, with a large number stored and available for use by identifying its reference number. This particular method for describing uncertainty was selected because of the relative ease of characterizing a broad range of forms and the ability to communicate with those not skilled in mathematics and statistics.[3–5]

The use of Monte Carlo modeling allows, within appropriate geographic segmentation, building end-use energy consumption (for each specific building type to be considered) and energy prices to be described in terms of histograms. In addition, current and new (i.e., resulting from technology investments) products can be characterized in terms of ranges of values (and the likelihood of achieving these values) associated with efficiency, price, and maintenance cost. The objective of the Monte Carlo approach is to allow a building to be selected and characterized by the random sampling of the consumption and energy price distributions and to then establish the user economics associated with the different product choices, which are based upon product attributes that are also established by the random sampling of the product attribute distributions. The result is the determination of the values of the decision attributes that are then used to establish the likelihood of product purchase and the ensuing product purchase decisions. This allows a large combination of consumption patterns (for the same building type), energy prices, and product attributes to be considered in the user-decision process that determines the market penetration rate.

In the MPM the Monte Carlo process proceeds as follows. A set of computations is performed outside the Monte Carlo loop. The values computed therefore remain constant during the Monte Carlo process. The main R loop is concerned with establishing values of global variables such as energy price, product efficiency, etc. Within the R loop is the r loop that considers the variability of parameters facing a decision maker. For example, at the time of a decision current energy price, efficiency, and installation cost are known, whereas future energy prices and maintenance costs are not known. The r loop is concerned with simulating the financial performance that will result from the different combinations of future prices and costs. This results in the set of financial performance measures or decision variables with a determination of probability of purchase for each run in the R loop. As a result, there will be a determination of probability of purchase of each product for a large number of buildings (as determined by the dimensionality of R) within each building type, geographic region, and year. The sample population is then extrapolated to the total population. This Monte Carlo process establishes expected values and standard deviations for product purchases and energy consumption and savings as a function of time. To accomplish this requires maintaining the sums and sums-of-the-squares (and not all calculated results) of all necessary variables.

The computational process can be further clarified by considering new buildings. Starting in the first year ($I = 1$), a building is selected and characterized in terms of its nominal end-use consumption. The end-use consumption of the building population is described by a set of end-use energy consumption histograms that are randomly sampled to obtain the specific energy consumption to be used for the selected sample. The energy price histograms are also sampled to establish the energy prices that are known to the decision maker. The next step is to select the attributes of the available envelope technologies. These attributes include cost, impact on end-use consumption, rate of change of impacts, level of standards, timing of standards, level of awareness of availability of the technology, etc. From this information the MPM calculates the expected payback, standard deviation of payback, and the relative magnitude of the investment of the new envelope technology (relative to the use of the current technology).

Using the decision database (which establishes the probability of purchasing the new technology in terms of the expected payback, risk, and relative magnitude decision variables), decisions are made with respect to the purchase of current or new envelope products. Once envelope purchase decisions have been made, end-use consumption is adjusted to reflect the actual product choices.

The next step is to select the attributes of the available end-use products. These attributes include cost, efficiency, maintenance cost, and associated rates of change, level of standards, timing of standards, level of awareness of the availability of the products, energy type, product competitions, etc. From this information the MPM calculates the expected payback, standard deviation of payback, and relative magnitude of investment of the new end-use products. Using the decision database (which establishes the probability of purchasing the new technology in terms of the expected payback, risk, and relative magnitude decision variables), decisions are made with respect to the purchase of current or new end-use products. Once end-use purchase decisions have been made, end-use consumption is adjusted for the start of the next year so as to reflect the actual product choices. The end-use consumption adjustments take into account the interrelationships between the end uses.

At this point all decisions have been made with respect to the original equipment (envelopes and end-use products) mix included in the new building. It is now necessary to track this new building through time to establish retrofit decisions and their timing. Two important assumptions must be noted: envelopes cannot be retrofitted (once selected, they remain in place and are not upgraded), and once a new product has been acquired for an end use, that end-use product is no longer considered as a candidate for upgrading. All energy consumption calculations are based upon the efficiency of the installed equipment that is a function of when the purchase decision was made. As time moves on, technology changes, and resulting values of decision variables, and hence future new building envelope decisions, will change. Energy consumption always depends on when the product was acquired and the level of efficiency in force at that time.

Product life characteristics are taken into account. An expected wear out life and standard deviation can be specified. Thus, as the MPM progresses with its tracking of products over time, the product life characteristics are taken into account. If a product has not failed (i.e., is still in service), then economic choice decisions are considered based upon annual cost savings and the cost of the new product. If the previously purchased product has failed, then a forced choice decision must be made with the decision considering the differences in annual operating costs and the relative magnitude of the required capital expenditures. (When considering existing buildings, the same analyses must be performed; however, an expected value and standard deviation can be specified for existing equipment age. This is taken into account when establishing whether or not an existing product is still operational. All computations then proceed in a manner similar to that already described.)

Upon completion of the analysis of a building, the MPM moves on to the next year for that building so that retrofit decisions can be considered. These decisions are always made in light of the product mix at the start of the new year. This product mix reflects all past purchase decisions and their timing. At the start of each year, adjustments are made to all end-use products to reflect the change in technology as a function of time. These new product attributes are then used when making retrofit decisions.

The new building retrofit decisions are based upon a comparison of the available new products and the previously selected products and their associated economics. The retrofit analysis considers the energy consumption of the installed product mix (which reflects the product attributes at the time that the product was selected) and the energy consumption that will result from the current level of new product technology. The retrofit decisions use a different set of decision data (discussed in following paragraphs) to establish whether or not the equipment mix will be altered or retrofit. If the equipment mix is altered, the MPM again adjusts end-use consumption to reflect the new equipment mix that then serves as the basis for making the next year's retrofit decisions.

At the completion of the last year's analysis (in other words, when a new building has been tracked through the complete time period covered by the analysis), the MPM moves on to the next year to start the process for considering new buildings added in the next year. This process continues until all new building introductions over the considered time frame have been evaluated.

Upon completion of the new building's portion of a simulation run, the MPM considers existing buildings. Existing buildings are considered in a manner similar to the new buildings with the following two notable exceptions: 1) the new building equipment mix (envelope and end-use equipment) is initialized with existing products; hence, only end-use equipment retrofits are considered as the randomly selected building is tracked through time. And 2) the time frame is traversed only once because existing buildings are not added over time as is the case of new buildings.

Upon completion of the analysis of each building within each year, results such as products purchased and energy consumption (by end use and fuel type) are saved (actually, tables of sums and sums-of-the-squares are appropriately incremented). Upon completion of all Monte Carlo simulation runs, expected values and standard deviations of quantities purchased and energy consumption are established.

To summarize, the process is to perform the following:

1) Select a building randomly from the population.
2) Characterize the attributes associated with the building.
3) Select building envelope technologies that will be included in the building.
4) Adjust end-use energy consumption to account for selected envelope technologies.
5) Select end-use product mix that will be incorporated into the building.
6) Adjust energy consumption for selected end-use product mix.
7) Trace the building through time over the total planning horizon establishing all end-use product retrofit purchase decisions and resulting consumption as they occur (each year).
8) Repeat this process a large number of times (i.e., selecting a large number of buildings).
9) Extrapolate results from the sample population to the total population.

The outlined methodology requires the development of a diverse set of input data. The data requirements can be categorized as follows:

1) Consumption forecasts (by end use and energy types) can be obtained from the EIA building-sector forecasts (Residential Energy Consumption Survey, DOE). Care must be taken to ensure that common policy scenarios are utilized.
2) Energy price and rate of change of price as a function of time and energy type are required.
3) Product competitions by end use and envelope and end-use product awareness as a function of time are necessary.
4) Level of standards and the timing of their implementation are required.
5) Product characterizations imply the specification of energy type, initial cost, efficiency, maintenance cost, timing, and the rate of change in efficiency as a function of time.
6) The relationship of building envelope products and end-use consumption must be specified as well as end-use consumption interactions.
7) Economic choice decision data must be provided on likely tradeoffs between decision variables and their impact on the likelihood of purchase. Reasonable data exist for commercial investment decisions.[2] Data must be developed for characterizing consumer purchase decisions. (This is described in the Appendix to this book.)

The MPM establishes energy consumption as a function of time by end use and fuel type. This is accomplished for a specified building/establishment type in a specified geographic area. Thus the MPM must be run a number of times to obtain total energy consumption and savings when multiple building/establishment types and geographic areas are to be considered.

New building product decisions are based upon a choice between competing current technology (existing or current) products and new products that result from technology investments and related policies. In other words, an implementation decision must be made, and the decision is concerned with which product should be acquired. New building product decisions are considered as probabilistic multiattribute decisions with the decision variables being expected payback period (the payback is of the new product relative to the current product), variability of the payback period (as measured by the standard deviation of payback period), and the relative cost of the competing products. A data set is provided, which relates probability of purchase (of the new product) to the multiple attributes of expected payback period, risk, and relative magnitude of investment. The MPM calculates values of expected payback period, risk, and relative magnitude of investment. These values are then used to interpolate the decision database

Building retrofit decisions are considered in a similar manner, but the choice is now whether to purchase a new product or to maintain the status quo. The same decision variables are considered but with the payback period being that of the new product relative to maintaining the status quo and the relative magnitude of purchase being the cost (including installation) of the new product relative to the household income. When commercial establishments are considered, the relative magnitude of investment is the cost of the new product relative to the establishment annual revenue. Probability of purchase and consequent decisions are established in the same manner as with the new buildings but using the appropriate data sets.

The decision model is only utilized when the decision maker is aware of the competing products and their attributes. Product awareness is specified and reflects both manufacturer activities as well as government information dissemination and technology programs and related policies. Altering awareness can be an important consideration when seeking possible alternatives to regulatory action pertaining to the setting of standards.

The possibility of fuel switching and multiple energy types are considered. Fuel type is part of the product identification data and the associated fuel cost, together with other product attributes and annual energy consumption, is used in the product economic choice decision analysis. Thus product choice decisions automatically indicate the energy type that will be utilized by the product choice. The resulting change in energy consumption and fuel type is then utilized to update energy consumption by end use.

The DOE developed technology can lead to products that will compete in the marketplace with multiple products. A single product can satisfy the same end use as multiple products, that is, heat pumps in competition with separate heating and cooling equipments. In addition, a single product can satisfy an end use while switching fuel type, for example, a gas cogeneration system producing electricity and hot water in competition with gas or electric hot water heating and purchase of electricity and/or gas.

Thus, product competitions among multiple energy types and multiple end uses are considered. To accomplish this, the information base associated with product characterization encompasses product competitions. The identification of product (including fuel type) competition is from the point of view of making economic choices; in other words, the definition of product competition type identifies the mix of products that are in competition with a single product.

C. Summary/Results/Status

The MPM was developed to provide a mechanism for forecasting the rate of penetration of building envelope and end-use equipment products into new building/establishment and retrofit residential and commercial markets and to establish the overall impacts on energy consumption. The MPM links product attributes to market penetration rates taking into account impacts of building envelope products on end-use consumption, interactions between the various end uses, interaction between current and future purchase decisions, the setting and timing of the introduction of standards, products that can affect multiple energy types, potential impacts of other technology transfer activities, and uncertainty perceptions.

The MPM forecasts energy consumption as a function of time in terms of residential and commercial purchase decisions. The MPM is predicated on the assumption that decisions to utilize new energy efficient products are based upon perceived product economics, but decisions must also take into account other economic and noneconomic factors. Thus, purchase decisions are described in probabilistic terms and are a function of multiple attributes. Because decisions are described in probabilistic terms, resulting forecasts are also described in probabilistic terms. It is thus possible to establish the chance that energy savings that can result from the investment in new technology, the imposition of standards, the undertaking of information dissemination programs, will exceed different levels.

The MPM takes into account 1) building envelope products and end-use products and their energy consumption interrelationships, 2) multiple end-use products and their energy consumption interrelationships, 3) characterization of program elements in terms of product attributes that will influence economic choices, 4) stochastic economic choice decision model, 5) energy consumption used in economic choice decisions based upon prior years' decisions, 6) the setting of standards, 7) retrofit and new construction together with geographic market segmentation, 8) fuel switching, 9) interrelated product decisions, and 10) multiple energy types.

The MPM is a user-friendly menu-driven PC-based system. It operates in the Windows environment and requires at least 2 MB of hard drive space. The input/output system operates in Excel 5.0 for Windows that interfaces (via PRN files) with a compiled FORTRAN program that performs the calculations on the data provided to the MPM. The Excel interface with its menus and on-screen instructions guides the user through the data screens for data entry, review and printing; activates the compiled FORTRAN program and pulls in the results to the report tables; and guides the user through the reports for review and printing.

The MPM has been vigorously tested in order to verify its internal validity. To test its efficacy, the MPM was used to forecast the market penetration rate of electric heat pumps into the residential market. This forecast (made in 1996) covered the period from 1983–1993 with an independent organization providing most of the data used in the analysis. The results of the forecast are illustrated in Fig. 6.2 and compared against actuals as obtained from the Residential Energy Consumption Survey (RECS), which indicates actual measured (i.e., via survey) results in the marketplace in the years 1984, 1987, 1990, and 1993. The results are based upon disaggregating the market into four geographic regions and considering 10 different product competitions within each region (i.e., heat pump in competition with 10 different heating and cooling systems). The MPM results are in reasonably good correspondence with the actuals. This single data point does not verify the model but certainly does not detract from its efficacy.

Though the model was specifically developed to assist with 1) formulating technology programs by providing an ability to relate technology program impacts on product attributes and product attributes on market penetration rates and 2) evaluating policy alternatives such as the setting of efficiency standards, which are aimed at altering manufacturing and/or consumer purchase decisions thereby enhancing energy conservation, the model has been used for neither. The reasons for this are, in the former application, the senior program planner and champion of the model development retired without having passed the baton, whereas in the latter case multiple internal organizations were involved each with vested interests in the continuation of the status quo with respect to the analysis of the appliance efficiency standards.

Though the model was developed for the preceding reasons, it provides a valuable linkage for management interested in understanding the likely impacts of design alternatives on product sales as a function of time, competition, government actions, and all or the other factors discussed earlier.

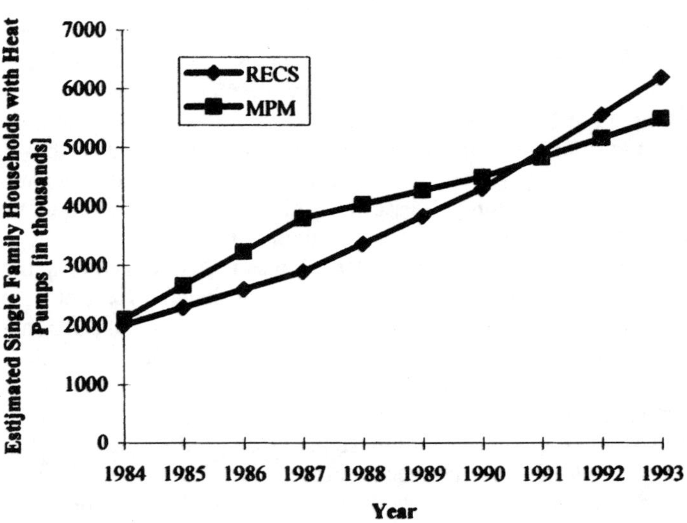

Fig. 6.2 Results of a demonstration of the use of the market penetration model: a forecast of electric heat pump market penetration compared with actuals (RECS).

II. Manufactured Housing: Product Design/Pricing/Investment Decisions

A. Introduction

The DOE's Industrialized Housing Program strives to improve quality built energy-efficient housing through the use of new manufacturing techniques. The University of Oregon reports "the factory-built process offers the promise of houses that are more efficient, of higher quality, and less costly. To ensure that this promise can be met, the U.S. industry must begin to develop and use new technologies, new design strategies, and new industrial processes."[6]

Promising technologies, such as preinsulated panels, integrated systems, multifunction housing components, and new manufacturing techniques are being developed by the University of Oregon and the Florida Solar Energy Center under the DOE Energy-Efficient Industrialized Housing Program.

Preinsulated panels have great potential for energy savings. Comparisons of utility bills of homes using conventional framing (i.e., stick construction) with homes using foam-core panels have revealed significant savings in energy costs in the foam core compared to the conventional homes. The Florida Solar Energy Center tested two homes, one built with conventional walls and the other with foam-core walls, and found an air-infiltration rate of the conventional home that was double that of the foam-core home. In one project the University of Oregon's Center for Housing Innovation demonstrated that a prototype low-income house could be built utilizing a highly energy efficient structure based upon the use of a stressed skin insulated core panel. (These panels use a sandwich lightweight core material faced with two layers of structural sheating.)

The energy savings potential of new factory-built building frame products requires that these products be utilized in new housing construction. This is likely to occur when the overall economics favor the factory-built building frame products relative to conventional stick construction and that this is believed (i.e., perceived as low risk). To understand the overall economics and likely choices that can be made for using particular building products, and to provide the means for manufacturers to evaluate the demand for new products and the financial viability of producing these products, the DOE supported the development of analytic techniques and tools that modeled both the demand for and the supply of manufactured housing products and explicitly considered uncertainty and resulting risk in the decision processes of both builders and manufacturers.

An approach was developed for estimating the demand for factory-built building frame products (sometimes referred to as industrialized or manufactured housing). The approach is aimed at estimating the demand for factory-produced panelized wood-frame construction and various panel construction products relative to conventional site-built (or stick-built) wood-frame construction. External and internal wall, roofing, and flooring products are considered.

The approach utilizes both supply and demand models for estimating manufacture prices and the effect of these prices on product demand. The supply-side model (*PSI-RISK*) takes into account the financial structure of a typical panel manufacturer with pricing established such that desired rate-of-return objectives are achieved considering pricing of competitive products. Prices established by this model, utilizing estimated demand, are then used in the demand model (HOUSE), which considers product awareness, construction time, building dimensions, number of new buildings, energy consumption, construction loan rate, mortgage rate, labor costs, material costs, transportation costs, importance of energy consumption, and other related factors. It also takes into account building-related product choices based upon both relative (stick relative to manufactured) expected costs and risks. This is included in a "probability of purchase" database. Results are estimates of market shares and level of demand (square feet of product) for the manufactured products. This level of demand can then be used in the supply-side model to obtain better estimates of prices, which can in turn be provided to the demand model. The process can be continued to iteratively develop an understanding of the likely consequences of panel pricing on the financial performance of a typical panel manufacturer and on the related product demand.

The developed approach can thus be used to evaluate 1) the demand for manufactured products in terms of the relative costs and risks associated with the manufactured product and the effect of energy savings and mortgage rates and other factors on builder decisions and 2) product attributes and the financial viability of product lines and related investments in producing the panel products.

B. Analysis

The objective was to develop a capability for estimating the demand for panel products in terms of new building starts and various construction costs and decision-maker attitudes (for example, the relative importance to the decision maker of energy savings and other costs in choosing between alternative construction techniques) and panel manufacturer design, production, and pricing decisions. The ultimate goal is to develop a better understanding of the specific factors, and tradeoffs, that are likely to be important in influencing decision-maker choices that will lead to increased market penetration rates for panel products and consequently will result in reduced heating and cooling energy consumption.

To accomplish this required the development of databases and mathematical models. Both manufacturer and builder databases were developed as a result of detailed information provided by both manufacturers and builders. (Actual collected data are not presented because, to a large extent, it is deemed to be proprietary.) The manufacturer database reflects the financial performance of typical panel manufacturers and related production decisions and product pricing. The builder database reflects builder perceptions with respect to construction costs when utilizing stick construction methods and panel construction methods. These databases reflect current industry practices but are not yet robust enough (because of data collection limitations) for the development of firm conclusions regarding market penetration rates and product pricing. (Budgetary constraints limited the number of organizations from which data could be obtained.)

It was necessary to develop both demand-side and supply-side models in order to understand the relationship of changes to these databases (and other related factors) in terms of their impact on the demand for panels. Separate models were developed and used interactively (through manual intervention) to develop the necessary relationships. These models and their use are described in the following paragraphs with details of the models provided in indicated references.

Because two separate models are utilized interactively, it is necessary to perform an initial synchronization prior to their utilization in analyzing panel demand as illustrated in Fig. 6.3. This synchronization process is aimed at tuning the models to provide results that are currently observed in today's marketplace (i.e., the demand model provides an estimate of panel demand commensurate with observed panel sales, and the supply model provides panel prices commensurate with current manufacturer prices). Once this is accomplished, it is then possible to vary technology, cost, and other assumptions and to observe the effect of these changes (as reflected through their impact on decision data) on both panel demand and panel manufacturer financial performance and panel pricing.

The synchronization of the supply and demand models is as follows (Fig. 6.3a):

Step 1: Data Collection. The data collection is concerned both with the establishment of current market supply and demand conditions (i.e., the product prices and sales volume that reflect current supply and demand) and with the determination of values of variables that will be used in the supply and demand models in the estimation of current pricing and sales volume.

Step 2: Adjustment of Input Data. The collected data are utilized in the demand-side model to establish an estimate of unit sales (i.e., the quantity demanded in the market place Q). Because the calculated demand and the actual demand (obtained through the data collection process) are likely to differ, it is necessary to selectively adjust the input data so that estimated and actual demand

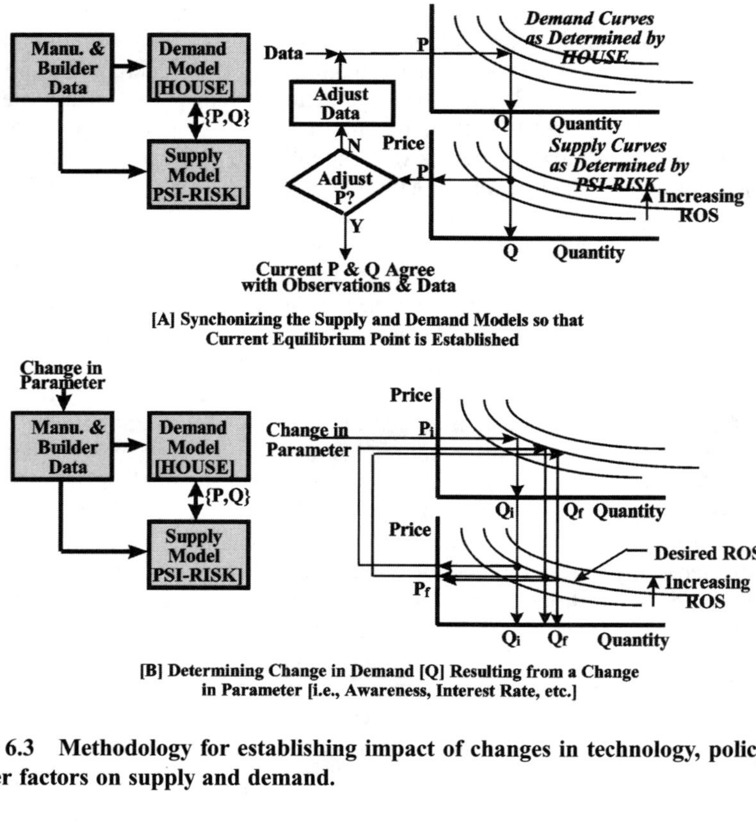

Fig. 6.3 Methodology for establishing impact of changes in technology, policy, and other factors on supply and demand.

are approximately equal and the current equilibrium point (i.e., current demand at current pricing) is established (*P,Q*). Because the collected data consist of both quantitative and qualitative data, with the latter converted into quantitative terms, the former's values are maintained and the latter's values are adjusted so as to achieve the equilibrium point.

Step 3: Establishment of Manufacturer Financial Objective. With the current equilibrium point (*P,Q*) established, the supply-side model is used to establish the return on sales (ROS) that is to be considered as the panel manufacturer rate-of-return objective. (Because panel manufacturers indicated that pricing is likely to be based upon the achievement of ROS objectives, ROS has been used as the rate of return objective.) Once this ROS is established, it is maintained as a constant throughout all subsequent analyses. In the following pages the ROS is based upon a detailed financial analysis of a typical panel manufacturer. Data used in this analysis are based upon industry-provided financial data.

When industry sales data are not forthcoming, an alternate course of action can be taken as follows:

Step 2A: Establishment of Manufacturer Financial Objective. An estimated manufacturer financial objective is determined by varying the ROS indicated to the supply model. Supply-side analyses are performed for each ROS establishing resulting product prices (i.e., those that would have to be charged to achieve each ROS). The return that yields the price closest to the actual manufacturer's price is then selected as the ROS objective. Once this ROS is established, it is maintained as a constant throughout all subsequent analyses. The ROS is based upon a detailed financial analysis of a typical panel manufacturer.

Step 3A: Adjustment of Input Data. The collected data are utilized in the demand-side model to establish an estimate of unit sales (i.e., the quantity demanded in the market place, *Q*). An estimate of unit sales is input to the supply model, and the model calculates a price for each product. This is the price that must be charged for the product in order for the ROS (as already estimated) to be achieved by the panel manufacturer. The new prices are then entered into the demand model. The demand model calculates a new quantity that would be sold at that new price. This quantity is entered into the supply model. The process continues until the "market" settles at an equilibrium price. This price is taken to be the base price. It is at this point that the models are considered to be in sync.

The following process is used when examining the implications of changes that can result from government and other actions (refer to Fig. 6.3b):

Step 1: Estimation of Changes. Once the supply and demand models are in sync, it is possible to estimate the impact that changes can have on the demand for panels and on the financial performance of panel manufacturers. The changes that are of concern might be the result of government or other actions (for example, increase in awareness of energy savings potential, improvement of manufacturing processes resulting in lower manufacturing costs, etc.) and require definition in terms of the specific variables considered in the demand-side and supply-side models.

Step 2: Establishment of Change in Demand. If the parameter to be considered for change is contained within the demand-side model (for example, builder awareness), then the demand-side model serves as the convenient point of departure for assessing the effect of the assumed change on the previously determined equilibrium demand (i.e., the calculation of a new value of Q with the starting value indicated in Fig. 6.3b as Q_i). This requires the use of the demand-side model with the data set used in the equilibrium analysis except for changes to identified variables. If the parameter to be considered for change is contained within the supply-side model (for example, reduced manufacturing labor cost), then the supply-side model serves as the convenient point of departure (i.e., go to step 3).

Step 3: Establishment of Panel Price Change. If the parameter to be considered for change is contained within the supply-side model (for example, reduced manufacturing labor cost), then the supply-side model serves as the convenient point of departure. Utilizing the supply-side model and the new value of Q as obtained from the demand-side model or utilizing the supply-side model directly to establish a revised price (P, with P_i being the initial or starting value of P as indicated in Fig. 6.1), establish a revised price (P) that will generate the same ROS as identified during the synchronization procedure us described.

Step 4: Establishment of a Revised Value of Demand. Utilizing the demand-side model and the revised price P, as established through the use of the supply-side model, establish a revised value of demand Q.

Step 5: Repetition of Steps 3 and 4. Repeat steps 3 and 4 until new values of price and quantity P_f, Q_f are established such that the changes in P and Q from one iteration to the next are deemed to be inconsequential. These new values represent the price and quantity values at a new equilibrium point with the changes being the direct result of the assumed action that is characterized by the changed data set. The new equilibrium point considers both price elasticity (as reflected by the builder decision database) and panel manufacturer business decisions.

1. Demand Model (HOUSE)[7]

The demand-side model, referred to as the HOUSE model, considers the decisions that can be made in the construction process given a set of product cost and other factor perceptions and allows the demand for factory-built envelope products to be estimated parametrically. The demand-side model does not address the supply-side and does not seek to develop a supply curve (i.e., cost as a function of quantity produced)—it uses price estimates obtained from a supply-side model. Thus the HOUSE model can be used to establish the form of the demand curve as a function of the price of factory-produced construction products and other related factors (for example, construction loan rates, importance of energy costs, energy prices, etc.).

The HOUSE model was developed to provide a means for estimating the demand for factory-produced panelized wood-frame construction, and various panel construction and related products relative to conventional site-built (or stick-built) wood-frame construction. External and internal wall, roofing and flooring products can be considered.

The developed demand-side model, HOUSE, considers a broad range of factors that describe the general market characteristics and are likely to affect decision-maker choices and markets. These include the following:

1) Number of buildings (of a generic type) constructed per year for which the manufactured housing products might be considered is a factor.

2) Building dimensions for each floor of either a single-or two-story building must be considered.
3) HVAC energy consumption for the stick-constructed building and consumption changes that can result from the use of factory-produced panelized and related external and internal wall, roofing, and flooring products need to be considered. The consumption changes can be described in terms of minimum, most likely, and maximum impacts as perceived by the decision maker.
4) The fraction of decision makers likely to consider energy savings in their product choices and the relative importance of their considerations is important.
5) Awareness (in the sense that the panel technology is known and is likely to be seriously considered as an alternative to stick construction) of the new products under consideration is a factor.
6) Effect of the new considered products on construction time is a factor.
7) Construction loan and mortgage rates must be considered.
8) Another factor is the multiple labor category labor costs for each of the considered external and internal walls, roofing, and flooring products. These labor costs can be described in terms of minimum, most likely, and maximum costs as perceived by the decision maker.
9) Multiple material category material costs for each of the considered external and internal walls, roofing and flooring products must be considered. These material costs can be described in terms of minimum, most likely, and maximum costs as perceived by the decision maker.
10) Transportation costs for the various considered materials can be explicitly considered if not included in the material prices.
11) Decision data that describe the probability of selecting a stick product in terms of the relative (i.e., ratio of stick to manufactured housing product) expected cost and relative risk need to be considered.

As a result of the consideration of the preceding factors, the demand-side model, **HOUSE**, develops expected labor, material, and energy costs and associated standard deviations for both the stick and manufactured housing products. (The supply-side model considers the effect of the demand and pricing on the financial viability of the firm.) These expected values and standard deviations are used to establish the ratio of expected buyer costs and the ratio of perceived risk (i.e., the ratio of the standard deviation of costs), which are then utilized to enter a decision database to establish the probability of selecting the stick construction approach (with 100 minus this amount being the probability of selecting the manufactured housing approach). [Actually, the utilized ratio is (expected stick cost + standard deviation of stick cost) ÷ (expected panel cost + standard deviation of panel cost).] This "selection probability" is considered as the panel product market share. This estimate of market share is then used to establish the demand for panel products. The market share and demand are established separately for both decision makers that do and do not consider energy consumption and related costs in their product selection decisions.

The basic concept for establishing market share is predicated upon the presumption that decision makers normally face choices and must make decisions that involve different perceptions with respect to expected costs and levels of risk and other often nonquantifiable factors. Thus, it is necessary to make decisions that involve tradeoffs between multiple attributes, both quantifiable and nonquantifiable. The market share computation acknowledges that builders, particularly when confronted with alternative construction products and techniques, must make construction-related decisions that involve tradeoffs between products and techniques that are perceived to have different expected costs and different levels of risk.

To take this into account, it is first necessary to establish builder perceptions that consider both the expected value and risk dimensions and then to establish how the tradeoffs between expected cost and risk affect decisions. The former is taken into account in the HOUSE model by allowing estimates to be made that explicitly take into account perceptions of uncertainty associated with impacts of new products on the reduction of HVAC consumption, labor costs, and material costs with the latter two covering both construction utilizing stick and panel construction methods. The latter is taken into account by utilizing a database that establishes the probability of making different choices when faced with mutually exclusive alternatives that have different expected values and perceived risk (see the Appendix).

The assumptions relating to the tradeoffs between relative expected cost and relative risk are illustrated in Fig. 6.4, where relative expected cost indicates the ratio of stick expected construction cost and panel expected construction cost (represented by the set of curves with the associated ratio of expected costs indicated in the data frame). Relative risk is the ratio of stick expected cost plus associated standard deviation of cost (i.e., the risk dimension) and panel expected

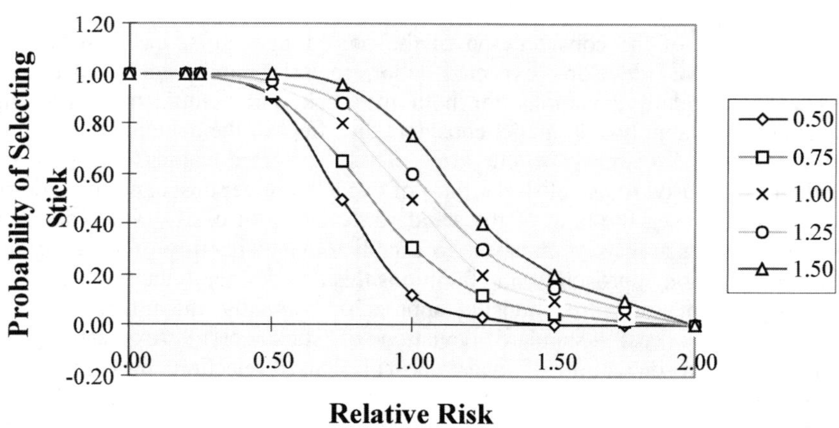

Fig. 6.4 Decision database for establishing the probability of selecting stick construction.

cost plus associated standard deviation of cost. This definition of relative risk allows both the relationship of stick and panel risk perceptions to be considered and the magnitude of these risks to be considered in relationship to the magnitude of the expected costs.

2. *Supply Model (PSI-RISK)*

The financial planning model, *PSI-RISK*, was developed as a general financial planning tool that specifically addresses uncertainty and risk in the business planning and evaluation process. *PSI-RISK* can be used to address a wide variety of financial planning and evaluation situations ranging from the planning and evaluation of either a new or ongoing business venture to the evaluation of pricing policies that aim to achieve a desired level of financial performance. (The original version of the model was developed more than 30 years ago to assist RCA corporate management in the capital budgeting process by explicitly considering uncertainty and risk. At about the same time Dupont developed a similar capability. These models and capabilities have been utilized successfully over the years.)[8–12] Through the provision of input data, *PSI-RISK* can be, as has been done for the supply-side analysis, used to consider an ongoing business and to establish product pricing in terms of rate of return objectives, production decisions, and competitive product pricing. *PSI-RISK*, which explicitly and quantitatively considers uncertainty, is a financial planning software tool that allows a business (such as a manufacturer of factory-produced panelized wood-frame construction products) to be modeled based on data provided to it and develops a pro forma income statement, cash-flow statement, and balance sheet. These are the basic financial documents that form the financial basis for decision making and are a major part of a business plan.

PSI-RISK is a general venture simulation model (operational in the DOS environment) that is capable of simulating the financial aspects of a wide range of multiproduct business ventures including both regulated and nonregulated. The basic cash-flow structure of these computations is illustrated in Fig. 6.5. A number of the key structural aspects of *PSI-RISK* that enable it to simulate a very diverse range of business ventures are highlighted here: 1) multiple time periods considered, 2) trends considered for following time periods, 3) large number of stored uncertainty profiles, 4) multiple products, 5) product interrelationships, 6) demand–price relationships (by product), 7) manufacturing cost–quantity relationships (by product), 8) choice of depreciation method (by capital expenditure), 9) inter-period correlation, 10) capital expenditure determination by production requirements, 11) consideration of initial corporate position, 12) interest rate a function of debt/equity ratio, 13) pricing policy established to achieve desired rate of return (with market constraints), 14) flexible tax structure, and 15) financial computation for different business situations that are a) part of Existing Corporation or b) a separate Corporation.

Data describing the projected market and competitive prices for as many as 20 products over a period of up to 10 years, manufacturing cost of sales and its elasticity, fixed and variable components of expenses, capital expenditures (both specifically identified and those determined to be required as a result of production

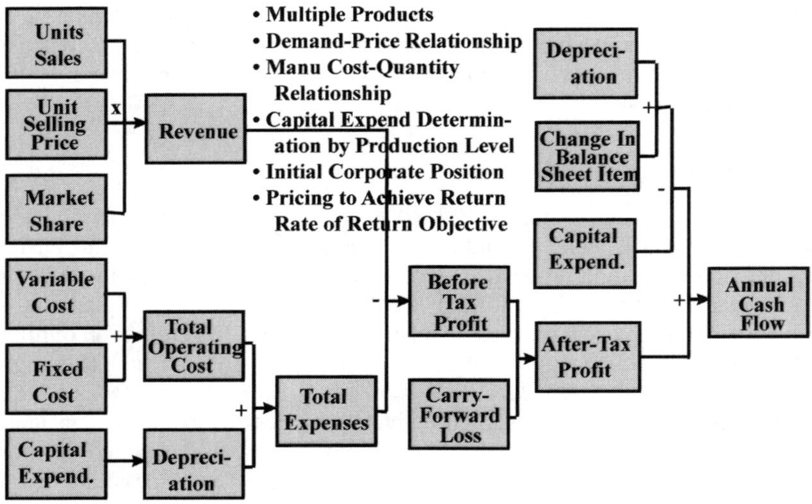

Fig. 6.5 Simplified cash-flow structure of *PSI-RISK*.

demands) and depreciation methods, business structure and pricing policy, tax structures, and balance-sheet items are entered into a menu-driven spreadsheet data input system. Data need only be provided within the ranges set by variables such as the number of years to be analyzed, number of products to be considered, number of capital expenditures to be considered, etc. Thus the data requirements are a function of the complexity of the business being analyzed. A FORTRAN-compiled program acquires the data set and calculates and prints pro forma income, cash-flow, and balance-sheet statements for the overall business, and individual income summaries for each product then. These statements can be reviewed to assess the prospective financial viability of the business modeled, product pricing and achievability of desired financial performance goals, etc., and necessary modifications can then be made to the business plan.

Although it can be used as a simple deterministic financial model, the *PSI-RISK* model is more complex and powerful than a simple financial planning model in several ways. It is designed as a stochastic simulation model using Monte Carlo simulation techniques to convert a combination of deterministic and probabilistic input data describing a business venture into risk profiles indicating the probability that various financial performance measures such as profit, cash flow, and return on assets will exceed different levels. The same model, using the same set of equations, can be used to also perform deterministic analyses. If a stochastic analysis is required, certain of the data can be entered in the form of a "range of uncertainty," whereby a minimum and maximum value are specified and an uncertainty profile describing the probability distribution of the data within the range of uncertainty is furnished. [To minimize data entry, the *PSI-RISK* Model contains a large number of stored uncertainty profiles (i.e., probability distributions) that can be referenced, or others can be created as the need arises]

Another useful feature of the *PSI-RISK* model is that it can either utilize specified product prices or it can derive product prices that will result in specified rates of return (either return on sales, assets or gross margin) within the constraints of competitive or substitute product prices. This feature plays a prominent role in the manufactured housing application. For each product an iteration and curve-fitting process are used to derive that price which will allow the desired rate of return to be achieved, subject to the market price constraints. Because of this capability, the *PSI-RISK* model can represent a price-setting firm or a price-taking firm. In the case of a price-setting firm, product selling price is computed to achieve a specified rate of return (either return on sales, assets, or gross margin). With a price-taking firm the firm must charge competitive prices (those that have been established by the price-setting firm or those current in the market), and rates of return for the firm are computed based on these prices.

The *PSI-RISK* model allows for the consideration of product interdependency relationships and relationships between unit sales and required production levels and associated capital expenditures. This allows the effects of variations in independent variables on dependent parameters to be automatically calculated by the model. Variations in prices for, or sales of, a product can affect sales of that or other products and therefore production levels. Changes in production level, in turn, can affect capital expenditures and manufacturing cost per unit.

These relationships are communicated to the model via the input data (for the current application these data are based upon information provided by panel manufacturers). Product interrelationships can be specified, whereby the sales of one product can be indicated as a function of the sales of another product sold by the firm (as would be the case with complimentary goods, such as razors and razor blades). Demand–price (or price elasticity) relationships for each product and manufacturing cost–quantity relationships (sensitivity of unit manufacturing cost to changes in production level) can be specified to the model. Capital expenditures can be specified as production independent and production dependent (which allows the *PSI-RISK* model to establish the magnitude and timing of expenditures depending on production level).

The model can represent various business situations, including a startup situation or a venture that is part of an existing corporation. In this latter case the venture can be considered either as a startup situation or as a continuation of current operations. The model has a flexible tax structure that calculates carry-forward losses and tax credits.

The following input data can be specified to the *PSI-RISK* Model via the set of spreadsheet data screens: 1) general data are time horizon of business, business type (separate entity or part of a larger organization), number of products, discount rates, timing of product introduction, uncertainty profiles, number of simulation runs, tax rates, investment tax credit rate, type of debt/equity computation (i.e., maintain constant debt/equity ratio), initial debt/equity position, and interest rates (cost of capital at different debt/equity ratios); 2) revenue-related data are unit sales, market share, distributor markup, product interrelationships, rate-of-return objectives and market constraints, price of competitive products, allowed or target rates of return, price elasticity data, uncertainty specifications (if desired) for unit sales, market share, unit price or price, and of competitive products; 3) cost of sales and expense data are cost of sales, sensitivity of cost of sales to production level, fixed and variable expense data items, and uncertainty specifications (if desired) for cost of sales and fixed and variable expense items; 4) capital expenditure data are unit independent capital expenditures, timing, depreciation type and life, applicability of investment tax credit, uncertainty specifications (if desired) for magnitude of expenditures, unit-dependent capital expenditures, production threshold level (production level at which a new production-level-related capital expenditure is required to be made), incremental production capability, expenditure to achieve incremental capability, depreciation type and life, and uncertainty specifications (if desired) for magnitude of expenditures; 5) balance-sheet-related data are debt/equity ratio, cash requirements, level of receivables, level of payables, level of inventory, other assets and liabilities, book value of existing (i.e., initial) capital assets, and depreciation type and life for existing capital assets.

After data are entered into the spreadsheet data program, a transfer file is created and sent to the FORTRAN-compiled program with a few simple keystrokes. Pro forma income, cash flow, and balance-sheet statements for the business are generated, which display after-tax profit, return on sales, return on assets, total assets and liabilities, debt, equity, debt/equity ratio, net cash flow, and indebtedness for each year of the analysis, as well as net present value (of annual cash flow) at several discount rates.

The consideration of production-dependent expenditures allows the *PSI-RISK* model to establish the magnitude and timing of expenditures that can be driven by level of production (the value of which is established by the HOUSE model). [These features, though important in general, were not of importance in the current analysis because of the specific capacity levels assumed and the level of changes encountered in the quantity (and resulting level of sales) resulting from price changes.] Another important feature is the ability to consider the initial position of the firm being modeled. *PSI-RISK* allows the initial corporate position to be specified in terms of current balance-sheet items. This includes cash, receivables, inventory, book value of capital assets (and remaining depreciation life and type of depreciation so that depreciation expenses can be computed), and debt/equity ratio. The model allows debt/equity ratio to be maintained as a constant over the time frame analyzed or allows equity to be maintained as a constant. Positive cash flows are used to reduce debt. When additional investment is required, it can be funded through additional debt (keeping equity constant) by increasing both debt and equity and maintaining a constant debt/equity ratio. When a constant debt/equity ratio is to be achieved, equity is not repurchased as positive cash flows occur, but distributions of earnings occur in a manner so as to maintain a near constant debt/equity ratio so that the cost of borrowing increases as the firm's financial position deteriorates.

To summarize, the *PSI-RISK* model is used to model a typical, or representative, price-setting panel manufacturer and to determine product pricing, which serves as input to the demand-side model. The judicious use of the House and *PSI-RISK* models provides a means for government agencies to evaluate the likely impact of technology programs and related policies on the demand for manufactured housing products, resulting energy savings, and impacts on product manufacturers. The models also provide the means for product manufacturers to evaluate the likely impacts of product attributes on demand, investment requirements, and resulting financial performance.

C. Summary/Results/Status

The manufactured housing models HOUSE and *PSI-RISK* were developed and used to demonstrate the ability to forecast the demand for manufactured housing products in terms of product attributes, to assess the effect of government technology programs and related government policies on the demand, and to provide a means for industry to plan and evaluate new product entries by analyzing demand and business structure on financial performance. This was accomplished successfully with the assistance of a panel manufacturer that provided product and financial data. This activity is continuing with the improvement of the user friendliness of the models in order to make them more useful to builders and manufactures. Currently, several panel manufacturers have expressed interest in acquiring and using the developed capabilities for their product planning and evaluation

III. Casino Gaming

Casinos offer an ever-increasing mix of table games and slot play. Until the early 1980s computer simulation was unknown to most casino management. Casino gaming decisions were based upon limited expected value analysis, informed judgment based upon experience, and folklore. Over the years there has been a great acceptance of the use of computers for database management (keeping track of player characteristics, awarding of complementaries, and win/loss history) and the development of casino management and evaluation information (for example, measuring table hold percentage as a function of shift, game, and table). [Hold "percentage" is the ratio of table win (as measured by the change in chip inventory) divided by the table drop (cash and markers placed in the table's drop box). In essence, hold percentage indicates the table, game, or casino rate of return from the players' investment and, as such, is a measure of performance.] Unfortunately, the acceptance of the use of computer simulation and analysis in support of management decisions has lagged considerably, but there have been a number of notable exceptions, several of which are discussed in the following pages. [Much of the analysis of the game of blackjack has been performed by those interested in improving player performance (i.e., the card counters).]

Blackjack, craps, and baccarat still remain the principal casino-offered table games. Little will be said about craps other than to say that what appears to be a rather simple game is actually an extremely complex game (particularly from the point of view of computer simulation modeling). Its importance in the generation of casino revenue however is waning with the aging of the World War II population. Recent experience in Atlantic City perhaps illustrates the general lack of appreciation of the game's complexities. Recently a number of the Atlantic City casinos attempted to utilize the pass-line odds bet [the rules of craps are presented in "Grand Gaming Guide," MGM Grand, Las Vegas, NV (A similar gaming guide can be obtained from any casino and presents the rules and procedures of the

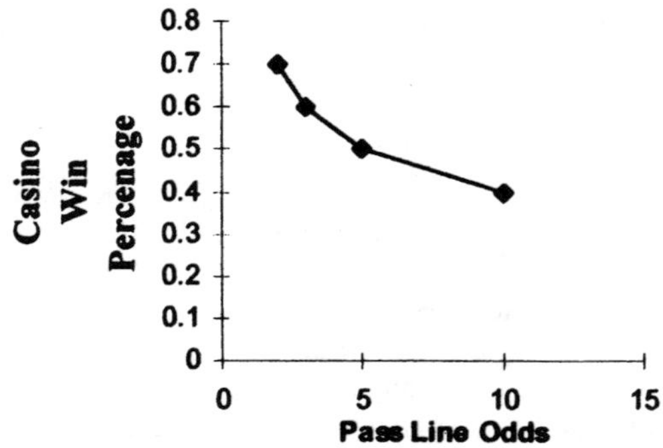

Fig. 6.6 Theoretical casino win percentage as a function of pass line odds (craps).

offered games)] as a marketing tool to improve market share. The effect of increasing the odds wager on the casino advantage (win percentage) is illustrated in Fig. 6.6. The Atlantic City craps performance is illustrated in Fig. 6.7 in terms of hold percentage. Of particular concern is the performance of the Sands casino, which introduced 5X-odds and then withdrew it, and Showboat, which also introduced 5X-odds. A little analysis could have predicted this deterioration in performance.

Mathematical modeling and computer simulation techniques have been used to help answer many questions relating to casino operations, management and related

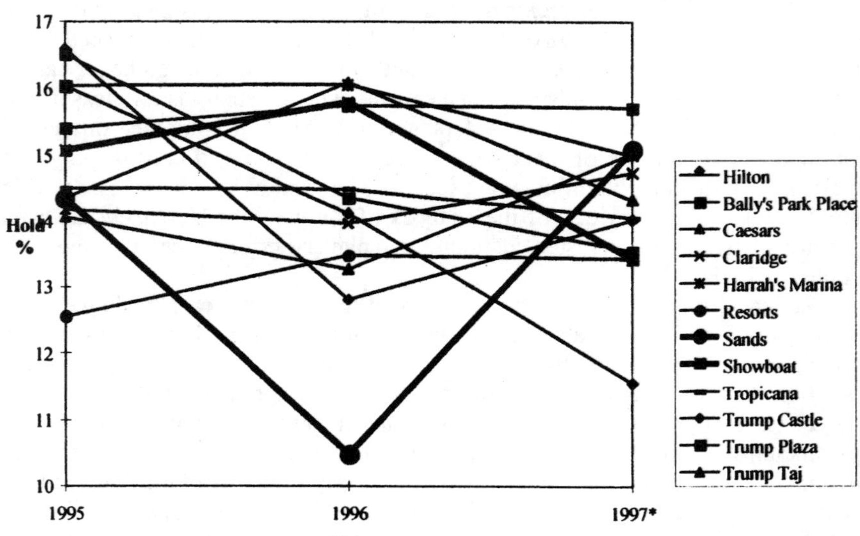

Fig. 6.7 Atlantic City casino experience (actual results) with craps hold percentage.

equipment and game design and uncertainty considerations have been accounted for in a number of different ways. These include such questions as: What effect will blackjack rule and procedure changes have on player performance (general public, basic strategy, and card counter players) and casino revenue? What effect will blackjack rule and procedure changes have on card counter (individual and team play) earnings potential? What complimentary policies should be initiated? When should casino policies be waived for a player, and what constraints should be placed on player actions in return for the special considerations? What will be the effect of new shuffle techniques? What will be the likely performance of a new piece of casino equipment, and how should it be designed to improve casino performance? These questions can be grouped into those that can be analyzed utilizing the long-term statistics of the games (for example, changing casino product attributes—rule and/or procedural changes—so as to improve overall casino performance), and those that require the consideration of the short-term statistics of the games (for example, negotiating with high rollers who will play for relatively short periods of time but will place extremely large wagers).

In the following paragraphs two specific gaming applications of mathematical modeling and computer simulation are described. The first is concerned with establishing the effect of blackjack rule changes on the players and on the casino with the joint objectives of maintaining casino financial viability while providing a "fair" game for the general public and basic strategy players. The second is concerned with simulations in support of real-time negotiations with high rollers who request special consideration (such as waiving table limits to allow wagers of tens to hundreds of thousands of dollars per hand) and are willing to accept other play constraints (such as duration of play and minimum and maximum bet size). The casino objective is to put together a complimentary package ("front-end" comps, discounting of losses, gifts, bonus if player achieves certain player bankroll objectives) that results in the player's play and achieves casino profitability and risk objectives.

IV. Casino Gaming: Decisions Based upon Long-Term Statistics

In the early 1980s casinos were legalized in New Jersey (Atlantic City), and a state regulating and control organization, the New Jersey Casino Control Commission (CCC), was established. The CCC was, amongst other things, tasked with establishing and maintaining a financially viable gaming industry while ensuring a "fair game" for the players and was given broad regulatory authority. This regulatory authority included the setting of gaming rules and procedures that were imposed on all of the casinos.

When the Atlantic City gaming industry started more than two decades ago, a set of rules was established by the CCC and agreed to by the casinos. This set of rules was deemed to be favorable to the casinos and yet offered a reasonably fair game to the players.

The blackjack rules used originally in Atlantic City were adopted from Las Vegas rules but also included an "early surrender" rule that was not in use in Las

Vegas. (The early surrender rule allowed a player to fold after seeing the dealer's face-up card while forfeiting half of the original wager.) Analysis, after the fact, showed the importance of the early surrender rule to basic strategy and card counter players. The key blackjack rules and procedures in place in the early 1980s were as follows: General rules included a) the use of six decks, b) a cut card placed 33% into the shoe (from the rear), c) blackjack wager odds of 3:2, and dealer sticks (hard and soft totals) at 17 or greater. Splitting included a) no resplitting, b) only one card drawn to split aces, and c) blackjack not possible after splitting. Doubling down was a) allowed on all hard and soft totals and b) allowed after splitting. Surrender included a) early surrender and b) 50% of wager given at surrender. Other rules were that a) insurance wagers were allowed and b) no dealer could shuffle at will.

Two problems arose shortly after the start of Atlantic City gambling. The first was the realization by some that the set of accepted rules provided basic strategy players with a slight advantage over the house [Basic strategy players follow a set of play decisions that are invariant with the history to the cards played at the table. The decision rules have been established based upon an expectation analysis assuming replacement (i.e., the probability of drawing any card remains constant).]; a totally unacceptable situation because this threatened the financial viability of the casinos. The second was the influx of card counters into the casinos with their significant advantage over the house. [Card counters are blackjack players that follow a set of play decisions that depend upon the history of the cards that have been played; in other words, they count the cards by at least keeping a running sum of the values of the cards that have been played (adding plus one for every low card and minus one for every high card) and vary their decisions accordingly. In addition, at high positive counts the wager amount is increased significantly, thus giving them significant leverage when the card count is in their favor (i.e., the probability of winning has increased). The decision rules have been determined by extensive simulation of the game of blackjack.] Because the set of rules and procedures were not developed with card counters in mind, the initial casino reaction was to exclude the play of card counters. (Allowing card counters to play and maintaining the financial viability of the casinos would have required a different set of rules and procedures. It was felt that to set different rules in order to allow the card counters to play would have serious adverse impacts on general public players and would ultimately lead to the elimination of blackjack in Atlantic City.) As a result, action was initiated to change the blackjack rules to eliminate the basic strategy player advantage and to then seek other rule and procedure changes that would eliminate the card counter advantage while having acceptable impacts on other players.

The CCC initiated the development of a Blackjack Policy Simulation Model[13] in order to demonstrate the effect of rule and procedure changes on the general public, basic strategy, and card counter players and on the financial performance of the casinos. The objective was to provide a sufficiently detailed analysis so that regulatory action could be taken to make the necessary changes in rules and procedures. At the time that this was undertaken, computer simulation was basically unknown to casino management.

The Blackjack Policy Simulation Model was developed specifically to provide a mechanism for analyzing the effect of rule and procedure changes on the casinos and on the players and to establish the feasibility of creating blackjack rules and procedures that would serve as an economically viable alternative to the exclusion of the card counting strategy players. The model was used to justify the elimination of the early surrender rule (player can surrender his/her hand after seeing the dealer face-up card and surrendering half of his/her wager) that was not available in other gaming jurisdictions and provided the basic strategy player a slight advantage over the casino. The model was also used to assess the impact of changes in player ability, mix of players, and betting strategies on casino performance. The model can also be used to assess the impact on casino performance of changes in the number of tables, mix of tables, speed of play, marketing strategies, level of productivity, and other factors under casino control.

The Blackjack Policy Simulation Model allowed the play of a large number of hands to be simulated for each rule or other change investigated (typically on the order of millions of hands were simulated for each analysis.). The results of the simulation analyses were in terms of both casino and player statistics. A typical set of casino statistics included expected value, standard deviation, and confidence intervals of the gross blackjack revenue per day and per hour for the casino and for each table type. Player types such as card counters, basic strategy players, and general public were simulated. The basic strategy and card counter decision strategies were obtained from published material. A unique approach was taken to establish general public "strategies." A group of casino experts was given the task of estimating the chance of different player decisions for every combination of player cards and totals. Thus the general public player became a probabilistic player with all play decisions being made by random sampling techniques after hands were "dealt" by the model to the players and the dealer.

The Model allowed the impact of a great number of rule and procedure changes to be examined. The rules cover splitting, doubling down, surrender insurance, and dealer reshuffle.

General rules cover the number of decks, placement of cut card, blackjack wager odds, dealer hard and soft sticking totals, player–dealer bust payments, player–dealer standoff payments, dealer announcement of dealer blackjack, player wager limitations, and table minima and maxima.

Splitting rules cover splitting/not splitting pairs, number of cards that may be drawn when splitting aces, and splitting on equal cards or equal values.

Doubling down rules cover double down/no double down, double down/no double down after splitting, allowable point counts for doubling down, amount of double-down maximum wager, and amount of additional split/double-down wager lost to dealer blackjack.

Surrender rules cover surrender/no surrender, surrender percentage, and surrender a function of dealer-up card.

Insurance rules cover insurance/no insurance, odds paid, maximum wager, and timing of insurance wager.

Dealer reshuffle (initial and relative wager size) rules cover wager size relative to previous wager, initial wager size, and dealer decisions random or certain.

Dealer reshuffle (based on true point count) rules cover point count limits and house card point values.

The rules and procedures that could be modeled are indicated and could be varied either individually or collectively and the impact on the casino and player determined. As rules were changed, the basic strategy and card counter decision strategy data were changed because there are different optimum strategies for each set of rules.

The data required for utilization of this Blackjack Policy Simulation Model are described here:

1) Number of player types include general public, strategy, and card counter.
2) General public (random) descriptor include distribution of wager size, insurance decision for each dealer-up card, chance of hard total decisions (stand, hit, double, and surrender) for each dealer-up card, chance of soft total decisions (stand, hit, double, and surrender) for each dealer-up card, and chance of pairs decisions (split/no split, stand, hit, double, and surrender) for each dealer-up card.
3) Strategy player's descriptive data include wager size determination (distribution or a function of point count), point count values of cards, approximation for true count, likelihood of error in incrementing point count, insurance decision procedure for each dealer-up card, hard total decision values and decision rules, soft total decision values and decision rules, and splitting pairs decision values and decision rules.
4) Environmental specification data include number of table types, table-related data, number of tables of each type, maximum/minimum wager sizes for each table type, utilization percentage for each table type, time to shuffle for each table type, time to play a round at each table type, and mix of players at each table type.

Typical obtained results are summarized in the following tables and figures. Investigations were concerned with elimination of the early surrender rule, no soft doubling, doubling on 9, 10, 11 only, dealer hits soft 17, and no doubling after splitting. In addition a number of procedural changes were analyzed including limiting entry in midshoe, limiting initial wager to table minimum when entering in midshoe (subsequent wagers within the same shoe are not restricted), limiting all wagers to table minimum when entering in midshoe (subsequent wagers within the same shoe are restricted to the table minimum), limiting bet multiples, and limiting table maximum/minimum ratios. In other words, a regulatory economic impact analysis was performed, which considered a number of regulatory alternatives before decisions were made and actions taken.

Table 6.1 illustrates the effect of the elimination of the early surrender rule on the different player categories. Elimination of the early surrender rule shifts the win percentage of the basic strategy player from a positive to a negative value — a necessary condition for the continued offering of the game of blackjack but does not solve the card counter problem. Figure 6.8 translates these win percentage impacts into casino revenue impacts as a function of the mix of players. There was considerable concern that the percentage of basic strategy players would increase over time because of the proximity of Atlantic City to major population centers.[14]

Because the elimination of the early surrender rule did not eliminate the significant advantage of the card counter over the casino, analyses were performed of other rule changes using the Blackjack Policy Simulation Model. Each time a rule was changed, optimum basic and card counter strategies were developed for

Table 6.1 Player performance: implications of the early surrender rule (player win percentage)[a]

Player category	Rule scenario: With early surrender	Rule scenario: Without surrender	Incremental change
General public			
Novice[b]			
Expected value	− 1.55	− 1.74	− 0.19
95% Confidence	− 1.43 to − 1.67	− 1.62 to − 1.86	− 0.02 to − 0.36
Experienced[b]			
Expected value	− 1.55	− 1.52	+0.03
95% Confidence	− 1.43 to − 1.67	− 1.40 to − 1.64	− 0.14 to +0.20
Basic strategy[c]			
Expected value	+0.16	− 0.54	− 0.70
95% Confidence	+0.08 to +0.24	− 0.46 to − 0.62	− 0.58 to − 0.82
Card counter			
Bet multiplier = 1			
Expected value	+0.76	+0.13	− 0.63
95% Confidence	+0.58 to +0.94	− 0.06 to +0.32	− 0.38 to − 0.88
Bet multiplier = 4			
Expected value	+4.38	+2.51	− 1.87
95% Confidence	+3.67 to +5.00	+1.86 to + 3.16	− 1.42 to − 2.32

[a]Player win percentage is the negative of the ratio of casino win/amount wagered.
[b]3.5 million hands.
[c]7.0 million hands.

the basic strategy and card counter players. A new general public strategy was also developed through the use of the casino gaming experts (i.e., assessments of the probability of player decisions given different card combinations and the specified rule change). The results of the analyses are summarized in Tables 6.2 and 6.3. Table 6.2 indicates the effect of the rule changes on the general public and basic strategy players. Table 6.3 indicates the effect of the rule changes coupled with various procedural changes on the card counters. The combined effect of limiting entry in midshoe coupled with limiting wager size when entering in midshoe is obvious. (Limiting entry in midshoe eliminated the advantage of "shadow counters." Shadow counters observe the table play, count the cards, but do not place wagers until the deck becomes rich for the player.)

Because the New Jersey Supreme Court decided that card counter play could not be excluded from the casinos, the reported work had a major impact and resulted in the rules and procedures that are currently used in Atlantic City. These rules and procedures make possible the play of the card counters by significantly reducing their revenue-generation capability. Of particular importance were the procedures of limiting entry in midshoe, limiting wager size if allowed entry in midshoe, and allowing the dealer to shuffle at will. (This latter procedure, though

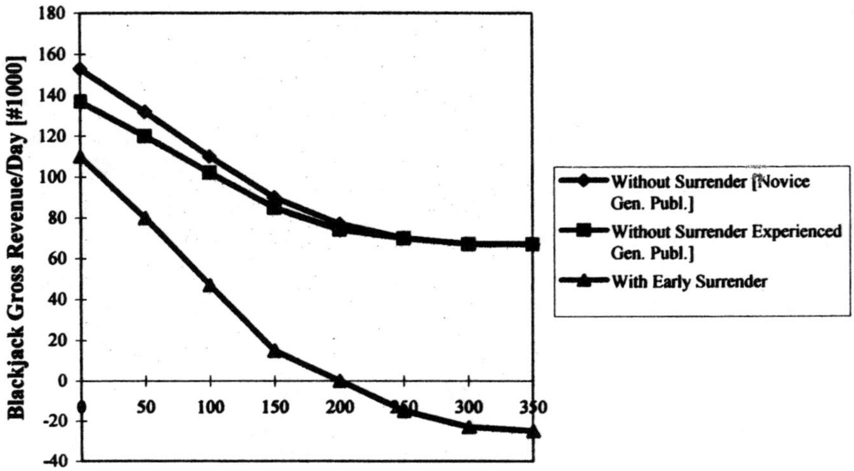

Fig. 6.8 Casino blackjack performance (for hypothetical casino).

of major effect, has the problem that it slows down the game and reduces revenue-generation time.)

The stochastic Blackjack Policy Simulation Model played a significant role in the formulation of gaming policy and the structuring of the design of a casino gaming product (blackjack). Particularly unique was the probabilistic description of general public play decisions through the use of blackjack experts that provided judgments as to the likelihood of player actions in terms of their options (i.e., stick, hit, double down, split, take insurance) and play constraints (i.e., the set of imposed rules). In other words, the product was designed specifically taking into account likely general public player decisions where it was not possible to define the player's utility function; it was necessary to describe player decisions as

Table 6.2 Player performance summary: impact of rule changes

	General public player		Basic strategy player	
Rule change	Expected win, %	95% Confidence	Expected win, %	95% Confidence
None (1985 AC Rules)	− 1.53	− 1.41 to − 1.65	− 0.43	− 0.31 to − 0.55
No soft double	− 1.64	− 1.52 to − 1.76	− 0.61	− 0.49 to − 0.73
Double on 9, 10, 11 only	− 1.59	− 1.47 to − 1.71	− 0.61	− 0.49 to − 0.73
Dealer hits soft 17	− 2.09	− 1.97 to − 2.21	− 0.77	− 0.65 to − 0.89
No double after split	− 1.87	− 1.75 to − 1.99	− 0.66	− 0.54 to − 0.78

Table 6.3 Summary of shadow card counter average win/hour[a]

	Rule change									
	Current Atlantic City rules		Double on 9, 10, 11		Dealer hits soft 17		No soft double		No double after split	
Procedural change	33%	50%	33%	50%	33%	50%	33%	50%	33%	50%
None	175[b]	97[b]	144[b]	78[b]	111[b]	57[b]	150[b]	81[b]	128[b]	71[b]
Limit entry in midshoe	—	—	—	—	—	—	—	—	—	—
Entry in midshoe:										
Limit initial wager	130	64	109	53	82	37	110	53	97	46
Entry in midshoe:										
Limit all wagers	18[c]	8[c]	14[c]	8[c]	11[c]	6[c]	15[c]	8[c]	13[c]	7[c]
Limit bet multiple (10X)	175[b]	97[b]	144[b]	78[b]	111[b]	57[b]	150[b]	81[b]	128[b]	70[b]
Limit table max/min ratio										
$100 Table ($2000 max)	175	97	144	78	111	57	150	81[b]	128[b]	70[b]
$50 Table ($1000 max)	175	97	144	78	111	57	150	81[b]	128[b]	70[b]
$25 Table ($500 max)	88	52	72	39	56	28	75	40	66	35
$10 Table ($200 max)	37	21	29	16	22	11	30	17	26	14

[a]Based upon 80 rounds/hour for 33% cut card and 75 rounds/hour for 50% cut card.

[b]Play at low-limit tables utilizing six decks.

[c]Play at $100 minimum tables.

probabilistic decisions and to include these probabilistic decisions within the simulation model and subsequent Monte Carlo simulation analysis.

The long-term analysis used a simulation model to analyze the impacts of blackjack rule and procedure changes on the players and on the casinos. The model simulated the play of the game by dealing the cards and making decisions. The analyses developed expectations: the expected value of casino win percentage, casino revenue, etc. The duration of play, that is, time, was of no concern as millions of hands were played per analysis. The risk dimension, that is, the variability of results, was not addressed.

V. Casino Gaming: Negotiations with High Rollers (Decisions Based upon Short-Term Play Statistics)

Casinos offer a range of table games including baccarat, blackjack, craps, and roulette to a broad range of players having differing playing skills, money-management systems, and bankrolls. A significant number of these players are considered as high rollers, who bring large bankrolls to the casino and request special considerations. These considerations include complimentaries in the form of free trips and gifts, discounting of player losses, and the awarding of a bonus if player win exceeds agreed-upon levels. Because the high-roller business is becoming increasingly competitive, casinos have demonstrated their concern by developing formal methods to improve their ability to attract high rollers with "special consideration" packages that meet casino profitability and risk objectives. Not adequately explicitly considering risk objectives in considering high-roller play can lead to disaster. Several years ago a Las Vegas casino allowed a high roller to wager $250,000 a hand while playing baccarat. A minimum play time constraint was apparently not agreed upon in advance of play and the player left after two days with between $18 and $28 million of the casino's money. What went wrong? The duration of play, as will be discussed in the following paragraphs, should have significantly exceeded the actual playing time in order to reduce casino risk to an acceptable level. In fact, the larger the allowed wager, the longer should be the required play time. Increasing the duration of play will reduce the chance that the casino will incur a large loss.

The statistical nature of the table games, and the somewhat unpredictable nature of the players, implies that casino win, and hence profitability, must be characterized in statistical terms, and casino profitability objectives should be specified in terms of desired level of profit and associated risk (i.e., the chance of achieving this profit). To address the issue of profit (or casino win or player bankroll) and its statistical nature, a baccarat simulation model that simulates the short term (i.e., hours to several tens of hours) play of Baccarat was developed.[15–17] (Simulation models have also been developed for the short-term play of blackjack and craps.) This model, referred to as the Baccarat Risk/Reward Model, develops the probability distribution of player or casino win as a function of time and as a function of player ability and money-management technique and other factors that are under the control of the casino. The Baccarat Risk/Reward Model is used to develop casino win statistics for specific or generic players and serves as the basis for developing the player and casino statistical data for the Risk/

Reward Decision Support System. The intent is to be able to establish the incentive package that should be offered to obtain the high-roller play and still meet casino profitability and risk objectives.

The Baccarat Risk/Reward Model simulates the play of the game according to the rules of baccarat and develops casino win statistics as a function of time that results from descriptions of player ability and money-management technique. The resulting casino win statistics are placed into a casino/player database that is accessed when negotiations are to be performed with the specified player. Upon identification of the player (which can be either a specific player or a "representative" or "generic" player) with whom negotiations are to be undertaken, the appropriate casino win statistics data are retrieved from the casino/player database, casino goals and requirements and negotiation attributes specified, and the player negotiations analysis performed.

The Risk/Reward Decision Support System, indicated in Fig. 6.9, is comprised of the following:

1) A Baccarat Risk/Reward Model is used to simulate the play of the game and develops casino win statistics in terms of specified player ability and money-management technique.
2) A casino/player database maintains casino win statistics for each player analyzed.
3) A casino/player Negotiations Model, based upon casino requirements, casino win statistics (as developed by the Baccarat Risk/Reward Model), and specified attributes of the negotiations "package," determines the required play duration which is necessary to achieve the specified casino win and risk objectives.

These are integrated together in the Risk/Reward Decision Support System so that the following can occur:

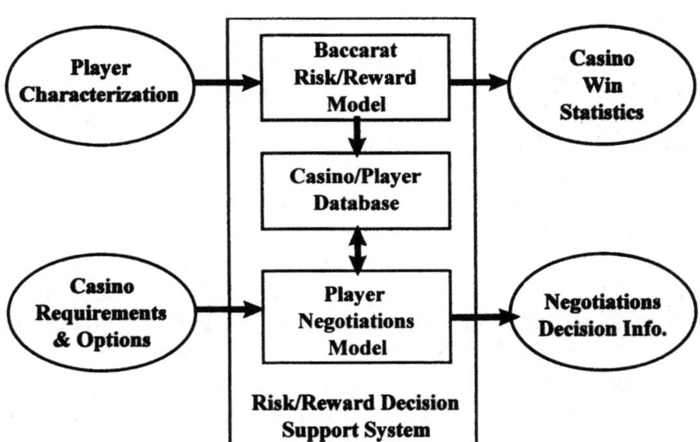

Fig. 6.9 Baccarat risk/reward decision support system.

1) Player's play can be described in terms of characterizations of play and wager (i.e., money management) decisions. (The characterization of the play normally is based upon routine floor operations associated with high-rollers. However, occasionally specific data collection procedures are developed. One such case involved a high roller baccarat play that was observed for several tens of hours of play with every card dealt, wager and play decision and outcome recorded. From these data it was possible to reasonably characterize the player's play and wager strategies.)
2) Casino win statistics can be developed for the identified players and maintained in a database.
3) Casino/player negotiations can be evaluated, by accessing the database, in terms of casino win requirements, casino negotiating attributes, and imposed player requirements.

Baccarat is an ancient game usually played with eight decks of cards with the cards dealt from a "shoe" one at a time. Those participating in the game place a wager on either the banker or the player. A wager can also be placed on a tie. Two hands (referred to as the bank and the player) of two cards each are dealt from the shoe. A third card can be drawn by the player's and/or the bank's hand according to a predefined set of rules for drawing a third card. The first hand represents those betting on the player, the other those betting on the Bank. The winning hand is the one that totals closest to nine. Tens, cards totaling 10, and picture cards have the value 0. Ace is counted as one, deuce as two, etc. If the cards total a two-digit number, only the unit's digit is considered. Bets are paid even money, but a 5% commission, or vigorish, is charged on winning banker bets. A tie bet can be made, which pays nine for one (i.e., 8 to 1).

The long-term casino statistical advantage is a function of the player wager strategy and differs for bets placed on the bank, player, and tie and for combination bets. Based upon the fact that the probability of the player's hand winning is 0.4461, the probability of the bank's hand winning is 0.4584, and the probability of a tie winning is 0.0955, the casino statistical advantage is as follows:

$$\text{Casino Statistical Advantage} = \begin{cases} 1.174\% \text{ for bets on bank} \\ 1.360\% \text{ for bets on player} \\ 14.050\% \text{ for bets on tie} \end{cases}$$

For combination bets the casino statistical advantage is a function of the amount wagered on the tie relative to the amount wagered on the bank or player.

To establish casino win statistics as a function of player decision characteristics, money management technique, bet size relative to bankroll, duration of play and other factors for the play of a finite number of hands—that is, long-term averages or expectations are not adequate to describe the casino statistics (thousands of hands are considered as finite, whereas millions of hands approach the infinite situation)—the Baccarat Risk/Reward Simulation Model was developed. The model simulates the play of a player over a period of time (typically 24 hours of play). Wagers are made, hands dealt, outcomes determined, and bankrolls are incremented (or decremented) for each hand played during the specified time period. The results are saved, and the play of another 24-hour period is simulated with all results saved. This process is repeated a large number of times with histograms of outcomes developed as a result of the large number of random play repetitions. These histograms form the basis of the risk/reward report, as illustrated in graphic form in Fig. 6.10. This is the key, as will be seen, to the negotiations process that explicitly allows casino acceptable level of risk to be taken into account.

Four snapshots in time are illustrated. As time progresses, expected player bankroll decreases while, because of the statistical nature of the table games, the spread (or standard deviation) of player bankroll increases and the probability of player bankroll exceeding a specified value decreases. Indicated in Fig. 6.10 is a value (the dashed line) below the initial or starting bankroll. This is the initial bankroll reduced by the amount that the casino must win from the player in order to cover all costs (complimentaries, discount on losses, bonus, gross contribution or direct cost of offering the game, and uncollectibles) associated with the player.

It is clear that the probability of achieving the desired win increases with time. Thus it is possible to specify, in addition to the required casino win, the risk that the casino is willing or able to take in achieving the win. The problem, expressed in a simplified fashion, is to establish the required player duration of play in order for the casino to recoup its investment in the player at a specified level of risk.

The baccarat Risk/Reward Simulation Model of the short-term play establishes the probability distribution of player bankroll, or casino win, as a function of time as illustrated in Fig. 6.10. Figure 6.11 presents an alternate depiction of the probability of player bankroll as a function of the duration of play—indicated are contours of constant probability. In other words, there is a 10% chance that player bankroll will exceed the values marked "10%." The effect of the level of acceptable risk on the casino achieving its desired level of win is illustrated. As time of play increases, these curves or contours of constant probability become downward

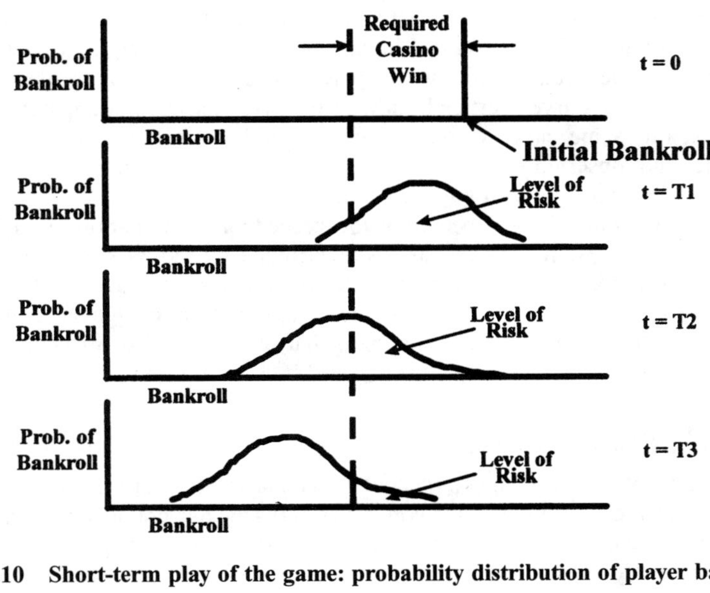

Fig. 6.10 Short-term play of the game: probability distribution of player bankroll.

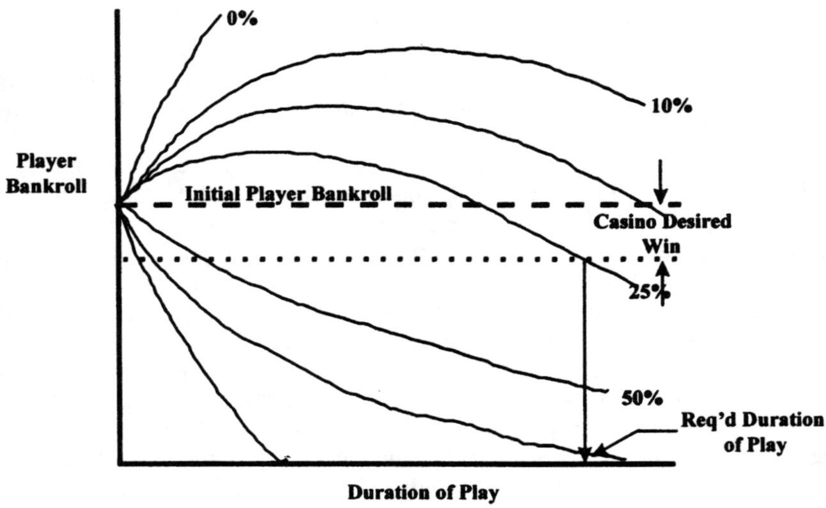

Fig. 6.11 Determination of required duration of play to meet casino objectives.

sloping. Also shown in Fig. 6.11 are the player initial bankroll and the level of casino desired win as established by the incentive package and casino profitability objectives.

The required win plus the level of comps plus the discount amount is the reduction in player bankroll that is required for the casino to accomplish its cost recovery and profit objectives. (Actually also added are the expected bonus and the expected level of uncollectibles.) If the casino is willing to accept a risk of 25% (i.e., there is an 75% chance of achieving its objectives), then the play must proceed until the 25% risk curve intersects the level of required casino win (indicated as the "req'd duration of play" as determined by projecting the intersection of the curves vertically downward). Not shown, but apparent, is that projecting this same curve vertically upward until it meets an indicated bonus threshold and observing the risk curve that passes through this intersection establishes the probability of paying a bonus.

The Negotiations Model performs the preceding computations based upon interpolating the risk/reward results (as determined by the Risk/Reward Model) and utilizing the casino specified profitability and risk objectives and specified negotiations package. The result is the determination of player constraints (duration of play and minimum bet size) that are required to meet the casino objectives.

A. Player Characterization

Before negotiations can be conducted with a player, the player analysis must be performed that results in the casino win statistics (vs the player under consideration) that become part of the casino/player database. Once the player has been incorporated into the database, negotiations can proceed as described earlier.

Player characterization consists primarily of describing the player's decision strategy (or player ability) and money management technique. Player ability is described in terms of the chance, or percentage of the time, that the player places wagers on the bank, player, tie, bank and tie, player and tie, and not wagering (i.e., sitting out play). A typical player characterization is illustrated in Fig. 6.12.

A player must also be described in terms of money management technique. A player can be described according to one of eight general money management techniques. The money management technique establishes the bet size that is to be used in the play of the current hand. The eight money management techniques available for consideration are referred to as 1) win progression (win), 2) win progression (loss), 3) percent of current bankroll, 4) hunch betting (average), 5) hunch betting (high roller), 6) loss progression (loss), 7) loss progression (recoup on win), 8) flat betting.

The money management technique establishes the bet size that is to be used in the play of the current hand. Depending upon the particular money management technique used, the bet size can be a function of one or more of the following factors: initial bankroll, current bankroll, outcome of previous hand(s), table limit, and relationship of current to initial bankroll (i.e., playing on casino money).

B. Player Analysis

The player analysis is concerned with converting the player characterization data into the casino win statistics as illustrated in Fig. 6.11. The play of the hand is simulated taking into account the cards that are played, the play decisions that are made, and the wager that is the result of the player's money management system. The play proceeds over time with appropriate results saved. The resulting win

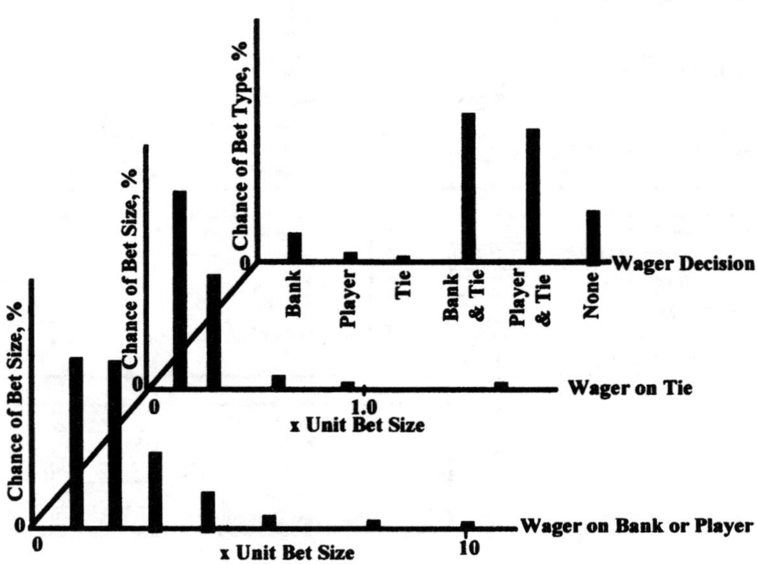

Fig. 6.12 Typical baccarat player characterization.

statistics are stored in the casino/player database such that player negotiations can then proceed.

C. Negotiations

Negotiations with a player can only be accomplished after the player ability and player money management technique data have been provided and the detailed player analysis accomplished. The detailed player analysis creates a casino win statistics spreadsheet that summarizes the casino win statistics (vs the specified player). This is automatically stored in the casino/player statistics database and remains in the database until specifically deleted. Negotiations can be performed with any player that has been previously stored in the database by simply scanning through the list of names and "clicking" on the desired player's name. This process automatically retrieves the casino statistics from the database and allows the negotiations to proceed. The process proceeds by providing the information as indicated on the screen illustrated in Fig. 6.13.

Having selected (identified) the player with whom negotiations are to be conducted, the negotiations screen indicates the player's name (including a player account or ID number) and the bet-to-bankroll ratio associated with the player. This bet-to-bankroll ratio can be changed slightly without having to reanalyze the player. (If large changes are to be considered, the player analysis must be redone using the new bet-to-bankroll ratio. This is necessary because the casino/player

Date last saved:	2/7/97	Baccarat Negotiation		Nominal
		Bet/Bankroll Ratio (%)	0.1	0.1
Hunch_Average		Unit Bet ($)	$2,000	
Maximum Time Considered	24	Bankroll ($)	$2,000,000	

Casino Objectives:		Casino Expenses	
Gross Contribution ($)	$20,000	3rd Party Commissions	
		Fixed ($)	$5,000
Chance of Achieving(%)	60	% of Loss	5
		Uncollectable(%)	10

Negotiating Points:		Player Requirements:	
Up front comp. ($)	$5,000		
Discount Level (%)	15	Unit Bet ($)	$2,000
Bonus Payment		Max Bet ($)	$20,000
Win Level ($)	$300,000	Duration of Play (hrs)	12.4
Bonus Amount($)	$100,000		

	Expected Win ($)	$43,258
Casino	Expected Handle ($)	$3,388,901
Impacts:	Expected Discounts ($)	$6,489
	Chance of Bonus (%)	3.6%
	Expected Commissions	$7,163

Fig. 6.13 Casino/player negotiations.

statistics are sensitive to the bet-to-bankroll ratio.) Specifying the bet-to-bankroll ratio and the unit bet size establishes the player bankroll. In addition to the bet-to-bankroll ratio, the casino negotiator may enter data relating to negotiating points, casino objectives, casino expenses, and player unit bet size.

The casino objectives that can be specified are **gross contribution** (i.e., the desired level of casino win in excess of the win necessary, to cover up-front comps, discounts, expected bonus, commissions, and losses deemed as uncollectibles) and the chance of achieving the gross contribution objective. The chance of achieving the gross contribution objective indicates the level of risk that the casino is willing to accept in order to gain the play of the high roller with the offered package of incentives (i.e., negotiating points) and identified player requirements. A 60% chance of achieving the gross contribution objective indicates that the casino is willing to play if it has a 60% chance of achieving (or exceeding) the indicated gross contribution. The greater the specified chance of achieving the gross contribution, the longer the required period of play. Casino expenses that can be considered are third-party commissions and player losses that might not be collectible.

The negotiating points that can be specified are the values of **up-front comps** (this includes the value of travel, lodging, gifts, etc.), **discount level** (the magnitude of the discount to be given on player losses), **bonus payment win level** (bonus can be awarded if the player wins more than the indicated amount after the required duration of play), and **bonus payment amount** (the amount of the bonus if the win level is achieved after the required duration of play).

The negotiations process can be repeated as many times as necessary (i.e., iterating to achieve a desired result or performing "what-if" analyses) by changing the negotiating points and/or casino objectives in order to 1) establish the constraints that should be placed on a player in terms of incentives (the negotiating points) and casino objectives, 2) develop insights into relationships between incentives and ability to achieve casino objectives, 3) establish the value of a player to the casino in terms of the incentive package that is likely to be necessary to obtain the play of the player, and 4) establish the feasibility of third-party commission arrangements.

D. Summary/Status

Techniques that allow analyses of the short-term play of table games and the subsequent real-time negotiations of high-roller incentive packages have been developed. These techniques result in the determination of the probability distribution of casino win as a function of time and allow the casinos to plan and evaluate complimentary policies that meet profitability and risk objectives. They allow uncertainty associated with player characterizations to be taken into account in the negotiations between casinos and high-rollers and significantly improves casino risk management.

The Risk/Reward Decision Support System was implemented for several Atlantic City and Las Vegas casinos that emphasize the high-roller markets. The casinos have utilized the methodology for negotiations with high rollers. To place the importance of these negotiations into proper perspective, the case of a baccarat high roller requesting special consideration bears consideration. Recently, a baccarat high roller requested that table limits be raised so that wagers of $250,000 per hand (there are approximately 75 hands per hour) could be placed. The player played for a weekend and left $18 to $28 million richer. This had a material effect on the casino quarterly financial reporting. The point is that inadequate duration of play (to offset the large wagers) resulted and the casino accepted an unduly high level of risk. Hopefully, more frequent use of analytic techniques will lead to fewer casino disasters.

References

[1]Princeton Synergetics, Inc., "A Stochastic Market Penetration Model for Building Envelope and End-Use Products," Battelle Pacific Northwest Labs., Contract No. 245163-A-U1, Washington, DC, April 1995.

[2]Greenberg, J.S., *Investment Decisions: The Influence of Risk and Other Factors*, American Management Association, New York, 1982.

[3]Edelman, F., and Greenberg, J.S., "Venture Analysis: The Assessment of Uncertainty and Risk," *Management: A Book of Readings*, edited by H. Koontz and C. O'Donnell, McGraw-Hill, 1972, pp. 200–207.

[4]Greenberg, J.S., "Risk Analysis," *Astronautics/Aeronautics*, AIAA, New York, Nov. 1974, pp. 48–57.

[5]Greenberg, J.S., "A Financial Risk Analysis of a District Heating Business Venture," *Proceedings of the 15th Intersociety Energy Conversion Engineering Conference*, Vol. 1, 1980, pp. 600–608.

[6]"Energy Efficient Industrialized Housing Research Program: Summary, FY 1989 Research Activities," Center for Housing Innovation, Univ. of Oregon and Florida Solar Energy Center, Feb. 1990.

[7]"The Economics of Industrialized Housing: A Parametric Approach for Estimating the Demand for Factory Built Envelope Products," Princeton Synergetics, Inc., July 1996.

[8]*Financial Planning Model: PSI-RISK*, Princeton Synergetics, Inc., NJ, March 1997.

[9]Greenberg, J.S., "Risk Analysis Results (A Space Processing Business Venture to Produce Monodispersed Polystyrene Latex Spheres)," NASA, July 1979.

[10]Greenberg, J.S., and Edelman, F., "Venture Analysis: The Assessment of Uncertainty and Risk," *Management: A Book of Readings*, edited by H. Koontz and C. O'Donnell, McGraw–Hill, New York, 1972.

[11]Greenberg, J.S., and Ross, D., "Manufacturer Impact Analysis of National Appliance Efficiency Standards," Operations Research Society of America/TIMS, Nov. 1985.

[12]Greenberg, J.S., and Gaelick, C., "Introduction of Fuel Efficient Technology into the Auto Industry: Financial Implications," Oak Ridge National Labs., Rept., Sept. 1986.

[13]Greenberg, J.S., Abram, P., and Clyne, F., "The Blackjack Policy Simulation Model (User's Manual)," ECON, Inc., Rept. 80-120-2, Princeton, NJ, Sept. 1980.

[14]Abram, P., "The ECON, Inc., Blackjack Policy Simulation Model and an Analysis of the Elimination of the Early Surrender Rule in Atlantic City," Gambling and Risk Taking Conf., 22 Oct. 1981.

[15]Greenberg, J.S., and Greenberg, R.S. "Baccarat: A Simulation Analysis," Operation Research Society of America/TIMS, Nov. 1984.

[16]Greenberg, J.S., "Simulation and the Casinos," *Proceedings of the Sixth National Conference on Gambling and Risk Taking*, edited by W. Eadington, Univ. of Nevada, 1985, pp. 146–182.

[17]Greenberg, J.S. "Play Constraints: The Result of Requests for Special Consideration," *Proceedings of the Sixth National Conference on Gambling and Risk Taking*, edited by W. Eadington, Univ. of Nevada, 1985, pp. 183–202.

Appendix: Estimating the Likelihood of Investment

SEVERAL surveys have been undertaken in order to understand private-sector investment decisions in terms of multiple financial performance attributes. A survey conducted in association with the American Management Association (AMA)[1] established the likelihood of private-sector investment in terms of perceived expected return on investment (ROI), risk (where risk is a measure of variability or unpredictability of ROI), expected payback period, and magnitude of the required investment relative to the budget under the control of the decision maker. In this case all investments were considered as independent. Because in many cases investment decisions are interrelated with many being mutually exclusive, another survey undertook to establish the probability of investment of mutually exclusive investments. This survey established the likelihood of investment as a function of relative risk and relative magnitude of investment.[2]

In the former survey data collection was accomplished through the AMA with vice presidents of finance and corporate planning and corporate controllers inputs solicited. The survey form consisted of a large number of graphics, with suitable explanatory material, as indicated in Fig. A.1. Each picture, depicting the ROI relative to the firm's cost of capital and risk, indicated five different investment situations. In successive pictures the expected ROI was shifted to the left and to the right. In addition, the set of situations was repeated for a number of different payback and relative magnitude of investment situations. The survey resulted in about a 15% response rate with most respondents offering to provide additional information. The result of the survey was the determination of the likelihood of investment as a function of expected ROI, risk, expected payback period, and relative magnitude of investment. (It was realized that the considered variables were not independent. However, survey testing resulted in the determination that these variables, whether independent or not, were explicitly considered in most investment decisions). Figure A.2 illustrates typical results obtained from the survey-generated data. Indicated is the expected rate of return that is required, as a function of expected payback period and level of perceived risk, in order to have a 50% chance of obtaining financing.

In the latter survey mutually exclusive investments were considered. A smaller population was surveyed but with a combination of telephone discussions and questionnaire. The survey background information clearly stated that the situation under consideration was such that a decision has been made to commit resources to a particular endeavor and that there existed multiple implementation alter-

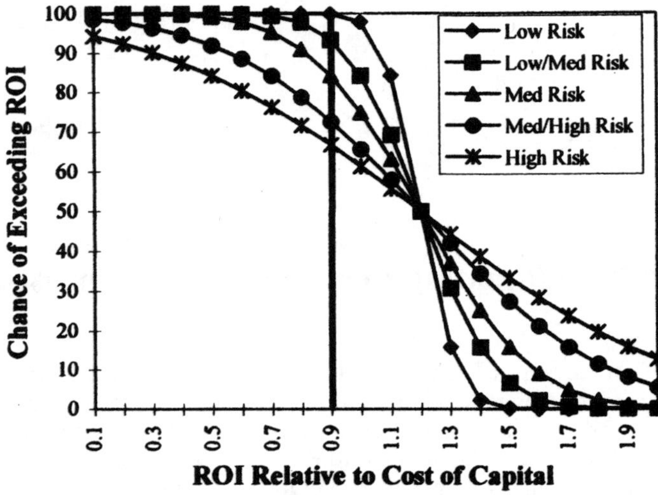

Chance of Investing In:

Low Risk [A]	**100**	**%**
Low/Med Risk [B]	**90**	**%**
Medium Risk [C]	**70**	**%**
Med/High Risk [D]	**50**	**%**
High Risk [E]	**10**	**%**

Fig. A.1 Determining the likelihood of investment when considering independent investments.

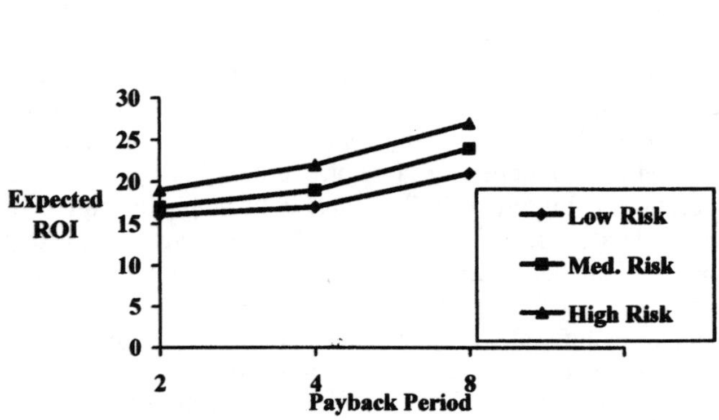

Fig. A.2 Determined relationship between expected payback period, risk, and required expected ROI to have a 50% chance of obtaining financing.

natives. The survey again presented a number of pictures, as indicated in Fig. A.3, depicting different investment situations with the respondent asked to indicate the investments that would be made as described by the graphics. Each picture indicated a basic situation A having a specified expected ROI and risk and five different alternatives (B:1 to B:5) having a different expected ROI and different levels of risk. The respondent was asked to indicate which of the alternatives would be selected. A number of similar graphics were used to establish the effect of the relative values of expected ROI and the level of risk associated with alternative A.

The survey provides robust insights into the likelihood of selecting between alternatives. An example of the use of the resulting likelihood of investment information is the estimation of space transportation service market share for providing launch services for communication satellite business ventures.

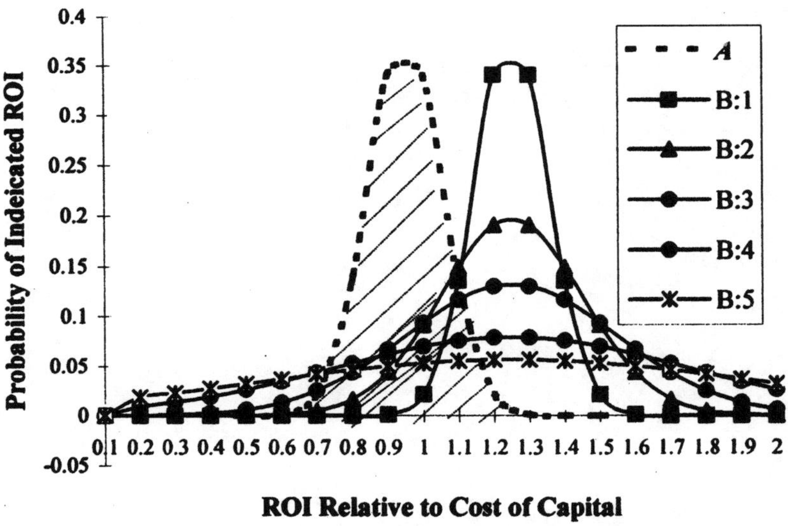

Would Select:

A	B	
	x	1
	x	2
x		3
x		4
x		5

Fig. A.3 Determining the likelihood of investment when considering mutually exclusive investments.

References

[1] Greenberg, J.S., *Investment Decisions: The Influence of Risk and Other Factors*, American Management Association, New York, 1982.
[2] Greenberg, J.S., "Pricing and Competitiveness of Present Space Transportation Services," Workshop on Space Launch Systems, Munich, June 1994.

Index

PROGRESS IN ASTRONAUTICS AND AERONAUTICS SERIES VOLUMES

***1. Solid Propellant Rocket Research (1960)**
Martin Summerfield
Princeton University

***2. Liquid Rockets and Propellants (1960)**
Loren E. Bollinger
Ohio State University
Martin Goldsmith
The Rand Corp.
Alexis W. Lemmon Jr.
Battelle Memorial Institute

***3. Energy Conversion for Space Power (1961)**
Nathan W. Snyder
Institute for Defense Analyses

***4. Space Power Systems (1961)**
Nathan W. Snyder
Institute for Defense Analyses

***5. Electrostatic Propulsion (1961)**
David B. Langmuir
Space Technology Laboratories, Inc.
Ernst Stuhlinger
NASA George C. Marshall Space Flight Center
J. M. Sellen Jr.
Space Technology Laboratories, Inc.

***6. Detonation and Two-Phase Flow (1962)**
S. S. Penner
California Institute of Technology
F. A. Williams
Harvard University

***7. Hypersonic Flow Research (1962)**
Frederick R. Riddell
AVCO Corp.

***8. Guidance and Control (1962)**
Robert E. Roberson
Consultant
James S. Farrior
Lockheed Missiles and Space Co.

***9. Electric Propulsion Development (1963)**
Ernst Stuhlinger
NASA George C. Marshall Space Flight Center

***10. Technology of Lunar Exploration (1963)**
Clifford I. Cumming
Harold R. Lawrence
Jet Propulsion Laboratory

***11. Power Systems for Space Flight (1963)**
Morris A. Zipkin
Russell N. Edwards
General Electric Co.

***12. Ionization in High-Temperature Gases (1963)**
Kurt E. Shuler, Editor
National Bureau of Standards
John B. Fenn, Associate Editor
Princeton University

***13. Guidance and Control–II (1964)**
Robert C. Langford
General Precision Inc.
Charles J. Mundo
Institute of Naval Studies

***14. Celestial Mechanics and Astrodynamics (1964)**
Victor G. Szebehely
Yale University Observatory

***15. Heterogeneous Combustion (1964)**
Hans G. Wolfhard
Institute for Defense Analyses
Irvin Glassman
Princeton University
Leon Green Jr.
Air Force Systems Command

***16. Space Power Systems Engineering (1966)**
George C. Szego
Institute for Defense Analyses
J. Edward Taylor
TRW Inc.

***17. Methods in Astrodynamics and Celestial Mechanics (1966)**
Raynor L. Duncombe
U.S. Naval Observatory
Victor G. Szebehely
Yale University Observatory

***18. Thermophysics and Temperature Control of Spacecraft and Entry Vehicles (1966)**
Gerhard B. Heller
NASA George C. Marshall Space Flight Center

***19. Communication Satellite Systems Technology (1966)**
Richard B. Marsten
Radio Corporation of America

*Out of print.

***20. Thermophysics of Spacecraft and Planetary Bodies: Radiation Properties of Solids and the Electromagnetic Radiation Environment in Space (1967)**
Gerhard B. Heller
NASA George C. Marshall Space Flight Center

***21. Thermal Design Principles of Spacecraft and Entry Bodies (1969)**
Jerry T. Bevans
TRW Systems

***22. Stratospheric Circulation (1969)**
Willis L. Webb
Atmospheric Sciences Laboratory, White Sands, and University of Texas at El Paso

***23. Thermophysics: Applications to Thermal Design of Spacecraft (1970)**
Jerry T. Bevans
TRW Systems

***24. Heat Transfer and Spacecraft Thermal Control (1971)**
John W. Lucas
Jet Propulsion Laboratory

***25. Communication Satellites for the 70's: Technology (1971)**
Nathaniel E. Feldman
The Rand Corp.
Charles M. Kelly
The Aerospace Corp.

***26. Communication Satellites for the 70's: Systems (1971)**
Nathaniel E. Feldman
The Rand Corp.
Charles M. Kelly
The Aerospace Corp.

***27. Thermospheric Circulation (1972)**
Willis L. Webb
Atmospheric Sciences Laboratory, White Sands, and University of Texas at El Paso

***28. Thermal Characteristics of the Moon (1972)**
John W. Lucas
Jet Propulsion Laboratory

***29. Fundamentals of Spacecraft Thermal Design (1972)**
John W. Lucas
Jet Propulsion Laboratory

***30. Solar Activity Observations and Predictions (1972)**
Patrick S. McIntosh
Murray Dryer
Environmental Research Laboratories, National Oceanic and Atmospheric Administration

***31. Thermal Control and Radiation (1973)**
Chang-Lin Tien
University of California at Berkeley

***32. Communications Satellite Systems (1974)**
P. L. Bargellini
COMSAT Laboratories

***33. Communications Satellite Technology (1974)**
P. L. Bargellini
COMSAT Laboratories

***34. Instrumentation for Airbreathing Propulsion (1974)**
Allen E. Fuhs
Naval Postgraduate School
Marshall Kingery
Arnold Engineering Development Center

***35. Thermophysics and Spacecraft Thermal Control (1974)**
Robert G. Hering
University of Iowa

***36. Thermal Pollution Analysis (1975)**
Joseph A. Schetz
Virginia Polytechnic Institute
ISBN 0-915928-00-0

***37. Aeroacoustics: Jet and Combustion Noise; Duct Acoustics (1975)**
Henry T. Nagamatsu, Editor
General Electric Research and Development Center
Jack V. O'Keefe, Associate Editor
The Boeing Co.
Ira R. Schwartz, Associate Editor
NASA Ames Research Center
ISBN 0-915928-01-9

***38. Aeroacoustics: Fan, STOL, and Boundary Layer Noise; Sonic Boom; Aeroacoustics Instrumentation (1975)**
Henry T. Nagamatsu, Editor
General Electric Research and Development Center
Jack V. O'Keefe, Associate Editor
The Boeing Co.
Ira R. Schwartz, Associate Editor
NASA Ames Research Center
ISBN 0-915928-02-7

***39. Heat Transfer with Thermal Control Applications (1975)**
M. Michael Yovanovich
University of Waterloo
ISBN 0-915928-03-5

---
*Out of print.

***40. Aerodynamics of Base Combustion (1976)**
S. N. B. Murthy, Editor
J. R. Osborn, Associate Editor
Purdue University
A. W. Barrows
J. R. Ward,
Associate Editors
Ballistics Research Laboratories
ISBN 0-915928-04-3

***41. Communications Satellite Developments: Systems (1976)**
Gilbert E. LaVean
Defense Communications Agency
William G. Schmidt
CML Satellite Corp.
ISBN 0-915928-05-1

***42. Communications Satellite Developments: Technology (1976)**
William G. Schmidt
CML Satellite Corp.
Gilbert E. LaVean
Defense Communications Agency
ISBN 0-915928-06-X

***43. Aeroacoustics: Jet Noise, Combustion and Core Engine Noise (1976)**
Ira R. Schwartz, Editor
NASA Ames Research Center
Henry T. Nagamatsu,
Associate Editor
General Electric Research and Development Center
Warren C. Strahle,
Associate Editor
Georgia Institute of Technology
ISBN 0-915928-07-8

***44. Aeroacoustics: Fan Noise and Control; Duct Acoustics; Rotor Noise (1976)**
Ira R. Schwartz, Editor
NASA Ames Research Center
Henry T. Nagamatsu,
Associate Editor
General Electric Research and Development Center
Warren C. Strahle,
Associate Editor
Georgia Institute of Technology
ISBN 0-915928-08-6

***45. Aeroacoustics: STOL Noise; Airframe and Airfoil Noise (1976)**
Ira R. Schwartz, Editor
NASA Ames Research Center
Henry T. Nagamatsu,
Associate Editor
General Electric Research and Development Center
Warren C. Strahle,
Associate Editor
Georgia Institute of Technology
ISBN 0-915928-09-4

***46. Aeroacoustics: Acoustic Wave Propagation; Aircraft Noise Prediction; Aeroacoustic Instrumentation (1976)**
Ira R. Schwartz, Editor
NASA Ames Research Center
Henry T. Nagamatsu,
Associate Editor
General Electric Research and Development Center
Warren C. Strahle,
Associate Editor
Georgia Institute of Technology
ISBN 0-915928-10-8

***47. Spacecraft Charging by Magnetospheric Plasmas (1976)**
Alan Rosen
TRW Inc.
ISBN 0-915928-11-6

***48. Scientific Investigations on the Skylab Satellite (1976)**
Marion I. Kent
Ernst Stuhlinger
NASA George C. Marshall Space Flight Center
Shi-Tsan Wu
University of Alabama
ISBN 0-915928-12-4

***49. Radiative Transfer and Thermal Control (1976)**
Allie M. Smith
ARO Inc.
ISBN 0-915928-13-2

***50. Exploration of the Outer Solar System (1976)**
Eugene W. Greenstadt
TRW Inc.
Murray Dryer
National Oceanic and Atmospheric Administration
Devrie S. Intriligator
University of Southern California
ISBN 0-915928-14-0

***51. Rarefied Gas Dynamics, Parts I and II (two volumes) (1977)**
J. Leith Potter
ARO Inc.
ISBN 0-915928-15-9

***52. Materials Sciences in Space with Application to Space Processing (1977)**
Leo Steg
General Electric Co.
ISBN 0-915928-16-7

---

*Out of print.

***53. Experimental Diagnostics in Gas Phase Combustion Systems (1977)**
Ben T. Zinn, Editor
Georgia Institute of Technology
Craig T. Bowman, Associate Editor
Stanford University
Daniel L. Hartley, Associate Editor
Sandia Laboratories
Edward W. Price, Associate Editor
Georgia Institute of Technology
James G. Skifstad, Associate Editor
Purdue University
ISBN 0-915928-18-3

***54. Satellite Communication: Future Systems (1977)**
David Jarett
TRW Inc.
ISBN 0-915928-18-3

***55. Satellite Communications: Advanced Technologies (1977)**
David Jarett
TRW Inc.
ISBN 0-915928-19-1

***56. Thermophysics of Spacecraft and Outer Planet Entry Probes (1977)**
Allie M. Smith
ARO Inc.
ISBN 0-915928-20-5

***57. Space-Based Manufacturing from Nonterrestrial Materials (1977)**
Gerald K. O'Neill, Editor
Brian O'Leary, Assistant Editor
Princeton University
ISBN 0-915928-21-3

***58. Turbulent Combustion (1978)**
Lawrence A. Kennedy
State University of New York at Buffalo
ISBN 0-915928-22-1

***59. Aerodynamic Heating and Thermal Protection Systems (1978)**
Leroy S. Fletcher
University of Virginia
ISBN 0-915928-23-X

***60. Heat Transfer and Thermal Control Systems (1978)**
Leroy S. Fletcher
University of Virginia
ISBN 0-915928-24-8

***61. Radiation Energy Conversion in Space (1978)**
Kenneth W. Billman
NASA Ames Research Center
ISBN 0-915928-26-4

***62. Alternative Hydrocarbon Fuels: Combustion and Chemical Kinetics (1978)**
Craig T. Bowman
Stanford University
Jorgen Birkeland
Department of Energy
ISBN 0-915928-25-6

***63. Experimental Diagnostics in Combustion of Solids (1978)**
Thomas L. Boggs
Naval Weapons Center
Ben T. Zinn
Georgia Institute of Technology
ISBN 0-915928-28-0

***64. Outer Planet Entry Heating and Thermal Protection (1979)**
Raymond Viskanta
Purdue University
ISBN 0-915928-29-9

***65. Thermophysics and Thermal Control (1979)**
Raymond Viskanta
Purdue University
ISBN 0-915928-30-2

***66. Interior Ballistics of Guns (1979)**
Herman Krier
University of Illinois at UrbanaChampaign
Martin Summerfield
New York University
ISBN 0-915928-32-9

***67. Remote Sensing of Earth from Space: Role of Smart Sensors (1979)**
Roger A. Breckenridge
NASA Langley Research Center
ISBN 0-915928-33-7

***68. Injection and Mixing in Turbulent Flow (1980)**
Joseph A. Schetz
Virginia Polytechnic Institute and State University
ISBN 0-915928-35-3

---
*Out of print.

***69. Entry Heating and Thermal Protection (1980)**
Walter B. Olstad
NASA Headquarters
ISBN 0-915928-38-8

***70. Heat Transfer, Thermal Control, and Heat Pipes (1980)**
Walter B. Olstad
NASA Headquarters
ISBN 0-915928-39-6

***71. Space Systems and Their Interactions with Earth's Space Environment (1980)**
Henry B. Garrett
Charles P. Pike
Hanscom Air Force Base
ISBN 0-915928-41-8

***72. Viscous Flow Drag Reduction (1980)**
Gary R. Hough
Vought Advanced Technology Center
ISBN 0-915928-44-2

***73. Combustion Experiments in a Zero-Gravity Laboratory (1981)**
Thomas H. Cochran
NASA Lewis Research Center
ISBN 0-915928-48-5

***74. Rarefied Gas Dynamics, Parts I and II (two volumes) (1981)**
Sam S. Fisher
University of Virginia
ISBN 0-915928-51-5

***75. Gasdynamics of Detonations and Explosions (1981)**
J. R. Bowen
University of Wisconsin at Madison
N. Manson
Universite de Poitiers
A. K. Oppenheim
University of California at Berkeley
R. I. Soloukhin
Institute of Heat and Mass Transfer, BSSR Academy of Sciences
ISBN 0-915928-46-9

***76. Combustion in Reactive Systems (1981)**
J. R. Bowen
University of Wisconsin at Madison
N. Manson
Universite de Poitiers
A. K. Oppenheim
University of California at Berkeley
R. I. Soloukhin
Institute of Heat and Mass Transfer, BSSR Academy of Sciences
ISBN 0-915928-47-7

***77. Aerothermodynamics and Planetary Entry (1981)**
A. L. Crosbie
University of Missouri-Rolla
ISBN 0-915928-52-3

***78. Heat Transfer and Thermal Control (1981)**
A. L. Crosbie
University of Missouri-Rolla
ISBN 0-915928-53-1

***79. Electric Propulsion and Its Applications to Space Missions (1981)**
Robert C. Finke
NASA Lewis Research Center
ISBN 0-915928-55-8

***80. Aero-Optical Phenomena (1982)**
Keith G. Gilbert
Leonard J. Otten
Air Force Weapons Laboratory
ISBN 0-915928-60-4

***81. Transonic Aerodynamics (1982)**
David Nixon
Nielsen Engineering & Research, Inc.
ISBN 0-915928-65-5

***82. Thermophysics of Atmospheric Entry (1982)**
T. E. Horton
University of Mississippi
ISBN 0-915928-66-3

***83. Spacecraft Radiative Transfer and Temperature Control (1982)**
T. E. Horton
University of Mississippi
ISBN 0-915928-67-1

***84. Liquid-Metal Flows and Magneto-hydrodynamics (1983)**
H. Branover
Ben-Gurion University of the Negev
P. S. Lykoudis
Purdue University
A. Yakhot
Ben-Gurion Universityof the Negev
ISBN 0-915928-70-1

*Out of print.

***85. Entry Vehicle Heating and Thermal Protection Systems: Space Shuttle, Solar Starprobe, Jupiter Galileo Probe (1983)**
Paul E. Bauer
McDonnell Douglas Astronautics Co.
Howard E. Collicott
The Boeing Co.
ISBN 0-915928-74-4

***86. Spacecraft Thermal Control, Design, and Operation (1983)**
Howard E. Collicott
The Boeing Co.
Paul E. Bauer
McDonnell Douglas Astronautics Co.
ISBN 0-915928-75-2

***87. Shock Waves, Explosions, and Detonations (1983)**
J. R. Bowen
University of Washington
N. Manson
Universite de Poitiers
A. K. Oppenheim
University of California at Berkeley
R. I. Soloukhin
Institute of Heat and Mass Transfer, BSSR Academy of Sciences
ISBN 0-915928-76-0

***88. Flames, Lasers, and Reactive Systems (1983)**
J. R. Bowen
University of Washington
N. Manson
Universite de Poitiers
A. K. Oppenheim
University of California at Berkeley
R. I. Soloukhin
Institute of Heat and Mass Transfer, BSSR Academy of Sciences
ISBN 0-915928-77-9

***89. Orbit-Raising and Maneuvering Propulsion: Research Status and Needs (1984)**
Leonard H. Caveny
Air Force Office of Scientific Research
ISBN 0-915928-82-5

***90. Fundamentals of Solid-Propellant Combustion (1984)**
Kenneth K. Kuo
Pennsylvania State University
Martin Summerfield
Princeton Combustion Research Laboratories, Inc.
ISBN 0-915928-84-1

***91. Spacecraft Contamination: Sources and Prevention (1984)**
J. A. Roux, Editor
University of Mississippi
T. D. McCay, Editor
NASA Marshall Space Flight Center
ISBN 0-915928-85-X

***92. Combustion Diagnostics by Nonintrusive Methods (1984)**
T. D. McCay, Editor
NASA Marshall Space Flight Center
J. A. Roux, Editor
University of Mississippi
ISBN 0-915928-86-8

***93. The INTELSAT Global Satellite System (1984)**
Joel Alper
COMSAT Corp.
Joseph Pelton
INTELSAT
ISBN 0-915928-90-6

***94. Dynamics of Shock Waves, Explosions, and Detonations (1984)**
J. R. Bowen
University of Washington
N. Manson
Universite de Poitiers
A. K. Oppenheim
University of California at Berkeley
R. I. Soloukhin
Institute of Heat and Mass Transfer, BSSR Academy of Sciences
ISBN 0-915928-91-4

***95. Dynamics of Flames and Reactive Systems (1984)**
J. R. Bowen
University of Washington
N. Manson
Universite de Poitiers
A. K. Oppenheim
University of California at Berkeley
R. I. Soloukhin
Institute of Heat and Mass Transfer, BSSR Academy of Sciences
ISBN 0-915928-92-2

***96. Thermal Design of Aeroassisted Orbital Transfer Vehicles (1985)**
H. F. Nelson, Editor
University of Missouri-Rolla
ISBN 0-915928-94-9

***97. Monitoring Earth's Ocean, Land, and Atmosphere from Space—Sensors, Systems, and Applications (1985)**
Abraham Schnapf
Aerospace Systems Engineering
ISBN 0-915928-98-1

---
*Out of print.

98. Thrust and Drag: Its Prediction and Verification (1985)
Eugene E. Covert
Massachusetts Institute of Technology
C. R. James
Vought Corp.
William F. Kimzey
Sverdrup Technology AEDC Group
George K. Richey
U.S. Air Force
Eugene C. Rooney
U.S. Navy Department of Defense
ISBN 0-930403-00-2

99. Space Stations and Space Platforms—Concepts, Design, Infrastructure, and Uses (1985)
Ivan Bekey
Daniel Herman
NASA Headquarters
ISBN 0-930403-01-0

***100. Single- and Multi-Phase Flows in an Electromagnetic Field: Energy, Metallurgical, and Solar Applications (1985)**
Herman Branover, Editor
Ben-Gurion University of the Negev
Paul S. Lykoudis, Editor
Purdue University
Michael Mond, Editor
Ben-Gurion University of the Negev
ISBN 0-930403-04-5

***101. MHD Energy Conversion: Physiotechnical Problems (1986)**
V. A. Kirillin, Editor
A. E. Sheyndlin, Editor
Soviet Academy of Sciences
ISBN 0-930403-05-3

***102. Numerical Methods for Engine-Airframe Integration (1986)**
S. N. B. Murthy, Editor
Purdue University
Gerald C. Paynter, Editor
Boeing Airplane Co.
ISBN 0-930403-09-6

***103. Thermophysical Aspects of Re-Entry Flows (1986)**
James N. Moss
NASA Langley Research Center
Carl D. Scott
NASA Johnson Space Center
ISBN 0-930403-10-X

***104. Tactical Missile Aerodynamics (1986)**
M. J. Hemsch, Editor
PRC Kentron, Inc.
J. N. Nielson, Editor
NASA Ames Research Center
ISBN 0-930403-13-4

***105. Dynamics of Reactive Systems Part I: Flames and Configurations; Part II: Modeling and Heterogeneous Combustion (1986)**
J. R. Bowen, Editor
University of Washington
J.-C. Leyer, Editor
Universite de Poitiers
R. I. Soloukhin, Editor
Institute of Heat and Mass Transfer, BSSR Academy of Sciences
ISBN 0-930403-14-2

***106. Dynamics of Explosions (1986)**
J. R. Bowen, Editor
University of Washington
J.-C. Leyer, Editor
Universite de Poitiers
R. I. Soloukhin, Editor
Institute of Heat and Mass Transfer, BSSR Academy of Sciences
ISBN 0-930403-15-0

***107. Spacecraft Dielectric Material Properties and Spacecraft Charging (1986)**
A. R. Frederickson
U.S. Air Force Rome Air Development Center
D. B. Cotts
SRI International
J. A. Wall
U.S. Air Force Rome Air Development Center
F. L. Bouquet
Jet Propulsion Laboratory, California Institute of Technology
ISBN 0-930403-17-7

***108. Opportunities for Academic Research in a Low-Gravity Environment (1986)**
George A. Hazelrigg
National Science Foundation
Joseph M. Reynolds
Louisiana State University
ISBN 0-930403-18-5

***109. Gun Propulsion Technology (1988)**
Ludwig Stiefel
U.S. Army Armament Research, Development and Engineering Center
ISBN 0-930403-20-7

*Out of print.

***110. Commercial Opportunities in Space (1988)**
F. Shahrokhi, Editor
K. E. Harwell, Editor
University of Tennessee Space Institute
C. C. Chao, Editor
National Cheng Kung University
ISBN 0-930403-39-8

***111. Liquid-Metal Flows: Magnetohydrodynamics and Application (1988)**
Herman Branover, Editor
Michael Mond, Editor
Yeshajahu Unger, Editor
Ben-Gurion University of the Negev
ISBN 0-930403-43-6

***112. Current Trends in Turbulence Research (1988)**
Herman Branover, Editor
Michael Mond, Editor
Yeshajahu Unger, Editor
Ben-Gurion University of the Negev
ISBN 0-930403-44-4

***113. Dynamics of Reactive Systems Part I: Flames; Part II: Heterogeneous Combustion and Applications (1988)**
A. L. Kuhl, Editor
R&D Associates
J. R. Bowen, Editor
University of Washington
J.-C. Leyer, Editor
Universite de Poitiers
A. Borisov
USSR Academy of Sciences
ISBN 0-930403-46-0

***114. Dynamics of Explosions (1988)**
A. L. Kuhl, Editor
R & D Associates
J. R. Bowen, Editor
University of Washington
J.-C. Leyer, Editor
Universite de Poitiers
A. Borisov
USSR Academy of Sciences
ISBN 0-930403-47-9

***115. Machine Intelligence and Autonomy for Aerospace (1988)**
E. Heer
Heer Associates, Inc.
H. Lum
NASA Ames Research Center
ISBN 0-930403-48-7

***116. Rarefied Gas Dynamics: Space Related Studies (1989)**
E. P. Muntz, Editor
University of Southern California
D. P. Weaver, Editor
U.S. Air Force Astronautics Laboratory (AFSC)
D. H. Campbell, Editor
University of Dayton Research Institute
ISBN 0-930403-53-3

***117. Rarefied Gas Dynamics: Physical Phenomena (1989)**
E. P. Muntz, Editor
University of Southern California
D. P. Weaver, Editor
U.S. Air Force Astronautics Laboratory (AFSC)
D. H. Campbell, Editor
University of Dayton Research Institute
ISBN 0-930403-54-1

***118. Rarefied Gas Dynamics: Theoretical and Computational Techniques (1989)**
E. P. Muntz, Editor
University of Southern California
D. P. Weaver, Editor
U.S. Air Force Astronautics Laboratory (AFSC)
D. H. Campbell, Editor
University of Dayton Research Institute
ISBN 0-930403-55-X

119. Test and Evaluation of the Tactical Missile (1989)
Emil J. Eichblatt Jr., Editor
Pacific Missile Test Center
ISBN 0-930403-56-8

***120. Unsteady Transonic Aerodynamics (1989)**
David Nixon
Nielsen Engineering & Research, Inc.
ISBN 0-930403-52-5

121. Orbital Debris from Upper-Stage Breakup (1989)
Joseph P. Loftus Jr.
NASA Johnson Space Center
ISBN 0-930403-58-4

122. Thermal-Hydraulics for Space Power, Propulsion and Thermal Management System Design (1990)
William J. Krotiuk
General Electric Co.
ISBN 0-930403-64-9

*Out of print.

123. Viscous Drag Reduction in Boundary Layers (1990)
Dennis M. Bushnell
Jerry N. Hefner
NASA Langley Research Center
ISBN 0-930403-66-5

***124. Tactical and Strategic Missile Guidance (1990)**
Paul Zarchan
Charles Stark Draper Laboratory, Inc.
ISBN 0-930403-68-1

125. Applied Computational Aerodynamics (1990)
P. A. Henne, Editor
Douglas Aircraft Company
ISBN 0-930403-69-X

126. Space Commercialization: Launch Vehicles and Programs (1990)
F. Shahrokhi
University of Tennessee Space Institute
J. S. Greenberg
Princeton Synergetics Inc.
T. Al-Saud
Ministry of Defense and Aviation
Kingdom of Saudi Arabia
ISBN 0-930403-75-4

127. Space Commercialization: Platforms and Processing (1990)
F. Shahrokhi
University of Tennessee Space Institute
G. Hazelrigg
National Science Foundation
R. Bayuzick
Vanderbilt University
ISBN 0-930403-76-2

128. Space Commercialization: Satellite Technology (1990)
F. Shahrokhi
University of Tennessee Space Institute
N. Jasentuliyana
United Nations
N. Tarabzouni
King Abulaziz City for Science and Technology
ISBN 0-930403-77-0

***129. Mechanics and Control of Large Flexible Structures (1990)**
John L. Junkins
Texas A&M University
ISBN 0-930403-73-8

130. Low-Gravity Fluid Dynamics and Transport Phenomena (1990)
Jean N. Koster
Robert L. Sani
University of Colorado at Boulder
ISBN 0-930403-74-6

131. Dynamics of Deflagrations and Reactive Systems: Flames (1991)
A. L. Kuhl
Lawrence Livermore National Laboratory
J.-C. Leyer
Universite de Poitiers
A. A. Borisov
USSR Academy of Sciences
W. A. Sirignano
University of California
ISBN 0-930403-95-9

132. Dynamics of Deflagrations and Reactive Systems: Heterogeneous Combustion (1991)
A. L. Kuhl
Lawrence Livermore National Laboratory
J.-C. Leyer
Universite de Poitiers
A. A. Borisov
USSR Academy of Sciences
W. A. Sirignano
University of California
ISBN 0-930403-96-7

133. Dynamics of Detonations and Explosions: Detonations (1991)
A. L. Kuhl
Lawrence Livermore National Laboratory
J.-C. Leyer
Universite de Poitiers
A. A. Borisov
USSR Academy of Sciences
W. A. Sirignano
University of California
ISBN 0-930403-97-5

134. Dynamics of Detonations and Explosions: Explosion Phenomena (1991)
A. L. Kuhl
Lawrence Livermore National Laboratory
J.-C. Leyer
Universite de Poitiers
A. A. Borisov
USSR Academy of Sciences
W. A. Sirignano
University of California
ISBN 0-930403-98-3

*Out of print.

***135. Numerical Approaches to Combustion Modeling (1991)**
Elaine S. Oran
Jay P. Boris
Naval Research Laboratory
ISBN 1-56347-004-7

136. Aerospace Software Engineering (1991)
Christine Anderson, Editor
U.S. Air Force Wright Laboratory
Merlin Dorfman, Editor
Lockheed Missiles & Space Company, Inc.
ISBN 1-56347-005-5

137. High-Speed Flight Propulsion Systems (1991)
S. N. B. Murthy
Purdue University
E. T. Curran
Wright Laboratory
ISBN 1-56347-011-X

138. Propagation of Intensive Laser Radiation in Clouds (1992)
O. A. Volkovitsky
Yu. S. Sedenov
L. P. Semenov
Institute of Experimental Meteorology
ISBN 1-56347-020-9

139. Gun Muzzle Blast and Flash (1992)
Günter Klingenberg
Fraunhofer-Institut für Kurzzeitdynamik, Ernst-Mach-Institut
Joseph M. Heimerl
U.S. Army Ballistic Research Laboratory
ISBN 1-56347-012-8

***140. Thermal Structures and Materials for High-Speed Flight (1992)**
Earl. A. Thornton
University of Virginia
ISBN 1-56347-017-9

141. Tactical Missile Aerodynamics: General Topics (1992)
Michael J. Hemsch
Lockheed Engineering & Sciences Company
ISBN 1-56347-015-2

142. Tactical Missile Aerodynamics: Prediction Methodology (1992)
Michael R. Mendenhall
Nielsen Engineering & Research, Inc.
ISBN 1-56347-016-0

143. Nonsteady Burning and Combustion Stability of Solid Propellants (1992)
Luigi De Luca, Editor
Politecnico di Milano
Edward W. Price, Editor
Georgia Institute of Technology
Martin Summerfield, Editor
Princeton Combustion Research Laboratories, Inc.
ISBN 1-56347-014-4

144. Space Economics (1992)
Joel S. Greenberg, Editor
Princeton Synergetics, Inc.
Henry R. Hertzfeld, Editor
HRH Associates
ISBN 1-56347-042-X

145. Mars: Past, Present, and Future (1992)
E. Brian Pritchard
NASA Langley Research Center
ISBN 1-56347-043-8

146. Computational Nonlinear Mechanics in Aerospace Engineering (1992)
Satya N. Atluri
Georgia Institute of Technology
ISBN 1-56347-044-6

147. Modern Engineering for Design of Liquid-Propellant Rocket Engines (1992)
Dieter K. Huzel
David H. Huang
Rocketdyne Division of Rockwell International
ISBN 1-56347-013-6

148. Metallurgical Technologies, Energy Conversion, and Magnetohydrodynamic Flows (1993)
Herman Branover
Yeshajahu Unger
Ben-Gurion University of the Negev
ISBN 1-56347-019-5

149. Advances in Turbulence Studies (1993)
Herman Branover
Yeshajahu Unger
Ben-Gurion University of the Negev
ISBN 1-56347-018-7

150. Structural Optimization: Status and Promise (1993)
Manohar P. Kamat
Georgia Institute of Technology
ISBN 1-56347-056-X

*Out of print.

151. Dynamics of Gaseous Combustion (1993)
A. L. Kuhl, Editor
Lawrence Livermore National Laboratory
J.-C. Leyer, Editor
Universite de Poitiers
A. A. Borisov, Editor
USSR Academy of Sciences
W. A. Sirignano
University of California
ISBN 1-56347-060-8

152. Dynamics of Heterogeneous Gaseous Combustion and Reacting Systems (1993)
A. L. Kuhl, Editor
Lawrence Livermore National Laboratory
J.-C. Leyer, Editor
Universite de Poitiers
A. A. Borisov, Editor
USSR Academy of Sciences
W. A. Sirignano, Editor
University of California
ISBN 1-56347-058-6

153. Dynamic Aspects of Detonations (1993)
A. L. Kuhl
Lawrence Livermore National Laboratory
J.-C. Leyer
Universite de Poitiers
A. A. Borisov
USSR Academy of Sciences
W. A. Sirignano
University of California
ISBN 1-56347-057-8

154. Dynamic Aspects of Explosion Phenomena (1993)
A. L. Kuhl
Lawrence Livermore National Laboratory
J.-C. Leyer
Universite de Poitiers
A. A. Borisov
USSR Academy of Sciences
W. A. Sirignano
University of California
ISBN 1-56347-059-4

155. Tactical Missile Warheads (1993)
Joseph Carleone, Editor
Aerojet General Corporation
ISBN 1-56347-067-5

156. Toward a Science of Command, Control, and Communications (1993)
Carl R. Jones
Naval Postgraduate School
ISBN 1-56347-068-3

***157. Tactical and Strategic Missile Guidance Second Edition (1994)**
Paul Zarchan
Charles Stark Draper Laboratory, Inc.
ISBN 1-56347-077-2

158. Rarefied Gas Dynamics: Experimental Techniques and Physical Systems (1994)
Bernie D. Shizgal, Editor
University of British Columbia
David P. Weaver, Editor
Phillips Laboratory
ISBN 1-56347-079-9

***159. Rarefied Gas Dynamics: Theory and Simulations (1994)**
Bernie D. Shizgal, Editor
University of British Columbia
David P. Weaver, Editor
Phillips Laboratory
ISBN 1-56347-080-2

160. Rarefied Gas Dynamics: Space Sciences and Engineering (1994)
Bernie D. Shizgal
University of British Columbia
David P. Weaver
Phillips Laboratory
ISBN 1-56347-081-0

161. Teleoperation and Robotics in Space (1994)
Steven B. Skaar
University of Notre Dame
Carl F. Ruoff
Jet Propulsion Laboratory, California Institute of Technology
ISBN 1-56347-095-0

162. Progress in Turbulence Research (1994)
Herman Branover, Editor
Yeshajahu Unger, Editor
Ben-Gurion University of the Negev
ISBN 1-56347-099-3

163. Global Positioning System: Theory and Applications, Volume I (1996)
Bradford W. Parkinson, Editor
Stanford University
James J. Spilker Jr., Editor
Stanford Telecom
Penina Axelrad, Associate Editor
University of Colorado
Per Enge, Associate Editor
Stanford University
ISBN 1-56347-107-8

164. Global Positioning System: Theory and Applications, Volume II (1996)
Bradford W. Parkinson, Editor
Stanford University
James J. Spilker Jr., Editor
Stanford Telecom
Penina Axelrad, Associate Editor
University of Colorado
Per Enge, Associate Editor
Stanford University
ISBN 1-56347-106-X

*Out of print.

165. Developments in High-Speed Vehicle Propulsion Systems (1996)
S. N. B. Murthy, Editor
Purdue University
E. T. Curran, Editor
Wright Laboratory
ISBN 1-56347-176-0

166. Recent Advances in Spray Combustion: Spray Atomization and Drop Burning Phenomena, Volume I (1996)
Kenneth K. Kuo
Pennsylvania State University
ISBN 1-56347-175-2

167. Fusion Energy in Space Propulsion (1995)
Terry Kammash
University of Michigan
ISBN 1-56347-184-1

168. Aerospace Thermal Structures and Materials for a New Era (1995)
Earl A. Thornton, Editor
University of Virginia
ISBN 1-56347-182-5

169. Liquid Rocket Engine Combustion Instability (1995)
Vigor Yang
William E. Anderson
Pennsylvania State University
ISBN 1-56347-183-3

170. Tactical Missile Propulsion (1996)
G. E. Jensen, Editor
United Technologies Corporation
David W. Netzer, Editor
Naval Postgraduate School
ISBN 1-56347-118-3

171. Recent Advances in Spray Combustion: Spray Combustion Measurements and Model Simulation, Volume II (1996)
Kenneth K. Kuo
Pennsylvania State University
ISBN 1-56347-181-7

172. Future Aeronautical and Space Systems (1997)
Ahmed K. Noor
NASA Langley Research Center
Samuel L. Venneri
NASA Headquarters
ISBN 1-56347-188-4

173. Advances in Combustion Science: In Honor of Ya. B. Zel'dovich (1997)
William A. Sirignano, Editor
University of California
Alexander G. Merzhanov, Editor
Russian Academy of Sciences
Luigi De Luca, Editor
Politecnico di Milano
ISBN 1-56347-178-7

174. Fundamentals of High Accuracy Inertial Navigation (1997)
Averil B. Chatfield
ISBN 1-56347-243-0

175. Liquid Propellant Gun Technology (1997)
Günter Klingenberg
Fraunhofer-Institut für Kurzzeitdynamik, Ernst-Mach-Institut
John D. Knapton
Walter F. Morrison
Gloria P. Wren
U.S. Army Research Laboratory
ISBN 1-56347-196-5

***176. Tactical and Strategic Missile Guidance Third Edition (1998)**
Paul Zarchan
Charles Stark Draper Laboratory, Inc.
ISBN 1-56347-254-6

177. Orbital and Celestial Mechanics (1998)
John P. Vinti
Gim J. Der, Editor
TRW
Nino L. Bonavito, Editor
NASA Goddard Space Flight Center
ISBN 1-56347-256-2

178. Some Engineering Applications in Random Vibrations and Random Structures (1998)
Giora Maymon
RAFAEL
ISBN 1-56347-258-9

179. Conventional Warhead Systems Physics and Engineering Design (1998)
Richard M. Lloyd
Raytheon Systems Company
ISBN 1-56347-255-4

180. Advances in Missile Guidance Theory (1998)
Joseph Z. Ben-Asher
Isaac Yaesh
Israel Military Industries—Advanced Systems Division
ISBN 1-56347-275-9

181. Satellite Thermal Control for Systems Engineers (1998)
Robert D. Karam
ISBN 1-56347-276-7

*Out of print.

182. Progress in Fluid Flow Research: Turbulence and Applied MHD (1998)
Yeshajahu Unger
Herman Branover
Ben-Gurion University of the Negev
ISBN 1-56347-284-8

183. Aviation Weather Surveillance Systems (1999)
Pravas R. Mahapatra
Indian Institute of Science
ISBN 1-56347-340-2

184. Flight Control Systems (2000)
Rodger W. Pratt, Editor
Loughborough University
ISBN 1-56347-404-2

185. Solid Propellant Chemistry, Combustion, and Motor Interior Ballistics (2000)
Vigor Yang, Editor
Pennsylvania State University
Thomas B. Brill, Editor
University of Delaware
Wu-Zhen Ren, Editor
China Ordnance Society
ISBN 1-56347-442-5

186. Approximate Methods for Weapons Aerodynamics (2000)
Frank G. Moore
ISBN 1-56347-399-2

187. Micropropulsion for Small Spacecraft (2000)
Michael M. Micci, Editor
Pennsylvania State University
Andrew D. Ketsdever, Editor
Air Force Research Laboratory, Edwards Air Force Base
ISBN 1-56347-448-4

188. Structures Technology for Future Aerospace Systems (2000)
Ahmed K. Noor, Editor
NASA Langley Research Center
ISBN 1-56347-384-4

189. Scramjet Propulsion (2000)
E. T. Curran, Editor
Department of the Air Force
S. N. B. Murthy, Editor
Purdue University
ISBN 1-56347-322-4

190. Fundamentals of Kalman Filtering: A Practical Approach (2000)
Paul Zarchan
Howard Musoff
Charles Stark Draper Laboratory, Inc.
ISBN 1-56347-455-7

191. Gossamer Spacecraft: Membrane and Inflatable Structures Technology for Space Applications (2001)
Christopher H. M. Jenkins, Editor
South Dakota School of Mines
ISBN 1-56347-403-4

192. Theater Ballistic Missile Defense (2001)
Ben-Zion Naveh, Editor
Azriel Lorber, Editor
Wales Ltd.
ISBN 1-56347-385-2

193. Air Transportation Systems Engineering (2001)
George L. Donohue, Editor
George Mason University
Andres Zellweger, Editor
Embry Riddle Aeronautical University
ISBN 1-56347-474-3

194. Physics of Direct Hit and Near Miss Warhead Technology (2001)
Richard M. Lloyd
Raytheon Electronics Systems
ISBN 1-56347-473-5

195. Fixed and Flapping Wing Aerodynamics for Micro Air Vehicle Applications (2001)
Thomas J. Mueller, Editor
University of Notre Dame
ISBN 1-56347-517-0

196. Physical and Chemical Processes in Gas Dynamics: Cross Sections and Rate Constants for Physical and Chemical Processes, Volume I (2002)
G. G. Chernyi
S. A. Losev
Moscow State University
S. O. Macharet
Princeton University
B. V. Potapkin
Kurchatov Institute
ISBN 1-56347-518-9

197. Physical and Chemical Processes in Gas Dynamics: Physical and Chemical Kinetics and Thermodynamics of Gases and Plasmas, Volume II
G. G. Chernyi
S. A. Losev
Moscow State University
S. O. Macharet
Princeton University
B. V. Potapkin
Kurchatov Institute
ISBN 1-56347-519-7

198. Advanced Hypersonic Test Facilities (2002)
Frank K. Lu, Editor
University of Texas at Arlington
Dan E. Marren, Editor
Arnold Engineering Development Center
ISBN 1-56347-541-3

*Out of print.

199. Tactical and Strategic Missile Guidance, Fourth Edition (2002)
Paul Zarchan
MIT Lincoln Laboratory
ISBN 1-56347-497-2

200. Liquid Rocket Thrust Chambers: Aspects of Modeling, Analysis, and Design
Vigor Yang, Editor
Pennsylvania State University
Mohammed Habiballah, Editor
ONERA
Michael Popp, Editor
Pratt & Whitney
James Hulka, Editor
Aerojet-General Corporation
ISBN 1-56347-223-6

201. Economic Principles Applied to Space Industry Decisions (2003)
Joel S. Greenberg
Princeton Synergetics, Inc.
ISBN 1-56347-607-X

*Out of print.